Gene Therapy Protocols

METHODS IN MOLECULAR MEDICINE™

John M. Walker, SERIES EDITOR

METHODS IN MOLECULAR MEDICINE™

Gene Therapy Protocols

2nd Edition

Edited by

Jeffrey R. Morgan

Brown University, Providence, RI

Humana Press ✳ Totowa, New Jersey

© 2002 Humana Press Inc.
999 Riverview Drive, Suite 208
Totowa, New Jersey 07512

www.humanapress.com

This publication is printed on acid-free paper. ∞

ANSI Z39.48-1984 (American National Standards Institute) Permanence of Paper for Printed Library Materials.

Cover design by Patricia F. Cleary.

Cover illustration: Most retrovirus particles are not able to infect a cell successfully. The barriers to infection (diffusion, half-life, and sparse cell coverage) are illustrated. *See* discussion on pp. 162–163.

For additional copies, pricing for bulk purchases, and/or information about other Humana titles, contact Humana at the above address or at any of the following numbers: Tel: 973-256-1699; Fax: 973-256-8341; E-mail: humana@humanapr.com or visit our website at http://humanapress.com

Photocopy Authorization Policy:

Printed in the United States of America.

10 9 8 7 6 5 4 3 2 1

Library of Congress Cataloging-in-Publication Data

Gene therapy protocols. -- 2nd ed. / edited by Jeffrey R. Morgan.
    p. ; cm. -- (Methods in molecular medicine ; 69)
    Includes bibliographical references and index.
    ISBN 0-89603-723-1 (hardcover; alk. paper) ISBN 0-89603-869-6 (comb binding, alk. paper)
    1. Gene therapy--Methodology. I. Morgan, Jeffrey Robert. II. Series.
    [DNLM: 1. Gene Therapy--methods. 2. Gene Transfer Techniques. 3. Genetic Vectors. QZ 50
G3213 2002]
    RB155.8 .G465 2002
    616'.042--dc21
                                                    2001026441

Preface

Efforts in gene therapy have grown dramatically in recent years. Basic research as well as clinical activity have made exciting progress and are beginning to offer renewed hope that gene therapy may be able to deliver novel approaches for the treatment of inherited as well as such acquired diseases as cardiovascular disease and cancer. With the sequencing of the human genome complete, we now have a comprehensive catalog of genes that further expands the potential role of gene therapy into such new fields as tissue engineering.

Central to gene therapy is the process of gene transfer; thus, advances in the technology of gene transfer are at the heart of this field's progress. Numerous technologies, based on a variety of methods (e.g., viral-mediated, physical/chemical), have been developed to achieve gene transfer. Some of the earliest methods, such as recombinant retroviruses, are still widely used, have undergone significant improvements, and have given rise to new vectors based on lentiviruses. Others, such as molecular conjugates, are exciting new methods that blur the distinctions between the fields of gene therapy and drug delivery. Regardless of the method chosen, gene therapy is performed in either one of two settings, namely ex vivo or in vivo. Individual gene transfer methods have inherent advantages and limitations in these settings and this is an important consideration in planning an overall strategy. Another, perhaps more important, distinction between gene transfer technologies is whether or not the genetic modification is permanent or temporary and this has a large impact on any new medical therapy envisioned. Each of these criteria plays a significant role as the field tries to weigh the advantages and disadvantages of each gene transfer technology and find a match with an intended medical application.

The field of gene therapy is advancing simultaneously on several fronts. The core technologies of gene transfer have undergone significant improvements in critical areas such as gene transfer efficiency, gene regulation, and vector production. Industry has joined academia, and a new emphasis is being placed on a quantitative understanding of gene transfer. And the applications of gene transfer continue to expand into new medical areas. In this new and entirely revised second edition, *Gene Therapy Protocols* presents detailed methods and protocols covering a comprehensive range of technologies and techniques used by leaders of the gene therapy field. The first part of the book covers the new molecular conjugates that show great promise for targeting gene transfer and regulating transgene expression. The second part of the book presents completely revised methods based on retroviruses and adenoviruses as well as new and promising methods based on lentiviruses and adeno-associated viruses. The third part of the book details protocols used in exciting applications of gene transfer in such areas as delivery of therapeutic proteins, vaccination, and tissue engineering.

This collection of protocols should make a valuable and indispensable resource for graduate students, postdoctoral fellows, as well as basic and clinical researchers in industry and academia.

Jeffrey R. Morgan

Contents

Contributors

RONALD V. ABRUZZESE • *Valentis, The Woodlands, TX*

HIROYUKI AIHARA • *Department of Nutrition and Physiological Chemistry, Osaka University Medical School, Suita, Osaka, Japan*

ROBIN R. ALI • *Molecular Immunology Unit, Institute of Child Health, London; Institute of Ophthalmology, London, United Kingdom*

STELIOS T. ANDREADIS • *Bioengineering Laboratory, Department of Chemical Engineering, University of New York at Buffalo, Amherst, NY*

ALFRED B. BAHNSON • *Department of Human Genetics, University of Pittsburgh, Pittsburgh, PA*

MARIUSZ G. BANASZCZYK • *The Immune Response Corporation, Carlsbad, CA*

JOHN A. BARRANGER • *Department of Human Genetics and Department of Molecular Genetics and Biochemistry, University of Pittsburgh, Pittsburgh, PA*

DONNA L. BASSETT • *The Immune Response Corporation, Carlsbad, CA*

KRISTIAN BERG • *Department of Biophysics, Institute for Cancer Research, The Norwegian Radium Hospital, Oslo, Norway*

GERALD W. BOTH • *CSIRO, North Ryde, Australia*

XANDRA O. BREAKEFIELD • *Molecular Neurogenetics Unit, Massachusetts General Hospital and Harvard Medical School, Boston, MA*

ALISON T. CARLO • *The Immune Response Corporation, Carlsbad, CA*

DENNIS J. CARLO • *The Immune Response Corporation, Carlsbad, CA*

LUNG-JI CHANG • *Powell Gene Therapy Center, Department of Molecular Genetics and Microbiology, University of Florida, Gainesville, FL*

K. REED CLARK • *Division of Molecular Medicine, Department of Pediatrics; Department of Molecular Virology, Immunology, and Molecular Genetics, Ohio State University, Columbus, OH*

CHRISTOPHER C. COFFIN • *The Immune Response Corporation, Carlsbad, CA*

LAUREN C. COSTANTINI • *Neuroregeneration Laboratory, Harvard Medical School; McLean Hospital, Belmont, MA*

MICHAEL A. CURRAN • *Program in Immunology, Stanford University, Stanford, CA*

JENNIFER L. CUSICK • *Massachusetts General Hospital, Harvard Medical School; Shriners Hospital for Children, Boston, MA*

PAMELA B. DAVIS • *Department of Pediatrics at Rainbow Babies and Children's Hospital, Case Western Reserve University, School of Medicine, Cleveland, OH*

MAHESH DE ALWIS • *Molecular Immunology Unit, Institute of Child Health, London; Institute of Ophthalmology, London, United Kingdom*

MICHAEL DEL TATTO • *Department of Pathology, The Miriam Hospital, Cell Based Delivery, Inc., Providence, RI*

LEISHA A. EMENS • *Department of Oncology, The Johns Hopkins University School of Medicine, Baltimore, MD*

GULSUN ERDAG • *Massachusetts General Hospital, Harvard Medical School; Shriners Hospital for Children, Boston, MA*

ELISABETH FEUDNER • *Molecular Immunology Unit, Institute of Child Health, London; Institute of Ophthalmology, London; University Eye Hospital, Tuebingen, Germany*

DAVID J. FINK • *Department of Molecular Genetics and Biochemistry, University of Pittsburgh, School of Medicine, Pittsburgh, PA*

ANTONIA FOLLENZI • *Laboratory for Gene Transfer and Therapy IRCC, Institute for Cancer Research and Treatment, University of Torino Medical School, Torino, Italy*

CORNEL FRAEFEL • *University Institute for Virology, Zurich, Switzerland*

PETER FREDERIK • *Center for Pharmacogenetics, School of Pharmacy, University of Pittsburgh, Pittsburgh, PA*

JOSEPH C. GLORIOSO • *Department of Molecular Genetics and Biochemistry, University of Pittsburgh, School of Medicine, Pittsburgh, PA*

WILLIAM F. GOINS • *Department of Molecular Genetics and Biochemistry, University of Pittsburgh, School of Medicine, Pittsburgh, PA*

ERIN K. GOUVEIA • *The Immune Response Corporation, Carlsbad, CA*

FRANK L. GRAHAM • *Departments of Biology and Pathology, McMaster University, Hamilton, Ontario, Canada*

KAREN E. HAMOEN • *Massachusetts General Hospital, Harvard Medical School; Shriners Hospital for Children, Boston, MA*

MAMORU HASEGAWA • *DNAVEC Research Inc., Tsukuba-shi, Japan*

MORITZ HILLGENBERG • *DeveloGen AG, NL Berlin, Berlin-Buch, Germany*

CHRISTIAN HOFMANN • *DeveloGen AG, NL Berlin, Berlin-Buch, Germany*

ANDERS HØGSET • *Department of Biophysics, Institute for Cancer Research, The Norwegian Radium Hospital, Oslo, Norway*

LEAF HUANG • *Center for Pharmacogenetics, School of Pharmacy, University of Pittsburgh, Pittsburgh, PA*

OLGA A. IGOUCHEVA • *Department of Dermatology and Cutaneous Biology, Department of Biochemistry and Molecular Pharmacology, Jefferson Institute of Molecular Medicine, Jefferson Medical College, Philadelphia, PA*

AKIHIRO IIDA • *DNAVEC Research Inc., Tsukuba-shi, Ibaraki, Japan*

OLE ISACSON • *Neuroregeneration Laboratory, Harvard Medical School; McLean Hospital, Belmont, MA*

ELIZABETH M. JAFFEE • *Department of Oncology, The Johns Hopkins University School of Medicine, Baltimore, MD*

YASUFUMI KANEDA • *Division of Gene Therapy Science, Graduate School of Medicine, Osaka, University, Suita Osaka, Japan*

DAVID M. KRISKY • *Department of Molecular Genetics and Biochemistry, University of Pittsburgh, School of Medicine, Pittsburgh, PA*

DANIEL KÜMIN • *DeveloGen AG, NL Berlin, Berlin-Buch, Germany*

CHIA-FENG KUO • *Department of Foods & Nutrition, Shih-Chen University, Institute of Bioagricultural Science, Academia Sinica, Taipei, Taiwan*

SONG LI • *Center for Pharmacogenetics, School of Pharmacy, University of Pittsburgh, Pittsburgh, PA*

CHARLES P. LOLLO • *Target Protein Technologies, San Diego, CA*

PETER LÖSER • *DeveloGen AG, NL Berlin, Berlin-Buch, Germany*

JEAN-PASCAL H. MACHIELS • *Laboratoire d'Oncologie Experimentale, Université Catholique de Louvain, Russels, Belgium*

FIONA C. MACLAUGHLIN • *Valentis, The Woodlands, TX*

AYALEW MERGIA • *Department of Pathobiology, College of Veterinary Medicine, University of Florida, Gainsville, FL*

AJAY R. MISTRY • *Molecular Immunology Unit, Institute of Child Health, London; Institute of Ophthalmology, London, United Kingdom*

JUN-ICHI MIYAZAKI • *Department of Nutrition and Physiological Chemistry, Osaka University Medical School, Suita, Osaka, Japan*

JEFFREY R. MORGAN • *Brown University, Providence, RI; Massachusetts General Hospital, Harvard Medical School and Shriners Hospital for Children, Boston, MA*

RICHARD A. MORGAN • *Clinical Gene Therapy Branch, National Human Genome Research Institute, National Institute of Health, Bethesda, MD*

PATRICIA M. MULLEN • *The Immune Response Corporation, Carlsbad, CA*

LUIGI NALDINI • *Laboratory for Gene Transfer and Therapy IRCC, Institute for Cancer Research and Treatment, University of Torino Medical School, Torino, Italy*

PHILIP NG • *Departments of Biology and Pathology, McMaster University, Hamilton, Ontario, Canada*

GARRY P. NOLAN • *Department of Molecular Pharmacology, Stanford University, Stanford, CA*

JEFFREY L. NORDSTROM • *Valentis, The Woodlands, TX*

ENRICO M. NOVELLI • *Department of Human Genetics and Department of Molecular Genetics and Biochemistry, University of Pittsburgh, Pittsburgh, PA*

JEONGHAE PARK • *Department of Pathobiology, College of Veterinary Medicine, University of Florida, Gainsville, FL*

ROBIN J. PARKS • *Ottawa Health Research Institute, Ottawa, Ontario, Canada*

COURTNEY POWELL • *Department of Molecular Pharmacology, Physiology and Biotechnology, Brown University; Department of Pathology, The Miriam Hospital, Providence, RI*

LINA PRASMICKAITE • *Department of Biophysics, Institute for Cancer Research, The Norwegian Radium Hospital, Oslo, Norway*

HINNE A. RAKHORST • *Massachusetts General Hospital, Harvard Medical School Shriners Hospital for Children, Boston, MA*

ULLA B. RASMUSSEN • *Laboratory of Molecular and Cellular Biology, TRANSGENE S.A., Strasbourg, France*

R. TODD REILLY • *Department of Oncology, The Johns Hopkins University School of Medicine, Baltimore, MD*

BRUCE SCHNEPP • *Children's Research Institute, Children's Hospital Inc., Columbus, OH*

KLAUS SCHUGHART • *Laboratory of Molecular and Cellular Biology, TRANSGENE S.A., Strasbourg, France*

JANET SHANSKY • *Cell Based Delivery, Inc., Providence, RI*

LOUIS C. SMITH • *Valentis, The Woodlands, TX*

NARASIMHACHAR SRINIVASAKUMAR • *Division of Hematology/Oncology, Department of Medicine, Vanderbilt University Medical Center, Nashville, TN*

WILLIAM P. SWANEY • *Department of Molecular Genetics and Biochemistry, University of Pittsburgh, Pittsburgh, PA*

YADI TAN • *Center for Pharmacogenetics, School of Pharmacy, University of Pittsburgh, Pittsburgh, PA*

ADRIAN J. THRASHER • *Molecular Immunology Unit, Institute of Child Health, London; Institute of Ophthalmology, London, United Kingdom*
KLAUS ÜBERLA • *Department of Molecular and Medical Virology, Ruhr University Bochum, Bochum, Germany*
HERMAN H. VANDENBURGH • *Department of Molecular Pharmacology, Physiology and Biotechnology, Brown University; Cell Based Delivery, Department of Pathology, The Miriam Hospital, Providence, RI*
JARMO WAHLFORS • *A.I. Virtanen Institute for Molecular Sciences, University of Kuopio, Kuopio, Finland*
CHERIE M. WALTON • *Division of Gastroenterology-Hepatology, Department of Medicine, University of Connecticut Health Center, Farmington, CT*
JENG HWAN WANG • *Department of Foods & Nutrition, Shih-Chen University, Institute of Bioagricultural Science, Academia Sinica, Taipei, Taiwan*
MARK WHITMORE • *Center for Pharmacogenetics, School of Pharmacy, University of Pittsburgh, Pittsburgh, PA*
DARREN P. WOLFE • *Department of Molecular Genetics and Biochemistry, University of Pittsburgh, School of Medicine, Pittsburgh, PA*
CATHERINE H. WU • *Division of Gastroenterology-Hepatology, Department of Medicine, University of Connecticut Health Center, Farmington, CT*
DONGPEI WU • *The Immune Response Corporation, Carlsbad, CA*
GEORGE Y. WU • *Division of Gastroenterology-Hepatology, Department of Medicine, University of Connecticut Health Center, Farmington, CT*
NING-SUN YANG • *Department of Foods & Nutrition, Shih-Chen University, Institute of Bioagricultural Science, Academia Sinica, Taipei, Taiwan*
KYONGGEUN YOON • *Department of Dermatology and Cutaneous Biology, Department of Biochemistry and Molecular Pharmacology, Jefferson Institute of Molecular Medicine, Jefferson Medical College, Philadelphia, PA*
HONG YU • *Department of Surgery/Vascular Division, Keck School of Medicine, University of Southern California, Los Angeles, CA*
ANNE-KATHRIN ZAISS • *Powell Gene Therapy Center, Department of Molecular Genetics and Microbiology, University of Florida, Gainesville, FL*
ASSEM G. ZIADY • *Department of Pediatrics at Rainbow Babies and Children's Hospital, Case Western Reserve University, School of Medicine, Cleveland, OH*

1

Poly-L-Lysine-Based Gene Delivery Systems

Synthesis, Purification, and Application

Charles P. Lollo, Mariusz G. Banaszczyk, Patricia M. Mullen, Christopher C. Coffin, Dongpei Wu, Alison T. Carlo, Donna L. Bassett, Erin K. Gouveia, and Dennis J. Carlo

1. Introduction

Nonviral gene delivery has great potential for replacement of recombinant protein therapy. In many cases, gene therapies would be a considerable improvement over existing therapies because of putative advantages in dosing schedule, patient compliance, toxicity, immunogenicity, and cost. Development of a nonviral gene delivery vehicle capable of efficient, cell-specific delivery will be a valuable addition to the clinical armamentarium.

The current situation has led to a focus on increasingly complex delivery systems as investigators try to achieve the delivery efficiency that viral systems already demonstrate. It will be very difficult to create a self-assembling gene delivery system that incorporates molecular mechanisms similar to those that allow viruses to trespass on vascular, cellular, and intracellular barriers and effectively deliver viral DNA to the nucleus of mammalian cells. However, much progress has been made with regard to production of uniform particles. Steric stabilization of materials in vascular compartments has been an area of intense investigation, and numerous strategies for surface modification of delivery vehicles have shown positive effects *(1–6)*. Incorporation of molecular components to accomplish receptor-mediated targeting, endosomal escape, and nuclear transport have all been attempted, with some success in vitro *(7,8)*.

From: *Methods in Molecular Medicine, Vol. 69, Gene Therapy Protocols, 2nd Ed.*
Edited by: J. R. Morgan © Humana Press Inc., Totowa, NJ

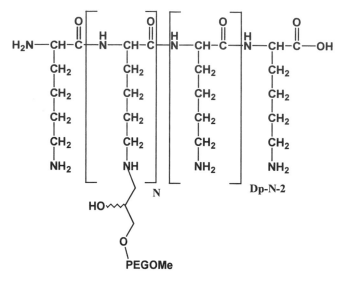

Fig. 1. Sample grafts.

1.1. Poly-L-Lysine

Poly-L-lysine (PLL) is a linear, biodegradable polymer that can be readily modified with a variety of chemical reagents to create novel conjugates with enhanced characteristics over those present in PLL *per se*. In the gene delivery arena, researchers have typically tried to mimic characteristics of proteins that enable viruses to deliver their DNA or RNA payload so efficiently. Thus, many synthetic chemists have focused on incorporating moieties that can facilitate cell-specific targeting, membrane penetration, and nuclear transport. Another common synthetic goal is to modify PLL so that it can protect the DNA payload effectively. More specifically, the intent is to diminish deleterious in vivo interactions such as immunogenicity, toxicity, adventitious binding, and uptake by the reticuloendothelial system. PLL can be grafted with various agents to alter polyplex performance characteristics depending on desired outcome and area of investigation. Cationic polymers other than PLL have also been modified and characterized in a similar fashion *(9–11)*.

1.2. Grafting

Grafts can consist of any natural or synthetic polymer, linear or branched, cyclic, heterocyclic, containing heteroatoms, or any combination of grafting molecules. The number of grafted chains can be varied to suit specific applications (**Fig. 1**).

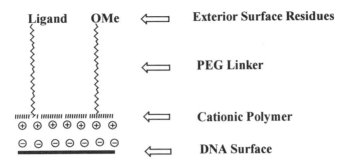

Fig. 2. Grafting of receptor ligands onto a cationic polymer.

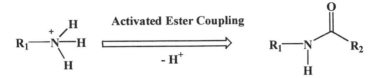

Fig. 3. Reaction of an activated ester with an amino group. An amide bond-linked conjugate is produced.

1.3. Ligand

To achieve cell-specific targeting, receptor ligands can be grafted onto PLL or other cationic polymers *(12–14)*. The preferred position of a ligand is on the exterior surface to ensure proper ligand recognition. However, it is conceivable that ligands may also be partially buried and subject to molecular mechanisms that expose them at an appropriate time *(15)*. Polymers, like polyethylene glycol (PEG), that are grafted onto surfaces form statistical clouds that are continually in flux. Therefore, simple covalent attachment of ligand onto the terminal end of a polymeric chain does not guarantee ligand recognition. The linker polymer, graft density, and chemistry will probably have to be optimized for individual cases (**Fig. 2**).

1.4. Graft Attachment

Nucleophilic substitution of activated esters is the most common chemistry to graft polymeric chains onto amino groups of proteins, cationic polymers, or more specifically PLL *(16)*. The reaction of an activated ester with an amino group produces an amide bond-linked conjugate and results in a net loss of charge on the conjugate (**Fig. 3**).

This loss of positive charge along the polymer chain significantly weakens the binding of conjugate to DNA. Conversely, chemistry that preserves the charge of the cationic domain is expected to have a lessened impact on DNA

Fig. 4. Reaction of an electrophilic reagent with an ε-amino group of PLL.

binding since the binding will be affected only by steric hindrance generated from the grafted moieties. For synthesis of our conjugates, we have chosen chemistries that preserve charges on the cationic domain and typically produce secondary and tertiary amines, and rarely quaternary ammonium species. All these amine species bear a positive charge at physiologic pH and consequently will bind to DNA electrostatically. The first method described below uses PEG–epoxide as the electrophilic reagent that reacts with ε-amino groups of PLL. The product of the reaction is a secondary amine with a racemic β-hydroxyl group (**Fig. 4**).

Grafts can be added successively if more than one feature is desired. Alternatively, the grafting molecule can be engineered to contain more than one functional domain.

1.5. Conjugate Synthesis, Purification, and Characterization

A variety of grafted PLL conjugates have been successfully synthesized *(17)*. These copolymers (e.g., poly-L-lysine-*graft*-R_1-*graft*-R_2-*graft*-R_3) can have a variety of molecules grafted on amino groups of cationic polymers in a stepwise synthesis. For example, PEG molecules can be grafted first (R_1), followed by introduction of other functional groups such as ligands (R_2), and finally fluorescent tags or other delivery-enhancing moieties (R_3). The synthesis of one grafted copolymer is described below in stepwise fashion. The procedure can be repeated to add other grafted domains.

2. Materials
2.1. Chemicals

1. Phosphate (J.T. Baker, Phillipsburg, NJ).
2. SP Sepharose FF resin (Amersham Pharmacia, Uppsala, Sweden).
3. NaOH (J.T. Baker).
4. NaCl (J.T. Baker).
5. PLL 10K (Sigma, St. Louis, MO).
6. Lithium hydroxide monohydrate (E.M. Science, Gibbstown, NJ).
7. Methanol (VWR Scientific Products, West Chester, PA).
8. BioCad 700E HPLC (PE Biosystems, Foster City, CA).
9. UV/VIS detector (PE Biosystems).
10. Glacial acetic acid (J.T. Baker).

11. PEG5K-epoxide (Shearwater Polymers, Huntsville, AL).
12. Sephadex G-25 fine resin (Amersham Pharmacia).
13. Trilactosyl aldehyde (Contract synthesis, e.g., SRI International, Menlo Park, CA).
14. Amino-PEG3.4k-amino-tBOC (Shearwater Polymers).
15. Sodium cyanoborohydride (Alfa Aesar, Ward Hill, MA).
16. Methyl iodide (Aldrich, Milwaukee, WI).
17. Trifluoroacetic acid (J.T. Baker).
18. Methylene chloride (VWR Scientific Products).
19. Succinimidyl bromoacetate (Molecular Biosciences, Boulder, CO).
20. Acetonitrile (J.T. Baker).

2.2. Materials for DNA Manipulation

1. Tris(hydroxymethyl)aminomethane (J.T. Baker).
2. EDTA (J.T. Baker).
3. Ethidium bromide (Sigma).

2.3. Materials for Animal Studies

1. Ketamine (Phoenix Pharmaceuticals, St. Joseph, MO).
2. Xylazine (Phoenix Pharmaceuticals).
3. Acepromazine (Fermenta Vet. Products, Kansas City, MO).
4. Potassium phosphate (J.T. Baker).
5. Triton X-100 (VWR Scientific Products).
6. Sigma Firefly luciferase L-5256 (BD Pharmingen, San Diego, CA).
7. 15-mL dounce homogenizer (Wheaton, Millville, NJ).

3. Methods

3.1. Synthesis of Poly-L-Lysine-graft-R_1-graft-R_2-graft-R_3 Copolymers

R_1 means PEG derivative and R_2 and R_3 no PEG derivative.

PLL-*graft*-PEG polymers can be prepared by reaction of a PEG-electrophile with ε-NH_2 lysine groups under basic conditions. For any specific copolymers, the ratio of activated PEG to poly-L-lysine, PEG size, and poly-L-lysine size can be varied as needed.

1. Poly-L-lysine 10K (600 mg, 0.06 mmol) and lithium hydroxide monohydrate (41 mg, 2.9 mmol) are dissolved in water (2 mL) and methanol (6 mL) in a siliconized glass flask.
2. Solid PEG5K-epoxide (600 mg, 0.12 mmol) is added to the flask, which is then sealed, and the solution is incubated at 65°C for 48 h.
3. After incubation, the solvent is removed *in vacuo*. The product is redissolved in a loading buffer (0.1 M sodium phosphate, pH 6, in 10% MeOH [v/v]).
4. The solution is loaded on a cation exchange column (SP Sepharose FF resin) attached to a high-performance liquid chromatography (HPLC) device (e.g., BioCad 700E), followed by an extensive washing step (up to 10 column volumes).

5. The product is eluted with 0.1 *N* NaOH in 10% MeOH solution. An in-line 214-nm UV/VIS detector is used to monitor the eluant, and fractions are collected in a standard manner.
6. Fractions containing the product are combined and neutralized, and the solvent is removed *in vacuo*.
7. The dried product, which contains inorganic salts, is redissolved in a minimum amount of 0.05 *M* acetic acid in 30% MeOH solution and separated over a G-25 column (Amersham Pharmacia Sephadex G-25 fine resin) using the same acetic acid solution.
8. The fractions are pooled and lyophilized. The average number of PEG moieties grafted onto each poly-L-lysine chain can be determined by ^{1}H nuclear magnetic resonance (NMR) *(18)*.

3.2. Synthesis of PL26k-graft-(ε-NH-CH$_2$CO-NH-PEG-ε-Trilactose-Ligand)$_{2.5}$

Stepwise grafting is one of the simplest ways to modulate properties of resulting copolymers. However, it does not provide easy means of incorporating targeting moieties at their optimal positions. The linker bearing the ligand should be at least as long (or longer) as other components grafted onto the cationic domain. Otherwise, the ligand could be buried and thus unavailable for binding interactions. Several heterobifunctional PEGs (abbreviated as X-PEG-Y) are commercially available in a 3.4-kDa size. These X-PEG-Y molecules can be used to connect ligands to cationic domains. An example of this type of synthesis is shown in **Fig. 5**.

1. Trilactosyl aldehyde (100 mg, 0.067 mmol) is stirred in water (0.5 mL) under argon.
2. Amino-PEG3.4k-amino-t-BOC (151 mg, 0.04 mmol) and lithium hydroxide (1.7 mg, 0.04 mmol) dissolved in methanol (1 ml) are then added to the trigalactosyl aldehyde solution and stirred under argon at 25°C for 30 min.
3. Two portions of sodium cyanoborohydride (6.2 mg, 0.1 mmol) are then added over a 24-h period.
4. Methyl iodide (568 mg, 4 mmol) is added and the solution stirred for 24 h.
5. The solution is then evaporated to dryness, and trifluoroacetic acid (0.7 mL) in methylene chloride (0.6 mL) is added.
6. The solvents are again evaporated to dryness and the residue redissolved in a mixture of methanol and water (3 mL).
7. The solution is adjusted to pH 9 with 10 *N* sodium hydroxide. Succinimidyl bromoacetate (118.5 mg, 0.5 mmol) is then added in acetonitrile (0.5 mL), and the mixture is stirred under argon at 25°C for 1 h.
8. The bromoacetyl intermediate is eluted over a Sephadex G-25 column in 0.05 *N* acetic acid.
9. The macromolecular fractions are combined and evaporated *in vacuo*.
10. Poly-L-lysine 26k (27.7 mg, 0.001 mmol) and lithium hydroxide (4.6 mg, 0.11

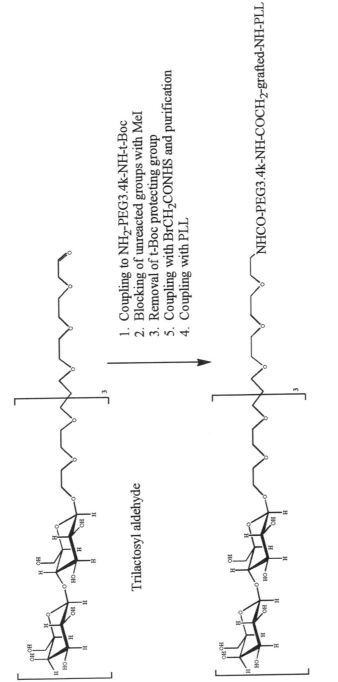

Fig. 5. An example of stepwise grafting.

Trilactosyl aldehyde

1. Coupling to NH₂-PEG3.4k-NH-t-Boc
2. Blocking of unreacted groups with MeI
3. Removal of t-Boc protecting group
5. Coupling with BrCH₂CONHS and purification
4. Coupling with PLL

NHCO-PEG3.4k-NH-COCH₂-grafted-NH-PLL

mmol) dissolved in methanol (1.5 mL) are added to the solution of iodoacetyl intermediate.

11. The reaction mixture is sealed and incubated overnight at 37°C.
12. The product is purified by SP Sepharose FF and Sephadex G-25 column chromatography.
13. The ratio of triantennary galactose/PEG/PLL is determined by ^{1}H NMR.

3.3. ^{1}H NMR Spectroscopy

1. Each polymer is first freeze-dried from D_2O and redissolved in D_2O for spectral analysis. This procedure minimizes the HOD peak and gives superior spectra.
2. ^{1}H NMR spectra are recorded on a high-resolution spectrometer (e.g., 300 MHz ARX-300 Bruker).
3. Chemical shifts are expressed in parts per million and referenced to the HDO signal at 4.7 ppm.
4. The integration ratio of PEG signal (3.68 ppm) to C_α-H of poly-L-lysine (4.2 ppm) is used to determine the composition of the copolymer.
5. The number of C_α-H protons per PLL molecule is calculated from the MW and known structure. For example, 10-kDa polylysine has 48 C_α-H.
6. The number of methylene protons ($-CH_2-$) per PEG molecule is calculated from the MW and known structure. For example, 5-kDa PEG has 454 methylene protons.
7. The number determined in **step 6** is divided by the number determined in **step 5** to yield the proton ratio expected for a 1:1 conjugation of PEG and PL.
8. The ratio computed in **step 4** is divided by the ratio computed in **step 7** to yield the average number of PEG grafts per PLL molecule.

3.4. Plasmid DNA

Preparation and purification of plasmid DNA is beyond the scope of this chapter, but a few salient points need to be made as to the use of plasmid DNA for polyplex formation and transfection studies. These remarks assume that the plasmid was constructed properly, contains the proper elements, and is known to express at reasonable levels in transfection assays in vitro. Plasmid DNA should be assayed by agarose gel electrophoresis with ethidium bromide staining to determine purity and relative amounts of linear and covalently closed circular forms including the super-coiled form. For best results, plasmid DNA used in transfection studies should be ≥90% in the covalently closed circular form. Plasmid DNA should be stored below 4°C in an appropriate buffer (e.g., 10 m*M* Tris(hydroxymethyl)aminomethane, 1 m*M* EDTA, pH 8.0). DNA preparations must be tested for endotoxin levels using the limulus amebocyte lysate assay (Bio-Whittaker, Walkersville, MD) or other methods *(19)*. Contamination should not exceed 10 endotoxin units per milligram of plasmid DNA.

3.5. Charge Ratio Determinations

Charge ratios (+/-) can be determined by several methods, and it is recommended that at least two independent methods be used to characterize conjugates. We recommend using a theoretical calculation based on composition combined with a fluorescence quenching assay.

3.6. Calculation Based on Composition

1. From the proton NMR data, calculate the expected molecular weight of the conjugate.
2. From the known composition of the conjugate, calculate the number of positive charges on each conjugate molecule.
3. Calculate the conjugate mass per positive charge (**step 1/step 2**).
4. The mean mass per unit negative charge for plasmid DNA is 330.
5. Conjugate mass per unit charge (**step 3**) divided by DNA mass per unit charge (330) is the theoretical mass ratio (R) to form a neutral polyplex.
6. To manufacture a polyplex at a given charge ratio, use the following equation: mass of conjugate = desired polyplex charge ratio × DNA mass × R

3.7. Fluorescence Quenching Assay

The binding abilities of polycationic polymers were examined using an ethidium bromide-based quenching assay.

1. Solutions (1 mL) containing 2.5 μg/mL ethidium bromide and 10 μg/mL DNA (1:5 molar ratio, EtBr/DNA phosphate) are prepared.
2. Highly concentrated aqueous conjugate solutions (≥1 mg/mL) are used to minimize the effect of dilution after multiple additions.
3. Fluorescence reading is taken of the DNA solution prepared in **step 1**, using a fluorometer with excitation and emission wavelengths at 540 and 585 nm, respectively.
4. Aliquots of the conjugate solution prepared in **step 2** are added incrementally to the DNA solution, and fluorescence readings are taken after each addition. Aliquots should be <10 μL and should contain enough conjugate to neutralize approximately 10% of the DNA charge.
5. Fluorescence reading after each addition is divided by fluorescence value for the DNA sample from **step 3** and multiplied by 100 to give a percent value. All readings have background subtracted.
6. Conjugate aliquots are added until no further change in fluorescence is achieved.
7. Results should be analyzed as the percentage of fluorescence relative to the control with no polycation.

3.8. Polyplex Formation

1. Polyplexes are typically formed at a 1.35± charge ratio and a final DNA concentration between 10 and 100 μg/mL (*see* **Note 1**).

2. An aqueous DNA solution is prepared at approximately twice the desired polyplex concentration.
3. An aqueous conjugate solution is prepared at approximately twice the desired polyplex concentration.
4. The 2× conjugate solution is added rapidly to the 2× DNA solution, and the solution is vigorously mixed.
5. Formulant is added if necessary.
6. Sufficient 5 M NaCl is added to achieve a final concentration of 150 mM.
7. The solution is vortexed briefly.
8. Filtration through a 0.2-µm filter is necessary for sterile applications.

3.9. Particle Size Analysis

Light scattering measurements of mean particle size and distribution of polyplex solutions can be determined on any of a variety of particle size analyzers, for example, a Brookhaven Instruments 90 Plus particle size analyzer equipped with a 50-mW, 532-nm laser or a Coulter N4 Plus PCS analyzer with 10-mW helium-neon 632.8-nm laser.

Reagents are filtered through a 200-nm surfactant-free cellulose acetate filter (NalgeNunc, Rochester, NY) prior to polyplex formation. Polyplex concentrations should be 30–75 µg/mL. Sample volume is 0.5–1 mL, and measurements are made in 4.5-mL methyl acrylate cuvettes (Evergreen Plastics, Los Angeles, CA). Results can be reported as effective diameter defined as the average diameter that is weighted by the intensity of light scattered by each particle.

It should be noted that the equations used to determine the effective diameter assume that the particles being measured are spherical. Typically, no correction is made to account for nonspherical particles, and since DNA condensed with PLL forms toroidal or rod-shaped particles, the measured effective diameters should be considered an approximation of the actual dimension of the polyplexes.

3.10. Luciferase Gene Expression Studies

A cohort of 8–10-week-old Balb/C mice is anesthetized with an 80-µL intramuscular injection of a cocktail containing 25 mg/mL ketamine (Phoenix Pharmaceuticals), 2.5 mg/mL xylazine (Phoenix Pharmaceuticals), and 5 mg/mL of acepromazine (Fermenta Vet. Products) in saline (*see* **Notes 2** and **3**).

After sedation, animals are injected in the tail vein with 0.2–0.5 mL of polyplex containing 15 µg pCMV-luciferase plasmid DNA (*see* **Notes 4** and **5**). Tuberculin syringes (1 mL; Becton-Dickinson, Franklin Lakes, NJ) can be used for administration of both anesthetic and polyplex.

Twenty-four hours after injection, the mice are euthanized by carbon dioxide inhalation. The livers are excised, homogenized with lysis buffer (100 m*M* potassium phosphate, 0.2% Triton X-100, pH 7.8), and analyzed against a luciferase standard curve (Sigma Firefly luciferase L-5256) using commercially available substrate solutions (BD Pharmingen) (*see* **Note 6**). Samples are read using a standard luminometer (e.g., Analytical Luminescence model #2010, BD Pharmingen).

4. Notes

1. Polyplexes can be formed at higher or lower ratios to meet specific needs or to test other protocols. Near neutral polyplexes are recommended for intravenous delivery. Polyplexes with high positive charge work best for in vitro work.
2. Animal studies can be done without anesthesia during administration, but our experience is that anesthetized animals generally give higher gene expression.
3. Anesthetic reagents from vendors are received at the following concentrations: 100 mg/mL ketamine, 20 mg/mL xylazine, and 10 mg/mL acepromazine. Prepare a stock solution for animal studies by combining 7.5 mL of ketamine, 3.8 mL of xylazine, 0.75 mL of acepromazine, and 17.95 mL of saline. This solution has the proper concentrations of each component such that 80 µL is suitable to anesthetize a mouse.
4. Solutions for intravenous injections should be at ambient temperature or body temperature whenever feasible. Cool or cold temperature solutions result in lower gene expression.
5. Rapid injections into the tail vein give the best results but are not truly representative of a clinically applicable method *(20,21)*.
6. A 15-mL Dounce homogenizer is used to grind each liver. The liver is rinsed with phosphate-buffered saline and weighed. The liver is then placed in a 15-mL Dounce homogenizer to which is added a volume of lysis buffer equal to liver weight multiplied by 10 (e.g., 1 g of liver would have 10 mL of lysis buffer). The liver is well homogenized, and the entire volume is centrifuged at 1000 rpm at 4°C for 15 min in a 15-mL conical tube. The fluid separates into a pellet, middle aqueous layer, and upper lipid layer. From the middle aqueous layer, 1.5 mL is aliquoted into an Eppendorf tube and recentrifuged for 5 min at 14,000 rpm. Three layers form again, and the middle aqueous layer is collected for assay.

References

1. Uster, P. S., Allen, T. M., Daniel, B. E., et al. (1996) Insertion of poly(ethylene glycol) derivatized phospholipid into pre-formed liposomes results in prolonged in vivo circulation time. *FEBS Lett.* **386,** 243–246.
2. Watrous-Peltier, N., Uhl, J., Steel, V., Brophy, L., and Merisko-Liversidge, E. (1992) Direct suppression of phagocytosis by amphipathic polymeric surfactants. *Pharm. Res.* **9,** 1177–1183.
3. Toncheva, V., Wolfert, M. A., Dash, P. R., et al. (1998) Novel vectors for gene delivery formed by self-assembly of DNA with poly(L-lysine) grafted with hydrophilic polymers. *Biochim. Biophys. Acta* **1380,** 354–368.

4. Lasic, D. D. and Needham, D. (1995) The "stealth" liposome: a prototypical bio-material. *Chem. Rev.s* **95,** 2601–2628.
5. Lollo, C. P., Kwoh, D. Y., Mockler, T. C., et al. (1997) Non-viral gene delivery: vehicle and delivery characterization. *Blood Coagul. Fibrinol.* **8,** S31–38.
6. Kwoh, D. Y., Coffin, C. C., Lollo, C. P., et al. (1999) Stabilization of poly-L-lysine/DNA polyplexes for in vivo gene delivery to the liver. *Biochim. Biophys. Acta* **1444,** 171–190.
7. Zanta, M. A., Belguise-Valladier, P., and Behr, J. P. (1999) Gene delivery: a single nuclear localization signal peptide is sufficient to carry DNA to the cell nucleus. *Proc. Natl. Acad. Sci. USA* **96,** 91–96.
8. Curiel, D. T., Wagner, E., Cotton, M., et al. (1992) High-efficiency gene transfer mediated by adenovirus coupled to dna-polylysine complexes. *Hum. Gene Ther.* **3,** 147–154.
9. Wolfert, M. A., Dash, P. R., Nazarova, O., et al. (1999) Polyelectrolyte vectors for gene delivery: influence of cationic polymers on biophysical properties of complexes formed with DNA. *Bioconjug. Chem.* **10,** 993–1004.
10. Choi, J. S., Joo, D. K., Kim, C. H., Kim, K., and Park, J. S. (2000) Synthesis of a Barbell-like triblock copolymer, poly(L-lysine) dendrimer-*block*-poly(ethylene glycol)-*block*-poly(L-lysine) dendrimer, and its self-assembly with plasmid DNA. *J. Am. Chem. Soc.* **122,** 474–480.
11. Yoshikawa, K., Yoshikawa, Y., Koyama, Y., and Kanbe, T. (1997) Highly effective compaction of long duplex DNA induced by polyethylene glycol with pendant amino groups. *J. Am. Chem. Soc.* **119,** 6473–6477.
12. Plank, C., Zatloukal, K., Cotton, M., Mechtler, K., and Wagner, E. (1992) Gene transfer into hepatocytes using asialoglycoprotein receptor mediated endocytosis of DNA complexed with an artificial tetra-antennary galactose ligand. *Bioconjug. Chem.* **3,** 533–539.
13. Perales, J. C., Grossman, G. A., Molas, M., et al. (1997) Biochemical and functional characterization of DNA complexes capable of targeting genes to hepatocytes via the asialoglycoprotein receptor. *J. Biol. Chem.* **272,** 7398–7407.
14. Wadhwa, M. S., Knoell, D. L., Young, A. P., and Rice, K. G. (1995) Targeted gene delivery with a low molecular weight glycopeptide carrier. *Bioconjug. Chem.* **6,** 283–291.
15. Harris, J. M. and Zalipsky, S., eds. (1997) *Poly(Ethylene Glycol) Chemistry and Biological Applications.* ACS, Washington, DC, pp. 170–181.
16. Hermanson, G. T. (1996) *Bioconjugate Techniques.* Academic, San Diego.
17. Banaszczyk, M. G., Lollo, C. P., Kwoh, D. Y., et al. (1999) Poly-L-lysine-*graft*-PEG comb-type polycation copolymers for gene delivery. *J.M.S. Pure Appl. Chem.* **A36(7&8),** 1061–1084.
18. Dust, J. M., Fang, Z., and Harris, M. (1990) Proton NMR characterization of poly(ethylene glycols) and derivatives. *Macromolecules* **23,** 3742–3746.
19. U.S. Department of Health and Human Services, Public Health Service, Food and Drug Administration. (1987) *Guideline on Validation of the Limulus Amebocyte*

Lysate Test as an End-Product Endotoxin Test for Human and Animal Parenteral Drugs, Biological Products, and Medical Devices. DHHS, Washington, DC.

20. Liu, F., Song, Y. K., and Liu, D. (1996) Hydrodynamics-based transfection in animals by systemic administration of plasmid DNA. *Gene Ther.* **6,** 1258–1266.

21. Zhang, G., Budker, V., and Wolff, J. A. (1999) High levels of foreign gene expression in hepatocytes after tail vein injections of naked plasmid DNA. Hum. Gene Ther. **10,** 1735–1737.

2

Targeted Gene Transfer to Liver Using Protein-DNA Complexes

Catherine H. Wu, Cherie M. Walton, and George Y. Wu

1. Introduction

The advantages of nonviral carriers are their ease of preparation and scale-up, capacity of DNA to be transferred, and safety in vivo. However, there also are disadvantages, including generally low efficiency and transience of transgene expression. To create more efficient systems, the use of approaches present in natural pathogens has been shown to be helpful. Based on an understanding of these natural components, ligand-polycation DNA delivery systems have been developed (1–3). In these systems, a DNA-binding polycation, such as polylysine (PL) was employed to compact plasmid DNA to a size that could be taken up by cells. To allow internalization by receptor-mediated endocytosis, cell binding ligands such as asialoglycoproteins for hepatocytes, anti-CD3 and anti-CD5 antibodies for T-cells, transferrin for some cancer cells, and hyaluronic acid polymers for endothelial cells have been covalently attached to polylysine.

Because the liver plays a central role in the metabolism and production of serum proteins, it is an important target organ for gene therapy. Metabolic diseases that result from a defect or deficiency of hepatocyte-derived gene products, as well as acquired diseases such as hepatocellular carcinomas and viral hepatitis, may also serve as targets for hepatic gene therapy. To be clinically useful, all require the development of delivery systems capable of efficiently introducing nucleic acids into the hepatocytes.

Parenchymal liver cells, hepatocytes, are useful target cells for gene delivery, as they are highly active metabolically, have a substantial blood supply

From: *Methods in Molecular Medicine, Vol. 69, Gene Therapy Protocols, 2nd Ed.*
Edited by: J. R. Morgan © Humana Press Inc., Totowa, NJ

and hepatocytes are the only cells that possess large numbers of high affinity cell-surface receptors that can bind asialoglycoproteins *(4)*.

Our early work in this area demonstrated that DNA could be delivered specifically to, and expressed in, the liver cells in vivo with an asialoglycoprotein-mediated system *(1,2)*. However, the efficiency in vivo has been poor. We have previously shown that incorporation of an endosome disruptive peptide into the delivery system could greatly increase the specific gene expression to liver in vivo. Recently, improvements have been undertaken to engender the DNA delivery system with high water solubility, serum stability and high gene expression efficiency. In some systems, polyethylene glycol (PEG) provides a biocompatible protective coating for the DNA complex. An endosomolytic peptide derived from Vesicular Stomatitis Viral G-Protein (VSV) or the bacterial protein, listeriolysin O (LLO), can be introduced to produce conjugates that can induce membrane changes at low pH allowing the internalized DNA to escape from lysosomal digestion. Finally, the targeting ligand itself can be converted to a DNA binding protein eliminating the need for a separate polycation.

2. Materials

2.1. Plasmid and Reporter Gene

A plasmid pCMVLuc containing a firefly luciferase gene driven by a cytomegalovirus (CMV) immediate early promoter was amplified in E. coli, isolated by alkaline lysis, and purified by cesium chloride gradient centrifugation. Ultra Pure cesium chloride was obtained from Life Technologies (Grand Island, NY).

2.2. Cells and Cell Culture

1. A hypersecretor strain of *Listeria monocytogenes* (gift of Dr. D. A. Portnoy, Stanford University).
2. Brain Heart Infusion media (Difco, Detroit, MI).
3. LB Broth (Life Technologies).
4. Huh7 human hepatoblastoma (asialoglycoprotein receptor positive) and SK Hep1 human hepatoma (asialoglycoprotein receptor negative) cells, grown to confluence in Dulbecco's modified Eagle's medium (DMEM) containing 10% fetal calf serum (Gibco/BRL, Grand Island, NY) under 5% CO_2 at 37°C.

2.3. Components of DNA Carriers Targetable to Liver

1. Polylysine (PL)(MW 3970) HBr.
2. Polyethylene glycol (PEG)-succinyl ester (MW 5000).
3. Potassium sulfate.
4. Dimethyl sulfoxide (DMSO).

5. Dithiothreitol (DTT).
6. Sodium chloride (NaCl).
7. Sodium acetate.
8. Lysine ester.
9. Sodium hydroxide (NaOH).
10. Sodium dodecyl sulfate (SDS).
11. Ammonium bicarbonate (NH_4HCO_3).
12. Ethylenediamine tetraacetic acid (EDTA).
13. Ethidium bromide.
14. Heparin.
15. Tetrahydrofuran (THF). **Items 1–15** from Sigma Chemical Co. (St. Louis, MO).
16. Succinimidyl 3-(2-pyridyldithio) propionate (SPDP).
17. 1-ethyl-3-(3-dimethylaminopropyl)-carbodiimide (EDC). **Items 16** and **17** from Pierce Chemical Co. (Rockford, IL).
18. Ultrapure agarose (Life Technologies, Grand Island, NY).
19. A vesicular stomatitis virus G peptide (VSV) of the following sequence: TIVFPHNQKGNWKNVPSNYHYCP.
20. Human asialoorosomucoid (ASOR). **Items 19** and **20** from Immune Response Corporation (Carlsbad, CA).
21. Dialysis membranes (12-14 kD exclusion limits; Spectra/Por, Spectrum Medical Industries, Houston, TX).
22. A S1Y30 spiral cartridge of 30,000 molecular weight cut off was purchased from Amicon Inc. (Beverly, MA).
23. A 10 cm DEAE Sephacel column.
24. PD-10 (diameter, 5 cm, containing Sephadex G-25 resin) desalting columns.
25. Whatman #1 paper. **Items 23–25** from Amersham Pharmacia Biotech (Piscataway, NJ).
26. TSK-GEL CM-650 S, 40–90 µm (Supelco, Inc.) was packed into a 2 × 10 cm column.
27. Bio-gel P-6 (Bio-Rad Lab.) was packed into a 2 × 50 column.
28. Syringe filters, 0.2 µ and 0.45 µm (Acrodisc, Gelman Sciences, Ann Arbor, MI).
29. A Waters HPLC system using a Shodex KW-804 column (300 × 8 mm; Waters Corporation, Milford, MA) was used for purification of some conjugates.

2.4. Animals

Balb C female mice (approx 20 g body weight; Charles River Laboratory, Wilmington, MA) were housed under controlled conditions of temperature and humidity, and fed normal chow ad libitum.

3. Methods

3.1. Synthesis of Asialoorosomucoid-Polylysine (AsOR-PL) Conjugates

1. Filter AsOR, 200 mg dissolved in 10 mL of water, through a 0.2-µm syringe filter, and adjust the solution to pH 7.4.

2. Dissolve PL, 160 mg, in 10 mL water, adjusted to pH 7.4 with 0.1 N NaOH.
3. Dissolve EDC, 92 mg, in 1 mL water and add directly to the AsOR solution.
4. Add the PL solution to the mixture and stir at 37°C for 24 h.
5. Dialzye the reaction mixture at 4°C through a membrane with 12–14 kDa exclusion limits against 20 L of water for 2 days.

3.1.1. Purification of Listeriolysin O (LLO)

A convenient pH-sensitive endosomolytic protein that has been found to enhance the efficiency of targeted gene delivery is listeriolysin O. This protein can be recovered from cultures of *L. monocytogenes* (*see* **Note 1**).

1. Inoculate a stab from a frozen culture of L. monocytogenes into 15 mL of Brain Heart Infusion medium and incubate with shaking overnight at 37°C.
2. Add the overnight culture to 1 L of Luria-Bertani (LB) broth, which is prewarmed to 37°C.
3. We recommend growing 6 L per purification batch, each grown for 15 h.
4. Remove bacteria by centrifugation at 10,000g for 15 min at 4°C.
5. Filter the supernatant through Whatman #1 filter paper, keeping the receiving flask on ice.
6. Apply 6 L of chilled supernatant to a CH_2 spiral cartridge concentrater with a S1Y30 spiral cartridge of 30,000 mol wt cut-off, concentrate to 500 mL. Add a total of 4 L of chilled water to the concentrator, and reconcentrate the entire volume to 500 mL to remove small proteins.
7. Apply the 500 mL of retentate to a 20-mL, 10-cm DEAE Sephacel column, which is equilibrated with 10 mM potassium phosphate, pH 6.8, and elute in a single pass-through.
8. Lyophilize the LLO sample, redissolve it in water, desalt it by application to a PD-10 desalting column, and elute it with 5 mL of water. Determine the protein peak by reading absorbencies at 280 nm.
9. Pool this peak and lyophilize it. Store samples either at –20°C as the lyophilized dry powder or redissolve in water and freeze *(5)*.

3.1.2. Synthesis of Asialoorosomucoid (AsOR)-PL-LLO Conjugates

1. Incubate 1 mg of ASOR-PL and LLO separately with 25 mM SPDP in dimethylsulfoxide (OMSO) for 30 min at 25°C.
2. Separate free SPDP from protein-linked by SPDP application to a PD-10 desalting column and elute with water.
3. Determine the concentration of SPDP linked to the proteins by measuring the release of 2-thione after reduction with 100 mM dithiothreitol (DTT) and reading the absorbance at 343 nm.
4. Activate the LLO-SPDP for coupling by reduction with 12 mg DTT in 100 mM NaCl, 100 mM Na acetate, pH 4.5.
5. Remove free DTT by application to a PD-10 desalting column and elute with water *(6)*.

3.1.3. AsOR-PL-LLO Complexes

1. Mix 1 mg of the SPDP-linked AsOR-PL with 0.1 mg CMV luc DNA and incubate for 30 min at room temperature in 0.15 M saline.
2. Add the AsOR-PL-SPDP-DNA complex to DTT-reduced LLO-SPDP at a 2:1 molar ratio.
3. Incubate the complex overnight at 4°C and filter through a 0.2 µm syringe filter *(6)*.

3.2. Synthesis of AsOR-Lysine Methyl Ester

To avoid the concentration of positively charged polycations such as polylysine, other derivatives of AsOR can be prepared in which the overall charge of the protein is made strongly positive. This can be accomplished by covalently coupling esters of dibasic amino acids.

1. Dissolve 10 mg AsOR in 1 mL water and filter through a 0.2-µm syringe-tip filter.
2. Dissolve 50 mg lysine ester in 1 mL water, add to the AsOR, and adjust the pH to 6.5 using 0.1 N NaOH.
3. Add to this mixture 4 mg EDC, dissolved in 1 mL water, and incubate with stirring at 37°C for 5 h, followed by dialysis of the reaction mixture through membranes with 12–14 kDa exclusion limits against 20 L of water at 4°C for 24 h.
4. Lyophilize the dialyzate and then redissolve it in 0.15 M NaCl (10 mg/mL), filter it through a 0.45-µm syringe-tip filter, and apply it on a Waters HPLC system using a Shodex KW-804 column (300 × 8 mm).
5. Inject samples of 250 µL and elute with 0.15 M NaCl at a flow rate of 0.12 mL/min.
6. Collect samples 0.6 mL each and monitor absorption at 280 nm. Analyze samples of the two peaks in the effluent by 12% sodium dodecyl sulfate (SDS) polyacrylamide gel electrophoresis.
7. Use the second peak for further conjugation, and lyophilize corresponding samples.

3.2.1. Preparation of AsOR-Lysine Ester-VSV Conjugates

The single cysteine residue in the VSVG peptide is used to couple peptides to AsOR-PL.

1. Dissolve 5 mg purified AsOR-lysine ester in 1 mL of phosphate-buffered saline (PBS, pH 7.4), and add 0.18 mL of freshly prepared SPDP (3.1 mg/ml in DMSO) to give a 30-fold molar excess of SPDP over AsOR-lysine ester.
2. Incubate the mixture with stirring for 1 h at 25°C.
3. Dialyze the reaction mixture in membranes with 12–14 kDa exclusion limits against 20 L water for 24 h at 4°C and lyophilize.

4. To remove any free SPDP, dissolve the conjugate in water and apply it to a PD-10 desalting column followed by elution with water.
5. Check the eluate for absorption at 280 nm and lyophilize *(7)*.

3.3. Synthesis of PEG-PL Conjugates

1. Dissolve polylysine (PL) 30 mg in 1.5 mL PBS (0.1 M, pH 7.2), adjust it to pH 7–8 with addition of 2 N NaOH.
2. Add 10 mg PEG-succinyl ester and incubate at 23°C for 5 h.
3. Dilute the reaction solution with 10 mL of water and chromatograph on an ion exchange column—TSK-GEL CM-650 (40–90 µm), 2 × 10 cm.
4. Elute the sample with 50 mL of water and then with 200 mL 0–5.0 M NaCl gradient, and then monitor the effluent by UV absorption at 230 nm.
5. Collect the second peak and freeze-dry it; dissolve the powder in 2 mL water.
6. Gel-filter the sample through a 2 × 50-cm Bio-gel P-6 column and elute with 0.2 M NH$_4$HCO$_3$.
7. Collect the first peak in 3 mL and freeze dry to give 10 mg white powder (*see* **Note 2**).

3.3.1. Synthesis of PEG-PL-VSV Conjugates

1. Dissolve PEG-PL, 8.5 mg, in 1.0 mL 0.2 M PBS (pH 7.2) and react with 1.2 mg SPDP in 0.2 mL tetrahydrofuran (THF).
2. After stirring at 23°C for 3 hs, gel-filter the product through a 2 × 50-cm Bio-gel P-6 column and elute with 0.2 M NH$_4$HCO$_3$ buffer.
3. Collect the first peak and lyophilize to give PEG-PL-dithiopyridyl (DTP).
4. Dissolve 7 mg of that product in 1.0 mL PBS (pH 7.2) and add to 0.1 mL 0.5 M EDTA; add to this 9.2 mg of VSV-G in 0.2 mL water and incubate at 23°C for 24 h.
5. Gel-filter the reaction solution through a Bio-gel P-6 column (2 × 50 cm) and elute with 0.2 M NH4HCO3.
6. Collect the first peak and lyophilize to give 6.0 mg powder.

3.3.2. Formation of PEG-PL-VSV/AsOR-PL-DNA Complexes

1. First mix 30 µg DNA in 1.0 mL 0.15 M saline with PEG-PL-VSV in 1.0 mL saline and then further complex it with AsOR-PL in 1.0 mL saline (*see* **Note 3**).
2. Filter the mixed complex formed through a 0.2-µm filter, and determine the DNA concentration by measuring UV absorption at 260 nm.
3. Store the filtered solution at 4°C for all further experiments.
4. Determine the number of amino groups in the conjugate by ninhydrin assay *(8)*, and from that, calculate the content of PL in the conjugate.
5. The average ratio of PEG to PL is determined to be 1.2:1.
6. Modify the conjugate with SPDP to introduce DTP groups for conjugating with VSV peptide.
7. The ratio of DTP to PEG-PL is determined to be 2.7. Monitor the conjugation of

PEG-PL-DTP with VSV by measuring the absorption at 343 nm, because of release of 2-mercaptopyridine *(9)*.

8. Determine the content of VSV in the conjugate from UV absorption at 280 nm; the ratio of components in the conjugate PEG-PL-VSV is determined to be 1.2:1:1.6 (*10*; *see* **Note 4**).

3.4. Measurement of DNA Binding and Compaction

To assess compaction of DNA after complexation with various conjugates, fluorescence of ethidium bromide excluded from DNA complexes was used. Fluorescence studies were performed using a Perkin-Elmer luminescence spectrometer at an excitation wavelength of 516 nm (slit width 6 nm) and an emission wavelength of 598 nm (slit width 10 nm).

1. Add ethidium bromide (1 mM final concentration) to 2.5 mL normal saline solution containing 30 µg DNA and determine a baseline fluorescence.
2. Add aliquots of conjugate slowly to a DNA solution containing 1 mM ethidium bromide, filter through 0.2 µm syringe filters, and measure the fluorescence. Maximum compaction of DNA by conjugate is the point at which no more changes in DNA fluorescence are observed.
3. Correct the fluorescence of the complexes for dilution as a result of the addition of the conjugate solutions, normalized to the fluorescence of free DNA before complexation, which is assigned a value of 100.

3.5. Particle Size and Zeta-Potential

1. To determine the effective hydrodynamic diameter and the net charge of DNA complexes, use a 90 Plus particle size analyzer (Brookhaven).
2. All DNA complexes are in normal saline, filtered through 0.2-µm syringe filters; perform measurements in triplicate at 23°C.

3.6. Stability of Complexes

1. Determine the stability of complexed DNA by incubation of DNA complexes with fresh rat serum or saline for 30 min at 37°C.
2. To determine the status of the DNA in the complexes, release DNA from the complexes by heparin (2000 U/mL) after 30 min of incubation.
3. Analyze all samples on 1% agarose gels.

3.7. Cell Transfections and Luciferase Activity Assays

1. Seed AsOR receptor-positive Huh7 and AsOR receptor-negative SK Hep1cells into 24-well plates.
2. After 20 h, remove media and add 0.5 mL of DMEM media containing 2.0 mM $CaCl_2$; then add 100 µL (1 µg DNA) of DNA as complexes.
3. To assess specificity, add a 100-fold molar excess of free AsOR over that calculated to be present in the complexes to complexes prior to administration to cells.

4. After incubation for 6 h, add 50 μL fetal bovine serum to each well, and incubate further for 20 h.
5. Remove the media, wash the cell layers with PBS, homogenized with 200 μL of tissue lysis buffer (Promega), and centrifuge at 8,000g for 5 min.
6. Mix 20 μL of supernatant solution with 50 μL luciferase substrate, and measure relative light units (RLUs) with a luminometer (Monolight 2001, Analytical Luminescence) for 30 s.
7. Perform all assays in triplicate, and express results as means + SD in units of RLU.

3.8. Liver-Directed Transfection to the Liver in Mice

1. Pass complexes containing 10 μg of pGL3CMVluc in 0.5 mL of 0.15 M NaCl through 0.2-μm syringe-tip filters and immediately inject into the tail veins of mice over 30 s (*see* **Note 5**).
2. After 24 h, and after 7 days sacrifice the animals, remove the livers, wash them with ice-cold PBS, and weigh them.
3. Remove a liver section (approx 100 mg), weigh it, and homogenize it in lysis buffer (100 mg/mL, Luciferase Assay System, Promega); determine liver luciferase activity by luminometry.
4. Perform a standard curve for luciferase using firefly luciferase (1 ng/mL, Analytical Luminescence Laboratory, Ann Arbor, MI) along with the test samples.

4. Notes

1. Care should be taken in the handling and disposal of *L. monocytogenes*, as it can be a significant pathogen in humans.
2. Purified conjugates are hydrolyzed in constant boiling HCl, and then amino acid analysis is performed to determine the ratios of components present. The total number of lysine residues minus the lysine residues expected from AsOR alone provides quantitation of the amount of lysine ester present. The number of aspartic acid residues is used to determine the amount of AsOR in each conjugate, and a lysine ester to AsOR molar ratio is calculated.
3. Concentrations of greater than 1 mg/mL DNA in complexes that lack PEG have an increased tendency to aggregate.
4. Incorporation of endosomolytic agents such as LLO or VSV peptides increases targeted gene expression by 100- to 1000-fold.
5. Tail vein injections in animals must be performed slowly (over the course of 30 s) *(8)*.

Acknowledgments

The secretarial assistance of Martha Schwartz is gratefully acknowledged. This work was supported in part by a grant from the National Institutes of Health, DK-42182 (to G.Y.W.), the Immune Response Corporation (to C.H.W.), and the Herman Lopata Chair for Hepatitis Research (to G.Y.W.).

References

1. Wu, G. Y. and Wu, C. H. (1987) Receptor-mediated in vitro gene transformation by a DNA carrier system. *J. Biol. Chem.* **262,** 4429–4432.
2. Wu, G. Y. and Wu, C. H. (1988) Receptor-mediated gene delivery and expression in vivo. *J. Biol. Chem.* **263,** 14621–14624.
3. Wagner, E., Zenke, M., Cotton, M., Beug, H., and Birnstiel, M. L. (1990) Transferrin-polycation conjugates as carriers for DNA uptake into cells. *Proc. Natl. Acad. Sci. USA* **87,** 3410–3414.
4. Wu, G. Y. and Wu, C. H. (1998) Receptor-mediated delivery of foreign genes to hepatocytes. *Adv. Drug Deliv. Rev.* **29,** 243–248.
5. Walton, C. M., Wu, C. H., and Wu, G. Y. (1999) A method for purification of listeriolysin O from a hypersecretor strain of *Listeria monocytogenes*. *Protein Expression Purif.* **15,** 243–245.
6. Walton, C. M., Wu, G. Y., and Wu, C. H. (1999) A DNA delivery system containing listeriolysin O results in enhanced hepatocyte-directed gene expression. *World J. Gastroenterol.* **5,** 465–469.
7. Schuster, M. J., Wu, G. Y., Walton, C. M., and Wu, C. H. (1999) A multicomponent DNA carrier with a vesicular stomatitis virus G peptide greatly enhances liver-targeted gene expression in mice. Bioconjug. Chem. **10,** 1075–1083.
8. Moore, S. (1968) Amino acid analysis: aqueous dimethyl sulfoxide as solvent for the ninhydrin reaction. *J. Biol. Chem.* **243,** 6281–6283.
9. Carsson, J., Drevin, D., and Axen, R. (1978) Protein thiolation and reversible protein-protein conjugation. N-succinimidyl 3-(2-pyridyldithio)-propionate, a new heterobifuctional reagent. *Biochem. J.* **173,** 723–737.
10. Zhong, B.-H., Wu, G. Y., and Wu, C. H. Progress towards a synthetic virus: a multicomponent system for liver-directed DNA delivery, in *Nonviral Vectors for Gene Therapy* (Findeis, M., ed.), Humana Press, Totowa, NJ, pp. 111–121.

3

Receptor-Directed Molecular Conjugates for Gene Transfer

Assem G. Ziady and Pamela B. Davis

1. Introduction

To circumvent the safety limitations of viral vectors and the cytotoxicity of liposomal carriers, several investigators have used receptor-targeted molecular conjugates to direct gene transfer into mammalian cells in vitro *(1–26)*, and in vivo *(27–40)*. This method has potential for human gene therapy, once it is perfected in animal models. DNA, noncovalently bound to a polycation polymer that is chemically conjugated to a ligand, can be bound to the cell surface and internalized. Various ligands have been used to target cell surface receptors for gene delivery. Some of these receptors (e.g., the asialoglycoprotein receptor [reviewed in **ref.** *41*]) are designed to traffic their cargo to degradation in the lysosomes, whereas other receptors recycle to the cell surface (e.g., transferrin receptor [reviewed in **ref.** *42*]) or transport their ligands across the cell (e.g., polymeric immunoglobulin receptor [reviewed in **ref.** *43*]). Success is enhanced if the receptor displays high specificity for a ligand but low selectivity for attached cargo, constitutive, abundant expression and the capability for bulk uptake. Receptor-directed molecular conjugates have advantages as gene therapy reagents. Receptor targeting confers specificity, immunogenicity is normally low for the polycation and DNA and variable for the ligand, and the packaging capacity is quite large *(44)*, allowing for the inclusion of native promoters or intronic sequences that will enhance gene expression *(45)*. To understand the methods used in receptor-mediated molecular conjugate gene transfer, we must first review the use of such receptors, which have provided specificity in the context of a noninfectious and nontoxic vector.

From: *Methods in Molecular Medicine, Vol. 69, Gene Therapy Protocols, 2nd Ed.*
Edited by: J. R. Morgan © Humana Press Inc., Totowa, NJ

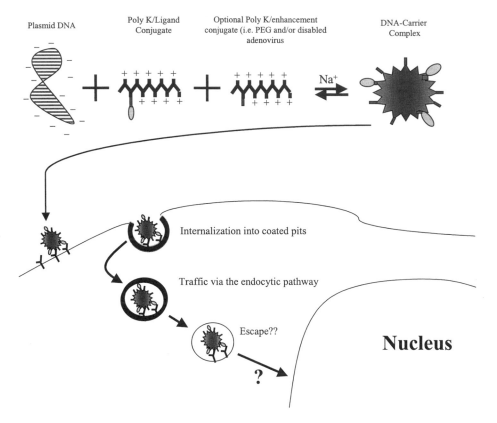

Fig. 1. *General scheme of receptor-targeted DNA complex construction and cellular internalization.* DNA complexes are formed by mixing plasmid DNA with the molecular conjugate under the proper salt conditions. Molecular conjugates consist of a polycation coupled to a receptor ligand. Polycations modified with enhancers (e.g., PEG, adenovirus, and so on) may also be included in the DNA complex. Once in contact with the cell surface receptor, the complex is internalized through the endocytic pathway and translocates to the nucleus by either an active or passive mechanism.

1.1. The Protein Carrier for DNA

Construction of the molecular conjugate begins with the selection of a suitable ligand to target a receptor on a specific cell type. Examples of such ligands include, mono- and disaccharides *(4–8,23,36)*, peptides/proteins *(2,9–13,37)*, glycoproteins *(1,27,28,38)*, lectins *(14)*, folate *(16)*, and antibodies *(3,15,17,25,26,32)*. **Figure 1** provides a general schematic for DNA complex construction and the receptor-mediated gene delivery process. Ligands are covalently linked, often using a linker reagent, to a polycation (e.g., poly-L-lysine), which in turn interacts electrostatically with the negatively charged phosphate backbone of the DNA of interest. Under appropriate conditions, this results in a complex that compacts DNA and protects it against degradation *(37,46)*, making it suitable for gene transfer. The level of substitution of the polycation with linker and ligand, as well as the length and type of polycation, markedly affects the efficiency of these complexes to transfer genes *(13,37,47)*.

Once inside the cell, these DNA complexes must translocate to the nucleus, where the DNA is transcribed. Some investigators suggest that this occurs following endosomal escape *(48)*, but this is based on pharmacologic data rather than direct observation. No published studies have yet examined the trafficking of receptor-targeted DNA complexes; however, investigators have incorporated endosomolytic agents *(49–53)* in the complex or in the media to disrupt the endocytic pathway and have proposed the use of nuclear targeting motifs *(54,55)* to improve nuclear entry. Both these approaches have increased transfection success in specific cell models.

1.2. Molecular Conjugate Condensation of DNA

In initial reports of studies that achieved gene transfer by targeting the asialoglycoprotein receptor *(1,28)*, little attention was paid to the size of the condensed DNA particles, which averaged 150–200 nm by electron microscopy (EM). However, later reports have underscored the importance of minimizing the size of the conjugate-DNA complexes for gene transfer, since endocytosed receptors may discriminate against ligands on the basis of size *(7,37,56)*. For example, Wagner and colleagues *(47)* have shown that condensation into toroid structures 80–100 nm in diameter improved transgene expression compared with larger DNA complexes.

Perales and colleagues *(7,56,57)* developed an alternative method of condensing DNA into very compact particles (12–30 nm in diameter, less than twice the minimum theoretical volume of the DNA) that are suitable for gene transfer. Unlike earlier methods, this technique allows for the stabilization of high concentrations of molecular conjugate DNA complexes, while avoiding the formation of positively charged DNA complexes that maybe inefficient in vivo. Key to this process is the gradual addition of small aliquots of the ligand-polycation conjugate to plasmid DNA over time under high salt conditions. By adjusting the sodium chloride concentration, unimolecular (with respect to DNA) complexes can be produced with a neutral or slightly negative zeta potential *(7)*. This helps avoid complement activation *(58)* and thus provides for a more efficient and safer vector for gene transfer in vivo.

The degree of DNA condensation in conjugate-DNA complexes depends on the concentration of sodium chloride, the length of the polycation polymer, and the degree of substitution of the polycation, as well as the size, sequence, and state of the DNA *(30,37)*. Different length polymers condense DNA under different conditions. Longer polymers require a higher salt concentration to form and maintain these DNA complexes, whereas shorter polymers require less salt *(7,37)*. The secondary structure of the polycation also influences the globular structure of the complexes and their efficacy *(7,37,47)*. Furthermore, the construction of molecular conjugates affects their binding to DNA. For instance, the interaction of the polycation with DNA is destabilized by excessive linker and ligand substitution *(13)*. Substitution usually eliminates positive charge on the polycation and thus lessens its affinity for DNA *(13)*. Steric hindrance by bulky ligands also results in less tightly packed complexes.

Wagner and colleagues *(47)* reported that molecular conjugates with fewer ligand moieties were more effective in delivering reporter genes by receptor-mediated endocytosis. Conjugates containing approximately 1 transferrin per 100 lysine residues resulted in maximal transgene expression in human erythroid cells; conjugates containing more or less ligand were suboptimal. Moreover, the partial replacement of the transferrin-based conjugate with free poly-L-lysine (poly K) produced smaller toroidal structures and improved transfection. In a more detailed analysis with larger ligands designed to target the serpin enzyme complex receptor (SEC-R), Ziady et al. *(13,37)* demonstrated that even less substitution of lysine residues produced optimal transfection complexes *(13,37)*. Both the rate of substitution and the polymer length affected DNA complex size, correlating with an effect on transfection efficiency. The polycation can be further substituted with other moieties such as polyethyleneglycol (PEG), so as to stabilize polycation DNA complexes *(59)*.

Condensation of plasmid DNA with polycations also protects it from degradation. DNA is susceptible to shearing by hydrodynamic forces and has a short half-life in the blood *(60)*. Tightly compacted DNA is thought to be more stable in the blood and, once internalized, in the cytoplasm *(14,46)*. Furthermore, condensed DNA is resistant to endonuclease digestion compared with free DNA *(31,37)*. These properties are probably important for duration of expression; presumably the longer the DNA survives, the longer its product will be expressed.

A number of different polycations have been used to transfer genes. Most commonly, poly K has been used to condense DNA, but poly-L-arginine *(61)*, poly-L-ornithine *(61)*, and polyethylenimine (PEI) *(62)* have also been tested. Plasmids compacted with poly-L-arginine were not expressed because of the tight interaction of poly-L-arginine with DNA that prevents transcription factors from interacting with the transgene even if it is delivered intact. Poly-L-ornithine was also inefficient in gene transfer *(61)*. Poly K and PEI remain the most efficient polymer cations used in molecular conjugate-mediated gene delivery. Since stretches of lysine occur in many nuclear localization signals *(54,55)*, it has been postulated that poly K may be more efficient in targeting

DNA to the nucleus once DNA complexes are internalized and have escaped the endosome. Also, poly-L-amino acid DNA complexes are relatively nonantigenic and are biodegradable *(63,64)*. Although it lacks these advantages, PEI has a superior ability to disrupt endosomes during acidification, causing rupture and more efficient release *(62)*. Other DNA binding molecules, such as histones *(6,47)* and protamines *(21,26)*, have also been used as DNA condensing agents.

1.3. Endosomal Escape and Export to the Nucleus

Although molecular conjugate-DNA complexes imitate the entry processes of some viruses, they lack their efficiency in escaping the endosomal compartment *(65)*. Szoka and colleagues *(66)* report that polycations that can be protonated may uncouple the endosomal proton pump, resulting in endosome lysis caused by an influx of water, but this process does not apply to all polycations. Other investigators have used a variety of agents to enhance endosomal escape, either incorporated into the molecular conjugate-DNA complexes or administered separately. The efficiency of gene transfer in many systems has been enhanced through disruption of the endocytic pathway by pharmacologic agents such as chloroquine *(16,49,65)*.

Whole adenovirus particles *(33–35,50,51)* have been coupled to molecular conjugate-DNA complexes to augment endosomal release. Acidification in the lysosomes results in conformational changes in the adenoviral capsid proteins that cause pore formation in the vesicle membrane and allows for the escape of its contents into the cytoplasm. In airway epithelial cells in vivo, adenovirus-linked poly K and transferrin-adenovirus-poly K delivered reporter genes through the luminal route *(40)*. Furthermore, the use of asialoorosomucoid-based molecular conjugates modified with an inactivated adenovirus particle resulted in high levels of expression in primary hepatocytes *(33,35)*. However, when a reagent that is itself capable of cell surface interaction is included in the complex, internalization of these intact viral conjugates might occur through the viral receptor and not through the intended targeting receptor.

Viral sequences, such as those implicated in the fusogenic activity of influenza hemagglutinin HA-2, were bound to transferrin-poly K DNA complexes by Wagner and colleagues *(52)*, resulting in a marked increase in the level of transgene expression in cell culture. Other peptides, like the fusion protein of the respiratory syncytial virus or synthetic endosomal release peptides, have also been suggested *(53)*. One major drawback of using viral particles and proteins, however, is their intense immunogenicity in vivo.

The ultimate target of molecular conjugate-DNA complexes is the nucleus, where transcription occurs. So far, no detailed reports have focused on nuclear entry, so it is unclear whether it occurs simply by mass action or whether there exists some specific uptake mechanism. Access of transgenes to the nucleus is favored by cell division, when the nuclear envelope disintegrates during mitosis. Gene transfer via the asialoglycoprotein receptor is increased when hepatic regeneration is induced by partial hepatectomy, which induces cell replication in a normally quiescent tissue *(27,28,38)*. However, condensed molecular conjugates can efficiently deliver genes to quiescent cells, such as airway epithelia *(3,32)*; thus, for at least some receptor-targeted systems, cell division is not required for gene expression.

An intriguing hypothesis is that the lysines on the poly K component of the molecular conjugate may target the attached DNA to the nucleus. Several viral proteins implicated in nuclear translocation have sequences rich in lysine, such as the amino acid sequence Phe-Lys-Lys-Lys-Arg-Lys-Val from the simian virus-40 large T-antigen *(54)* or Lys-Lys-Lys-Tyr-Lys-Leu-Lys from the human immunodeficiency virus type-1 *(55)*. Enhanced nuclear localization may help explain why poly K-containing molecular conjugates have been more efficient in gene transfer than other poly-L-amino acid polycations. However, other investigators suggest that poly K itself provides little, if any, nuclear targeting.

Once delivered to the nucleus, different DNA plasmids result in varying expression patterns. The intensity and duration of transgene expression depend on the tissue or cell type transduced, the promoter and whether it can be extinguished, the nature of the transgene, and the relative survival of the recipient cells as well as other factors discussed above. For instance, the expression of certain reporter genes (e.g., bacterial β-galactosidase) can induce a cytotoxic lymphocyte response *(67)*, so that the transfected cell is eliminated. Degradation of foreign DNA within the cell will also limit expression. Endotoxin contamination of plasmid preparations damages cells and can affect expression *(68)*. Several investigators have designed episomal DNA vectors that extend the survival of the transgene. Self-replicating episomal vectors that contain a viral origin of replication allow transgenes to persist in dividing cells *(69)*. In addition, artificial chromosomes *(70)* may allow for regulated permanent expression but face formidable problems of delivery. Thus, once the receptor-mediated molecular conjugate has accomplished its function of delivering its cargo (i.e., the DNA) across the target cell membrane and into its nucleus, the stability and design of the DNA become crucial.

1.4. Targeting Cell Surface Receptors

The strength of receptor-mediated gene transfer is its selective nature. Indiscriminant transgene delivery and expression may be disadvantageous and are thus undesirable. **Table 1** lists a number of the cell surface receptors that have been targeted for molecular conjugate gene delivery. We will discuss only

TABLE 1
Receptors Targeted with Molecular Conjugate Vectors

Receptor	Ligand	Trafficking	Target cells	Successful transfection	Cotransfer elements used	Reference
Asialogly-coprotein	Asialoorsomucoid galactose	Lysosomal	Hepatocytes	Moderate in vitro Low in vivo	Endosomolytic agents (unless partial hepatectomy)	1,4–7,27–31, 33,35,38,39
Transferrin	Transferrin	Recycled	Ubiquitous	Moderate in vitro Minimal in vivo	Endosomolytic agents/adenovirus particles	2,20,21,24, 34,40,44,47, 49–53,71
Polymeric immuno-globulin	Anti-pIgR antibody	Transcytotic	Respiratory and intestinal epithelia, hepatocytes	Moderate in vitro and in vivo	None	3,32,72
Serpin enzyme complex	Peptide ligands	Lysosomal	Hepatocytes, glia, neurons, macrophages, respiratory epithelia	High in vitro Moderate in vivo	None	12,13,37
Epidermal growth factor (EGF)	EGF and anti-EGF-R antibody	Lysosomal/recycled	Ubiquitous	Moderate in vitro	Adenovirus particles	25
Integrin	Peptide ligand	Lysosomal	Ubiquitous	Moderate in vitro	None	11
Folate	Folate	Lysosomal	Ubiquitous	Moderate in vitro	None	16

(continued)

TABLE 1 (continued)

Receptor	Ligand	Trafficking	Target cells	Successful transfection	Cotransfer elements used	Reference
Mannose	Mannose	Lysosomal	Macrophages	Moderate in vitro Low in vivo	None	8,23,36
Cell surface gylcocalyx	Lectins	Lysosomal/ recycled	Ubiquitous	Moderate in vitro	None	14
Surfactant A	Surfactant protein A	Lysosomal	Respiratory epithelia and alveoli	Moderate in vitro	None	10
c-kit	Anti-CD3 antibody	Lysosomal	Hematopoietic	Moderate in vitro stem cells	None	9
Carbohydrate	Anti-T$_n$ antibody	Lysosomal	Carcinoma cells and lymphocytes	Moderate in vitro	None	19
CD3	Steel factor (SLF)	Lysosomal	Lymphocytes	Moderate in vitro	Adenovirus particles	17

those tested in vivo as well as in vitro, although the others are listed in the table. In 1987, Wu and Wu. *(1)* described a soluble DNA carrier that targeted the asialoglycoprotein receptor, an integral membrane glycoprotein on hepatocytes that clears galactosylated (or partially degraded) glycoproteins from the blood for lysosomal degradation (reviewed in **ref. *41***). Treatment of orosomucoid with neuraminidase exposes the terminal galactose, and this protein was covalently linked to poly K, which was complexed with a plasmid encoding chloramphenicol acetyltransferase. The resulting complex successfully targeted the livers of rats, although protracted (weeks) expression only occurred if animals underwent partial hepatectomies at the time of injection *(27,28,38)*. This receptor-targeted molecular conjugate was used in subsequent studies, but variability was greater and the low-level gene expression observed was transient. In Nagase analbuminemic rats, systemic injection of targeted complexes containing a chimeric gene encoding human albumin after partial hepatectomies resulted in expression for as long as 4 weeks *(28)*.

Wilson and co-workers *(29)* used similar molecular conjugates to deliver a gene encoding the low-density lipoprotein (LDL) receptor to the livers of Watanabe rabbits, a model for familial hypercholesterolemia. LDL receptor mRNA was detected 1 day after administration, but not at 3 days, and total cholesterol levels in the blood of transfected rabbits were reduced by 30% 2 days after treatment but returned to pretreatment levels 5 days after transfection. Some repeat injections failed to produce expression. Stankovics and colleagues *(39)* reported that asialoorosomucoid-poly K molecular conjugates delivered the methylmalonyl coenzyme A (CoA) mutase gene to the liver in quantities that may be therapeutic in patients with methylmalonic aciduria, an inborn error of metabolism, but expression lasted <2 days, and repeated injections produced an antibody response against the ligand. Perales and colleagues demonstrated that a molecular conjugate consisting of a poly K chemically linked to α-D-galactopyranosyl phenylisothiocyanate could introduce functional genes to hepatocytes in vitro *(7)* and in vivo *(30)*. Human factor IX cDNA was specifically introduced into the livers of adult animals and was expressed for weeks after administration. Thus, in general, asialoglycoprotein receptor-directed molecular conjugates give transient or low-level gene expression in vivo, and the larger ligands for the receptor have proved to be immunogenic.

Another target for gene transfer has been the transferrin receptor, a dimeric glycoprotein 180 kDa in size (reviewed in **ref. *42***). This receptor binds to its natural ligand, transferrin, rapidly internalizes, and then recycles the ligand back to the cell surface. The transferrin receptor is present in many cells, including erythroblasts, hepatocytes, and tissue macrophages. This endocytotic pathway has also been exploited to deliver drugs and toxins to tumor cells in vitro *(20)*. Bernstiel and associates *(2,21,24)* reported that expression plasmids targeted to the transferrin receptor were efficiently delivered to avian and human erythroid cells in vitro in a receptor-specific fashion and that transgene expression was augmented by treatment with lysosomotropic agents, such as chloroquine. The transferrin receptor was also capable of delivering DNA plasmids as large as 48 kB *(44)*, overcoming one of the limitations of virus packaging systems. The size of the DNA complexes formed with the transferrin-based conjugates was critical for gene transfer, and complexes <100 nm in diameter were better than larger preparations.

In vivo, systemic injection with transferrin-directed DNA complexes failed to produce significant transgene expression in tissues, but local injection into the liver resulted in high levels of reporter gene expression *(51)*. Zatloukal and colleagues *(71)* used the transferrin-based conjugate to deliver the interleukin-2 gene to murine melanoma cells and obtained high levels of the cytokine. Airway epithelia of intact animals have also been transfected using human transferrin-poly K and transferrin-adenovirus-poly K molecular conjugates *(40)* designed to exploit the endosomolytic property of adenovirus. Intratracheal instillation of DNA bound to these conjugates resulted in transient low-level expression of the reporter gene, which peaked 1 day after transfection and returned to pretreatment levels by 7 days.

The lung is an attractive organ for gene therapy. Davis and co-workers *(3,32,72)* targeted the polymeric immunoglobulin receptor (pIgR), which is a bulk flow receptor for dimeric IgA and polymeric IgM expressed in human respiratory epithelium and the serous cells of the submucosal glands (reviewed in **ref.** *43*). These cells express the cystic fibrosis transmembrane conductance regulator, so the pIgR may be an attractive target for the treatment of cystic fibrosis *(32)*. In animals, the polymeric immunoglobulin receptor introduced expression plasmids to airway epithelial cells when the pIgR-directed molecular conjugates were injected into the systemic circulation. (The receptor is predominantly expressed on the basolateral surface.) Expression was transient, lasting <12 days after injection, and repeated injection provoked a neutralizing serologic response directed against the Fab portion of the complexes *(72)*.

Tissue macrophages have been targeted by way of the mannose receptor both in vitro and in vivo. This receptor is abundantly expressed by a variety of macrophage subtypes and internalizes glycoproteins with mannose, glucose, fucose, and N-acetylglucosamine residues in exposed, nonreducing positions, for lysosomal degradation (reviewed in **ref.** *73*). Systemic administration of expression plasmids complexed to the mannose-terminal glycoprotein molecular conjugates resulted in successful delivery to the reticuloendothelial organs in adult mice *(36)*. However, transfection efficiency was low, and transgene expression, which peaked 4 days after administration, was transient.

Ziady et al. *(13,37)* have used SEC-R to examine a number of characteristics of receptor-mediated molecular conjugate gene transfer. SEC-R, originally described as a binding site on human hepatoma cells and blood monocytes, recognizes a sequence in α_1-antitrypsin that is exposed only when it is complexed with a serine protease such as neutrophil elastase or modified by either metalloelastase or by the collaborative action of active oxygen intermediates and neutrophil elastase (reviewed in **ref. 74**). The receptor is present on such cell types as mononuclear phagocytes, neutrophils, myeloid cell lines U937 and HL60, the human intestinal epithelial cell line CaCo2, mouse fibroblast L cells, the rat neuronal cell line PC12, and the human glial cell line U373MG. Using synthetic peptides based in sequence on α_1-antitrypsin, Ziady and colleagues *(12)* targeted reporter genes specifically to receptor-bearing cells. Further studies detailed the effects of substitution of poly K with receptor ligands on expression *(13)*. Using sparsely substituted poly K of various lengths it was possible to extend or shorten the duration of expression as well as affect the intensity of expression in vitro and in vivo *(13,37)*.

Thus, investigators have targeted a variety of cell surface receptors for molecular conjugate-mediated gene delivery with considerable success in cell culture and in animals. In the consideration of a suitable receptor, a number of criteria concerning the type of receptor should be met, including abundance and selectivity of cargo. Design of the ligand should be simple and reproducible. Since the ligand is the most immunogenic and/or toxic portion of receptor-targeted molecular conjugates, manipulation of ligand size, structure, and design may reduce these undesirable effects.

2. Materials
2.1. Ligand Design and Production

The choice of ligand depends on the receptor to be targeted. Peptide display technology may also be used to identify a ligand that will be internalized by the cells of interest. Preferably, ligands should have high-affinity for the receptor and low immunogenicity, initiate minimal cell signaling, and be easy to couple (reliably and efficiently) to the polycation.

2.2. Molecular Conjugate Construction

Given the variety of options, we will describe the basic construction of SEC-R-directed molecular conjugate lacking endosomolytic, nuclear localizing, or complex stabilizing enhancements, as a prototype for conjugate preparation.

1. Poly K (Sigma, St. Louis, MO) or other nonlipid polycation in water (see **Notes 1** and **2**).
2. Sulfo *LC*-SPDP (Pierce, Rockford, IL) or other hetero- or homobifunctional linker (*see* **Note 3**) dissolved in a solvent lacking phosphate (*see* **Note 2**).
3. Polypropylene 0.5–2.0 mL presiliconized microcentrifuge tubes that are RNase/

DNase free and sterile (National Scientific Supply, San Rafael, CA) or similar tubes (*see* **Notes 4** and **5**).
4. Purified receptor ligand (>98% pure; *see* **Note 6**) dissolved in a nonphosphate-containing solvent (*see* **Note 1**).
5. Phosphate-buffered saline (PBS): 1.0% NaCl, 0.025% KCl, 0.14% Na_2HPO_4, 0.025% KH_2PO_4 (all w/v), pH 7.4; sterile.
6. Although they are not necessary, enhancement moieties such as reactive PEG and/or biotinylated viral particles may be used (*see* **Note 7**).
7. Sterile dialysis tubing of the appropriate molecular weight cutoff (*see* **Note 8**).

2.3. Analysis of Receptor-Targeted Conjugates

1. Lyophilizer (for example, Freezemobile II by Virtis) and appropriate lyophilizing equipment.
2. Deuterated water (Sigma).
3. 300 MHz or higher nuclear magnetic resonance (NMR) spectrometer (for example, the Varian Unity Plus 600 NMR; *see* **Note 9**).
4. A VIS/UV light spectrophotometer (for example, Beckman DU-64).
5. Ellman's reagent (Pierce) or other linker analysis reagent (*see* **Note 10**).
6. Polyacrylamide and conventional sodium dodecyl sulfate-polyacrylamide gel electrophoresis (SDS/PAGE) equipment (*see* **Note 11**).
7. Fast Protein Liquid Chromatography (FPLC™) or equivalent equipment.

2.4. DNA Plasmid Preparation

Plasmid DNA quality is crucial to the quality and efficiency of receptor-targeted DNA complexes. DNA should be purified by double CsCl gradient centrifugation *(75)* or an equivalent high-quality method. The choice of DNA will depend on the experimental goal. Primarily, reporter genes should be used to assess efficacy of gene transfer to provide an understanding of the parameters for using the optimal molecular conjugate and dose for therapeutic genes. DNA size can vary, as discussed above. However, although molecular conjugates can compact large DNA molecules, the larger the plasmid, the larger the complex size, so minimizing plasmid size may be advantageous.

2.5. Molecular Conjugate Condensation of DNA

1. Expression plasmid DNA (*see* **Note 12**).
2. Receptor ligand-polycation conjugate (molecular conjugate).
3. 5 *M* stock of sterile RNase/DNase-free NaCl.
4. Polypropylene ultraclear 0.5–2.0 mL presiliconized microcentrifuge tubes that are RNase/DNase free and sterile (National Scientific Supply) or similar tubes (*see* **Note 4**).
5. Microcentrifuge tube shaker (for example, Janke and Kunkle IKA VIBRAX).
6. Sterile dialysis tubing of the appropriate molecular weight cutoff (*see* **Note 8**).

7. Polyethersulfone filtration membrane (Whatman, Fairfield, NJ) or other similar filtration membrane (*see* **Note 4**).

2.6. Analysis of Receptor-Targeted DNA Complexes

Although analysis is not necessary for the formation of efficient receptor-targeted DNA complexes, particle size and structure are key parameters for gene transfer, so monitoring is strongly recommended. It is desirable to use more than one technique in assessing complex structure to confirm observations. Of course, no clinical or preclinical trials should be done without such analysis.

1. A laser cytometer or similar device capable of dynamic light scattering (*see* **Note 13**).
2. A VIS/UV light spectrophotometer (for example, Beckman DU-64).
3. Carbon type-B electron: 400–1000-mesh copper or similar micrograph grids.
4. 2×2-cm ultraclean mica wafer or similar substrate for atomic force microscopy.
5. Uranyl acetate (Polysciences, Warrington, PA).
6. Conventional agarose and agarose gel electrophoresis equipment.
7. Transmission electron microscope (for example, JEOL-100C).
8. Scanning atomic force microscope (for example, Nanoscope III with a SPARC 10 Sun Microsystems workstation).

3. Methods
3.1. Generation of Receptor-Targeted Molecular Conjugates

Selection of the polycation portion of the molecular conjugate is determined by the investigators' preferences. Poly K and PEI are the most promising polycations for DNA condensation. As discussed above, many different materials and methods are employed by investigators to form molecular conjugates, but the basic principles regarding the rate of substitution of the polycation and size of the polycation will always apply. We describe the use of heterobifunctional linkers to conjugate SEC-R receptor-targeted peptide ligands to poly K to illustrate construction of a conjugate. These molecular conjugates are efficient in vitro *(12,13)* and in vivo *(37)*, and are good models for conjugations of sulfhydryl-containing ligand and primary amines on polycations. Enhancements such as PEGylation *(59)* or conjugation to viral particles *(33–35,50,51)* may be carried out, but it should be noted that excessive modification can interfere with DNA condensation. **Figure 2** demonstrates this method of polycation to ligand coupling. Since lysine residue substitution is an important parameter in conjugation, modifier concentrations are expressed as percent of lysine residues, not poly K molecules. All solutions must be sterile.

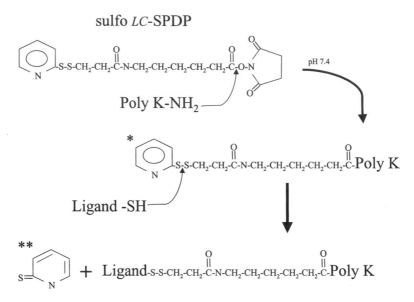

Fig. 2. *Chemical coupling of poly K to a sulfhydral-containing ligand with sulfo LC-SPDP.* Poly K can be chemically coupled to a sulfhydral-containing ligand using the heterobifunctional linker sulfo *LC*-SPDP. This scheme is representative of molecular conjugate construction. After the linker is reacted with the poly K, the resulting compound can be examined by proton NMR. Since the aromatic ring (single asterisk) produces a proton spectrum distinct from lysine, accurate measurement of linker substitution of poly K is possible. Once the ligand is coupled to the linker, this ring is released (double asterisk) and can be monitored by OD. However, if the ligand is small and also produces a distinct proton spectrum, NMR analysis is more accurate.

1. Produce peptide ligands ($M_r <$ 5 kDa) by conventional solid-phase synthesis to contain a cysteine at the N terminus.
2. Incubate poly K (approx 252 lysines, average M_r = 53.7 kDa) with the heterobifunctional crosslinking reagent sulfo *LC*-SPDP. Add 42 μL of 1 m*M* sulfo *LC*-SPDP in water to 12 mg poly K (1000-fold molar excess of lysine to sulfo *LC*-SPDP or 0.1% linker) in 0.1 PBS, pH 7.4, at room temperature for 30 min (*see* **Note 2**).
3. Dialyze reaction mixture exhaustively in 10,000 M_r cutoff dialysis tubing at room temperature against PBS to remove unreacted sulfo *LC*-SPDP and low molecular weight reaction products.
4. Following dialysis, a portion of the sample may be set aside to assess exact concentrations of postdialysis modified poly K, purity, and linker coupling efficiency by NMR (*see* **Notes 9** and **14**) or other analysis.
5. To drive the disulfide reaction to completion, add ligand (10–100-fold molar excess over coupled linker) to the modified poly K in PBS (pH 7.4) and allow the reaction to proceed at room temperature for 24 h (*see* **Note 14**).

6. Exhaustively dialyze the conjugate in 10,000 M_r cutoff dialysis tubing at room temperature against ultrapure water to remove unreacted ligand and low molecular weight reaction products. If the ligand used is not easily separated from the conjugate by dialysis (for example, if the ligand size is too close to the size of the conjugate), then high-performance liquid chromatography (HPLC) or an equivalent technique may be used (*see* **Note 15**)

7. A portion of the sample may be separated after dialysis for analysis by NMR (*see* Notes 9 and **16**) or other methods.

8. Store the remaining sample in aliquots at –80°C to avoid freeze/thaw damage.

3.2. Molecular Conjugate Analysis

Because of the repetitive nature of the polycation and the small size of the ligand used (*see* **Note 9**), it is possible to use NMR to verify the degree of substitution of the poly K as well as the concentration and purity of the components of the molecular conjugate. Furthermore, linkers such as sulfo *LC*-SPDP usually contain aromatic rings that produce resonances distinct from the polycation, allowing for monitoring of each step of the conjugation process. Ligands and polycations that do not lend themselves to such analysis may be examined by conventional SDS-PAGE and spectrophotometric methods *(3)*.

1. Dialyze an aliquot (4 mg in respect to poly K) of the conjugate exhaustively against water in M_r cutoff 5000–10,000 dialysis tubing at room temperature.

2. Lyophilize from water and then from D_2O.

3. Resuspend in 0.75 mL of 99.99% D_2O.

4. Obtain proton NMR spectra at 300–600 MHz using standard proton parameters, previously described *(13)*. Spectra acquisitions typically require between 0.5 and 16 h.

5. Chemical shifts should be referenced to the residual deuterated water resonance at approximately 4.8 ppm.

6. Aliquots of dialysis bath water as well as dialysis bag wash should also be lyophilized and examined by NMR to verify the absence of contaminants.

7. Integration of spectra provides the molar ratios of the components of the molecular conjugate and may be used to calculate the efficiency for each of the conjugation steps.

8. A known concentration of a simple compound, such as acetate, with a proton spectrum is that nonoverlapping with the spectra produced by the polycation or ligand may be added to determine the exact concentration of the molecular conjugate.

3.1. Receptor-Targeted DNA Complex Production

As discussed above, two main methods have been used to produce receptor-targeted DNA complexes. Initial reports *(1)* described a technique capable of compacting 200–500 µg of DNA with short-length poly K ($M_r = 3800$) in 1 mL low-salt (approx 150 m*M* NaCl) solution. Condensation with longer polymers with this method results in precipitation unless the molecular conjugate is modified with PEG, or else a smaller concentration of DNA (30–60 μg/mL) must be used. Furthermore, this technique produced a heterogeneous population of large DNA complexes averaging 150–200 nm in diameter *(1,27,28)*. Filtration could isolate the smaller particles from this mixed population, but it reduced the effective DNA concentration. An alternative technique described by Perales and colleagues *(56)* avoids some of these limitations. This method allows for compaction of DNA at high concentrations (up to 12 mg/mL) with any size polycation. In addition, smaller particles are produced, averaging 20–30 nm in diameter, which is advantageous in vivo, as discussed above. Here we describe the basis for each method using the molecular conjugate constructed in **Subheading 3.1.** to condense DNA into charge neutral particles, although a 1:1 DNA to conjugate charge ratio is not necessary for particle production. However, positively charged particles can be internalized nonspecifically because of interaction with the negatively charged cell membrane. Furthermore, excessive positive charge will activate the complement cascade in vivo. Therefore, we recommend production of neutral charge particles.

3.3.1. Low-Salt Condensation

1. Add 2.3 μg of molecular conjugate in 250 μL of water to 5 μg of plasmid DNA in 250 μL of 0.3 *M* NaCl while agitating in ultraclear microcentrifuge tubes.
2. Higher NaCl concentrations can be used with higher concentrations of DNA, followed by stepwise dialysis to bring the salt concentration down to saline levels.
3. Incubate mixture at room temperature for 30 min.
4. The mixture can be filtered through a 0.2-μm polyethersulfone filter to remove large aggregates.
5. An aliquot of the mixture (50–100 μL) should be retained for analysis.
6. These complexes are reasonably stable and can be stored at 4°C for 1 week.

3.3.2. High Salt Condensation

1. To 200 μg plasmid DNA in 500 μl 0.4 *M* NaCl, add 10 μL of molecular conjugate (80 μg in 500 μL 0.4 *M* NaCl) every 3 min under constant vortexing at room temperature in ultraclear nonstick microcentrifuge tubes.
2. Continue additions until total amount of conjugate is added (about 2.5 h).
3. After the addition of the carrier to the DNA is complete, aggregates should be visible in solution.
4. Adjust the sodium chloride concentration by slowly adding small aliquots of 5 *M* NaCl (approx 2–5 μL) until the rise in ionic strength dissociates aggregated DNA complexes, and the turbidity of the solution clears.
5. The final volume of the DNA complex solution should contain 0.8–1 μg plasmid

DNA/5 μL (1:0.40 w/w DNA to peptide-poly K conjugate ratio) in 1.0–1.1 *M* NaCl for 53.7-kDa polymers.

6. Controls often include DNA condensed in the same method with unconjugated poly K and naked DNA.
7. The mixture can be filtered through a 0.2-μm polyethersulfone filter to remove large aggregates, but this is not as important as with complexes formed with the low salt method.
8. Retain an aliquot of the mixture (50–100 μL) for analysis. Unless modified for stability (for example, with PEG) these complexes should be used within 1 h of condensation.

3.4. Analysis of Receptor-Targeted DNA Complexes

Since the quality of the DNA complexes is essential to efficacy, analysis of the purity, structure, and size of these particles is recommended. Conventional agarose gel electrophoresis retardation assays have been used by a number of investigators, as rough measures of proper condensation. As DNA charge is neutralized by the molecular conjugate, its electrophoretic mobility is diminished. Complete retardation should indicate charge neutrality, whereas reversed migration toward the negative electrode indicates excess positive charge. Dynamic light scattering can also be used to estimate the size of constructed DNA complexes, but this technique has limited accuracy when measuring very small particles dispersed among a heterogeneous population of larger aggregates of varying shapes. Two techniques, EM and atomic force microscopy (AFM), have become widely used by investigators to assess complex size and structure. Transmission EM is a powerful, and well-established method for examining these complexes with high resolution *(7,12)*. AFM, a newer technique, further allows for examination of particles in solution, thus providing information on hydrated complex structure *(76)*.

3.4.1. EM Analysis of DNA Complexes

1. Add a 10-μL aliquot of diluted DNA complex solution (1:10 dilution) to a 1000-mesh electron microscope carbon grid, immediately after DNA condensation. High salt concentrations can sometimes destroy the carbon micrographs. Thus, although it is not absolutely required, we recommend diluting the DNA complex solution.
2. Then blot grids and fix them in methanol or ethanol.
3. Stain the grids with a drop (10–15μL) of 0.04% uranyl acetate.
4. Let grids dry for 5–10 min. at room temperature in a clean, dust-free area.
5. Grids can be sputter-coated with platinum to allow rotary shadowing on samples to provide information of the three-dimensional structure of the DNA complexes.
6. Examine samples using a JEOL-100C transmission electron microscope.
7. Solutions used in complex formation as well as blank wafers should be examined to ensure absence of contaminants.

3.4.2. AFM Analysis of DNA Complexes

1. Prepare samples on suitable substrate surfaces (for example, mica chips for neutral charge particles).
2. For analysis of dried samples, add a drop of the solution (1:100 dilution) to the surface of a 2 × 2-cm mica wafer immediately after formation of DNA complexes, and let dry for 3 h in a clean, dust-free area.
3. Lightly rinse dried wafers with double-deionized water to remove salt crystals.
4. For analysis of hydrated samples, add 10–20 µL to the fluid chamber of an AFM, and allow sample to settle and bind the substrate surface at the bottom of the chamber for 1 h.
5. Set the microscope so that the cantilever deflection maintains an applied force less than 10 nN; the force of adhesion is about 30 nN.
6. Adjust feedback gain and scanning speed to minimize errors caused by temporal response limitations.
7. Transfer images in binary format to a SPARC 10 Sun Microsystem workstation, where they can be converted to gray scale and analyzed.
8. Solutions used in complex formation as well as blank wafers should be examined to ensure absence of contaminants.

4. Notes

1. Solutions involved in the condensation process should be phosphate free, since negatively charged phosphates interfere with the polycation-DNA interaction.
2. Polycations other than poly K may be used for DNA compaction. Modification of these polymers will differ depending on their chemical properties. Also, most commercial poly K is heterodispersed; for better defined complexes, as for preclinical trials, may be purified or custom synthesized.
3. A number of chemical linkers are available for specific or nonspecific linkage of various moieties. These may be used according to preference as long as the basic premises of polycation modification are applied.
4. Polycation-DNA complexes and especially the polycation prior to condensation tend to adhere to uncoated surfaces. Thus, nonstick containers and filters should be used throughout the preparation to minimize sample loss.
5. Since most preparations are carried out at room temperature, it is important to use sterile containers that are DNase free, to avoid bacterial contamination and/or DNA degradation.
6. The purity of the ligand is essential to the success of chemical coupling and gene transfer. Contaminants will reduce the effective concentration of the active ingredient in receptor targeting and may lead to nonreceptor-specific internalization.
7. Molecular conjugate enhancers such as PEG and/or biotinylated viral particles may be coupled to the conjugate at low levels to avoid interference with DNA condensation.
8. Azide, used in many instances to maintain dialysis bag sterility, must be removed so as to not contaminate samples that will be applied to cells later in vitro or in vivo.

9. In coupling small moieties (e.g., ligand, PEG, and so on) to the polycation, NMR can be used to monitor the conjugation process if these moieties and the coupling reagents produce resonances distinct from the polymer.

10. Many protein-modifying chemicals, such as Ellman's reagent, may be used in spectrophotometric analysis of the extent of coupling of the DNA carrier.

11. Poly K has an excess positive charge and a high pK_a and thus will not enter SDS-PAGE gel unless coupled to a large, more neutrally charged protein. Other polycations such as protamine and histone may enter the gel but may travel in an unexpected manner because of charge.

12. All DNA preparations should be purified by double-CsCl gradient centrifugation or an equivalent method and stored at –20°C. DNA purity is crucial for proper condensation and thus for gene transfer. Furthermore, contamination of DNA with bacterial toxins may be toxic to cells during transfection.

13. Variation in solubility and a wide range of different size particles in solution may result in inaccurate measurements by dynamic light scattering. Thus, these measurements should be confirmed with other techniques such as EM or AFM.

14. Dialysis bath water and solutions used in molecular conjugate construction should be examined by NMR to assess the presence of contaminants.

15. Because of the stickiness of the conjugate, the proper column should be used to allow for total retrieval of the sample.

16. The same aliquots examined by NMR may also be examined by other techniques, such as SDS-PAGE, and chemical analysis.

References

1. Wu, G. Y. and Wu, C. H. (1987) Receptor-mediated *in vitro* gene transformation by a soluble DNA carrier system. *J. Biol. Chem.* **262,** 4429–4432.

2. Zatoukal, K., Wagner, E., Cotten, M., et al. (1992) Transferrinfection: a highly efficient way to express gene constructs in eukaryotic cells. *Ann. N.Y. Acad. Sci.* **660,** 136–153.

3. Ferkol, T., Kaetzel, C. S., and Davis, P. B. (1993) Gene transfer into respiratory epithelial cells by targeting the polymeric immunoglobulin receptor. *J. Clin. Invest.* **92,** 2394–2400.

4. Monsigny, M., Roche, A. C., Midoux, P., and Mayer, R. (1994) Glycoconjugates as carriers for specific delivery of therapeutic drugs and genes. *Adv. Drug. Delivery Rev.* **14,** 1–24.

5. Midoux, P., Mendes, C., Legrand, A., et al. (1993) Specific gene transfer mediated by lactosylated poly-L-lysine into hepatoma cells. *Nucleic Acids Res.* **21,** 871–878.

6. Chen, J., Stickles, R. J., and Daiscendt, K. A. (1994) Galactosylated histone-mediated gene transfer and expression. *Hum. Gene Ther.* **5,** 429–435.

7. Perales, J. C., Grossmann, G. A., Molas, M., et al. (1997) Biochemical and functional characterization of DNA complexes capable of targeting genes to hepatocytes via the asialoglycoprotein receptor. *J. Biol. Chem.* **272,** 7398–7407.

8. Erbacher, P., Bousser, M. T., Raimond, J., et al. (1996) Gene transfer by DNA/

glycosylated polylysine complexes into human blood monocyte derived macrophages. *Hum. Gene Ther.* **7,** 721–729.

9. Schwarzenberger, P., Spence, S. E., Gooya, J.M., et al. (1996) Targeted gene transfer to human hematopoietic progenitor cell lines through the c-kit receptor. *Blood* **87,** 472–478.

10. Ross, G. F., Morris, R. E., Ciraolo, G., et al. (1995) Surfactant proteinA-polylysine conjugates for delivery of DNA to airway cells in culture. *Hum. Gene Ther.* **6,** 31–40.

11. Hart, S. L., Harbottle, R. P., Cooper, R., et al. (1995) Gene delivery and expression mediated by an integrin-binding peptide. *Gene Ther.* **2,** 552–554.

12. Ziady, A., Perales, J. C., Ferkol, T., et al. (1997) Gene transfer into hepatocyte cell lines via the serpin enzyme complex (SEC) receptor. *Am. J. Physiol.* **273,** G545–G552.

13. Ziady, A., Ferkol, T., Gerken, T., et al. (1998) Ligand substitution of receptor targeted DNA complexes affects gene transfer into hepatoma cells. *Gene Ther.* **5,** 1685–1697.

14. Batra, R. K., Wang-Johanning, F., Wagner, E., Garver, R. I., and Curiel, D. T. (1994) Receptor-mediated gene delivery employing lectin-binding specificity. *Gene Ther.* **1,** 255–260.

15. Rojanasakul, Y., Wang, L. J., Malanga, C. J., Ma, J. K. H., and Liaw, J. H. (1994) Targeted gene delivery to alveolar macrophages via Fc receptor-mediated endocytosis. *Pharm. Res.* **11,** 1731–1736.

16. Gottschalk, S., Christiano, R. J., Smith, L. C., and Woo, S. L. (1994) Folate receptor mediated DNA delivery into tumor cells: potosomal disruption results in enhanced gene expression. *Gene Ther.* **1,** 185–191.

17. Buschle, M., Cotten, M., Kirlappos, H., et al. (1995) Receptor-mediated gene transfer into human T lymphocytes via binding of DNA/CD3 antibody particles to the CD3 T cell receptor complex. *Hum. Gene Ther.* **6,** 753–761.

18. Cotten, M., Wagner, E., and Birnstiel, M. L. (1993) Receptor-mediated transport of DNA into eukaryotic cells. *Methods Enzymol.* **217,** 618–644.

19. Thurher, M., Wagner, E., Clausen, H., et al. (1990) Carbohydrate receptor mediated gene transfer to human T leukemic cells. *Glycobiology.* **4,** 429–435.

20. Wagner, E., Curiel. D. T., and Cotten, M. (1994) Delivery of drugs, proteins, and genes into cells using transferrin as a ligand for receptor-mediated endocytosis. *Adv. Drug. Del. Rev.* **14,** 113–135.

21. Wagner, E., Zenke, M., Cotten, M., Beug, H., and Birnstiel, M. L. (1990) Transferrin-polycation conjugates as carriers for DNA uptake into cells. *Proc. Natl. Acad. Sci. USA* **87,** 3410–3414.

22. Gottschalk, S., Sparrow, J. T., Hauer, J., et al. (1996) Synthetic vehicles for efficient gene transfer and expression in mammalian cells. *Gene Ther.* **3,** 448–457.

23. Ferkol, T., Perales, J. C., Mularo, F., and Hanson, R. W. (1996) Receptor-mediated gene transfer into macrophages. *Proc. Natl. Acad. Sci. USA* **93,** 101–105.

24. Zenke, M., Steinlein, P., Wagner, E., et al. (1990) Receptor-mediated endocytosis of transferrin-polycation conjugates: an efficient way to introduce DNA into hematopoietic cells. *Proc. Natl. Acad. Sci. USA* **87,** 3655–3659.

25. Chen, J., Gamau, S., Takayanagi, A., and Shimizu, N. (1994) A novel gene delivery system using EGF receptor-mediated endocytosis. *FEBS Lett.* **338,** 167–169.
26. Chen, S. Y., Zani, C., Khouri, Y., and Marasco, W. A. (1995) Design of a genetic immunotoxin to eliminate immunogenicity. *Gene Ther.* **2,** 116–123.
27. Wu, C. H., Wilson, J. M., and Wu, G. Y. (1989) Targeting genes: delivery and persistent expression of a foreign gene driven by mammalian regulator elements *in vivo. J. Biol. Chem.* **264,** 16985–16987.
28. Wu, G. Y., Wilson, J. M., Shalaby, F., et al. (1991) Receptor-mediated gene delivery *in vivo*: partial correction of genetic analbuminemia in Nagase rats. *J. Biol. Chem.* **266,** 14338–14342.
29. Wilson, J. M., Grossman, M., Wu, C. H., et al. (1992) Hepatocyte-directed gene transfer *in vivo* leads to transient improvement of hypercholesterolemia in low-density lipoprotein receptor-deficient rabbits. *J. Biol. Chem.* **267,** 963–967.
30. Perales, J. C., Ferkol, T., Beegen, H, Ratnoff, O. D., and Hanson, R. W. (1994) Gene transfer *in vivo*: Sustained expression and regulation of genes introduced into the livers by receptor targeted uptake. *Proc. Natl. Acad. Sci. USA* **91,** 4086–4090.
31. Chowdhury, N. R., Wu, C. H., Wu, G. Y., et al. (1993) Fate of DNA targeted to the liver by asialoglycoprotein receptor-mediated endocytosis *in vivo. J. Biol. Chem.* **268,** 11265–11271.
32. Ferkol, T., Perales, J. C., Kaetzel, C. S., et al. (1995) Gene transfer into airways in animals by targeting the polymeric immunoglobulin receptor. *J. Clin. Invest.* **95,** 493–502.
33. Christiano, R. J., Smith, L. C., and Woo, S. L. (1993) Hepatic gene therapy: adenovirus enhancement of receptor-mediated gene delivery and expression in primary hepatocytes. *Proc. Natl. Acad. Sci. USA* **90,** 2122–2126.
34. Cotten, M., Wagner, E., Zatloukal, K., and Birnstiel, M. L. (1993) Chicken adenovirus (CELO) particles augment receptor-mediated DNA delivery to mammalian cells and yield exceptional levels of stable transformants. *J. Virol.* **67,** 3777–3785.
35. Christiano, R. J., Smith, L. C., Kay, M. A., Brinkley, B. R., and Woo, S. L. (1993) Hepatic gene therapy: efficient gene delivery and expression in primary hepatocytes utilizing a conjugated adenovirus DNA complex. *Proc. Natl. Acad. Sci. USA* **90,** 11548–11552.
36. Ferkol, T., Mularo, F., Hilliard, J., et al. (1998) Transfer of the gene encoding human alpha$_1$ antitrypsin into pulmonary macrophages *in vivo. Am. J. Respir. Cell Mol. Biol.* **18,** 591–601.
37. Ziady, A., Ferkol, T., Dawson, D. V., Perlmutter, D. H., and Davis, P. B. (1999) Chain length of the polymer portion of receptor-targeted DNA complexes modulates gene transfer both *in vitro* and *in vivo. J. Biol. Chem.* **8,** 4908–4916.
38. Ferkol, T., Lindberg, G. L., Perales, J. C., et al. (1993) Regulation of the phosphoenolpyruvate carboxykinase/human factor IX gene introduced into the livers of adult rats by receptor-mediated gene transfer. *FASEB J.* **7,** 1081–1091.
39. Stankovics, J., Crane, A. M., Andrews, E., et al. (1994) Overexpression of human

methylmalonyl CoA mutase in mice after *in vivo* gene transfer with asialoglycoprotein/polylysine/DNA complexes. *Hum. Gene Ther.* **5,** 1095–1104.

40. Gao, L., Wagner, E., Cotten, M., et al. (1993) Direct *in vivo* gene transfer to airway epithelium employing adenovirus-polylysine-DNA complexes. *Hum. Gene Ther.* **4,** 17–24.

41. Stockert, R. J. (1995) The asialoglycoprotein receptor: relationships between structure, function, and expression. *Physiol Rev.* **75,** 591–609.

42. Ponka, P. and Lok, C. N. (1999) The transferrin receptor: role in health and disease. *Int. J. Biochem. Cell Biol.* **31,** 1111–1137.

43. Kaetzel, C. S., Blanch, V. J., Hempen, P. M., et al. (1997) The polymeric immunoglobulin receptor: structure and synthesis. *Biochem. Soc. Trans.* **25,** 475–480.

44. Cotten, M., Wagner, E., Zatloukal, K., et al. (1992) High-efficiency receptor-mediated delivery of small and large (48 kilobase) gene constructs using endosome-disruption activity of defective or chemically inactivated adenovirus particles. *Proc. Natl. Acad. Sci. USA* **89,** 6094–6098.

45. Yew, N. S., Wysokenski, D. M., Wang, K. X., et al. (1997) Optimization of plasmid vectors for high-level expression in lung epithelial cells. *Hum. Gene Ther.* **8,** 575–584.

46. Adami, R.C., Collard, W. T., Gupta, S. A., et al. (1998) Stability of peptide-condensed plasmid DNA formulations. *J. Pharm. Sci.* **87,** 678–683.

47. Wagner, E., Cotten, M., Foisner, R., and Birnstiel, M. T. (1991) Transferrin-polycation-DNA complexes: the effect of polycations on the structure of the complex and DNA delivery to cells. *Proc. Natl. Acad. Sci. USA* **88,** 4255–4259.

48. Miller, N. and Vile, R. (1995) Targeted vectors for gene therapy. *FASEB J.* **9,** 190–199.

49. Cotten, M., Laengle-Rouault, F., Kirlappos, H., et al. (1990) Transferrin-polycation mediated introduction of DNA into human leukemic cells: stimulation by agents that affect survival of transfected DNA or modulate transferrin receptor levels. *Proc. Natl. Acad. Sci. USA* **87,** 4033–4037.

50. Curiel, D. T., Agarwal, S., Wagner, E., and Cotten, M. (1991) Adenovirus enhancement of transferrin-polylysine-mediated gene delivery. *Proc. Natl. Acad. Sci. USA* **88,** 8850–8854.

51. Cotten, M. (1995) The entry mechanism of adenovirus and some solutions to the toxicity problems associated with adenovirus-augmented, receptor-mediated gene delivery, in *The Molecular Repertoire of Adenoviruses III* (Doerfler, W. and Böhm, P. eds.), *Curr. Top. Microbiol. Immunol.* **199,** 283–295.

52. Wagner, E., Plank, C., Zatloukal, K., Cotten, M., and Birnstiel, M. L. (1992) Influenza virus hemaglutinin HA-2 N-terminal fusogenic peptides augment gene transfer by transferrin-polylysine-DNA complexes: toward a synthetic virus-like gene-transfer vehicle. *Proc. Natl. Acad. Sci. USA* **89,** 7934–7938.

53. Plank, C., Oberhauser, B., Mechtler, K., Koch, C., and Wagner, E. (1994) The influence of endosome-disruptive peptides on gene transfer using synthetic virus-like gene transfer systems. *J. Biol. Chem.* **269,** 12918–12924.

54. Kalderon, D., Richardson, W. D., Markham, A. F., and Smith, A. E. (1984) Se-

quence requirements for nuclear location of simian virus 40 large-T antigen. *Nature* **311,** 33–38.

55. Bukrinsky, M. I., Haggerty, S., Dempsey, M. P., et al. (1993) A nuclear localization signal within HIV-1 matrix protein that governs infection of non-dividing cells. *Nature* **365,** 666–669.

56. Perales, J. C., Ferkol, T., Molas, M., and Hanson, R. W. (1994) Receptor-mediated gene transfer. *Eur. J. Biochem.* **226,** 255–266.

57. Perales, J. C., Ferkol, T., Molas, M., and Hanson, R. W. (1994) An evaluation of receptor-mediated gene transfer using synthetic DNA-ligand complexes. *Eur. J. Biochem.* **226,** 255–266.

58. Plank, C., Mechtler, K., Szoka, F., and Wagner, E. (1996) Activation of the complement system by synthetic DNA complexes: a potential barrier for intravenous gene delivery. *Hum. Gene Ther.* **7,** 1437–1446.

59. Katayose, S. and Kataoka, K. (1997) Water-soluble polyion complex associates of DNA and poly(ethylene glycol)-poly(L-lysine) block copolymer. *Bioconjug Chem.* **8,** 702–707.

60. Emlen, W. and Mannik, M. (1984) Effect of DNA size and strandedness on the *in vivo* clearance and organ localization of DNA. *Clin. Exp. Immunol.* **56,** 185–192.

61. Bond, V. C. and Wold B. (1987) Poly-L-ornithine-mediated transformation of mammalian cells. *Mol. Cell. Biol.* **7,** 2286–2293.

62. Boussif, O., Lezoualc'h, F., Zanta, M. A., et al. (1995) A versatile vector for gene and oligonucleotide transfer into cells in culture and *in vivo*: polyethylenimine. *Proc. Natl. Acad. Sci. USA* **92,** 7297–7301.

63. Maurer, P. H. (1962) Antigenicity of polypeptides (poly-alpha amino acids). *J. Immunol.* **88,** 330–345.

64. Kantor, F. S., Ojeda, A., and Benacerraf, B. (1962) Studies on artificial antigens: antigenicity of DNP-polylysine and DNP copolymer of lysine and glutamic acid in guinea pigs. *Hum Exp. Med.* **117,** 55–69.

65. Michael, S. I. and Curiel, D.T. (1994) Strategies to achieve targeted gene delivery via the receptor-mediated endocytosis pathway. *Gene Ther.* **1,** 223–232.

66. Haensler, J. and Szoka, F. C. Polyamidoamine cascade polymers mediate efficient transfection of cells in culture. *Bioconjug. Chem.* **4,** 372–379.

67. Brubaker, J. O., Thompson, C. M., Morrison, L. A., et al. (1996) Th1-associated immune response to beta-gal expressed by a replication-defective herpes simplex virus. *J. Immunol.* **157,** 1598–1604.

68. Cotten, M., Baker, A., Saltik, M., Wagner, E., and Buschle, M. (1994) Lipopolysaccharide is a frequent contaminant of plasmid DNA preparations and can be toxic to primary human cells in the presence of adenovirus. *Gene Ther.* **1,** 239–246.

69. Cooper, M. J. (1996) Noninfectious gene transfer and expression systems for cancer gene therapy. *Semin. Oncol.* **23,** 172–187.

70. Huxley, C. (1994) Mammalian artificial chromosomes: a new tool for gene therapy. *Gene Ther.* **1,** 7–12.

71. Zatoukal, K., Schneeberger, A., Berger, M., et al. (1995) Elicitation of a systemic

and protective anti-melanoma immune response by an IL-2-based vaccine: assessment of critical cellular and molecular parameters. *J. Immunol.* **154,** 3406–3419.

72. Ferkol, T., Pellicena-Palle, A., Eckman, E., et al. (1996) Immunologic responses of gene transfer via the polymeric immunoglobulin receptor in mice. *Gene Ther.* **3,** 669–678.
73. Stahl, P. D. and Ezekowitz, R. A. (1998) The mannose receptor is a pattern recognition receptor involved in host defense. *Curr. Opin. Immunol.* **10,** 50–55.
74. Perlmutter, D. H. (1994) The SEC receptor: a possible link between neonatal hepatitis in α_1-antitrypsin deficiency and Alzheimer's disease. *Pediatr. Res.* **36,** 271–277.
75. Maniatis, T., Frinsch, E. F., and, Stambrook, J. (1989) *Molecular Cloning: A Laboratory Manual.* Cold Spring Harbor Laboratory Press, Cold Spring Harbor, NY.
76. Yengerov, Y. Y. and Semenov, T. E. (1992) Electron microscopy of DNA complexes with synthetic oligopeptides. *Electron Microsc. Rev.* **5,** 193–207.

4

Gene Transfer into Muscle by Electroporation In Vivo

Jun-ichi Miyazaki and Hiroyuki Aihara

1. Introduction

Among the nonviral techniques for gene transfer in vivo, the direct injection of plasmid DNA into muscle is especially simple, inexpensive, and safe. However, applications of this method have been limited by the relatively low expression levels of the transferred gene. Recently, we investigated the applicability of in vivo electroporation for gene transfer into muscle, using plasmid DNA expressing a cytokine as the vector. The results demonstrated that gene transfer into muscle by electroporation in vivo is far more efficient than simple intramuscular DNA injection and provides a potential approach toward systemic delivery of cytokines, growth factors, and other serum proteins for human gene therapy.

1.1. Naked DNA Injection into Muscle

Plasmid DNA injected into skeletal muscle is taken up by muscle cells, and the genes in the plasmid are expressed for more than 2 months thereafter *(1–9)*, although the transfected DNA does not usually undergo chromosomal integration *(1,2)*. However, the relatively low expression levels attained by this method have limited its applications for uses other than as a DNA vaccine *(4)*. There are a number of reports analyzing the conditions that affect the efficiency of gene transfer by intramuscular DNA injection and assessing the fine structures of expression plasmid vectors that may affect expression levels *(5,10)*. For example, Vitadello et al. *(6)* showed that regenerating muscle produced 80-fold or more protein than did normal muscle, following injection of an expression plasmid. Muscle regeneration was induced by treatment with

From: *Methods in Molecular Medicine, Vol. 69, Gene Therapy Protocols, 2nd Ed.*
Edited by: J. R. Morgan © Humana Press Inc., Totowa, NJ

cardiotoxin or bupivacaine *(6,7)*. By combining a strong promoter and bupivacaine pretreatment, we previously demonstrated that intramuscular injection of an interleukin-5 (IL-5) expression plasmid results in IL-5 production in muscle at a level sufficient to induce marked proliferation of eosinophils in the bone marrow and eosinophil infiltration of various organs in mice *(8)*. Recently, it was also reported that a single intramuscular injection of an erythropoietin expression plasmid produced physiologically significant elevations in serum erythropoietin levels and increased hematocrits in adult mice *(9)*. Hematocrits in these animals remained elevated at >60% for at least 90 days after a single injection. However, improvements to this method have not been sufficient to extend its application to human gene therapy.

1.2. In Vivo Electroporation

Electroporation has been widely used to introduce DNA into various types of cells in vitro. Recently, it has been reported that gene transfer by electroporation in vivo, that is, DNA injection followed by the application of electric fields, is highly effective for introducing DNA into mouse skin *(11)*, chick embryos *(12)*, rat liver *(13)*, and murine melanoma *(14)*. Based on these findings, we investigated the applicability of in vivo electroporation for gene transfer into muscle, using plasmid DNA carrying the gene for IL-5, which is involved in the growth and differentiation of B-cells and eosinophils *(15)*. Fifty micrograms each of a control plasmid or an IL-5 expression plasmid were injected into the bilateral tibialis anterior muscles of mice. Two electrode needles were inserted into the muscle to encompass the DNA injection sites, and electric pulses were delivered using an electric pulse generator. Then six 50-ms-long pulses of 50 V each were delivered to each injection site at a rate of 1 pulse/s. Five days later, serum IL-5 levels were measured by enzyme-linked immunosorbent assay (ELISA). In the mice injected with control plasmid, IL-5 levels were below the detection limit of the assay (<10 pg/mL), irrespective of the administration of electric pulses. IL-5 levels in the mice injected with IL-5 expression plasmid were 0.2 ng/mL without electropulsation, but 12 ng/mL with electropulsation.

Histochemical analysis of muscles injected with a lacZ expression plasmid showed that in vivo electroporation markedly increased both the number of muscle fibers taking up plasmid DNA and the copy number of plasmids introduced into muscle cells. These results clearly indicate that in vivo electroporation is a much more effective means of introducing DNA into muscle than is simple intramuscular DNA injection *(16)*. Recently, we also demonstrated that the muscle-targeted transfer of an erythropoietin expression plasmid into rats by in vivo electroporation produced significant elevations in serum erythropoietin levels and increased hematocrits *(17)*.

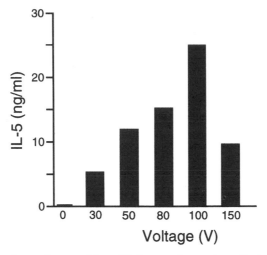

Fig. 1. Voltage dependence of the efficiency of gene transfer by electroporation in vivo. The bilateral tibialis anterior muscles were injected with 50 μg each of pCAGGS-interleukin-5 (IL-5) plasmid DNA. Electric pulses of the indicated voltages were delivered to the DNA injection site. Solid bars indicate serum IL-5 levels 5 days after electroporation. Each value represents the mean IL-5 concentration from three mice.

1.3. Parameters Affecting the Efficiency of Gene Transfer into Muscle by Electroporation

Although one hypothesis suggests that electropores are created by electric pulses, few data describe the structural effects of electropermeabilization. The transmembrane potential difference at 200–250 mV is believed to lead to transient permeabilization of the membrane, which results in the exchange of molecules across the membrane. Conditions affecting the efficiency of gene transfer have been studied in detail for in vitro electroporation *(18)*. Typically, DNA is mixed with cells suspended in phosphate-buffered saline (PBS) at 4°C or lower, and a single electric pulse lasting 20–100 ms at 200–700 V/cm is applied. Nonetheless, there have been no reports of systematic studies of the optimum conditions required for in vivo electroporation. However, square pulses are believed to be better than exponentially decaying pulses.

We examined various conditions for in vivo electroporation in muscle. To optimize the voltage of the electric pulses used for electroporation in vivo, we compared the serum IL-5 levels of mice subjected to electroporation at various electrode voltages. In this experiment, the pulse length (50 ms), number of pulses (6), amount of DNA injected (50 μg/site), and DNA concentration (1.5 μg/μL in saline), which could all affect the efficiency of gene transfer, were fixed. Immediately after DNA injection, electrode needles with a 5-mm gap

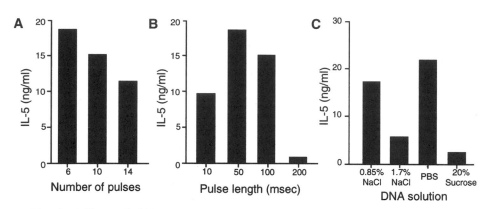

Fig. 2. Effects of different numbers of pulses (**a**), pulse lengths (**b**), and solution types (**c**) on the efficiency of gene transfer by electroporation. The bilateral tibialis anterior muscles were injected with 50 μg each of pCAGGS-IL-5 plasmid DNA at 1.5 μg/μL in saline (**a, b**) or in other solution types as indicated (**c**). Electric pulses of 100 V were delivered to the DNA injection sites. Serum samples were obtained 5 days after electroporation and measured for IL-5 by ELISA. Each value represents the mean IL-5 concentration from three mice.

were inserted to encompass the DNA injection sites, and electric pulses were administered at different voltages. Five days later, serum IL-5 levels were measured. As shown in **Fig. 1**, the serum IL-5 levels increased nearly proportionally with the voltage up to 100 V. However, administration of voltages higher than 100 V resulted in a marked decrease of the serum IL-5 levels, probably because of damage to the muscle cells. Optimal gene expression was achieved at 100 V, resulting in a serum IL-5 concentration of 25 ng/mL. The serum IL-5 concentration at 0 V was <0.1 ng/mL. Therefore, gene transfer by electroporation in vivo is more efficient than the traditional method of intramuscular DNA injection by more than two orders of magnitude.

We examined the effect of different pulse lengths, numbers of pulses, DNA concentrations, volumes of injection fluid, and solution types on the efficiency of gene transfer by electroporation. Some of these studies are shown in **Fig. 2**. The optimal conditions were as follows: pulse length, 50 ms; number of pulses, 6; DNA concentration, >0.5 μg/mL; DNA amount, >50 μg/site; and solution type, PBS. In these experiments, electric pulses were applied to the muscle immediately after DNA injection. However, our study showed that electric pulses may be applied up to 30 min after DNA injection without affecting the efficiency of gene expression. It should be noted that the optimal conditions for in vivo electroporation will be very different if this method is applied to other animals. Optimizing these parameters should permit the efficiency of gene transfer in other species to be improved.

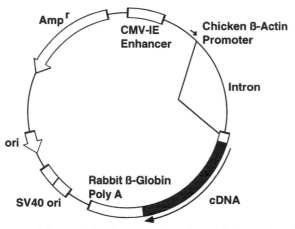

Fig. 3. Structure of the pCAGGS expression plasmid. Plasmid pCAGGS has the cytomegalovirus immediate early (CMV-IE) enhancer-chicken β-actin hybrid (CAG) promoter and a 3'-flanking sequence and a polyadenylation signal of the rabbit β-globin gene.

We next examined whether the direction of the electric field relative to that of muscle fibers affects the efficiency of gene transfer. After intramuscular DNA injection, electric pulses of 60 V were delivered in either a longitudinal or a transverse direction relative to the muscle fibers. To fit the tibialis anterior muscles between a pair of electrodes, we used electrodes with a 3-mm gap. Five days later, serum IL-5 levels were measured to evaluate the efficiency of DNA transfer. There were no significant differences between electric fields that were applied longitudinally (31 ng/mL) or transversely (40 ng/mL) to the muscle fibers. However, before this topic is closed, the following issues must be considered. Because muscle cells are oblong, the number of cells lying between the electrodes is much larger in a transverse orientation than in a longitudinal orientation. Therefore, more cells should be transfected in a transverse orientation. However, a higher voltage might be required for efficient electroporation in this orientation, since there are more cell membranes to be permeabilized between the electrodes. Further studies are required to determine how the direction of the electric field affects the efficiency of gene transfer. Because the tibialis anterior muscles of mice are small and it is difficult to insert a pair of electrodes into them in a transverse orientation, electric pulses were delivered in a longitudinal direction in the other experiments.

1.4. Expression Plasmid Used in Gene Transfer into Muscle

We used the pCAGGS plasmid *(19)* as an expression vector (**Fig. 3**). This vector contains the CAG (cytomegalovirus immediate early enhancer–β-actin

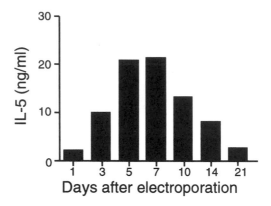

Fig. 4. Time course of interleukin-5 (IL-5) expression after transfer of pCAGGS-IL-5 plasmid by electroporation in vivo. The bilateral tibialis anterior muscles were injected with pCAGGS-IL-5 plasmid DNA. Electric pulses of 100 V were delivered to the DNA injection sites. Serum samples were obtained on the indicated days after electroporation and measured for IL-5 by ELISA. Each value represents the mean IL-5 concentration from three mice.

hybrid) promoter, which is especially active in muscle. Other expression plasmids may be used if their promoter is very active in muscle cells, e.g., the VR1255 vector developed by Norman et al. *(10)* for muscle-targeted expression contains the cytomegalovirus (CMV) promoter, enhancer, intron, as well as the kanamycin resistance gene.

For gene therapy, it is sometimes desirable to regulate the expression of foreign genes introduced by in vivo electroporation. This may be attained by using a tetracycline (Tet)-inducible gene expression system, in which two expression units are required: one to express a Tet-regulatable chimeric transactivator and the other to express the gene of interest under the promoter regulated by the transactivator *(20)*. Recently, it was reported that pharmacologic regulation through the Tet-inducible system allows the regulation of serum erythropoietin and hematocrit levels *(21)*. This method is expected to provide a potentially safe and low-cost treatment for serum protein deficiencies.

1.5. Duration of Expression

The time course of gene expression by electroporation in vivo was determined by following the serum IL-5 levels after electroporation at 100 V. As shown in **Fig. 4**, the serum IL-5 levels reached a maximum 5–7 days after electroporation and gradually decreased thereafter, followed by a decrease to approximately 10% of the peak value by 3 weeks after electroporation. We also examined the duration of expression when a rat erythropoietin expression plasmid was introduced into the thigh muscle of rats by in vivo electroporation. The serum erythropoietin levels reached a maximum 7 days after electroporation and gradually decreased thereafter. However, the erythropoietin levels were still approximately 30% of their peak value 1 month after electroporation. At present, we do not know the reason for this difference in the duration of expression between IL-5 and erythropoietin. It cannot be explained by the species difference between rats and mice, because other groups have reported long-term expression of luciferase in mouse muscle after DNA transfer by electroporation *(22)*. Alternatively, local inflammation possibly caused by IL-5 expression might have reduced the duration of expression.

Recently, Vicat et al. *(23)* showed that electroporation with high-voltage and short-pulse (900 V/100 μs) currents provides high-level and long-lasting gene expression in muscle. It will be necessary to study the relation between the conditions for in vivo electroporation and the time course of gene expression in detail.

1.6. Advantages of In Vivo Electroporation

Gene transfer by electroporation, which uses plasmid DNA as the vector, has several advantages over transfer using viral vectors. A large quantity of highly purified plasmid DNA is easily and inexpensively obtained. Gene transfer can be repeated without apparent immunologic responses to the DNA vector. Although the gene expression is usually transient, there is less likelihood of recombination events with the cellular genome, eliminating the risk of insertional mutagenesis that is associated with the use of viral vectors. Since there are fewer size constraints than with current viral vectors, plasmid vectors can carry larger genes. It may also be possible to transfer a mixture of two or more different plasmid constructs into muscle by electroporation. Finally, new DNA plasmid constructs can be rapidly made and tested. Furthermore, gene transfer by electroporation also has advantages over other methods of gene transfer using nonviral vectors. Intramuscular electrotransfer strongly decreases the interindividual variability in expression usually observed after plasmid DNA injection into muscle *(22)*. Moreover, plasmid DNA can be readily used without any modification, such as conjugation with liposomes.

1.7. Possible Application of In Vivo Electroporation

Electroporation-mediated gene transfer has been used effectively in the muscles of mouse, rat, rabbit, and monkey *(22)*, and it has been applied to gene transfer into cardiac muscle *(24)*. Thus, this method should have broad applications in physiologic and pharmacologic studies using experimental animals. It is likely that further improvement of this method will provide a new approach to efficient DNA vaccination and gene therapy for human diseases. Among the potentially treatable human diseases are various autoimmune diseases, chronic inflammatory disorders, infections, malignancies, and acquired or inherited serum protein deficiencies. This method may also be applied to the constitutive overexpression of vascular endothelial growth factor (VEGF) or hepatocyte growth factor (HGF) to induce therapeutic angiogenesis in patients with critical limb ischemia.

2. Materials

2.1. Experimental Animals

1. Mice: We used 8-week-old female C57BL/6J mice purchased from CLEA Japan (Osaka, Japan). Mice of other ages or strains or rats may be treated similarly.
2. Anesthetic: 50 mg/mL pentobarbital sodium solution (Nembutal; Abbott Lab., North Chicago, IL) was diluted to 6 mg/mL with a diluent containing 40% (v/v) propylene glycol and 10.5% (v/v) ethanol.

2.2. Plasmid DNA

The plasmid vector must include an expression unit that is active in striated muscles. We have successfully used the pCAGGS vector (**Fig. 3**) *(19)* (*see* **Subheading 1.4.**). To assess the efficiency of gene transfer, the following two constructs were convenient: pCAGGS-IL-5 *(8)* and pCAGGS-lacZ *(16)*, which were constructed by inserting mouse IL-5 cDNA and the *E. coli* lacZ gene, respectively, into the unique *Eco*RI site between the CAG promoter and a 3'-flanking sequence of the rabbit β-globin gene of pCAGGS. These plasmid vectors can be provided by J. M. upon request.

The pCAGGS plasmid is based on pUC13, a high-copy-number plasmid, and is easily grown in *E. coli* HB101 or other strains. Plasmid DNA is extracted by the alkaline lysis method and purified by two cycles of ethidium bromide-CsCl equilibrium density gradient ultracentrifugation (*see* **Note 2**). Plasmid DNA is further purified by isopropanol precipitation, phenol and phenol/chloroform extraction, and ethanol precipitation. DNA is dissolved in pure water, and its quantity and quality are assessed by optical density at 260 and 280 nm. Prior to injection, DNA is diluted to its final concentration, 1–1.5 μg/μL in PBS (137 mM NaCl, 2.68 mM KCl, 8.1 mM Na$_2$HPO$_4$, 1.47 mM KH$_2$PO$_4$, pH 7.4). Because the salt concentration seems to affect the efficiency of gene transfer, the final DNA solution is made by adding 1 vol of 10× PBS to 9 vol of DNA solution diluted with water.

2.3. Intramuscular DNA Injection and Electroporation

1. Insulin syringe with a 27-gage needle.
2. Electrodes consisting of a pair of stainless steel needles, 5 mm in length and 0.4 mm in diameter, fixed with a distance (gap) between them of 5 mm (**Fig. 5**).

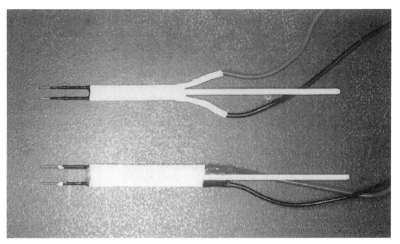

Fig. 5. Appearance of electrodes consisting of a pair of stainless steel needles, 5 mm in length and 0.4 mm in diameter, fixed with a distance (gap) between them of 3 mm (top) or 5 mm (bottom).

   These electrodes can be obtained from TR Tec (Tokyo, Japan; Fax: +81-3-3944-6196).
3. Electric pulse generator (Electro Square Porator T820; BTX, San Diego, CA) connected to a switch box (MBX-4; BTX), which can produce square waves, i.e., the voltage remains constant during the pulse duration. Electric pulses can be monitored by a graphic pulse analyzer (BTX400; BTX).

2.4. Assessment of the Efficiency of Gene Transfer

1. CsCl-purified preparations of pCAGGS-IL-5 and pCAGGS-lacZ plasmid DNA at a concentration of 1.5 µg/µL in PBS.
2. Murine IL-5 ELISA kit (Endogen, Woburn, MA).
3. 4% paraformaldehyde in PBS.
4. 40 mM X-gal (5-bromo-4-chloro-3-indolyl-β-D-galactopyranoside) in dimethyl-sulfoxide (DMSO). This is diluted to 1 mM in PBS before use for staining.
5. O.C.T. compound (Miles, Elkhart, IL).
6. Dry ice-acetone.
7. Cryostat.
8. Slide glasses coated with 3-aminopropyltriethoxysilane (Sigma, St. Louis, MO).
9. 1.5% glutaraldehyde in PBS.
10. Eosin.

3. Methods

   In the following section, we describe the method of gene transfer into tibialis anterior muscles of adult mice by in vivo electroporation. It will be necessary to modify this method to use it in other muscles or other species.

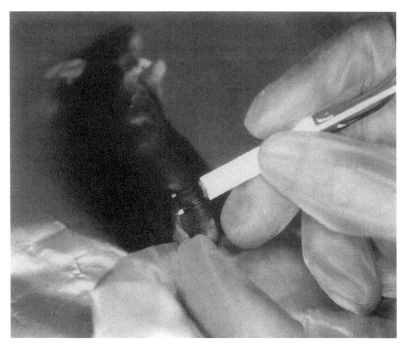

Fig. 6. Insertion of electrodes. A pair of electrode needles are inserted into the right tibialis anterior muscle to a depth of 5 mm to encompass the DNA injection sites.

3.1. Intramuscular DNA Injection

1. Anesthetize mice by intraperitoneal injection of 0.1 mL/g body weight of 6 mg/mL pentobarbital sodium solution.
2. Inject the tibialis anterior muscles with 50 μg of purified closed circular plasmid DNA at 1.5 μg/μL in PBS using an insulin syringe with a 27-gage needle (*see* **Notes 3** and **4**).

3.2. Electroporation In Vivo

1. Insert a pair of electrode needles into the muscle to a depth of 5 mm to encompass the DNA injection sites (**Fig. 6**) (*see* **Note 5**). Push on the chamber resistance switch and monitor the resistance value. If the value is 1–2 kΩ, it shows that the electrodes are correctly inserted into the muscle. Otherwise, change the insertion site of the electrodes.
2. Deliver three 50-msec–long electric pulses using an electric pulse generator (*see* **Note 6**) followed by three more pulses of the opposite polarity to each injection site at a rate of 1 pulse/s. The shape of the pulses can be monitored using a graphic pulse analyzer.

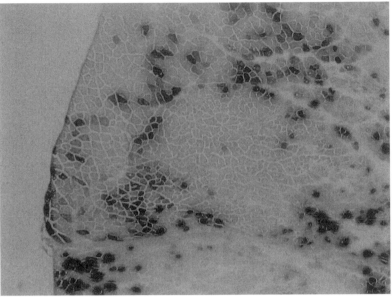

Fig. 7. Histochemical staining for β-galactosidase activity in the muscle after gene transfer of pCAGGS-lacZ DNA with electropulsation. The tibialis anterior muscle was injected with 50 µg of pCAGGS-lacZ plasmid DNA and treated with electric pulses of 100 V. Five days later, the muscle was excised. Transverse sections of the muscle sample were stained for β-galactosidase activity and counterstained with eosin (original magnification, ×40).

3.3. Assessment of the Efficiency of Gene Transfer

Before introducing the gene of interest by electroporation, it is important to test the effectiveness of the experimental procedures using a positive control. This may be done using a plasmid that expresses some cytokine, e.g., IL-5, or β-galactosidase (*see* **Note 7**).

3.3.1. IL-5 Expression

1. Inject the bilateral tibialis anterior muscles of anesthetized mice with 50 µg each of pCAGGS-IL-5 plasmid DNA at a concentration of 1.5 µg/µL in PBS, and deliver electric pulses at 100 V, as described in **Subheading 3.2.**
2. Five days after injection, obtain serum samples from the tail vein of the mice.
3. Assay the serum samples for IL-5 using an ELISA kit (Endogen), according to the supplier's instructions.

3.3.2. β-Galactosidase Expression

1. Inject the bilateral tibialis anterior muscles of anesthetized mice with 50 µg each of pCAGGS-lacZ plasmid DNA at a concentration of 1.5 µg/µL in PBS, and deliver electric pulses at 100 V as described in **Subheading 3.2.**

2. Four or 5 days after injection, sacrifice the mice by cervical dislocation.
3. Fix the tibialis anterior muscles in cold 4% paraformaldehyde in PBS for 3 h, and then wash in PBS for 1 h.
4. To detect *E. coli* β-galactosidase activity in whole muscle, stain the muscle sample at 37°C for 18 h in the presence of 1 mM X-gal. Otherwise, go to **step 5**.
5. For transverse sections, embed the muscle in O.C.T. compound and freeze in dry ice-acetone.
6. Slice serial sections (15-μm thick) with a cryostat and place on slide glasses coated with 3-aminopropyltriethoxysilane.
7. Fix the slices in 1.5% glutaraldehyde for 10 min at room temperature, then wash three times in PBS.
8. Incubate the samples at 37°C for 3 h in the presence of 1 mM X-gal.
9. Counterstain the muscle sections with eosin.
10. Observe the sections with a microscope for X-gal staining (**Fig. 7**).

4. Notes

1. Prior to the in vivo experiment, the DNA construct should be tested for transgene expression in vitro using a myoblast cell line, such as C2C12, which can be transfected easily by the lipofection method.
2. The plasmid DNA preparation should be pure. Contaminating impurities may cause local immunologic reactions, which may lead to early loss of gene expression or affect the experimental results. Two cycles of ethidium bromide-CsCl equilibrium density gradient ultracentrifugation are recommended.
3. Prior to the actual experiments, we recommend that you confirm that you can reproducibly inject the tibialis anterior muscles of anesthetized mice using some kind of dye.
4. The tibialis anterior muscle is relatively small, and the maximal volume of DNA solution that can be injected into it is <50 μL.
5. Other types of electrodes may be used: external plate-type electrodes have been successfully used in our laboratory (unpublished data).
6. We have used needle-type electrodes fixed with a distance (gap) of 5 mm. This distance should be altered to take into account the size of target muscle. The electric field strength (voltage/distance), but not the voltage itself, is assumed to be directly related to the efficiency of electropermeabilization. Therefore, if the electrodes have a gap of 10 mm instead of 5 mm, the voltage of the electric pulses must be doubled to obtain the same strength of electric field. However, higher voltages may cause very high local current around the electrodes, which may lead to irreversible damage to muscle and other tissues.
7. It was reported that X-gal histochemistry following gene transfer of constructs encoding lacZ may underestimate the anatomic extent of gene expression *(25)*.

References

1. Wolff, J. A., Malone, R. W., Williams, P., et al. (1990) Direct gene transfer into mouse muscle in vivo. *Science* **247**, 1465–1468.

2. Wolff, J. A., Ludtke, J. J., Acsadi, G., Williams, P., and Jani, A. (1992) Long-term persistence of plasmid DNA and foreign gene expression in mouse muscle. *Hum. Mol. Genet.* **1,** 363–369.

3. Davis, H. L., Whalen, R. G., and Demeneix, B. A. (1993) Direct gene transfer into skeletal muscle in vivo: factors affecting efficiency of transfer and stability of expression. *Hum. Gene Ther.* **4,** 151–159.

4. Davis, H. L., Michel, M.-L., and Whalen, R. G. (1995) Use of plasmid DNA for direct gene transfer and immunization. *Ann. N.Y. Acad. Sci.* **772,** 21–29.

5. Wolff, J. A., Williams, P., Acsadi, G., Jiao, S., Jani, A., and Chong, W. (1991) Conditions affecting direct gene transfer into rodent muscle in vivo. *Biotechniques* **11,** 474–485.

6. Vitadello, M., Schiaffino, M. V., Picard, A., Scarpa, M., and Schiaffino, S. (1994) Gene transfer in regenerating muscle. *Hum. Gene Ther.* **5,** 11–18.

7. Wells, D. J. (1993) Improved gene transfer by direct plasmid injection associated with regeneration in mouse skeletal muscle. *FEBS Lett.* **332,** 179–182.

8. Tokui, M., Takei, I., Tashiro, F., et al. (1997) Intramuscular injection of expression plasmid DNA is an effective means of long-term systemic delivery of interleukin-5. *Biochem. Biophys. Res. Commun.* **233,** 527–531.

9. Tripathy, S. K., Svensson, E. C., Black, H. B., et al. (1996) Long-term expression of erythropoietin in the systemic circulation of mice after intramuscular injection of a plasmid DNA vector. *Proc. Natl. Acad. Sci. USA* **93,** 10876–10880.

10. Norman, J. A., Hobart, P., Manthorpe, M., Felgner, P., and Wheeler, C. (1997) Development of improved vectors for DNA-based immunization and other gene therapy applications. *Vaccine* **15,** 801–803.

11. Titomirov, A. V., Sukharev, S., and Kistanova, E. (1991) In vivo electroporation and stable transformation of skin cells of newborn mice by plasmid DNA. *Biochim. Biophys. Acta* **1088,** 131–134.

12. Muramatsu, T., Mizutani, Y., Ohmori, Y., and Okumura, J. (1997) Comparison of three nonviral transfection methods for foreign gene expression in early chicken embryos in ovo. *Biochem. Biophys. Res. Commun.* **230,** 376–380.

13. Heller, R., Jaroszeski, M., Atkin, A., et al. (1996) In vivo electroinjection and expression in rat liver. *FEBS Lett.* **389,** 225–228.

14. Rols, M.-P., Delteil, C., Golzio, M., et al. (1998) In vivo electrically mediated protein and gene transfer in murine melanoma. *Nature Biotechnol.* **16,** 168–171.

15. Takatsu, K. (1992) Interleukin-5. *Curr. Opin. Immunol.* **4,** 299–306.

16. Aihara, H. and Miyazaki, J. (1998) Gene transfer into muscle by electroporation in vivo. *Nat. Biotechnol.* **16,** 867–870.

17. Maruyama, H., Sugawa, M., Moriguchi, Y., et al. (2000) Continuous erythropoietin delivery by muscle-targeted gene transfer using in vivo electroporation. *Hum. Gene Ther.* **11,** 429–437.

18. Wolf, H., Rols, M. P., Boldt, E., Neumann, E., and Teissie, J. (1994) Control by pulse parameters of electric field-mediated gene transfer in mammalian cells. *Biophys. J.* **66,** 524–531.

19. Niwa, H., Yamamura, K., and Miyazaki, J. (1991) Efficient selection for high-expression transfectants with a novel eukaryotic vector. *Gene* **108,** 193–199.
20. Liang, X., Hartikka, J., Sukhu, L., Manthorpe, M., and Hobart, P. (1996) Novel, high expressing and antibiotic-controlled plasmid vectors designed for use in gene therapy. *Gene Ther.* **3,** 350–356.
21. Rizzuto, G., Cappelletti, M., Maione, D., et al. (1999) Efficient and regulated erythropoietin production by naked DNA injection and muscle electroporation. *Proc. Natl. Acad. Sci. USA* **96,** 6417–6422.
22. Mir, L. M., Bureau, M. F., Gehl, J., et al. (1999) High-efficiency gene transfer into skeletal muscle mediated by electric pulses. *Proc. Natl. Acad. Sci. USA* **96,** 4262–4267.
23. Vicat, J. M., Boisseau, S., Jourdes, P., et al. (2000) Muscle transfection by electroporation with high-voltage and short-pulse currents provides high-level and long-lasting gene expression. *Hum. Gene Ther.* **11,** 909–916.
24. Harrison, R. L., Byrne, B. J., and Tung, L. (1998) Electroporation-mediated gene transfer in cardiac tissue. *FEBS Lett.* **435,** 1–5.
25. Couffinhal, T., Kearney, M., Sullivan, A., Silver, M., Tsurumi, Y., and Isner, J. M. (1997) Histochemical staining following LacZ gene transfer underestimates transfection efficiency. *Hum. Gene Ther.* **8,** 929–934.

5

Viral Liposomes

Preparation and Use

Yasufumi Kaneda

1. Introduction

With the aim of developing successful human gene therapy, numerous viral and nonviral (synthetic) methods of gene transfer have been developed *(1,2)*, each method having limitations as well as advantages. To develop in vivo gene transfer vectors with high efficiency and low toxicity, several groups have attempted to overcome the limitations of one vector by combining them with the strengths of another.

1.1. Development of HVJ-Liposomes

Our basic concept is the construction of novel, hybrid-type liposomes with functional molecules inserted into them *(3,4)*. Based on this concept, DNA-loaded liposomes were fused with ultraviolet (UV)-inactivated hemagglutinating virus of Japan (HVJ; Sendai virus) to form HVJ liposomes (approximately 400–500 nm in diameter). These viral liposomes bind to cell surface sialic acid receptors and fuse with the cell membrane to introduce DNA directly into the cytoplasm without degradation (**Fig. 1**). The HVJ-liposomes can encapsulate DNA smaller than 100 kb. RNA, oligodeoxynucleotides, proteins, and drugs can also be enclosed and delivered to cells. HVJ liposomes are useful for in vivo gene transfer *(5)*. When HVJ-liposomes containing the LacZ gene were injected directly into one rat liver lobe, approximately 70% of cells expressed LacZ gene activity, and no pathologic hepatic changes were observed *(6)*. One advantage of HVJ liposomes is allowance for repeated injections. Gene trans-

From: *Methods in Molecular Medicine, Vol. 69, Gene Therapy Protocols, 2nd Ed.*
Edited by: J. R. Morgan © Humana Press Inc., Totowa, NJ

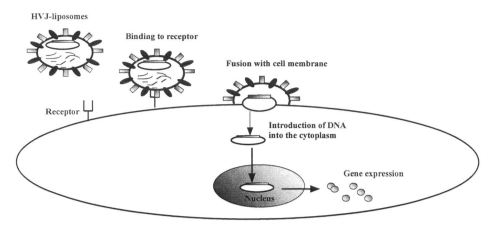

Fig. 1. Gene transfer by HVJ liposomes. HVJ liposomes bind to cell surface sialic acid receptors and associate with lipids in the lipid bilayer to induce cell fusion. By the fusion of the envelope of HVJ liposomes with cell membrane, DNA in the HVJ liposomes can be introduced directly into the cytoplasm.

fer to rat liver cells was not inhibited by repeated injections. After repeated injections, anti-HVJ antibody generated in the rat was not sufficient to neutralize HVJ liposomes. Cytotoxic T-cells recognizing HVJ were not detected in rats transfected repeatedly with HVJ liposomes *(6)*.

1.2. Improvements in HVJ-Liposomes

The HVJ-liposome gene delivery system has several advantages, but improvement has been needed before use in humans. To increase the efficiency of gene delivery, we investigated the lipid components of liposomes *(7)*. Our conclusions were threefold: the most efficient gene expression occurred with a phosphatidylcholine, phosphatidylethanolamine, and sphingomyelin molar ratio of 1:1:1; anionic HVJ liposomes should be prepared using phosphatidylserine (PS) as the anionic lipid; and the ratio of phospholipids to cholesterol should be 1:1. Accordingly, we developed new anionic liposomes called HVJ artificial viral envelope (AVE) liposomes. The lipid components of AVE liposomes are very similar to the HIV envelope and mimic the red blood cell membrane *(8)*. HVJ-AVE liposomes have yielded gene expression in liver and muscle 5–10 times higher than that observed with conventional HVJ liposomes *(7)*. As shown in **Fig. 2**, HVJ-AVE liposomes were the most effective for gene transfer to mouse skeletal muscle in various nonviral gene transfer methods. HVJ-AVE liposomes were also very effective for gene delivery to isolated rat heart via the coronary artery. LacZ gene expression was observed in the entire heart, whereas expression was not observed with empty HVJ-AVE liposomes

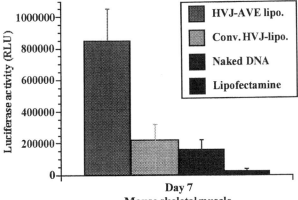

Fig. 2. Luciferase gene was introduced into mouse skeletal muscle using various nonviral gene delivery systems including HVJ-AVE liposome, conventional HVJ-liposome, Lipofectamine® (Gibco, BRL), and naked DNA. Luciferase activity in the muscle (*n* = 4) was assayed on day 7 after gene transfer. Mean values and standard deviations are shown. RLV, relative light units.

(9). The safety of HVJ liposomes has been tested and evaluated in monkeys. There were no significant pathologic signs after injection of HVJ liposomes into skeletal muscle or the saphenous vein of cynomolgus monkeys. Messenger RNAs for fusion proteins of HVJ were not detected in monkey tissues after the injection.

Another improvement was construction of cationic-type HVJ liposomes using cationic lipids. Of the cationic lipids, positively charged 3β-(N-[N',N'-dimethylaminoethane]-carbamoyl) cholesterol hydroxide (DC) *(10)* has been the most efficient for gene transfer. For luciferase expression, HVJ-cationic DC liposomes were 100 times more efficient than conventional HVJ-anionic liposomes *(7)*. Although it has been very difficult to transfer genes to bone marrow and spleen cells using conventional HVJ liposomes, HVJ-cationic liposomes have been shown to be effective for gene transfer to both types of cells. However, when introduced into mouse muscle or liver, total luciferase expression after transfection with HVJ-cationic liposomes was shown to be 10–150 times lower than that with conventional anionic HVJ liposomes *(7)*, which were less efficient for in vitro transfection. AVE liposomes were modified further to create AVE+DC10 (containing 10% PS and 10% DC), AVE+DC20 (containing 10% PS and 20% DC), and AVE-PS (containing neither PS nor DC) liposomes.

We examined in vivo gene transfection efficiency with these liposomes after conjugation with the HVJ envelope. AVE yielded the highest luciferase

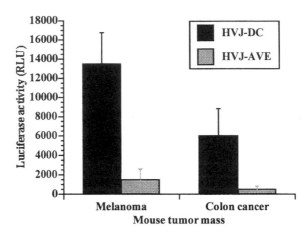

Fig. 3. Luciferase gene was transferred to mouse tumor masses ($n = 5$; melanoma and colon cancer) using HVJ-DC liposomes (cationic) or HVJ-AVE liposomes (anionic), and luciferase gene expression was analyzed on day 1 after the transfer. Mean values and standard deviations are shown. RLV, relative light units.

expression in liver. AVE-PS and AVE+DC10 liposomes, which have a net neutral charge, showed intermediate luciferase activities. AVE+DC20 liposomes, which have an excessive amount of cationic lipid, yielded luciferase activities similar to those of HVJ-DC liposomes. However, we recently found HVJ-cationic liposomes to be more effective in some cases for in vivo gene transfer. High expression of the LacZ gene was obtained in restricted regions of chick embryos after injection of HVJ-cationic liposomes *(11)*, whereas HVJ-anionic liposomes were ineffective. In addition, when HVJ-cationic liposomes containing the LacZ gene were administered to rat lung with a jet nebulizer, more efficient gene expression in the epithelium of the trachea and bronchus was observed compared with that found with HVJ-anionic liposomes *(12)*. HVJ-cationic liposomes were also much more effective for gene transfer to tumor masses (**Fig. 3**) or disseminated cancers *(3,13)* in an animal model compared with HVJ-AVE liposomes.

Therefore, HVJ-anionic and -cationic liposomes can complement each other, and each liposome should be used for proper targeting.

Methods to prepare HVJ-anionic and -cationic liposomes as well as applications to gene transfer both in vitro and in vivo are described below.

2. Materials
2.1. Preparation of HVJ

1. Seed of HVJ: 100-µL aliquots of the chorioallantoic fluid containing HVJ (Z strain) in 10% dimethylsulfoxide, stored in liquid nitrogen.

2. Polypeptone solution: 1% polypeptone, 0.2% NaCl, pH 7.2) and balanced salt solution (BSS); 137 mM NaCl, 5.4 mM KCl, 10 mM Tris-HCl, pH7.6), sterilized by autoclaving and stored at 4°C.
3. Embryonated chick eggs: 10–14 days after fertilization.
4. Incubater: Temperature and moisture set at 36.5°C and at 30–40%, respectively.
5. Centrifuge tubes including 50-mL conical tubes (Becton-Dickinson, Lincoln Park, NJ), 35-mL centrifuge tubes (Beckman, Tokyo, Japan) and 10-mL ultra-centrifuge tubes (Hitachi, Tokyo, Japan) sterilized.
6. Photometer (Spectrophotometer DU-68; Beckman).
7. Low-speed centrifuge (05PR-22; Hitachi, Tokyo, Japan).
8. Centrifuge with JA-20 rotor (J2-HS; Beckman).

2.2. Preparation of Lipid Mixtures

1. Chromatographically pure bovine brain phosphatidylserine-sodium salt (PS) (cat. no. 83032L; Avanti Polar Lipids, Birmingham, AL), dioleoyl-L-α-phosphatidyle-thanolamine (DOPE; P-5078; Sigma, St. Louis, MO), sphingomyelin (Sph) (S-0756; Sigma), egg yolk phosphatidylcholine (PC; P-2772; Sigma), DC (C2832; Sigma), and cholesterol (Chol; C-8667; Sigma), stored at –20°C.
2. Glass tubes (24-mm caliber and 12 cm long), custom-made (Fujiston 24/40; Iwaki Glass, Tokyo, Japan). Immerse the fresh tubes in saturated KOH-ethanol (180 g KOH in 500 mL ethanol) solution for 24 h, rinse with distilled water, and heat at 180°C for 2 h before use.
3. Rotary evaporator with water bath (type SR-650; Tokyo Rikakikai Tokyo, Japan).
4. Vacuum pump with pressure gage (type Asp-13; Iwaki Glass).

2.3. Preparation of HVJ Liposomes

1. Plasmid DNAs, purified by a column procedure (Qiagen, Germany). Dissolve the preparations in BSS; the final concentration of DNA should be >1 mg/mL, and should be stored at –20°C.
2. Cellulose acetate membrane filters (0.45 μm, [cat. no. 2053-025] and 0.20 μm [cat. no. 2052-025]; Iwaki Glass), used for sizing liposomes.
3. BSS and 30% (w/v) sucrose in BSS, sterilized by autoclaving and stored at 4°C.
4. Water bath (Thermominder Jr 80; TAITEC, Saitama, Japan) for preparing liposomes.
5. Water bath shaker (Thermominder; TAITEC) for fusing liposomes with HVJ.
6. Ultracentrifuge with an RPS-40T rotor (55P-72, Hitachi) for purifying HVJ liposomes.
7. Ultraviolet crosslinker (Spectrolinker XL-1000, Spectronics) for inactivating HVJ.

3. Methods

A flow chart for the preparation of HVJ liposomes is shown in **Fig. 4**.

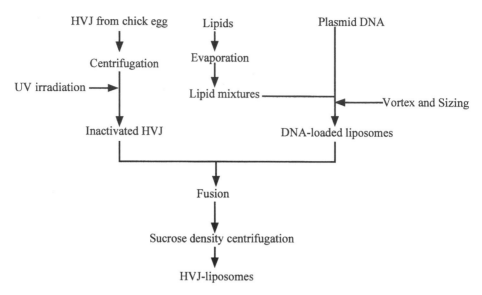

Fig. 4. Flow chart for the preparation of HVJ liposomes.

3.1. Preparation of HVJ in Eggs

1. Thaw the seed quickly and dilute to 1000 times with polypeptone solution. Keep the diluted seed at 4°C before proceeding to the next step.
2. Observe embryonated eggs under illumination in a dark room, and mark an injection point at about 0.5 mm above the chrioallantoic membrane. Dissenfect the eggs with tincture of iodine and puncture at the point marked.
3. Infect the diluted seed (0.1 mL) into each egg using a 1-mL disposable syringe with a 26-gage needle. Insert the needle vertically so as to stab the chorioallantoic membrane.
4. After inoculation of the seed, cover the hole in the egg with melted paraffin. Then incubate the eggs for 3 days at 36.5°C in 30–40% moisture.
5. Chill the eggs at 4°C for >6 h before harvesting the virus.
6. Partially remove the egg shell and remove the chorioallantoic fluid to an autoclaved bottle using a 10-mL syringe with an 18-gage needle. The virus in the fluid stored at 4°C is stable at least for 3 months.

   Steps 2, 3 and **6** can be carried out at room temperature.

3.2. Purification of HVJ from Chorioallantoic Fluid

1. Transfer 200 mL of the chorioallantoic fluid into four 50-mL disposable conical tubes, rotate at 1000g for 10 min at 4°C in a low-speed centrifuge.
2. Then aliquot the supernatant into six tubes (JS-20; Beckman) and centrifuge at 27,000g for 30 min at 4°C.
3. Add about 5 mL of BSS to the pellet in one of the tubes, and keep the materials at 4°C overnight.

4. Gently suspend the pellets, collect in two tubes, and centrifuge as described in **step 2**. Keep the resultant pellet in each tube at 4°C in 5 mL of BSS for more than 8 h.
5. Gently suspend the pellets and rotate at 1000*g* in a low-speed centrifuge.
6. Remove the supernatant to an aseptic tube and store at 4°C.
7. Indicate virus titer by measuring the absorbance at 540 nm of the 10× diluted supernatant using a photometer. An optical density at 540 nm corresponds to 15,000 hemagglutinating units (HAU), which correlates well with fusion activity. The supernatant prepared as above usually shows 20,000–30,000 HAU/mL. An aseptically prepared virus solution maintains the fusion activity for 3 weeks.

3.3. Preparation of Lipid Mixture

1. Dissolve dry reagents of DOPE (12.2 mg), Sph (11.8 mg), and Chol (24.0 mg) in 3870 μL of chloroform. Add 130 μL of PC (13.0 mg) chloroform solution to the 3870-μL lipid solution (*see* **Note 1**). This 4000-μL lipid solution is called a basal mixture for liposomes. The basal mixture is ready for preparation of the anionic or cationic liposomes described below, or it can be stored at –20°C after infusing nitrogen gas.
2. To prepare the anionic lipid mixture, add 10 mg of PS to the basal mixture. To obtain the cationic lipid mixture, add 6 mg of DC to the mixture.
3. Aliquot the lipid solution of 0.5 mL into eight glass tubes. Keep the tubes on ice or –20°C in nitrogen gas before evaporation. Evaporate the lipid solution as soon as possible.
4. Connect the tube to a rotary evaporator. Immerse the tube in a water bath at the tip, and set the water bath at a temperature of 40°C.
5. Evaporate the organic solvent in a rotary evaporator in a vacuum. The usual drying period is about 5–10 min. Lipids appropriate for liposome preparation are those that have stuck inside the tube in a thin layer. Those that have accumulated at the bottom of the tubes are inappropriate (*see* **Note 2**).

3.4. Preparation of HVJ Liposomes Containing DNA

1. Add plasmid DNA (200 μg) in 200 μL BSS to a lipid mixture in the glass tube prepared as above and agitate intensely by vortexing for 30 s followed by incubation at 37°C for 30 s. Repeat this cycle eight times. By this method, plasmid DNA is enclosed at a ratio of 10–30% in anionic liposomes or 50–60% in cationic liposomes.
2. For preparation of sized unilamellar liposomes, filter the liposome suspension with 0.45-μm pore size cellulose acetate filter and then with 0.2-μm filter. Sizing by an extruder with polycarobanate filters is better for preparing sized liposomes.
3. In the meantime, inactivate the HVJ virus and keep on ice (*see* **Note 3**). Add 15,000 HAU of the HVJ virus to the liposome suspension and leave the tube on ice for 5–10 min (*see* **Note 4**). Then incubate sample at 37°C for 1 h with shaking (120/min) in a water bath.
4. Add 7 mL 30% sucrose solution to a centrifuge tube and overlay the HVJ lipo-

some mixture on it. Separate the HVJ-liposome complexes from the free HVJ by sucrose gradient density centrifugation at 62,000g for 90 min at 4°C.

5. Stop the centrifuge and gently remove only the conjugated liposomes. The final volume of the HVJ-liposome suspension should be approximately 1 mL.

6. Only in the case of HVJ-anionic liposomes, after centrifugation, collect the complexes, add 4 vol of chilled BSS, spin at 27,000g for 30 min, and suspend the pellet in 0.5–1.0 mL of appropriate buffer (*see* **Note 5**).

3.5. Applications

3.5.1. Transfer of DNA into Cultured Cells

1. HVJ-cationic liposomes should be used for in vitro gene transfer because HVJ-cationic liposomes are approximately 100 times more efficient in gene transfer to cultured cells than HVJ-anionic liposomes *(7)*. Add 10 µL of 1-mL HVJ-cationic liposome suspension to 10^5 cells in serum-containing culture medium *(14)*.

2. Incubate the cells with the liposomes at 37°C for 2 h. Then change to fresh medium and continue the culture (*see* **Note 6**).

3.5.2. Gene Transfer In Vivo by HVJ Liposomes

1. For gene transfer to tissues, HVJ-anionic liposomes are recommended. The liposomes are useful for gene transfer to liver, skeletal muscle, heart, lung, artery, brain, spleen, eye, and joint space of rodent, rabbit, dog, lamb, and monkey. For example, to introduce DNA into rat liver, inject 2–3 mL of HVJ-anionic liposomes into the portal vein using a 5-mL syringe with a butterfly-shaped needle *(15,16)* or directly into the liver under the perisplanchnic membrane using a 5-mL syringe with a 27-gage needle *(17,18)*. For gene transfer into rat kidney, inject 1 mL of anionic HVJ-liposome suspension into the renal artery *(19,20)*. For gene transfer to rat carotid artery, fill a lumen of a segment of the artery with 0.5 mL anionic HVJ-liposome complex for 20 min at room temperature using a cannula *(21)*.

2. For gene transfer to tumor masses or disseminated tumors, direct injection of cationic HVJ liposomes (0.1–0.5 mL) is recommended.

4. Notes

1. Lipid vials should be left at room temperature for about 30 min before opening the lids. Many lipids are highly hygroscopic.

2. The lipid mixture in the glass tube can be stored at –20°C in nitrogen gas for 1 month, after evaporation.

3. UV-inactivated HVJ can be stored for >6 months in 10% DMSO at –80°C. Do not store it at 4°C for more than 1 day.

4. Reconstituted fusion liposomes can be prepared using isolated fusion proteins derived from HVJ instead of inactivated whole viral particles *(23)*.

5. HVJ liposomes can be stored for 3 weeks at 4°C and for more than 3 months with 10% DMSO at the final concentration in a freezer (below –20°C).
6. Gene transfer efficiency of HVJ liposomes is greatly affected by the fusion activity of the HVJ envelope. The hemagglutinating ability should be frequently checked by hemagglutination of chick red blood cells *(22)*.

References

1. Mulligan, R. C. (1993) The basic science of gene therapy. *Science* **260,** 926–932.
2. Ledley, F. D. (1995) Non-viral gene therapy: the promise of genes as pharmaceutical products. *Hum. Gene Ther.* **6,** 1129–1144.
3. Kaneda, Y. (1998) Fusigenic Sendai-virus liposomes: a novel hybrid type liposome for gene therapy. *Biogenic Amines* **14,** 553–572.
4. Kaneda, Y., Saeki, Y., Morishita, R. (1999) Gene therapy using HVJ-liposomes; the best of both worlds. *Mol. Med. Today* **5,** 298–303.
5. Dzau, V. J., Mann, M., Morishita, R., and Kaneda, Y. (1996) Fusigenic viral liposome for gene therapy in cardiovascular diseases. *Proc. Natl. Acad. Sci. USA* **93,** 11421–11425.
6. Hirano, T., Fujimoto, J., Ueki, T., et al. (1998) Persistent gene expression in rat liver *in vivo* by repetitive transfections using HVJ-liposome. *Gene Ther.* **5,** 459–464.
7. Saeki, Y., Matsumoto, N., Nakano, Y., et al. (1997) Development and characterization of cationic liposomes conjugated with HVJ (Sendai virus): reciprocal effect of cationic lipid for *in vitro* and *in vivo* gene transfer. *Hum. Gene Ther.* **8,** 1965–1972.
8. Chander, R. and Schreier, H. (1992) Artificial viral envelopes containing recombinant human immunodeficiency virus (HIV) gp160. *Life Sci.* **50,** 481–489.
9. Sawa, Y., Kaneda, Y., Bai, H.-Z., et al. (1998) Efficient transfer of oligonucleotides and plasmid DNA into the whole heart through the coronary artery. *Gene Ther.* **5,** 1472–1480.
10. Goyal, K. and Huang, L. (1995) Gene therapy using DC-Chol liposomes. *J. Liposome Res.* **5,** 49–60.
11. Yamada, G., Nakamura, S., Haraguchi, R., et al. (1997) An efficient liposome-mediated gene transfer into the branchial arch, neural tube and the heart of chick embryos: a strategy to elucidate organogenesis. *Cell. Mol. Biol.* **43,** 1165–1169.
12. Yonemitsu, Y., Kaneda, Y., Muraishi, A., et al. (1997) HVJ (Sendai virus)-cationic liposomes: a novel and potentially effective liposome-mediated gene transfer technique to the delivery to the airway epithelium. *Gene Ther.* **4,** 631–638.
13. Mabuchi, E., Shimizu, K., Miyao, Y., et al. (1997) Gene delivery by HVJ-liposome in the experimental gene therapy of murine glioma. *Gene Ther.* **4,** 768–772.
14. Nishikawa, T., Edelstein, D., Du, X.L., et al. (2000) Normalization of mitochondrial superoxide production blocks three pathways of hyperglycaemic damage. *Nature* **404,** 787–790.

15. Kaneda, Y., Iwai, K., and Uchida,T. (1989) Increased expression of DNA cointroduced with nuclear protein in adult rat liver. *Science* **243,** 375–379.
16. Kaneda, Y., Iwai, K., and Uchida, T. (1989) Introduction and expression of the human insulin gene in adult rat liver. *J. Biol. Chem.* **264,** 12126–12129.
17. Kato, K., Nakanishi, M., Kaneda, Y., Uchida, T., and Okada, Y. (1991) Expression of hepatitis B virus surface antigen in adult rat liver. *J. Biol. Chem.* **266,** 3361–3364.
18. Tomita, N., Morishita, R., Higaki, J., et al. (1996) *In vivo* gene transfer of insulin gene into neonatal rats by HVJ-liposome method resulted in sustained transgene expression. *Gene Ther.* **3,** 477–482.
19. Tomita, N., Higaki, J. , Morishita, R., et al. (1992) Direct in vivo gene introduction into rat kidney. *Biochem. Biophys. Res. Commun.* **186,** 129–134.
20. Isaka, Y., Fujiwara, Y., Ueda, N., et al. (1993) Glomerulosclerosis induced by *in vivo* transfection with TGF-β or PDGF gene into rat kidney. *J. Clin. Invest.* **92,** 2597–2601.
21. Morishita, R., Gibbons, G., Kaneda, Y., Ogihara, T., and Dzau, V. (1993) Novel and effective gene transfer technique for study of vascular renin angiotensin system. *J. Clin. Invest.* **91,** 2580–2585.
22. Okada, Y. and Tadokoro, J. (1962) Analysis of giant polynuclear cell formation caused by HVJ virus from Ehrlich's ascites tumor cells. *Exp. Cell Res.* **26,** 98–128.
23. Suzuki K, Nakashima H, Sawa Y, et al. (2000) Reconstituted fusion liposomes for gene transfer *in vitro* and *in vivo*. *Gene Ther. Reg.* **1,** 65–77.

6

LPD Nanoparticles—Novel Nonviral Vector for Efficient Gene Delivery

Yadi Tan, Mark Whitmore, Song Li, Peter Frederik, and Leaf Huang

1. Introduction

1.1. Composition

Liposome-Polycation-DNA (LPD) nanoparticles is a novel nonviral vector developed in our laboratory for efficient systemic gene delivery. Currently there are two LPD formulations in use, differing in cationic liposome composition. One is composed of cationic lipid dioleoyl-trimethylammonium propane (DOTAP) and cholesterol at molar ratio of 1:1. The other is composed of cationic lipid 3ß(N-(N', N'-dimethylaminoethane) carbamoyl) cholesterol (DC-Chol) and neutral lipid dioleoylphosphatidylethanolamine (DOPE) at a molar ratio of 6:4. The polycation component is added for the condensation of plasmid DNA. Polylysine first was used (1), and later it was changed to protamine sulfate for improved activity (2). The optimal composition for DOTAP/cholesterol LPD is 1200 nmol DOTAP/1200 nmol cholesterol/60 µg protamine sulfate/100 µg plasmid DNA, which has a charge ratio of 4:1 (+:–) between DOTAP and DNA and a 1/1 charge ratio between protamine and DNA. The optimal composition for DC-Chol/DOPE LPD is 60 nmol total lipids (36 nmol DC-Chol and 24 nmol DOPE)/80 µg protamine/100 µg plasmid DNA.

1.2. Preparation, Physical Property, and Storage

The LPD nanoparticles are prepared by adding an equal volume of DNA solution to an aqueous mixture of cationic liposome and protamine sulfate in a dropwise manner. The resulting tertiary complex contains a highly condensed

From: *Methods in Molecular Medicine, Vol. 69, Gene Therapy Protocols, 2nd Ed.*
Edited by: J. R. Morgan © Humana Press Inc., Totowa, NJ

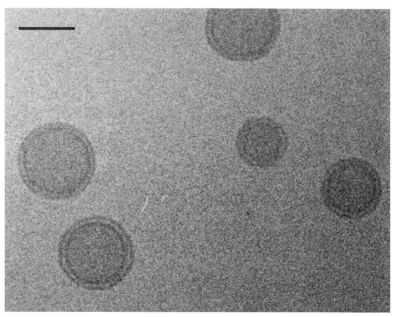

Fig. 1. Cryoelectron micrograph of LPD. Scale bar = 100 nm.

DNA core surrounded by lipid bilayers (**Fig. 1**). LPD nanoparticles range from 30–200 nm in diameter, with an average of about 100 nm. It can be stored as solution at 4°C for at least 4 weeks without losing activity. Alternatively, it can be stored for more than a year at room temperature as a lyophilized powder and is fully active after reconstitution *(3)*.

1.3. Biologic Activity

When DOTAP/cholesterol LPD containing a marker gene is injected from the tail vein of a mouse, transgene expression is found within 2 h in all major organs except the brain. The highest transgene activity is found in the lung (**Fig. 2**), predominantly in the pulmonary endothelial cells *(4)*. Transgene expression is also found in metastatic tumor cells in the lung *(5)*. The transgene expression peaks at about 8–24 h after injection and gradually declines to lower levels within 48–72 h.

When a tumor suppresser gene (Rb) was delivered with LPD nanoparticles, it was observed that metastatic tumor cells in the lung underwent spontaneous apoptosis. This led to a significant reduction in the percentage of mice with lung metastasis after LPD-Rb gene but not LPD-control gene treatment *(5)*.

DC-Chol/DOPE LPD has been optimized for brain gene transfer by intracranial injection. A high level of transgene expression in the brain was detected for more than 10 months in rodents and at least at 1 month in primates. Two

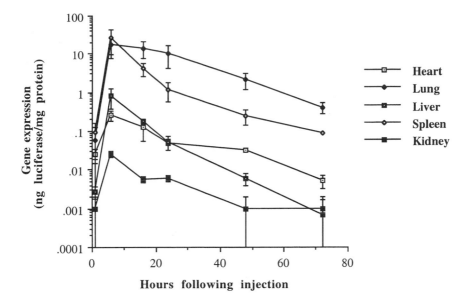

Fig. 2. Time-course of transgene expression following intravenous injection of LPD.

children with Canavan's disease were given intraventricular administration of DC-Chol/DOPE LPD carrying the ASPA gene, and both showed clinical improvements *(6)*.

1.4. Toxicity

Intracranial injection of LPD has been shown to be safe in animals and in human patients without any noticeable side effects *(6)*. However, systemic administration of LPD or liposome-plasmid DNA complex can trigger a proinflammatory cytokine response, which is dose dependent *(7)*. Although the induction of cytokines (tumor necrosis factor-α [TNF-α], interleukin-12 [IL-12], IL-6, interferon-γ [IFN-γ], etc.) may be beneficial for the treatment of tumors *(8,9)*, these cytokines are toxic to the injected animals at high doses. Furthermore, these cytokines inhibit transgene expression *(7,10)*. Recent studies have revealed that the induction of proinflammatory cytokine is largely due to the unmethylated CpG dinucleotides that are present in the bacterial plasmid DNA *(7)*. Ways to overcome this problem include modification of plasmid DNA to reduce the number of CpG motifs *(11)*, use of a polymerase chain reaction (PCR) fragment instead of the intact plasmid *(12)*, and the use of a general immunosuppressant such as dexamethasone *(10)*.

2. Materials

2.1. Preparation of LPD Complex

1. 50-mL polypropylene conical tube.
2. 1.5-mL Eppendorf tube.
3. Dextrose (Sigma).
4. Sterile water (autoclaved and 0.2-μm-pore sterile-filtered)
5. Cationic liposomes, prepared as described in the protocol (*see* **Subheading 3.2**).
   a. 10 mg/mL (14.3 m*M*) DOTAP/Chol liposome
   b. 2 m*M* DC-Chol/DOPE liposome (*see* **Note 2**).
6. 1 mg/mL plasmid DNA in sterile water (*see* **Note 3**).
7. 10 mg/mL protamine sulfate USP (Elkins-Sinn, Cherry Hill, NJ).
8. 5X Dextrose (26.0% [w/v]), 100 mL. Dissolve 13 g dextrose in approximately 40 mL sterile water, and then bring to 50 mL with sterile water. Sterile filter the solution through a 0.2 μm membrane and store at room temperature for up to 1 year.

2.2. Liposome Preparation by Thin-Film Hydration followed by Extrusion

1. N_2 gas tank.
2. Vacuum dessicator.
3. Bath sonicator.
4. 30.0-mL Corex glass centrifuge tube (or any clean glass tube).
5. Sterile water.
6. 25.0 mg/mL DOTAP stock solution in chloroform (Avanti Polar Lipids, Alabaster, AL).
7. 20.0 mg/mL cholesterol stock solution in chloroform.
8. 2.0 mg/mL DC-Chol stock solution (3.72 m*M* DC-Chol) in chloroform.
9. 20.0 mg/mL DOPE stock solution in chloroform (Avanti Polar Lipids, Alabaster, AL).
10. LiposoFast™ extruder, with 1.0-mL syringes (Avestin, Ottawa ON, Canada).
11. 1.0, 0.4, and 0.1 μm Nuclepore® polycarbonate membrane filters (Corning®, available from VWR).
12. Cholesterol (20 mg/mL in chloroform), 5 mL Weigh 100 mg cholesterol into a glass test tube previously rinsed 3× with chloroform and dried. Dissolve in approximately 4 mL chloroform. Bring to 5 mL with chloroform. Store at –20°C in capped and sealed glass container.

3. Methods

3.1. Preparation of LPD Complex

1. Volumes and quantities are for 100 μg of plasmid DNA in a final volume of 600 μL (DOTAP/Chol liposome) or 400 μL (DC-Chol/DOPE liposomes). For larger doses, volumes and quantities should be proportionately scaled up (*see* **Fig. 3** for illustration of the steps).

Protamine Sulfate   Cationic liposomes

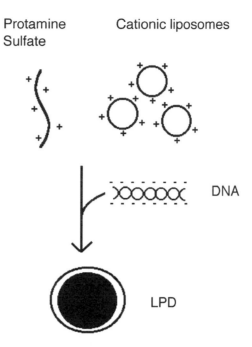

DNA

LPD

Fig. 3. Preparation of liposome-polycation-DNA (LPD).

2. a. For DOTAP/Chol liposomes prepare liposome/polycation (LP) solution in a 50-mL conical tube by adding 60 μL 5X dextrose, 148.2 μL sterile water, 86 μL DOTAP liposomes, and 6 μL protamine sulfate. Mix by vortexing at medium speed for 10 s.

   b. For DC-Chol/DOPE liposomes, prepare LP solution in a 50-mL conical tube by adding 40 μL 5X dextrose, 122 μL sterile water, 30 μL DC-Chol/DOPE liposomes, and 8 μL protamine sulfate. Mix by vortexing at medium speed for 10 s.

3. a. Make up a DNA solution (D solution for DC-Chol/DOPE liposomes) in an eppendorf tube by adding 60 μL 5X dextrose, 140 μL sterile water, and 100 μL plasmid DNA. Gently tap tube to mix. *Do not vortex.*

   b. Make up a DNA solution (D solution for DOTAP-chol liposomes) in an eppendorf tube by adding 40 μL 5X dextrose, 60 μL sterile water, and 100 μL plasmid DNA. Gently tap tube to mix. *Do not vortex.*

4. While gently swirling LP solution, slowly add D solution dropwise. It takes about 5–10 s to add 300–400 μL of D solution. A tip with a wide opening to transfer the D solution is recommended, such as the 1000-μL pipet tip or plastic transfer pipet (*see* **Note 4**).

5. Incubate the complex at room temperature for 10–15 min prior to injection to allow for complex maturation.

3.2. Preparation of Liposomes by Thin-Film Hydration followed by Extrusion

Thin-film hydration followed by extrusion is one of many methods that can be used to prepare liposomes. This method was selected because it provides a quick and easy protocol for the preparation of a concentrated and homogeneous suspension of small unilammellar liposomes suitable for laboratory-scale animal experiments. The method involves forming a thin film of dried lipids on the bottom of a glass tube. The lipids are then hydrated in a desired aqueous solution (e.g., water or 5.2% dextrose) to form a suspension of mixed, multilammelar liposomes. Extrusion is a method of liposome preparation whereby lipid suspensions are pushed through polycarbonate filters of desired pore size. The shear force produced by passage of the lipids through the pores results in the formation of liposomes with unilamallar lipid bilayers. The size of the resultant liposomes can be dictated, in part, by choosing the appropriate pore size of the filters used during extrusion. It should be noted that for large-scale preparation of liposomes, microfluidization is the method of choice *(13)*.

The preparation of DC-Chol/DOPE liposome (2 mM DC-Chol) DOTAP/ Chol liposome (14.3 mM DOTAP = 10 mg/mL DOTAP) stock solutions is described below. A cationic liposome preparation with a 3:2 molar ratio of DC-Chol/DOPE represents the optimal formulation to be used in LPD for intracranial injection. The optimal liposome preparation to be used in LPD for systemic injection is composed of a 1:1 molar ratio of DOTAP/cholesterol. Both cationic liposome stock solutions are stable for several months when stored at 4°C.

3.2.1. Preparation of Thin Film

1. Rinse a 30-mL Corex tube 3× with chloroform. There is no need to dry the residual chloroform since it will be evaporated in **step 3**.
2. Mix desired amount of lipids in the glass tube.
    a. For 2 mL of 10 mg/mL DOTAP/cholesterol liposomes (1:1 molar ratio), mix 0.8 mL DOTAP (mol wt 698.6) stock solution and 0.55 mL cholesterol (MW386.7) stock solution.
    b. For 2 mL of 2 μmol/mL DC-chol/DOPE liposomes (3:2 molar ratio), add 1.07 mL DC-Chol stock solution and 0.1 mL DOPE stock solution and mix.
3. In a chemical hood, evaporate chloroform to form a thin lipid film on the glass by blowing N$_2$ gas down the side of the tube while rotating the tube. This process should be done in a chemical hood. The chloroform fume is hazardous. To speed up the process, the tubes can be dipped in a warm water bath.
4. Dry film to completion by vacuum desiccation for 2–3 h (*see* **Note 5**). To prevent loss of lipid film under vacuum, cover tube with aluminum foil and poke small holes in the foil with a needle.

3.2.2. Hydration of the Thin Film

1. Add 2 mL of sterile water to film.
2. Suspend lipids in solution by vortexing for 15 s on maximum speed several times. Lipids may remain on the side of the tube, appearing as a solid, white precipitate. Several quick (10–15 s) bursts in a bath sonicator will help to suspend the lipids in solution.
3. Incubate suspension either at room temperature for 2–3 h or at 4°C overnight to allow for complete hydration of lipids. It is our experience that the longer the hydration, the easier the extrusion. Overnight hydration is recommended for common practice, and 48 h of hydration is also acceptable.

3.2.3. Extrusion

1. Vortex the lipid suspension and heat it for 5–10 min in a 65°C water bath. Should you observe any lipid aggregates, sonicate the suspension in a bath sonicator until all aggregates disappear, and then return to water bath.
2. Place two 1.0-µm polycarbonate filters in the extruder.
3. Heat the extruder to 65°C for 5 min. Heating the lipid suspension and the extruder to 65°C maintains the lipids in a liquid state. This allows for easier extrusion and improved lipid mixing.
4. Extrude the lipid dispersion by passing the suspension through the extruder 5×. Return the extruder to the water bath. Following extrusion, the lipid suspension should be transformed from an opaque, cloudy solution to a transluscent suspension.
5. Repeat **steps 2–4** using 0.4- and 0.1-µm polycarbonate filters sequentially to obtain small unilamellar liposomes with a mean diameter of 100–200 nm.
6. Store liposomes at 4°C.

4. Notes

1. A variety of cationic liposome formulations are commercially available for lipofection of cells in tissue culture. Optimal formulations are highly cell line dependent and typically do not correlate with optimal formulations used for gene transfer of corresponding tissues in vivo. Thus, for in vitro transfection, the reader is advised to test a variety of the commercially available liposomes following the manufacture's instructions.
2. The DOTAP/Chol liposome is made of a 1:1 molar ratio of DOTAP and cholesterol. Each DOTAP molecule carries a positive charge, and cholesterol is neutral. In gene transfer studies, one often needs to calculate the amount of DOTAP according to different charge ratio requirement. Thus, for convenience, the DOTAP/Chol liposome solution is labeled as 10 mg DOTAP/mL; it also contains 5.5 mg Chol/mL. For DC-Chol/DOPE liposomes, the stock solution is 2 mM of total lipids, i.e, 1.2 mM of DC-Chol and 0.8 mM of DOPE.
3. Plasmid DNA shall be highly purified and endotoxin free (e.g., using Double-CsCl gradient purification or Endofree™ plasmid/cosmid purification kits avail-

able from QIAGEN®). In addition, it is *required* that plasmid DNA (or other forms of DNA) be dissolved in water or 5.2% dextrose solution. This is because salt interferes with the charge-charge interaction between DNA and cationic lipid or polymer and causes the formation of aggregates.

4. White, string-like precipitates may form during mixing. This is most commonly caused by salt in the solutions (TE, NaCl, etc.) The precipitates appear to be toxic to mice following iv injection. Proper contour of the mixing vessel (i.e., the 50-mL conical tube), swirling of the LP solution, and dropwise addition of the D solution help to prevent the formation of the precipitate. If the precipitate forms, it is *not* advised to proceed with injection. Instead, try preparing the complex again. With practice, precipitate formation can routinely be avoided. Again, never vortex the complex since it contains plasmid DNA.

5. For the preparation of liposomes by the thin-film hydration followed by extrusion method, one should allow 3–4 h for the formation of the dry film, 2–3 h or overnight for hydration of the lipids, and 1–2 h for extrusion. For preparation of LPD, one should allow 1/2–1 h.

References

1. Gao, X. and Huang, L. (1996) Potentiation of cationic liposome mediated gene delivery by polycations. *Biochemistry* **35,** 1027–1036.
2. Sorgi, F. L., Bhattacharya, S., and Huang, L. (1997) Protamine sulfate enhances lipid mediated gene transfer. *Gene Ther.* **4,** 961–968.
3. Li, B., Li, S., Tan, Y., et al. (2000) Lyophilization of cationic lipid-protomine-DNA (LPD) complexes. *J. Pharm. Sci.* **89,** 355–364.
4. Li, S. and Huang, L. (1997) In vivo gene transfer via intravenous administration of cationic lipid/protamine/DNA (LPD) complexes. *Gene Ther.* **4,** 891–900.
5. Nikitin, A. Y., Juarez, M. I., Li, S., Huang, L., and Lee, W. H. (1999) RB-mediated suppression of spontaneous multiple endocrine neoplasia and lung metastases in Rb+/- mice. *Proc. Natl. Acad. Sci USA* **96,** 3916–3921.
6. Leone, P., Janson, C.G., Bilianuk, L., et al. (2000) Aspartoacylase gene transfer to the mammalian central nervous system with therapeutic implications for Canavan disease. *Ann. Neurol.* **48,** 27–38.
7. Li, S., Wu, S.P., Whitmore, M., et al. (1999) Effect of immune response on gene transfer to the lung via systematic administration of cationic lipidic vectors. *Am. J. Physiol. Lung Cell. Mol. Physiol.* **276,** L796–L804.
8. Hofland, H. and Huang, L. (1995) Inhibition of human ovarian carcinoma cell proliferation by liposome-plasmid DNA complex. *Biochem. Biophys. Res. Commun.* **207,** 492–507.
9. Whitmore, M., Li, S., and Huang, L. (1999) LPD lipopolyplex initiates a potent cytokine response and inhibits pulmonary tumor growth. *Gene Ther.* **6,** 1867–1875.
10. Tan, Y., Li, S., Pitt, B., and Huang, L. (1999) The inhibitory role of CpG immunostimulatory motifs in cationic lipid vector-mediated transgene expression in vivo. *Hum. Gene Ther.* **10,** 2153–2161.

11. Yew, N. S., Zhao, H., Wu, I. H., et al. (2000) Reduced inflammatory response to plasmid DNA vectors by elimination and inhibition of immunostimulatory CpG motifs. *Mol. Ther.* **1,** 255–262.

12. Hofman, C. R., Dileo, J. P., Li, Z., Li, S., and Huang, L. Efficient *in vivo* gene transfer by PCR amplified fragment with reduced inflammatory activity. *Gene Ther.*, **8**, 71–74.

13. Sorgi, F. L. and Huang, L. (1996) Large scale production of DC-Chol liposomes by microfluidization. *Int. J. Pharmaceut.* **144,** 131–139.

7

Solvoplex Synthetic Vector for Intrapulmonary Gene Delivery

Preparation and Use

Klaus Schughart and Ulla B. Rasmussen

1. Introduction

For in vivo gene transfer, synthetic (nonviral) vectors are thought to have several advantages compared with viral vectors: they can accommodate large-size DNA molecules (therapeutic genes including their endogenous regulatory regions), be used under reduced confinement conditions, be administered repeatedly, and be modified with appropriate ligands that allow specific cell targeting. Until now, many different types of organic compounds have been tested as synthetic gene delivery vectors in animal models by various routes of administration (*reviewed in* **refs.** *1–13*).

Intrapulmonary gene delivery represents an important route of administration for treatment of monogenic or multifactorial diseases like cystic fibrosis and asthma. Furthermore, gene delivery via the intrapulmonary route would be an attractive noninvasive delivery method for systemic expression of therapeutic proteins by targeting alveolar cells, which then express and secrete the gene products into the bloodstream. Different synthetic vectors have been studied for their potential use as gene transfer vectors to achieve expression after intrapulmonary delivery; some of them have been evaluated in clinical studies (*reviewed in* **refs.** *14–17*).

Recently, we described a new type of synthetic vector, named solvoplex, which is based on organic solvents and which can greatly enhance gene delivery to the lung epithelium after intrapulmonary delivery *(18)*. Solvoplexes rep-

From: *Methods in Molecular Medicine, Vol. 69, Gene Therapy Protocols, 2nd Ed.*
Edited by: J. R. Morgan © Humana Press Inc., Totowa, NJ

RLU/mg

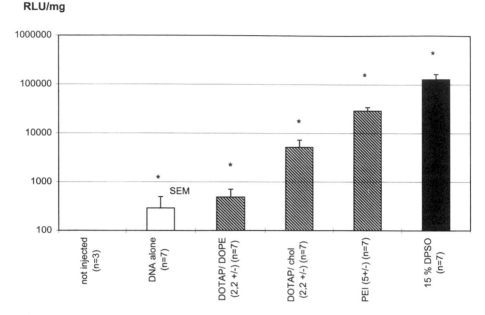

Fig. 1. DPSO solvoplexes are more efficient than DNA alone, DOTAP lipoplexes, or PEI polyplexes. Female B6SJLF1 mice were injected IT with solvoplexes, DNA alone, lipoplexes, or polyplexes (25 μg/50 μL of plasmid pTG11033 in each case; *for details, see* **ref. 18**), and luciferase expression in lung homogenates was determined 1 day after injection. As control, naive mice (not injected) were analyzed. (+/−) refers to the N/P molar ratio of lipid or polycation to DNA in lipo- and polyplexes *(21)*. Mean values were calculated and statistical analyses were performed using the Mann-Whitney rank sum test. *, $p < 0.04$ in a pairwise comparison of the lipoplexes and polyplexes to DPSO solvoplexes. *n*, number of mice analyzed in each group. (From **ref. 18** with permission.)

resent highly efficient gene transfer vectors to target the lung airway epithelium in mice and rats after intrapulmonary delivery, they are easy to prepare and stable for weeks, and they can be readministered. Here we describe the preparation and use of solvoplex vectors for preclinical studies in rodents.

1.1. Characteristics of Solvoplex Vectors

Solvoplex vectors represent easy to prepare mixtures of plasmid DNA and organic solvents. Among several polar aprotic solvents tested for efficacy of intrapulmonary gene delivery after intratracheal (IT) injection in mice, di-*n*-propylsulfoxide (DPSO), dimethylsulfoxide (DMSO), tetramethylurea (TMU), and butylmethylsulfoxide (BMSO) solvoplexes yielded the highest luciferase reporter gene expression in the lung. The best results were obtained with

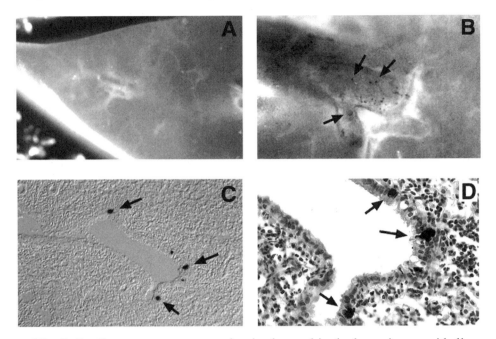

Fig. 2. LacZ reporter gene expression is observed in the lung airway epithelium after IT injection of solvoplexes. Lungs were prepared 1 day after IT injection of female B6SJLF1 mice, stained with X-Gal, and cut into large pieces; then the cut open surface was photographcd undcr a dissection microscope. **(A)** Noninjected control mice. **(B)** Mice injected with 15% DPSO solvoplexes (50 µg/50 µL of pTG11034 plasmid DNA). Lungs from 15% DPSO solvoplex-injected mice were embedded in paraffin, sectioned, mounted, and viewed under the light microscope using **(C)** Nomarski or **(D)** brightfield illumination after H&E staining. Arrows indicate examples of LacZ-positive (blue-stained) cells. (From **ref.** *18* with permission.)

solvoplexes containing 15% DPSO *(18)*. At this concentration, up to 400-fold increased expression levels were obtained when compared with DNA alone ("naked DNA") (**Fig. 1**). In addition, diethylsulfoxide (DESO), methylethylsulfoxide (MESO), propylenglycol (PPG), tetramethylensulfoxide (TEMSO), and sulfolane resulted in lower levels of gene expression but still higher than with DNA alone. Compared with dioleoyl-trimethylammonium propane/dioleoylphosphatidylethanolamine (DOTAP/DOPE), DOTAP/cholesterol lipoplexes, or polyethyleneimine (PEI) polyplexes, increases of 260-, 25-, or 4-fold, respectively, were observed with DPSO solvoplexes (**Fig. 1**). Thus, solvoplexes represent more efficiently performing vectors for intrapulmonary gene delivery than many of the currently used synthetic vectors.

Also in rats, IT delivery of 15% DPSO solvoplexes resulted in gene expres-

RLU/mg

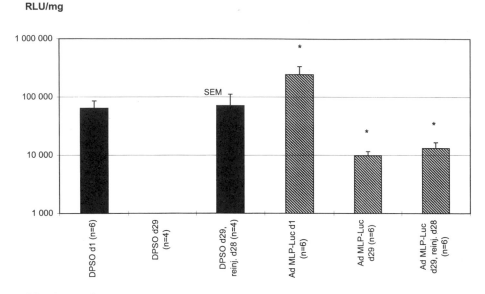

Fig. 3. Readministration of solvoplexes results in gene expression levels similar to single administration. Female B6SJLF1 mice were injected IT with solvoplexes (50 µg/25 µL of plasmid pTG11033) or with 1 ¥ 10E^8 PFU AdTG8509 adenovirus (Ad) and analyzed for luciferase expression 1 or 29 days later. Two groups were reinjected with solvoplexes or with adenovirus at day 28, and 1 day later, luciferase expression in lung homogenates was determined. Mean values were calculated, and statistical analyses were performed using the Mann-Whitney rank sum test. *, $p < 0.01$ in a pairwise comparison of expression levels of adenovirus-injected mice analyZed after 29 days and reinjected mice, with mice analyzed at day 1. *n*, number of mice analyzed in each group. SEM: standard error of mean. (From ref. *18* with permission.)

sion in the lung *(18)*. An important parameter for the clinical and preclinical use of synthetic vectors is the stability of complexes upon storage. When DPSO solvoplexes were kept at –20°C or at room temperature, no degradation of DNA was observed within a period of 15 days, and no loss in transfection efficiency of the DNA was noted *(18)*. Furthermore, reporter gene expression in the lung could be observed after delivery of solvoplexes as a spray using a hand-held, high-pressure microsprayer device (IA-1C Intrapulmonary Aerosolizer; PennCentury) *(18)*.

1.2. Gene Expression in the Lung Airway Epithelium

Individual cells in the lung expressing a foreign gene can be localized after solvoplex delivery of LacZ or GFP reporter genes. For example, in mice injected with 15% DPSO solvoplexes, LacZ-positive cells were detected in the epithelium of the proximal regions of the lung airways (**Fig. 2**).

1.3. Readministration of Solvoplex Vectors

Gene expression after solvoplex delivery was highest on day 1 after injection and declined rapidly until day 7, when expression levels were only slightly above background. Thus, when prolonged gene expression is needed, readministration will be required. We observed that, in contrast to first-generation adenovirus vectors, after a second administration of 15% DPSO solvoplexes, expression levels could be obtained that were equivalent to the levels seen after the first injection (**Fig. 3**).

1.4. Toxicity of Solvoplexes

Administration of solvoplexes resulted in pathologic changes in the lung at 1 day after injection that were not significantly different from the changes observed after delivery of DNA alone. Seven days after injection, no or only very weak residual inflammation was seen for solvoplexes and DNA alone. Changes in relative body weight or serum transaminase levels were not significantly different between injected and noninjected mice. It should also be noted that the amount of DPSO applied in our protocol is 5000-fold below the median lethal dose (LD_{50}) determined for mice after intraperitoneal injection *(19)*.

1.5. Mechanism of Transfection

The mechanism of transfection of airway epithelial cells by solvoplexes is currently unknown. It is conceivable that organic solvents change the composition of the cell membrane by altering its fluidity or the properties and distribution of receptors and that such changes enhance DNA uptake. Also, it is possible that the basolateral cell surfaces of epithelial cells in the lung become exposed in the presence of solvent because the solvent weakens cell-cell interactions, thus allowing DNA uptake also via the basolateral membrane. In addition to effects at the cellular level, the conformation of DNA may be altered in the presence of solvents in such a way that its uptake is facilitated.

1.6. Conclusion and Perspectives

Our results demonstrate the use of solvoplexes as an attractive alternative and complementary gene delivery vehicle to the currently available gene transfer vectors, especially for preclinical studies. They result in high levels of gene expression in the lung airway epithelium, and they are easy to prepare, stable over time, and of moderate toxicity; they can also be readministered. However, before solvoplexes could be used in the clinic, several issues, especially those concerning potential local pathogenic effects and improved delivery methods, must be addressed in more detail.

2. Material

2.1. Solutions

1. Phosphate buffered saline (PBS): 2 mM KH$_2$PO$_4$, 8 mM Na$_2$HPO$_4$, 150 mM NaCl.
2. X-Gal staining solution: 5 mM K$_3$Fe(CN)$_6$, 5 mM K$_4$Fe(CN)$_6$, 2 mM MgCl$_2$, 1 mg/mL X-Gal, 0.02% NP-40, 0.01% Na-deoxycholate in PBS.
3. Anesthetic mixture for mice: 5.2 mL 0.15 M NaCl, 1 mL Imalgene 1000 (ketamine; Merial, Lyon, France), 0.5 mL 2% Rompun (xylazine; Bayer, Leverkusen, Germany).
4. Anesthetic mixture for rats: 5 mL Imalgene 1000 (ketamine), 1.5 mL 0.5% Vetranquil (acepromazine; Sanofi-Santé-Nutrition Animale, Libourne, France).

2.2. Solvents for Solvoplexes

1. TMU was obtained from Sigma (L'Isle d'Abeau Chesnes, France; cat. no. T3875).
2. BMSO was synthesized by oxidation of methylbutylsulfide (Lancaster, Strasbourg, France; cat. no. 12039) with NaIO$_4$ (Sigma; cat. no. S1147).
3. DMSO was purchased as tissue culture grade from Sigma (cat. no. D2650).
4. DPSO could not be obtained commercially in a pure enough form and was thus synthesized by oxidation of di-n-propylsulfide with NaIO$_4$ (*see* **Subheading 3.2.**).

2.3. Expression Plasmids

The expression plasmid pTG11033 contained the luciferase gene under control of the IE1-CMV promoter, intron 1 of the HMGCoAR gene, and the SV40 polyA signal. Plasmid pTG11034 contained the bacterial LacZ gene with a nuclear localization signal under control of the IE1-CMV promoter, the rabbit ß$_1$-globin ivs2 intron, and SV40 polyA. Plasmid pEGFP-C1 (Clontech, Palo Alto, CA) contained the GFP gene under control of the IE1-CMV promoter.

2.4. Mice and Rats

Female C57BL/6 and B6SJLF1 mice (8–10 weeks old) and female Sprague-Dawley (OFA-SD) rats (approx 100 g) were obtained from authorized suppliers (Iffa Credo, l'Arbresle, France; Charles River, St. Germain sur l'Arbresle France; Janvier, Le Genest, France) as special pathogen-free (SPF) grade animals (*see* **Note 1**). They were housed under SPF conditions and according to the French regulations for animal experimentation. After arrival, animals were allowed at least 1 week to recover from transport and adapt to the environment before the beginning of the experiments.

3. Methods

3.1. Preparation of DNA

Plasmid DNA is prepared by the alkaline lysis method *(20)* from cultures grown in Erlenmeyer flasks or fermenters. Plasmid DNA is then precipitated by isopropanol, and the pellet is washed with 70% ethanol. Crude plasmid preparations are then purified by two subsequent CsCl density gradient centrifugations *(20)*.

1. Harvest the plasmid band from the second CsCl density gradient (about 8 mL of solution).
2. Extract ethidiumbromide 4–5× with 5 mL butanol (saturated with water/CsCl).
3. Dialyze against TE.
4. Digest RNA by adding, for 8 mL of plasmid solution, 10 µL of an RNAase solution (10 mg/mL in water).
5. Incubate for 30 min at 37°C and add 80 µL 10% sodium dodecyl sulfate (SDS) and 10 µL of proteinase K (10 mg/mL in water).
6. Incubate at 37°C for 2 h and stop reaction by incubation at 65°C for 5 min.
7. Extract plasmid solution twice with dichlormethan.
8. Precipate DNA by adding 1/10 vol of 3 M Na-acetate, pH 5.0, and 2.5 vol of 99.9% ethanol; wash pellet 3X with 70% ethanol.
9. Dissolve DNA in TE to a final concentration of 1 mg/mL or less.
10. Extract endotoxins by adding 1/9 vol of 3 M Na-acetate, pH 5.0 and then 1/9 vol of 10% TX114 (Sigma; cat. no. T-7003) in 0.3 M Na-acetate, pH 5.0, vortex for 1 min, incubate on ice for 10 min, vortex for 3 s, incubate at 42°C for 5 min, centrifuge for 10 min at 37°C at 2200g, and transfer aqueous phase to new tube; repeat twice.
11. Precipitate DNA by adding 2.5 vol ethanol, incubate for 2 h at –20°C, wash the pellet 3× with 70% ethanol,and dry the pellet for 5 min in a lyophylizator.
12. Dissolve DNA in TE.
13. Determine DNA concentration and purity by measuring OD260 and OD280 *(20)*.
14. Determine endotoxin concentration by using the LAL Coatest Endotoxin kit (Biogenic, Maurin, France). For use in animals, endotoxin concentrations should be <5 entropy units (eu)/mL.
15. Store DNA at –20°C until use.

3.2. Synthesis of DPSO

1. Cool 25 g NaIO$_4$ (117 mmol) in 245 mL water to 0°C.
2. Keep at 0°C and slowly add while stirring 13.2 g (15.85 mL) di-*n*-propylsulfide (111 mmol; Aldrich, l'Isle d'Abeau Chesnes, France; cat. no. P5,428-0).
3. Follow the progress of the reaction by thin-layer chromatography (R_f = 0.41; silica plates [Aldrich; cat. no. Z12,269-6]: layer 250 µm, particle size 5–17 µm, pore size 60 A; solvent: 90/10 CH$_2$Cl$_2$/CH$_3$OH; detection I$_2$, KMnO$_4$).
4. Filter the crude product on Celite (Sigma; cat. no. C8656) and wash Celite with

300 mL water, 200 mL CH_2Cl_2 (Merck, Fontenay-sous-Bois, France; cat. no. 1.06050.1000).

5. Extract eluate twice with water/CH_2Cl_2.
6. Wash aqueous phase with CH_2Cl_2 and combine organic phases.
7. Wash organic phase with H_2O saturated with $Na_2S_2O_3$ (Prolabo, Fontenay-sous-Bois, France; cat. no. 28130.292) and rinse with CH_2Cl_2.
8. Evaporate the solvent by drying over anhydrous Na_2SO_4.
9. Purify DPSO by silica gel chromatography (Kieselgel 60; Merck; column i.d. 4 cm, height 25 cm; eluent: CH_2Cl_2/CH_3OH 98/2).
10. Distill DPSO by vacuum distillation at 0.2 mm Hg (boiling point: 63°C; yield: 12.3 g, i.e., 91.7 mmol, i.e., 83%).

3.3. Preparation of Solvoplexes

All solutions and the final solvoplex mixture need to be prepared sterile.

1. Calculate the amount of DNA needed (final concentration 2 mg/mL; 25 µL per mouse).
2. Precipitate DNA by adding 0.1 vol of 3 M Na-acetate, pH 5.0 and 4 vol ethanol, and incubate for 2 h at –20°C.
3. Wash the pellet with ethanol/H_2O (80:20 vol), and let dry for 1 h at room temperature.
4. Calculate amount of solvent and buffer (20 mM HEPES, pH 7.5) required for solvoplex (*see* **Note 2**).
5. Dissolve DNA pellet in calculated amount of buffer (20 mM HEPES, pH 7.5).
6. Add calculated amount of solvent.
7. Mix by pipetting.
8. Inject into animals within next 2 h (*see* **Note 3**).

3.4. Intratracheal Injection

3.4.1. Mice

1. Anesthetize mice by intraperitoneal injection of 100 µL of anesthetic mixture (*see* **Subheading 2.1.** and **Note 4**).
2. Fix the mouse by taping the legs on a support (i.e., piece of styrofoam).
3. Disinfect the skin with 70% ethanol.
4. Make an incision in the skin above the trachea.
5. Expose the trachea by dilating the muscles and connective tissue with a fine curved forceps; be careful not to damage the esophagus or nerves and veins surrounding the trachea.
6. Insert a 1 mL 27G 3/8 gage injection needle (Becton Dickinson, LePont de Claix, France) between two neighboring cartilage rings with the tip pointing in the direction of the lung; make sure that the tip of the needle is well inside the trachea.
7. Slowly inject a volume of 25–50 µL into the trachea (*see* **Note 5**).

8. Carefully watch the breathing rhythm of the animal during this period and arrest the injection if the animal seems to be suffocating (*see* **Note 6).**
9. After injection, close the incision by suture, i.e., Ethibond 3/0 (Johnson and Johnson, Brussels, Belgium)

3.4.2. Rats

Treatment is the same as described above for mice, except that they are injected intraperitoneally with 110 μL/100 g of a different anesthetic mixture (*see* **Subheading 2.1.**). We injected volumes up to 150 μL of solvoplex solution into rat tracheas. Again, since solvoplexes are more viscous than purely aqueous solutions, special attention needs to be given to the breathing rhythm of the animal.

3.5. Analysis of Luciferase Expression

1. Sacrifice mice at the desired time point after injection (usually 1–2 days; *see* **Note 7**).
2. Prepare lungs and trachea separately (*see* **Note 8**).
3. Freeze them immediately in liquid nitrogen and store at –80°C until further use.
4. Disrupt the tissues by adding 500 μL (lung) or 200 μL (trachea) luciferase lysis buffer (Promega, Charnonnieres, France) and homogenize for 2× 30 s (Polytron homogenizer, Kinematica, Littau, Switzerland).
5. Freeze/thaw (liquid nitrogen, water bath) the homogenate 3×.
6. Remove cell debris by centrifugation.
7. Determine luciferase activity in a luminometer (Microlumat LB 96P; Berthold, Evry, France) by following the supplier's instructions for the luciferase assays (Promega, Charnonnieres, France).
8. Determine protein content by the BCA protein assay (Pierce, Montlucon, France) from an aliquot of the extract supernatant.
9. Calculate luciferase activity as relative light units (RLU)/min/mg (protein) (*see* **Note 9**)

3.5. Histologic Analyses

3.5.1. LacZ Reporter Gene Expression

1. Remove lungs in PBS.
2. Wash in PBS and fix in 2% formaldehyde (for at about 1 h at room temperature; *see* **Note 10**).
3. Wash lungs in PBS.
4. Incubate in X-Gal staining solution for 2 days at 37°C.
5. Fix tissues for 12–24 h in 5% formaldehyde at 4°C.
   For direct observation:
6. Place lungs in Petri dishes with PBS.
7. Using a dissection microscope, cut lungs open with scalpel to expose airways.
8. Directly observe and photograph lungs under a dissection microscope.
   For tissue sections:

9. Dehydrate organs or tissue pieces and embed in paraffin (Histoplast; Life Science International, Eragny, France).

10. Cut 8-μm sections with a microtome, dewax sections, perform histologic staining if desired, and mount them in Mowiol (Calbiochem/Novabiochem, La Jolla, CA) or Eukitt (Labonord, Villeneuve d'Ascq, France; *see* **Note 11**).

3.5.2. GFP Reporter Gene Expression

1. Deeply anesthetize mice.
2. Perfuse with PBS and then with 4% formaldehyde.
3. Prepare organs and fix them in 4% formaldehyde for 30–60 min at room temperature.
   For direct observation:
4. Place lungs in Petri dishes with PBS.
5. Using a dissection microscope, cut lungs open with scalpel to expose airways.
6. Directly observe and photograph under a fluorescence microscope using the fluorescein isothiocyanate (FITC) filter set(*see* **Note 12**).
   For sections (cryosections):
7. Equilibrate fixed organs or tissue pieces in 30% sucrose solution (4°C, overnight).
8. Freeze in OCT (Tissue Tek, Bayer Diagnostics, Munich, Germany).
9. Cut with cryostat (5-μm sections).
10. Fix sections in methanol (–20°C, 20 min) and mount in Mowiol.
11. Observe and photograph sections under the fluorescence microscope using the FITC filter set.

4. Notes

1. We used mice from 8 to 12 weeks of age, and we did not observe significant age-related differences in expression levels. Most of our data were generated with C57BL/6 and B6SJLF1 mice, and some experiments were performed with CBA, C3H/He, and BALB/c mice. We did not observe significant differences between mouse strains. However, we preferred to use B6SJLF1 hybrid mice because of the hybrid but identical background in all individuals and because F1 hybrids are more robust than inbred mice.

2. Best results were obtained with 15% DPSO and DMSO solvoplexes. DNA remains in solution at higher solvent concentrations, and formulations containing 20% DPSO resulted in higher expression levels in some mice. However, variations between individual mice were much more pronounced. This effect is probably because of the increased viscosity of solvoplex solutions at high DPSO concentrations and the consequent difficulty reproducibly delivering them to the lung.

3. Even though no change in the DNA or its capability to transfect was observed, we usually prepare the solvoplexes just prior to injection.

4. In principle, different anesthetics may be used, but be aware that they may influence the results. We did not perform an evaluation of different anesthetics.

5. Injection of the lowest volumes possible, i.e., 25 μL is recommended.
6. Solvoplexes containing >10% of solvent are more viscous than water and need to be injected very slowly. If you are not very experienced with IT injections, the results may vary considerably between individual animals in the beginning.
7. Maximal expression levels were observed 1 day after injection. The peak of expression may be slightly earlier than 24 h; we did not investigate this in detail.
8. After solvoplex injection, we observed expression in both the lung and trachea, but expression levels were variable for the trachea. In the trachea, similar expression levels were seen for solvoplexes as for DNA alone, whereas in the lung, expression was much higher after solvoplex injection (**Fig. 1**). Thus, if one is interested in lung gene transfer, the lung should always be studied separately from the trachea.
9. RLU can be converted into the corresponding mass (ng) of luciferase protein produced by preparing standard curves with purified commercially available luciferase. In this case, it is necessary to generate a standard curve at the same time as samples are being analyzed. However, be aware that the conversion depends on the purity of the luciferase protein batch used for the standard curve and that this can vary between batches and suppliers. Therefore, even though the conversion of RLU to ng luciferase protein allows a better comparison of expression levels between laboratories, it is not an "absolute" value.
10. For better fixation of the tissues, mice can be deeply anesthesized and perfused with PBS and then 2% formaldehyde; isolated organs can be fixed in this solution for 30–60 min at room temperature.
11. Histologic staining of all sections is not recommended, because the LacZ staining could be masked. We prefer to leave one section unstained and to counterstain the neighboring section with haematoxylin and eosin (Merck) for visualization of morphologic structures.
12. Long working distance objectives are required.

5. Acknowledgments

We acknowledge the excellent technical help of D. Ali Hadji, J. Kintz, M. de Meyer, H. Schultz, M. Dupont, F. Mischler, D. Elmlinger, C. Mougin, N. Accart, and V. Nourtier; Dr. B. Cavallini and I. Renardet for preparing plasmid DNAs; and Drs. R. Bischoff, F. Perraud, O. Boussif, N. Silvestre, Y. Cordier, A. Pavirani, and H.V.J. Kolbe for their contributions to the development and characterization of solvoplex vectors. We thank Dr. M. Courtney for critical comments on the manuscript. This work was supported in part by the AFM (Association Française contre les Myopathies) and the AFLM (Association Française de Lutte contre la Mucoviscidose).

References

1. Abdallah, B., Sachs, L., and Demeneix, B. (1995) Non-viral gene transfer: applications in developmental biology and gene therapy. *Biol. Cell.* **85,** 1–7.

2. Felgner, P. L., Tsai, Y. L., Sukhu, L., et al. (1995) Improved cationic lipid formulations for in vivo gene therapy. *Ann. N.Y. Acad. Sci.* **772**, 126–139.

3. Hagstrom, J. (2000) Self-assembling complexes for in vivo gene delivery. *Curr. Opin. Mol. Ther.* **2**, 143–149.

4. Hutchins, B. (2000) Characterization of plasmids and formulations for non-viral gene therapy. *Curr. Opin. Mol. Ther.* **2**, 131–135.

5. Lollo, C., Banaszczky, M., and Chiou, H. (2000) Obstacles and advances in non-viral gene delivery. *Curr. Opin. Mol. Ther.* **2**, 136–142.

6. Lee, R. J. and Huang, L. (1997) Lipidic vector systems for gene transfer. *Crit. Rev. Ther. Drug Carrier Syst.* **14**, 173–206.

7. Li, S. and Huang, L. (2000) Nonviral gene therapy: promises and challenges. *Gene Ther.* **7**, 31–34.

8. Luo, D. and Saltzman, W. M. (2000) Synthetic DNA delivery systems. *Nat. Biotechnol.* **18**, 33–37.

9. Kabanov, A., Felgner, P., and Seymour, L. (1998) *Self-Assembling Complexes for Gene Delivery*. John Wiley & Sons, Chichester, UK.

10. Mahato, R. I., Rolland, A., and Tomlinson, E. (1997) Cationic lipid-based gene delivery systems: pharmaceutical perspectives. *Pharm. Res.* **14**, 853–859.

11. Rolland, A. P. (1998) From genes to gene medicines: recent advances in nonviral gene delivery. *Crit. Rev. Ther. Drug Carrier Sys.* **15**, 143–198.

12. Schaffer, D. and Lauffenberger, D. (2000) Targeted synthetic gene delivery vectors. *Curr. Opin. Mol. Ther.* **2**, 155–161.

13. Scherman, D., Bessodes, M., Cameron, B., et al. (1998) Application of lipids and plasmid design for gene delivery to mammalian cells. *Curr. Opin. Biotechn.* **9**, 480–485.

14. Alton, E. W., Geddes, D. M., Gill, D. R., et al. (1998) Towards gene therapy for cystis fibrosis: a clinical progress report. *Gene Ther.* **5**, 291–292.

15. Korst, R. J., Mcelvaney, N. G., Chu, C. S., et al. (1995) Gene therapy for the respiratory manifestations of cystic fibrosis. *Respir. Crit. Care Med.* **151**, S75–S87.

16. Flotte, T. (1999) Gene therapy for cystic fibrosis. *Curr. Opin. Mol. Ther.* **1**, 510–615.

17. Rosenfeld, M. A. and Collins, F. S. (1996) Gene therapy for cystis fibrosis. *Chest* **109**, 241–252.

18. Schughart, K., Bischoff, R., Rasmussen, U. B., et al. (1999) Solvoplex: a new type of synthetic vector for intrapulmonary gene delivery. *Hum. Gene Ther.* **10**, 2891–2905.

19. Tatken, R. L. and Lewis Sr., J. R. (1983) *Registry of Toxic Effects of Chemical Substances*. Tractor Jitco, Rockville, MD.

20. Sambrook, J., Fritsch, E., and Maniatis, T. (1989) *Molecular Cloning: A Laboratory Manual*. Cold Spring Harbor Laboratory Press, Cold Spring Harbor, NY.

21. Zanta, M. A., Boussif, O., Adib, A., and Behr, J. P. (1997) In vitro gene delivery to hepatocytes with galactosylated polyethylenimine. *Bioconjug. Chem.* **8**, 839–844.

8

Gene Correction Frequency by Chimeric RNA-DNA Oligonucleotide Using Nuclear Extracts

Olga A. Igoucheva and Kyonggeun Yoon

1. Introduction

Site-specific correction of single-base mutations in target DNA sequences has been developed using an RNA-DNA hybrid oligonucleotide (RDO) to correct or cause a specific point mutation in episomal and genomic DNA *(1–14)*. In its original design, an RDO was composed of two strands, one strand containing both RNA and DNA (RNA-containing strand) and the other strand fully complementary to the RNA-containing strand but consisting of all DNA (DNA-containing strand; **Fig. 1**). The RNA was incorporated because an RNA-DNA hybrid duplex is more active than a DNA duplex in homologous recombination by the RecA and Rec2 proteins *(15,16)*. To render the oligonucleotide resistant to the RNaseH, ribose sugars were 2'-*O*-methylated, and to protect the 5' and 3' ends, a five GC clamp was placed at each end of the molecule. This original RDO molecule was empirically designed and has been modified subsequently. Recent RDO designs include a complete sequence complementarity of the RNA-containing strand to target DNA, a replacement of the central five DNA residues with 2'-*O*-methyl RNA, a placement of mismatch in the DNA-containing strand, and a chemical modification of the hairpin loops. These modifications result in a 3- to 10-fold increase in gene correction activity *(17,18)*. These results suggest that two different functions could be imparted to two strands of RDO: the RNA-containing strand stabilization of a labile intermediate and the DNA-containing strand for strand pairing and directing DNA repair activity.

From: *Methods in Molecular Medicine, Vol. 69, Gene Therapy Protocols, 2nd Ed.*
Edited by: J. R. Morgan © Humana Press Inc., Totowa, NJ

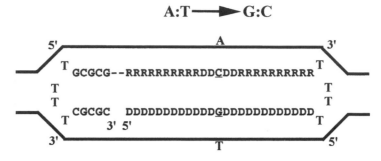

Fig. 1. The structure of RDO. RDO consists of RNA- and DNA-containing strands linked by a double T-loop and a 5-bp GC clamp. The RNA-containing strand has 10 2'-O-methyl RNA residues (R) flanking either side of a 5-residue DNA (D) stretch, which contains the base change desired for the mutation. The sequence of the RNA-containing strand is complementary to the targeted DNA over a stretch of 25 residues except for a single mismatch (A:C) to the targeted base to be altered. The DNA-containing strand is complementary to the RNA-containing strand.

1.1. Standardized System for Measuring Gene Conversion

The general application of RDO technology has been hampered by the lack of a standardized system to measure gene conversion in a particular cell type in a rapid and reproducible manner. For this purpose, we have constructed a shuttle vector, pCH110-G1651A, to measure the targeted gene correction of the mutant *E. coli* β-galactosidase gene *(19)*. The mutant plasmid contained a single-base-point mutation at position 1651 (G→A) in the plasmid pCH110, which encodes the full-length *E. coli* β-galactosidase gene under the control of the SV40 early promoter. This mutation causes a Glu→Lys substitution in the Mg^{2+} binding site (amino acid 461) and results in a complete loss of catalytic activity. A shuttle vector whereby a gene product is expressed in both mammalian and bacterial cells makes it possible to transfer the conversion event mediated by the RDO in mammalian cells to bacteria for further analysis (**Fig. 2**). The advantage of this scheme is that, in bacteria, gene correction events can be rapidly and unambiguously scored, as blue colonies. We also established an in vitro system in which nuclear extracts from mammalian cells were shown to process RDO-mediated gene correction of the mutant β-galactosidase plasmid *(19)*. This in vitro reaction was utilized to compare frequencies of gene conversion among cell lines *(19)* and to optimize the design of RDO *(17)*.

1.2. Variation in Gene Correction Frequency among Different Cell Types

We found a large variation in gene conversion frequencies among different cell types *(5,19)*. Neither uptake nor nuclear stability of the oligonucleotide

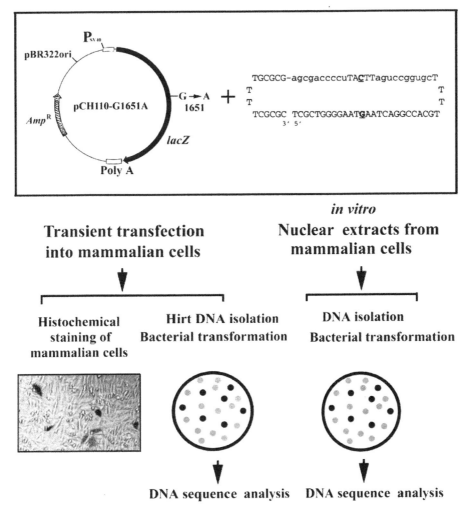

Fig. 2. Strategy of gene correction by RDO, an episomal transfection, and in vitro reaction with mammalian nuclear extracts, using the shuttle plasmid pCH110-G1651A containing a point mutation (G→A) in the *E. coli* β-galactosidase gene.

appears to be a limiting factor in our experiments, as measured by a radiolabeled or a fluorescein-conjugated oligonucleotide *(5)*. Variable frequencies among different cell types are not surprising since each cell type is likely to have different rates of recombination and mismatch repair. However, it has been difficult to predict whether RDO technology can be applied to a particular cell type.

Using the in vitro reaction, we measured the gene correction frequency of the mutant β-galactosidase shuttle vector. The frequency of gene conversion of

Table 1
Frequency of In Vitro Gene Correction by RDO in Different Cell Types

Cell type	RDO	Molar ratio or RDO/mutant plasmid	No. of Lac⁻ colonies	No. of Lac⁺ colonies	% Lac⁺/Lac⁻ [a]
CHO-K1	β-Gal B	1000	4.0×10^3	5	0.1
HeLa	β-Gal B	1000	34×10^3	1	0.003
HaCaT	β-Gal B	1000	28×10^3	1	0.004
PKC	β-Gal B	1000	110×10^3	2	0.002
p53$^{-/-}$	β-Gal B	1000	6.9×10^3	53	0.8
p53$^{+/+}$	β-Gal B	1000	26×10^3	0	<0.004
p53$^{+/+}$[b]	β-Gal B	1000	0.6×10^3	5	0.8
DT40	β-Gal B	1000	15×10^3	12	0.1

[a] Frequency of gene correction was measured by averaging three bacterial transformations of DNA prepared from each in vitro reaction by nuclear extracts from CHO-K1, HeLa, HaCaT, human primary keratinocytes (PKC), p53$^{-/-}$ and p53$^{+/+}$ mouse embryonic fibroblasts, and DT40 cells. One tenth-of the recovered DNA from each experiment was transformed into electrocompetent P90C bacteria and plated into 10 LB dishes containing 100 µg/mL of X-Gal and 50 µg/mL of ampicillin. The number of blue colonies was divided by the total number of colonies. The same molar ratio of β-Gal A (1000) was used in all experiments as a negative control and did not produce any blue colony among 10^4 colonies generated from each experiment.

[b] In vitro reaction carried out by nuclear extracts isolated from p53$^{+/+}$ cells after 15 passages

different cell types is shown in **Table 1**. Although an absolute gene correction frequency measured by the in vitro reaction is lower than that of transfection, it can be used to compare RDO-mediated gene correction activity among different cell types in a relative manner. For example, the embryonic fibroblasts from the p53$^{-/-}$ mouse showed much higher gene correction than that of isogenic p53$^{+/+}$ cells, indicating the importance of p53 in gene conversion *(19)*. Homologous recombination activity has previously been shown to be dependent on the status of p53 protein *(20,21)*. In addition, nuclear extracts made from the chicken B cell line (DT40) exhibited a high frequency of gene correction. DT40 cells were shown to have a higher homologous recombination rate than any other tissue culture cells *(22–24)*. Taken together, these results suggest that recombination is a rate-limiting step in gene conversion by RDO in cells with competent mismatch repair activities. Thus, the highly sensitive and convenient assay utilizing *E. coli* β-galactosidase is useful not only to optimize RDO structure but also to compare gene correction frequencies among different cell types.

In another in vitro approach, a point or a frameshift mutation in the drug-resistant gene (tetracycline or kanamycin) was corrected by cell-free extracts from HuH-7 cells in the presence of RDO at a frequency of approx 0.1%, scored by bacterial colonies growing in the drug selection *(25)*. Levels of gene correction by RDO decreased in HuH-7 cell extracts when the anti-hMSH2 antibody was added in the reaction, supporting the involvement of the hMSH2 mismatch repair protein. General application of RDO for site-specific gene correction or mutation should obviously benefit from such mechanistic studies.

2. Materials
2.1. Cell Culture

1. Chicken B-cell line (DT40): These cells are split 1/20 every day and maintained at a cell density lower than 1×10^6/mL in RPMI-1640 medium containing 10% fetal bovine serum (FBS), 1% chicken serum, and 0.05 mM 2-mercaptoethanol. The DT40 cells can be obtained from American Type Culture Collection (Rockville, MD; ATCC cat. no. CRL-2111).
2. Cell types to be tested: Maintain cells in an appropriate growth medium at log phase.
3. RPMI-1640 medium containing 2 mM glutamine.
4. FBS.
5. Chicken serum.
6. 2-Mercaptoethanol.

2.2. Nuclear Extract Isolation

1. Harvested cells.
2. Prechilled phosphate-buffered saline (PBS): 1% NaCl, 0.025% KCl, 0.14% Na_2HPO_4, 0.025% KH_2PO_4 (all w/v), pH 7.2; sterile.
3. Prechilled hypotonic buffer: 10 mM HEPES, pH 7.9, 1.5 mM $MgCl_2$, 10 mM KCl, 0.2 mM phenylmethylsulfonyl fluoride (PMSF; add just before using), and 0.5 mM dithiothreitol (DTT; add just before using).
4. Prechilled low-salt buffer: 20 mM HEPES, pH 7.9, 25% glycerol, 1.5 mM $MgCl_2$, 0.02 M KCl, 0.2 mM EDTA, 0.2 mM PMSF (add just before use), and 0.5 mM DTT (add just before use).
5. Prechilled high-salt buffer: 20 mM HEPES, pH 7.9, 25% glycerol, 1.5 mM $MgCl_2$, 1.2 M KCl, 0.2 mM EDTA, 0.2 mM PMSF (add just before use), and 0.5 mM DTT (add just before use).
6. Prechilled dialysis buffer: 20 mM HEPES, pH 7.9, 20% glycerol, 100 mM KCl, 0.2 mM EDTA, 0.2 mM PMSF (add dropwise just before use), and 0.5 mM DTT (add just before use).
7. 50-mL graduated conical polypropylene centrifuge tubes and 1.5-mL Eppendorf tubes.
8. Glass Dounce homogenizer (type B pestle) or 1-mL syringe (needle G21).
9. Slide-A-Lyzer Dialysis Cassettes (10,000 mol wt; Pierce, Rockford, IL).
10. Spectrophotometer.
11. Coomassie protein reagents kit (Pierce).

2.3. In Vitro Reaction

1. Nuclear extracts (*see* **Subheading 2.2.**).
2. DNA samples: supercoiled pCH110-G1651A plasmid DNA and RDO.
3. Reaction buffer: 30 mM HEPES, pH 7.8, 7 mM MgCl$_2$, 4 mM adenosine triphosphate (ATP), 200 μM each of cytidine triphosphate (CTP), guanosine triphosphate (GTP), uridine triphosphate (UTP), 100 μM each deoxy-ATP, deoxy-GTP, deoxy-CTP, deoxythymidine triphosphate (dTTP), 40 mM creatine phosphate, 100 μg/mL creatine phosphokinase, and 15 mM sodium phosphate (pH 7.5).
4. Phenol/chloroform/isomyl alcohol.
5. 100% and 70% ethanol.

2.4. Microbiology Reagents

1. Single colony of *E. coli* P90C and CC106 strains.
2. Luria-Bertani (LB) medium: 10 g tryptone, 5 g yeast extract, 5 g NaCl /L, pH 7.0.
3. Half-salt LB medium: 10 g tryptone, 5 g yeast extract, 2.5 g NaCl /L, pH 7.0.
4. 5× M63 medium: 10 g (NH$_4$)$_2$SO$_4$, 68 g KH$_2$PO$_4$, 2.5 mg FeSO$_4$ • 7H$_2$O/L, pH 7.0.
5. LB plates containing 50 μg/mL ampicillin and 100 μg/mL 5-bromo-4-chloro-3-indolyl-β-D-galactoside (X-Gal) (*see* **Note 1**).
6. 1 mM HEPES buffer, pH 7.5.
7. M63 minimal medium plates containing 0.4% D-(+)-lactose.
8. M63 minimal medium plates containing 0.4% glucose.
9. Spectrophotometer.
10. Ice-cold H$_2$O.
11. Ice-cold 10% glycerol.
12. SOC medium: 0.5% yeast extract, 2% tryptone, 10 mM NaCl, 2.5 mM KCl, 10 mM MgCl$_2$, 20 mM MgSO$_4$, 20 mM glucose.
13. Beckman JS-4.2 rotor or equivalent and adapters for 50-mL narrow-bottomed tubes.
14. Electroporation apparatus.
15. Chilled electroporation cuvets with two flat-topped electrode bosses separated by 0.1 cm.

3. Methods
3.1. Preparation of Nuclear Extracts

Nuclear extracts are prepared from the cells in a log phase growth by the standard method *(26)* with a slight modification. DT40 cells can be utilized as a positive control, since it showed a higher gene correction frequency than other cells tested *(19)*.

1. Grow DT40 cells in the 250-mL flasks in RPMI containing 10% FBS, 1% chicken serum, until they are 3–5 × 10^7/150 mL.

2. Collect cells from the spinner culture. Centrifuge 5×10^7 cells/150 mL in a 50-mL graduated conical centrifuge tubes (one tube for 45–50 mL of cell culture) for 10 min at 1850g (*see* **Note 2**).
3. Measure the packed cell volume (PCV) and resuspend cells in ice-cold PBS, approximately 5× (v/v) of the PCV. Centrifuge cells for 10 min at 1850g at 4°C.
4. Aspirate PBS and rapidly resuspend the cell pellets in an ice-cold hypotonic buffer, approximately 5× (v/v) of the PCV. Centrifuge cell pellets for 5 min at 1850g at 4°C.
5. Aspirate the solution and resuspend cells for the second time in an ice-cold hypotonic buffer, approximately 3× (v/v) of the original PCV (*see* **step 3**).
6. Allow the cells to swell on ice for 15–20 min (cells should swell twofold).
7. Homogenize the cell pellets with 20–25 slow up-and-down strokes using a 1-mL syringe with a needle (gage 21). At this stage, check for cell lysis under a microscope using Trypan blue.
8. Centrifuge for 15–20 min at 3300g.
9. Collect the supernatant containing cytoplasmic extracts and, if needed, save it for further analysis.
10. Measure the packed nuclear volume (PNV) from **step 9**.
11. Resuspend the nuclei in the low-salt buffer in a half-volume of the PNV (*see* **step 10**).
12. Add high-salt buffer dropwise the in a half-volume of the PNV (*see* **step 10**). Homogenize the nuclei as in **step 7** if necessary. Allow nuclei to extract for 30–45 min with continuous gentle shaking (use a rotating platform) at 4°C.
13. Centrifuge nuclei for 45–60 min at 14,000g at 4°C.
14. Collect the supernatant (nuclear extract), transfer it to the new tube, and keep it at 4°C.
15. Prepare a Slide-A-Lyzer cassette for dialysis by immersing it in dialysis buffer for 1–2 min. Then, remove excess liquid by tapping the edge of the cassette gently on a paper towel. Do not blot the membrane.
16. Fill a syringe (use a new one) with nuclear extract solution and inject the sample slowly. Dialyze against a large excess of dialysis buffer for 8–12 h at 4°C.
17. Remove the extracts from the cassette and collect the supernatant by centrifugation for 10–15 min at 14,000g at 4°C.
18. Determine the protein concentration using the Coomassie protein reagent kit. Measure the absorbance at 595 nm. Usually, 20 µg of protein is obtained from 1 million cells.
19. Aliquot extracts into tubes, freeze by submerging in liquid nitrogen, and store at –80°C (*see* **Note 3**).

3.2. In vitro Reaction between the Mutant β-Galactosidase Plasmid and RDO Catalyzed by Nuclear Extracts

The following reaction condition was optimized for the DT40 nuclear extracts. Because each cell extract has a different level of nuclease and gene correction activity, it may be necessary to optimize the reaction condition by varying the amount of nuclear extracts, the ratio of the plasmid DNA and oligonucleotide, and the time of reaction (*see* **Note 4**).

1. Prepare the reaction mixture containing a 100 ng supercoiled plasmid, pCH110-G1651A, and 3.0 µg RDO in the reaction buffer (*see* **Subheading 2.3.**). Prior to the reaction, heat the RDO for 5 min at 95°C and then chill it for 2 min on ice. This step is designed to fold the RDO into a double hairpin structure.
2. Add 50 µg of the nuclear extracts and adjust the reaction volume to 100 µL of DNase-free H_2O. Mix very gently by pipetting.
3. Incubate at 37°C for 1–3 h.
4. The DNA can be extracted first with an equal volume of phenol/chloroform (1:1) and then with chloroform, and precipitated by adding a 1/10 vol of 3 M NaOAc, pH 5.2, and 2.5 vol of 100% ethanol at –20°C overnight.
5. Wash the DNA pellet once with 70% ethanol and resuspend it in 5–10 µL of DNase-free water. Store at –20°C for further analysis into bacteria.

3.3. Gene Correction by RDO in Eukaryotic/Mammalian Cells Measured by Bacterial Transformation

The frequency of correction by RDO is measured by bacterial transformation of DNA isolated after in vitro reaction into bacterial strain P90C or CC106. Since a large number of bacterial colonies are generated and screened rapidly, the frequency of correction can be measured with increased accuracy. The P90C (*ara∆[lac proB]$_{XIII}$*) has a deletion in the entire *lac* operon *(27)*. Transformation of plasmid containing a functional or a mutant β-galactosidase gene into P90C results in either blue or white colonies on the X-Gal plate, respectively. The CC106 strain, a derivative of P90C, carries a mutation at amino acid 461 (Glu→Lys) of the *lac Z* gene in a F' *lacI⁻ Z⁻ proB⁺* episome *(28)*. When a functional β-galactosidase gene is supplied, the CC106 strain can metabolize lactose but not the P90C strain. Therefore, when DNA is transformed into the CC106 competent cells, a large number of colonies (over a million colonies) can be plated onto one minimum plate containing 0.4% lactose since only Lac⁺ colonies will grow. Therefore, the CC106 strain can be used to measure a lower frequency in comparison with the P90C strain. However, we found that the plating efficiency of CC106 on a minimum plate is 2–5 times lower than that of P90C, and we used P90C for most measurements. Both P90C and CC106 strains were generous gifts from Dr. Cupples.

3.3.1. Preparation of the P90C Competent Cells

1. Propagate the P90C bacteria culture from glycerol stock by streaking on a fresh LB medium plate.
2. Pick a single colony from the plate for the starter culture and grow it in 5 mL half-salt media overnight.

3. Prepare 1 L of a half-salt LB medium and pour into two Fernbach flasks (500 mL in each). Autoclave and shake at 37°C.
4. Remove 2 mL of LB medium from **step 3** and use it as a reference solution.
5. Inoculate Fernbach flasks with 2.5 mL of the overnight starter culture. Use sterile techniques.
6. Turn on the spectrophotometer and set the wavelength at 600 nm.
7. Grow bacteria until the absorbance at 600 nm reaches 0.45–0.55. This takes approximately 2 h. Absorbance should be read 1 h after adding the overnight culture, and then every 30 min.
8. When the absorbance reaches 0.45, chill flasks in ice/water bath for 15 min. After this step, work fast and keep cells cold (work on ice).
9. Transfer cells to cold 250-mL autoclaved centrifuge bottles. Weigh and balance appropriately.
10. Centrifuge for 10 min at 4000g.
11. Pour off the supernatant and resuspend pellets in a 25 mL of cold 1 mM HEPES buffer, pH 7.5. Gently resuspend the pellets using a pipet, then add 200 mL of 1 mM HEPES buffer, pH 7.5, to each bottle, and mix by gently inverting bottle.
12. Weigh and balance appropriately.
13. Centrifuge for 10 min at 4000g.
14. Immediately pour off supernatant. Pellets are very loose at this point. Add 25 mL of cold 1 mM HEPES buffer, pH 7.5. Resuspend pellets in a 100 mL of cold 1 mM HEPES buffer, pH 7.5, by gentle swirling. Add 100 mL of cold 1 mM HEPES buffer, pH 7.5, and mix by gently inverting the bottle.
15. Weigh and balance, and then centrifuge for 10 min at 4000g.
16. Quickly decant the supernatant. Pellets are very loose at this point. Resuspend the pellets in 5 mL of cold 10% (w/w) glycerol solution by gentle swirling. Transfer cells to one centrifuge bottle. Add 5 mL of cold 10% (w/w) glycerol in water to each bottle, swirl gently, and transfer solutions to the bottle containing cells.
17. Centrifuge for 10 min at 4000g. During centrifugation, prepare the dry ice/ethanol bath, label 0.5-mL Eppendorf tubes, and chill them on ice.
18. Immediately pour off the supernatant. Pellets are very loose at this point. Resuspend the pellets in an equal volume of cold 10% (w/w) glycerol in water. The pellet volume from a 500-mL culture is approximately 0.5 mL (1 L yields 1 mL). Resuspend pellets by gentle swirling.
19. Aliquot cells into cold 0.5-mL eppendorf tubes (100 μL each).
20. Freeze tubes in dry ice/ethanol bath for 5 min and store cells in a –80°C freezer. Properly stored, cells should be competent for transformation for at least 2–3 months. Transformation efficiency of the P90C competent cells should be greater than 5×10^8 transformants/μg of the control pUC19 plasmid DNA.

3.3.2. Preparation of the CC106 Competent Cells

1. Propagate the CC106 bacteria culture from glycerol stock by streaking on the M63 minimal plate containing glucose as a carbon source.
2. Pick a single colony from the plate for the overnight starter culture and grow in 5 mL 1× M63 minimal media with glucose as a carbon source.

3. Perform **steps 3–20** from **Subheading 3.3.1.**
4. Transformation efficiency of CC106 competent cells should be greater than 5×10^8 transformants/µg of the control pUC19 plasmid DNA.

3.3.3 Transformation and Plating of Bacteria for Quantitation of Correction Frequency

1. Prepare a dry ice/ethanol bath and maintain at –80°C.
2. Chill the electroporation cuvets and 0.5 mL Eppendorf tubes on ice.
3. Remove the competent cells (P90C or CC106) from the –80°C freezer.
4. Add 20 µL of cold 10% glycerol solution to an Eppendorf tube, then add 20 µL of the competent cells, and mix by flicking the tube gently.
5. Immediately refreeze unused competent cells in a dry-ice/ethanol bath for 5 min and put them back in the –80°C freezer for storage. Do not repeat this step more than three times, since the efficiency of transformation decreases upon repetitive freezing and thawing (*see* **Note 5**).
6. Add approx 10–20% of DNA recovered from the in vitro reaction (*see* **Subheading 3.2.**) to the competent cells/glycerol mixture. In most cases, plating of 10^5–10^6 clones, which can be generated from 0.2–2 ng of the recovered DNA after in vitro reaction, is sufficient to determine the frequency of gene correction. Incubate for 5 min on ice.
7. Transfer the solution to a 0.1-cm cuvet and flick the cuvet to mix.
8. Let stand for 5 min on ice.
9. Electroporate DNA at 25 µF and 250 W.
10. Immediately add 900 µL of SOC medium at room temperature, and transfer the electroporated mixture to a 3-mL snap cap tube.
11. Shake the transformed cells at 225 rpm (Incubator shaker series 25, New Brunswick Scientific, Edison, NJ) at 37°C for 1 h.
12. Plate cells appropriately and place plates upside down in the bacterial incubator overnight.
13. For DNA transformed into P90C cells, the frequency of correction is measured by dividing the number of blue colonies by the total number of colonies. To avoid overcrowding of bacterial colonies for each sample, approx 1000–2000 colonies should be plated per LB dish containing 50 µg/mL of ampicillin and 100 µg/mL of X-gal.
14. For DNA transformed into CC106 cells, over a million colonies can be plated onto one minimum plate containing 0.4% lactose. Prior to plating, spin the CC106 transformed cells and gently resuspend the pellet in 1x M63 medium containing 0.4% lactose. Repeat the wash, and resuspend the pellet in 100–150 µL of 1× M63 medium. Save 10 µL and plate the entire transformation mixture onto one to two minimum plates containing 0.4% lactose. The number of colonies grown on a lactose plate (Lac$^+$) should be normalized by the total number of colonies, which can be measured by plating serial dilutions of the saved 10 µL transformed CC106 cells in a minimum plate containing 0.4% glucose (*see* **Note 6**).

4. Notes

1. All plates containing X-Gal must be wrapped in aluminum foil and kept at 4°C because X-Gal is a light-sensitive reagent. Plates can be used within 1 month.
2. During nuclear extract preparation, all procedures must be performed at 4°C, preferably in a cold room using precooled buffers and equipment. Carry out all centrifugations at 4°C.
3. After isolation, the nuclear extract should be aliquoted and stored at –80°C. For the best results, each aliquot should be used once. Repeated freezing and thawing will decrease the efficiency.
4. An in vitro reaction may need to be optimized for each cell type by varying the amount of nuclear extracts, the ratio of the plasmid DNA and oligonucleotide, and the time of reaction.
5. Electroporation is currently the most efficient method for transformation of bacteria with plasmid DNA. Frozen competent cells should be aliquoted and stored at –80°C. For best results, each aliquot should be used once. Repeated freezing and thawing will decrease the efficiency by twofold.
6. For the M63 minimal medium plates, optional X-Gal and ampicillin can be added.

References

1. Yoon, K., Cole-Strauss, A., and Kmiec, E. B. (1996) Targeted gene correction of episomal DNA in mammalian cells mediated by a chimeric RNA/DNA oligonucleotide. *Proc. Natl. Acad. Sci. USA* **93**, 2071–2076.
2. Cole-Strauss, A., Yoon, K., Xiang, Y., et al. (1996) Correction of the mutation responsible for sickle cell anemia by a chimeric RNA/DNA oligonucleotide. *Science* **273**, 1386–1389.
3. Kren, B. T., Cole-Strauss, A., Kmiec, E. B., and Steer, C. J. (1997) Targeted nucleotide exchange in the alkaline phosphatase gene of HuH-7 cells mediated by a chimeric RNA/DNA oligonucleotide. *Hepatology* **25**, 1462–1468.
4. Xiang, Y., Cole-Strauss, A., Yoon, K., Gryn, J., and Kmiec, E. B. (1997) Targeted gene conversion in a mammalian CD34+-enriched cell population using a chimeric RNA/DNA oligonucleotide. *J. Mol. Med.* **75**, 829–835.
5. Santana, E., Peritz, A. E., Iyer, S., Uitto, J., and Yoon, K. (1998) Different frequency of gene targeting events by the RNA-DNA oligonucleotide among epithelial cells. *J. Invest. Dermatol.* **111**, 1172–1177.
6. Alexeev, V. and Yoon, K. (1998) Stable and inheritable changes in genotype and phenotype of albino melanocytes induced by an RNA-DNA oligonucleotide. *Nat. Biotechnol.* **16**, 1443–1346.
7. Bandyopadhyay, P., Ma, X., Linehan-Stieers, C., Kren, B. T., and Steer, C. J. (1999) Nucleotide exchange in genomic DNA of rat hepatocytes using RNA/DNA oligonucleotides. Targeted delivery of liposomes and polyethyleneimine to the asialoglycoprotein receptor. *J. Biol. Chem.* **247**, 10163–10172.
8. Kren, B. T., Metz, R., Kumar, R., and Steer, C. J. (1999b) Gene repair using chimeric RNA/DNA oligonucleotides. *Semin. Liver Dis.* **19**, 93–104.
9. Kren, B. T., Parashar, B., Bandyopadhyay, P., et al. (1999) Correction of the UDP-

glucuronosyltransferase gene defect in the Gunn rat model of Crigler-Najjar syndrome type I with a chimeric oligonucleotide. *Proc. Natl. Acad. Sci. USA* **96**, 10349–10354.

10. Beetham, P. R., Kipp, P. B., Sawycky, X. L., Arntzen, C. J., and May, G. D. (1999) A tool for functional plant genomics: chimeric RNA/DNA oligonucleotides cause in vivo gene-specific mutations. *Proc. Natl. Acad. Sci. USA* **96**, 8774–8778.

11. Zhu, T., Peterson, D. J., Tagliani, L., et al. (1999) Targeted manipulation of maize genes in vivo using chimeric RNA/DNA oligonucleotides. *Proc. Natl. Acad. Sci. USA* **96**, 8768–8773.

12. Alexeev, V., Igoucheva, O., Domashenko, A., Cotsarelis, G., and Yoon, K. (2000) Localized in vivo genotypic and phenotypic correction of the albino mutation in skin by RNA-DNA oligonucleotide. *Nat. Biotechnol.* **18**, 43–47.

13. Rando, T. A., Disatnik, M.-H., and Zhou, L. Z.-H. (2000) Rescue of dystrophin expression in *mdx* mouse muscle by RNA/DNA oligonucleotides. *Proc. Natl. Acad. Sci. USA* **97**, 5363–5368.

14. Zhu, T., Mettenburg, K., Peterson, D. J., Tagliani, L., and Baszczynski, C. L. (2000) Engineering herbicide-resistant maize using chimeric RNA/DNA oligonucleotides. *Nat. Biotechnol.* **18**, 555–558.

15. Kotani, H., Germann, M. W., Andrus, A., et al. (1996) RNA facilitates RecA-mediated DNA pairing and strand transfer between molecules bea–ing limited regions of homology. *Mol. Gen. Genet.* **250**, 626-634.

16. Kmiec, E. B., Cole, A., and Holloman, W. K. (1994) The REC2 gene encodes the homologous protein of *Ustilago maydis*. *Mol. Cell Biol.* **14**, 7163–7172.

17. Igoucheva, O. and Yoon, K. (2000) Improvement of RNA-DNA oligonucleotide design by using mammalian nuclear extracts. *Gene Ther. Reg.* **1**, 1–11.

18. Gamper, H. B., Cole-Strauss, A., Metz, R., et al. (2000) A plausible mechanism for gene correction by chimeric oligonucleotides. *Biochemistry* **39**, 5808–5816.

19. Igoucheva, O. A., Peritz, A. E., Levy, D., and Yoon, K. (1999) A sequence-specific gene correction by an RNA-DNA oligonucleotide in mammalian cells characterized by transfection and nuclear extract using a lacZ shuttle system. *Gene Ther.* **6**, 1960–1971.

20. Mekeel, K. L., Tang ,W., Kachnic, L. A., et al. (1997) Inactivation of p53 results in high rates of homologous recombination. *Oncogene* **14**, 1847–1857.

21. Thyagarajan, B., McCormick-Graham, M., Romero, D. P., and Campbell, C. (1996) Characterization of homologous DNA recombination activity in normal and immortal mammalian cells. *Nucleic Acids Res.* **24**, 4084–4091.

22. Buerstedde, J. and Takeda, S. (1991) Increased ratio of targeted to random integration after transfection of chicken B cell lines. *Cell* **67**, 179–188.

23. Takata, M., Sabe, H., Hata, A., et al. (1994) Tyrosine kinase Lyn and Syk regulate B cell receptor-coupled Ca^{2+} mobilization through distinct pathways. *EMBO J.* **13**, 1341–1349.

24. Yanez, R. J. and Porter, A. C. (1998) Therapeutic gene targeting. *Gene Ther.* **5**, 149–159.

25. Cole-Strauss, A., Gamper, H., Holloman, W. K., et al. (1999) Targeted gene repair directed by the chimeric RNA/DNA oligonucleotide in a mammalian cell-free extract. *Nucleic Acids Res.* **27,** 1323–1330.

26. Dignam, J. D., Lobovitz, R. M., and Roeder, R.G. (1983) Accurate transcription initiation by RNA polymerase II in a soluble extract from isolated mammalian nuclei. *Nucleic Acids Res.* **11,** 1475–1489.

27. Cupples, C. G. and Miller, J. H. (1988) Effects of amino acid substitutions at the active site in *E. coli* β-galactosidase. *Genetics* **120,** 637–643.

28. Cupples, C. G. and Miller, J. H. (1989) A set of *lacZ* mutations in *Escherichia coli* that allow rapid detection of each of the six base substitutions. *Proc. Natl. Acad. Sci. USA* **86,** 5345–5349.

9

Regulated Expression of Plasmid-Based Gene Therapies

Ronald V. Abruzzese, Fiona C. MacLaughlin, Louis C. Smith, and Jeffrey L. Nordstrom

1. Introduction
1.1. Intramuscular Plasmid Delivery

Following direct intramuscular injection of plasmids, transgene expression follows a biphasic pattern in which expression levels are maintained for approximately 1 month, and then decline to trace levels that persist for 1 year or more *(1,2)*. However, even during the first month, transgene expression levels are too variable and too low to provide consistent therapeutic levels of circulating proteins. Thus, most intramuscular plasmid-based gene therapy efforts have focused on indications that require short-term, low-level, local expression. Examples include induction of collateral vessel growth following injection of vascular endothelial growth factor (VEGF) plasmids into regions of ischemic muscle *(3)* and induction of humoral and cell-based immunity following intramuscular injection of plasmid-based genetic vaccines *(4)*. The use of protective, interactive, noncondensing (PINC^{TM}) polymers, such as poly(vinyl) derivatives *(5)* or nonionic block copolymers made of ethylene oxide and propylene oxide monomers *(6)*, has led to the development of formulations that provided higher, more persistent transgene expression compared with plasmids in saline. For example, a single dose of an insulin-like growth factor-I (IGF-I) plasmid formulated in poly(vinylpyrrolidone) caused reinnervation of motor end plates and increased muscle fiber size when delivered to paralyzed adult rat laryngeal muscle *(7)*.

From: *Methods in Molecular Medicine, Vol. 69, Gene Therapy Protocols, 2nd Ed.*
Edited by: J. R. Morgan © Humana Press Inc., Totowa, NJ

1.2. Intramuscular Plasmid Delivery with In Vivo Electroporation

The application of an electric field in the form of low-voltage, long-duration pulses enables the transport of macromolecules (i.e., plasmids) across a cell membrane. Upon reaching a critical voltage, cell membrane breakdown occurs, pores form, and plasmids are transported into the cytoplasm of cells *(8–10)*. When intramuscular plasmid delivery is combined with in vivo electroporation, levels of transgene expression are increased by 2–3 orders of magnitude, allowing levels as high as 2 µg/mL of expressed protein to be attained *(11–17)*. Transgene expression levels are reproducible and may persist at stable levels for at least 1 year in mice. Using low-voltage, long-duration pulses delivered by electrodes that are used noninvasively (caliper electrodes applied to the entire hindlimb) or invasively (two wires applied to exposed muscle tissue), sustained levels of several circulating proteins (secreted alkaline phosphatase, erythropoietin, factor IX) have been attained in mice.

1.3. Rationale for Regulated Gene Therapy

The ability to attain long-term, high-level transgene expression raises new challenges, since the expression of all endogenous genes is controlled by regulatory mechanisms. An effective system for transgene regulation will minimize the frequency and severity of side effects by maintaining circulating protein levels within the therapeutic window. If serious adverse effects due to excessive transgene expression are encountered, the regulation system can serve as a "safety switch" to terminate transgene expression rapidly. In addition, an effective transgene regulation system may permit the circulating levels of the therapeutic protein to be raised or lowered according to therapeutic need, thereby maximizing an individual patient's benefit.

1.4. Ligand-Dependent Transgene Regulation Systems

Several systems for ligand-dependent transgene regulation have been employed in gene therapy studies in animals. The tetracycline-dependent system is based on the responsiveness of an *E. coli* repressor protein to doxycycline and related antibiotics *(18)*. The rapamycin-dependent system utilizes the drug-induced heterodimerization properties of the immunophilins, FKBP and FRAP *(19)*. The ecdysone-dependent system is based on the ligand responsiveness of heterodimers formed from mammalian retinoid X receptors and modified *Drosophila* or *Bombyx* ecdysone receptors *(20)*. The mifepristone-dependent GeneSwitch™ system exploits the responsiveness of a mutant progesterone receptor to synthetic antiprogestins *(21,22)*. The tetracycline-, rapamycin-, and ecdysone-dependent systems have been primarily applied to viral-based gene therapy *(23–26)*. By contrast, the mifepristone-dependent

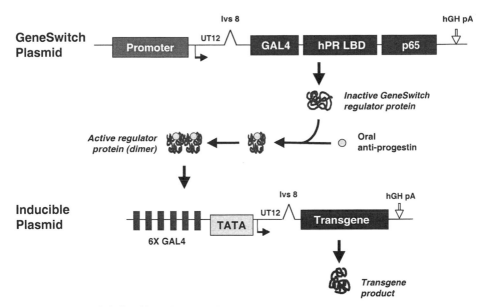

Fig. 1. Model for ligand-dependent regulation of transgene expression by the GeneSwitch system. Various promoters (CMV, chicken skeletal α-actin, or auto-inducible promoters) can be used to drive expression of the GeneSwitch regulator protein. The promoter of the inducible EPO plasmids contains six GAL4 sites linked to a TATA box. UT12 is derived from the 5' UTR of CMV I/E transcripts, IVS8 is a synthetic intron, and hGH pA is the human growth hormone poly(A) site. The GeneSwitch regulator protein is a chimera of the GAL4 DNA binding domain (GAL4), truncated human progesterone receptor ligand-binding domain (hPR-LBD), and the p65 subunit of human NF-κB (p65). It is synthesized in an inactive form. Binding of a synthetic antiprogestin drug (e.g., mifepristone) triggers a conformational change that leads to activation and dimerization. Activated homodimers bind to GAL4 sites in the inducible promoter and stimulate transcription of the transgene.

GeneSwitch system, which functions effectively in the context of viral vectors *(27,28)*, has also been utilized for plasmid-based gene therapy applications *(29,30)*.

1.5. Plasmid-Based GeneSwitch System for Transgene Regulation

A two-plasmid GeneSwitch system is illustrated in **Figures 1** and **2**. One plasmid encodes the inducible transgene of interest and is controlled by an inducible promoter that consists of six GAL4 sites linked to a minimal TATA box promoter. The GAL4 binding sites are 17-bp palindromic sequences that are found in the promoter regions of galactose-regulatable genes of yeast but appear to be absent in eukaryotic genomes.

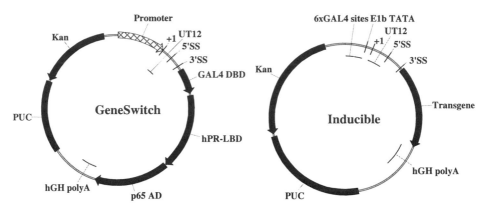

Fig. 2. Maps for the GeneSwitch and inducible transgene plasmids. +1 indicates the start of transcription. UT12 is the 5' UTR, 5'SS and 3'SS refer to the splice sites of the synthetic intron, IVS8, and hGH polyA is the human growth hormone poly(A) site. PUC is the modified ColE1 origin of replication derived from pUC18. Kan is the kanamycin resistance gene derived from transposon Tn5.

The second plasmid encodes the GeneSwitch regulator protein. This chimeric protein is based on the unique properties of a truncated human progesterone receptor ligand-binding domain (hPR-LBD), which has lost the ability to be activated by natural progestin ligands but gained the ability to be activated by synthetic antiprogestin ligands, such as mifepristone. To function as a specific transgene regulator, DNA binding domain from the yeast GAL4 protein was joined to the N-terminus of the hPR-LBD, and transactivation domain from the p65 subunit of human NF-κB was joined to its C-terminus *(22,27,29,30)*. The resultant chimeric protein is approx 85% human in sequence. When produced in transfected cells, the GeneSwitch regulator protein remains intracellular and probably forms inert complexes with heat shock proteins and immunophilins *(31)*. Following binding of the inducing drug (mifepristone), the GeneSwitch regulator protein undergoes a conformational change and becomes activated, dissociates from the components of the inert complex, forms homodimers, binds to the GAL4 sites in the inducible promoter, and then induces transcription of the inducible transgene (**Fig. 1**). The concentration of mifepristone required for half-maximal induction of transgene expression in transfected tissue culture cells is approximately 10^{-11} M (**Fig. 3**).

The plasmid-based GeneSwitch system has been used to regulate the expression of secreted alkaline phosphatase, erythropoietin, and VEGF in a ligand-dependent manner in mice following intramuscular administration with in vivo electroporation *(29,30)*. A single oral dose of mifepristone yielded peak transgene expression within 24 h that returned to basal levels by 72–96 h. The

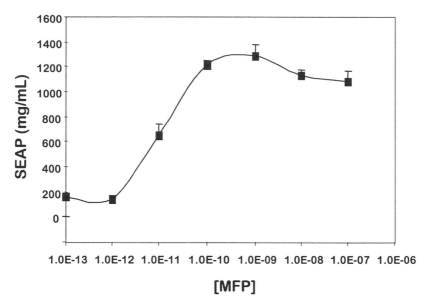

Fig. 3. Representative mifepristone (MFP) dose-response curve generated in transiently transfected HeLa cells using a 1:1 ratio of GeneSwitch and inducible plasmids. Twenty-four hours after transfection, media containing MFP at concentrations between 10^{-13} and 10^{-7} M were added to the cells. Twenty-four hours later, transgene expression was measured. Peak levels of expressed protein were obtained at 10^{-10} M. SEAP, human placental secreted alkaline phosphatase

oral mifepristone dose that yielded half-maximal induction of transgene expression in mice was 0.03 mg/kg body weight *(29)*. In comparison, for use as an abortifacient in humans, mifepristone is administered at doses that are 100-fold higher (3–10 mg/kg body weight), and it requires the coadministration of a prostaglandin, misoprostol **(32).** Thus, it is anticipated that the doses of mifepristone (MFP) required for regulated gene therapy will be very low and well tolerated.

The identity of the promoter that drives expression of the gene for the GeneSwitch protein influences the performance of the system. If the highly active, constitutive, cytomegalovirus (CMV) promoter is utilized, detectable levels of transgene expression occur even in the absence of inducing ligand *(29)*. If an autoinducible promoter, consisting of four GAL4 sites linked to a minimal thymidine kinase promoter, is utilized, the tightness of regulation is improved by over 1 order of magnitude. This promoter confers the ability to regulate the synthesis of the GeneSwitch regulator protein itself in a mifepristone-dependent manner. Thus, high-level induction of transgene expression by mifepristone is achieved through a series of self-amplifying in-

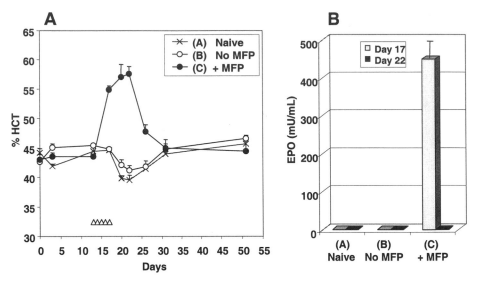

Fig. 4. Ligand-dependent regulation of erythropoietin (EPO) expression and hematocrit (HCT) in mice following intramuscular delivery of plasmids with electroporation. **(A)** Induction of hematocrit. **(B)** Induction of EPO expression. An inducible human EPO plasmid and a muscle-specific GeneSwitch plasmid were mixed in a 1:1 ratio and formulated in 6 mg/mL sodium poly-L-glutamate at a final DNA concentration of 1.0 mg/mL. A plasmid dose of 150 µg was delivered on day 0 to both tibialis and gastrocnemius muscles (25 µg per tibialis, 50 µg per gastrocnemius) of C57Bl/6 mice with electroporation (caliper electrodes, 2 pulses, 375 V/cm, 25 ms). The number of animals in each group was five. Group A animals (naïve) received no DNA. Group B and C animals received the plasmids for the EPO/GeneSwitch system. Group B animals (no mifepristone [MFP]) were not treated with MFP. Group C animals (+ MFP) were administered oral doses of MFP (0.3 mg/kg) on 5 consecutive days, beginning on day 13 (indicated by the open triangles). Hematocrit levels were measured at various times, and serum EPO levels were measured on days 17 and 22. All values are expressed as mean ± SEM.

duction cycles *(30)*. If a muscle-specific (chicken α-actin) promoter, which is approx 1% as active as the CMV promoter, is utilized, extremely tight transgene regulation in vivo also is achieved (**Fig. 4**).

2. Materials

2.1. Plasmids

Both of the plasmids for the GeneSwitch system have backbones that contain a modified ColE1 origin of replication, derived from pUC18, and a kanamycin resistance gene, derived from transposon Tn5. Small-scale lots of plasmid (<10 mg) are prepared from transformed cultures of *E. coli* (strain DH5α; GIBCO BRL, Rockville, MD) grown in 1 L of Terrific Broth media supplemented with 100 μg/mL kanamycin overnight at 37°C in 2.8-L fluted shaker flasks with shaking at 250 rpm in an orbital shaker. Cells are pelleted by centrifugation in 500-mL bottles at 2500*g* for 20 min. Plasmid is then isolated according to Qiagen's protocol for Endo Free Giga prep kits and suspended in 3 mL of QN buffer. The average yield is 10 mg, and endotoxin levels are less than 5 endotoxin units/mg. Larger scale lots of plasmid (>200 mg) are prepared following bacterial growth in 4.5–10 L fermentors, alkaline lysis, and purification by ion exchange chromatography, according to a proprietary process developed by Valentis, Inc. (*see* **Note 1**).

2.2. Plasmid Formulation

1. Sterile glass vials (The West Company, Lionville, PA).
2. 5 *M* NaCl (cat. no. S0155;Spectrum, New Brunswick, NJ).
3. Poly-L-glutamic acid, sodium salt (mol wt 15,000–50,000, cat. no. P4761; Sigma, St. Louis, MO). Prepare a stock solution of 25–50 mg/mL. The dry powder is stored in a dessicator at –20°C. A stock solution of 25 mg/mL is prepared in water, then stored at –4°C for up to one month.
4. 1 *M* Tris-HCl, pH 7.5, is from Life Technologies, Rockland, MD (cat. no. 15567-027). Prepare a 100 m*M* stock solution.

2.3. Inducing Ligand

Mifepristone (2.5 mg; Sigma) is dissolved in 0.5 mL of absolute ethanol, and then 29.5 mL of sesame oil (Sigma) is added, followed by gentle mixing by inversion. The final concentration is 0.083 mg/mL. The tube is protected from light by wrapping in aluminum foil and stored at –20°C.

2.4. Animals

We have used male or female CD-1 mice (6–8 weeks old, weight range 29–31 g) or female C57Bl/6 mice (6–8 weeks old, weight range 19–21 g). Animals are housed in a ventilated caging system (Thoren Caging Systems, Hazelton, PA) in a facility with 12-h light and dark cycles and fed RodentDiet 5001 (PMI Nutrition International, Brentwood, MO). Animals are acclimated for at least 3 days prior to experimentation. The combination anesthesia ketamine (74 mg/mL), xylazine (3.7 mg/mL), and acepromazine (0.73 mg/mL) is administered at a dose of 1.8–2.0 mL/kg. All procedures conform to state and federal guidelines.

2.5. Intramuscular Injection and In Vivo Electroporation

1. Insulin syringes (3/10 cc; Becton Dickinson, Franklin Lakes, NJ).
2. 70% isopropyl alcohol (Sigma).

3. Heating pad (GayMar T/pump, GayMar Industries, NY).
4. Electro Square Porator™, model T820 (BTX, San Diego, CA).
5. Enhancer 400 Graphic Pulse Display (BTX).
6. Stainless steel caliper electrodes, 1.5 or 2.0 cm² (BTX).

2.6. Assays on Serum or Tissue Homogenates

1. Blood is collected by retroorbital bleeding using disposable Pasteur pipets (cat no. 22-230-482, Fisher Scientific; Pittsburgh, PA).
2. Microtainer serum separator tubes (cat. no. VT5960 [VWR]; Becton Dickinson, Franklin Lakes, NJ).
3. Cryovial, polypropylene vials (2 mL) with screw cap lids (cat. no. 10832; BIOSPEC Products, Bartlesville, OK), for storing muscle tissue samples and preparing homogenates.
4. Silicon beads (cat. no. 11079125z; BIOSPEC).
5. Mini Bead Beater 8 (cat no. 693, BIOSPEC).
6. Reporter lysis buffer, 5× (cat. no. E3971; Promega, Madison, WI).
7. BCA Protein Assay Reagents A and B, and albumin standard (2 mg/mL; Pierce, Rockford, IL).

3. Methods
3.1. Formulation of Plasmid

Plasmids may be formulated in saline or sodium poly-L-glutamate (*see* **Note 2**). For delivery by in vivo electroporation, formulation of plasmids in poly-L-glutamate yields transgene expression levels that are up to 10-fold higher than plasmids in saline (unpublished data).

3.1.1. Plasmids in Saline

1. Combine the desired amount of GeneSwitch and inducible transgene plasmids in a 1:1 (w/w) ratio in a sterile glass vial.
2. Add sterile water and mix gently (with a pipet tip).
3. Add sterile 5 M NaCl for a final concentration of 0.15 M and mix gently (*see* **Note 3**).
4. Add sterile 100 mM Tris-HCl, pH 7.5, for a final concentration of 5 mM.

3.1.2. Plasmids in Poly-L-Glutamate

1. Combine the desired amount of GeneSwitch and inducible transgene plasmids in a 1:1 (w/w) ratio in a sterile glass vial.
2. Add sterile water with gentle mixing.
3. Add sterile poly-L-glutamate with gentle mixing for a final concentration of 6 mg/mL.
4. Add 5 M NaCl with gentle mixing for a final concentration of 0.15 M (*see* **Note 3**).
5. Add sterile 100 mM Tris-HCl, pH 7.5, for a final concentration of 5 mM.

3.2. Intramuscular Plasmid Delivery with In Vivo Electroporation

1. Anesthetize the mice just prior to the procedure by injecting the combination anesthetic intraperitoneally at a dose of 1.8–2.0 mL/kg body weight. The anesthesia typically lasts for 30 min.
2. Shave the hindlimb with a clipper or razor and cleanse the skin by swabbing with 70% isopropyl alcohol.
3. Inject the plasmid solution into the skeletal muscle of choice. We routinely orient the injection needle parallel to the muscle fibers. The injection volume for the tibialis cranialis muscle is 25 µL. The injection volume for the gastrocnemius muscle is 50 µL, administered in 25-µL volumes to each muscle belly (*see* **Note 4**).
4. Place the leg firmly, but without compression, between the stainless steel electrode plates. Too much pressure will result in reduced levels of gene expression and potential damage. For electroporation of the tibialis muscle alone, ensure that the muscle lies in the middle of the plate. For electroporation of the gastrocnemious muscle, ensure that the fleshy part of the muscle is firmly gripped between the electrode plates. For electroporation of tibialis and gastrocnemius muscles, place the leg as centrally as possible between the two plates of the electrode.
5. Measure the distance between the electrode plates. The average diameter of the tibialis and gastrocnemius muscles of CD-1 mice (29–31 g in weight), is 0.4 cm.
6. Two minutes after plasmid injection, initiate the electric pulsing regimen (2 square wave pulses of 375-V/cm intensity and 25-ms duration). To achieve this voltage intensity for a hindlimb that is 0.4 cm in diameter, a voltage setting of 150 V is required. Upon administration of each pulse, the toes will separate and the lower leg may twitch.
7. Place the animal on a 37°C heating pad for recovery. The lower limbs may appear to be weak immediately following the procedure but will completely recover within 1 or 2 h. Some edema may be observed immediately after the procedure and may last for up to 24 h.

3.3. Administration of Inducing Ligand

To deliver a dose of 0.3 mg/kg body weight to a 30-g mouse, 110 µL of the mifepristone suspension (0.083 mg/mL) is administered orally (by gavage) or intraperitoneally. Administration of the ligand to mice by gavage is performed with a Popper & Sons animal feeding/intubation needle (curved, 18 gage) (VWR cat. no. 20068-640). Intraperitoneal administration is performed with a 3-mL Luer lock syringe with an 18-gage needle. Ligand can be administered daily or on an intermittent schedule, as desired. Peak transgene expression is typically observed after 24 h (*see* **Note 5**).

3.4. Assays for Transgene Products in Serum or Tissue Homogenates

3.4.1. Assay for a Secreted Gene Product (see Notes 6–8)

1. Collect blood by retroorbital methods.
2. Separate the serum (after a 30-min incubation at room temperature) by centrifugation at $2500g$ for 5–10 min at room temperature.
3. Assay serum aliquots by the assay method of choice (e.g., by enzyme-linked immunosorbent assay (ELISA).

3.4.2. Assay for a Nonsecreted Gene Product

1. Excise the muscle tissue, place into a capped polypropylene tube with silica beads (8–10 beads), snap-freeze in liquid nitrogen, and lyophilize overnight. Store the samples at $-80°C$.
2. Remove the tubes from the freezer and pulverize the tissue by placing the tubes on a bead beater, located in a $4°C$ cold room, for 2 min.
3. Add 1 mL of 0.5× reporter lysis buffer and bead-beat for an additional 3 min.
4. Centrifuge the muscle suspension at $16,000g$ for 10 min at $4°C$ and collect the supernatant. Assay aliquots for the desired transgene protein product. Aliquots may be diluted in lysis buffer or in the sample dilution buffer supplied with the analyte kits.
5. For determination of total protein of tibialis or gastrocnemius muscles, dilute an aliquot of the muscle supernatant 1:10 with sterile water and add 10 µL of the diluted extract to each well. Add 200 µL of the BCA reagent, and proceed according to the manufacturer's protocol.

4. Notes

1. Plasmids for the autoinducible GeneSwitch system are available from Invitrogen (Carlsbad, CA). These plasmids differ from those shown in **Figure 2** in two ways. The Invitrogen plasmid contains a bacterial gene for ampicillin resistance, rather than kanamycin resistance. It also contains an additional gene, one that is driven by the SV40 promoter and codes for hygromycin or zeocin resistance, which permits the selection of stable transfectants. For regulatory reasons, plasmids that confer kanamycin resistance are used for gene therapy applications *(33)*.
2. Poly-L-glutamate is an anionic polymer that enhances the uptake and retention of plasmids delivered to skeletal muscle in combination with in vivo electroporation (unpublished data). It may enhance uptake by saturating nonspecific DNA binding sites on the cell surface and/or in the extracellular matrix. It may also inhibit nucleases, protecting the plasmid from degradation.
3. Once prepared, plasmid suspensions can be used immediately but may be stored at $4°C$. Total plasmid concentration should not exceed 3 mg/mL. It is prudent to prepare the minimum required volumes for the experiment plus an excess of 30%. When preparing formulations for the plasmid-based GeneSwitch system, first calculate the volume of each plasmid stock solution that is needed for a 1:1 (w/w)

ratio and the desired total DNA concentration. We have prepared formulations with final DNA concentrations that range from 0.02–3.0 mg/mL. Second, calculate the volumes of poly-L-glutamate, NaCl, and Tris-HCl. Third, calculate the volume of water that adjusts the solution to the final desired volume. For example, the volumes needed to prepare 1.0 mL of two plasmids (1:1 mixture, w/w) formulated at a total DNA concentration of 1.5 mg/mL in 6 mg/mL poly-L-glutamate are 187.5 µL of GeneSwitch plasmid (4.0 mg/mL stock); 250 µL of inducible transgene plasmid (3.0 mg/mL stock); 242.5 µL water; 240 µL poly-L-glutamate (25 mg/mL stock); 30 µL 5 M NaCl; 50 µL 100 mM Tris-HCl. It is prudent to verify that the final solution has an osmolality of 290 ± 10 mOsm/kg.

4. The total plasmid dose may be varied as desired. The highest dose we have used in a 20-g mouse is 150 µg (25 µg per tibialis, 50 µg per gastrocnemius), which is 7.5 mg/kg body weight. The lowest dose we have used is 1 µg (delivered to a single tibialis), which is 0.05 mg/kg.

5. We currently prefer to use a muscle-specific promoter in the GeneSwitch plasmid, since it produces the GeneSwitch regulator protein at lower levels and restricts its expression to muscle cells. This provides lower background levels of transgene expression, yet still provides the capability of a robust induction of transgene expression in vivo. The ability of a muscle-specific GeneSwitch system to regulate erythropoietin (EPO) expression in mice is shown in **Figure 4**. Plasmids (1:1 mixture of inducible EPO plasmid and muscle-specific GeneSwitch plasmid) were delivered with electroporation at a high dose (approx 7.5 mg/kg body weight). No EPO expression and, more importantly, no increase in hematocrit, was detected in the absence of mifepristone administration. Administration of oral doses of mifepristone caused EPO expression to be induced to >400 mU/mL. Within 5 days of the last mifepristone dose, EPO expression returned to undetectable levels. The 5-day mifepristone treatment triggered a substantial increase in hematocrit, with peak levels (57%) occurring 6 days following the final mifepristone dose. By day 30, 2 weeks after the final mifepristone dose, hematocrit returned to baseline levels, which was maintained at normal levels for an additional 20 days. The data demonstrate the ability of the muscle-specific GeneSwitch system to provide strict, ligand-dependent regulation of transgene expression in mice.

6. The GeneSwitch systems for regulating erythropoietin and VEGF perform effectively in vivo at 1:1 ratios of the GeneSwitch and inducible plasmids. However, it is difficult to predict precisely the tightness of regulation that will be obtained with different transgenes, since sequence elements within the transgene itself can influence the strength of the inducible promoter. Thus, it may be necessary to modify the ratios of the two plasmids to refine the desired response.

7. Delivery of plasmids to muscle by in vivo electroporation is associated with a transient period (approximately 30 days in length) during which localized regions of muscle damage and inflammation are evident. After approximately 30 days, the muscle tissue is normal in appearance, except for some evidence of recent myotube regeneration.

8. The two components of the system, the GeneSwitch gene and the inducible transgene, can be combined as a cassette on one DNA molecule. When the single cassette was inserted into an adenovirus vector, highly effective regulation of growth hormone expression was observed in mice *(27)*. However, when a single cassette is inserted into a plasmid, we typically observe a decline in the magnitude of drug-dependent transgene regulation due to increased basal expression of the inducible transgene. Thus, tight, drug-dependent regulation of transgene expression is best achieved with two separate plasmids.

References

1. Wolff, J. A., Ludtke, J. J., Acsadi, G., Williams, P., and Jani, A. (1992) Long-term persistence of plasmid DNA and foreign gene expression in mouse muscle. *Hum. Mol. Genet.* **1,** 363–369.
2. Doh, S. G., Vahlsing, H. L., Hartikka, J., Liang, X., and Manthorpe, M. (1997) Spatial-temporal patterns of gene expression in mouse skeletal muscle after injection of lacZ plasmid DNA. *Gene Ther.* **4,** 648–663.
3. Takeshita, S., Tsurumi, Y., Couffinahl, T., et al. (1996) Gene transfer of naked DNA encoding for three isoforms of vascular endothelial growth factor stimulates collateral development in vivo. *Lab Invest.* **75,** 487–501.
4. Ulmer, J. B., Donnelly, J. J., Parker, S. E., et al. (1993) Heterologous protection against influenza by injection of DNA encoding a viral protein. *Science* **259,** 1745–1749.
5. Mumper, R. J., Wang, J., Klakamp, S. L., et al. (1998) Protective interactive noncondensing (PINC) polymers for enhanced plasmid distribution and expression in rat skeletal muscle. *J. Control. Release* **52,** 191–203.
6. Lemieux, P., Guerin, N., Paradis, G., et al. (2000) A combination of poloxamers increases gene expression of plasmid DNA in skeletal muscle. *Gene Ther.* **7,** 986–991.
7. Shiotani, A., O'Malley, B. W. Jr., Coleman, M. E., Alila, H. W., and Flint, P. W. (1998) Reinnervation of motor endplates and increased muscle fiber size after human insulin-like growth factor I gene transfer into the paralyzed larynx. *Hum. Gene Ther.* **9,** 2039–2047.
8. Neumann, E., Kakorin, S., and Toensing, K. (1999) Fundamentals of electroporative delivery of drugs and genes. *Bioelectrochem. Bioenerg.* **48,** 3–16.
9. Mir, L. M. and Orlowski, S. (1999) Mechanisms of electrochemotherapy. *Adv. Drug Deliv. Rev.* **35,** 107–118.
10. Somiari, S., Glasspool-Malone, J., Drabick, J. J, et al. (2000) Theory and in vivo application of electroporative gene delivery. *Mol. Ther.* **2,** 178–187.
11. Smith, L. C. and Nordstrom, J. L. (2000) Advances in plasmid gene delivery and expression in skeletal muscle. *Curr. Opin. Mol. Ther.* **2,** 150–154.
12. Aihara, H. and Miyazaki, J.-I. (1998) Gene transfer into muscle by electroporation in vivo. *Nat. Biotechnol.* **16,** 867–870.
13. Mathiesen, I. (1999) Electropermeabilization of skeletal muscle enhances gene transfer in vivo. *Gene Ther.* **6,** 508–514.

14. Bettan, M., Emmanuel, F., Darteil, R., et al. (2000) High-level protein secretion into blood circulation after electric pulse-mediated gene transfer into skeletal muscle. *Mol. Ther.* **2,** 204–210.

15. Rizzuto, G., Cappelletti, M., Maione, D., et al. (1999) Efficient and regulated erythropoietin production by naked DNA injection and muscle electroporation. *Proc. Natl. Acad. Sci. USA* **96,** 6417–6422.

16. Mir, L. M., Bureau, M. F., Gehl, J., et al. (1999) High-efficiency gene transfer into skeletal muscle mediated by electric pulses. *Proc. Natl. Acad. Sci. USA* **96,** 4262–4267.

17. Vicat, J. M., Boisseau, S., Jourdes, P., et al. (2000) Muscle transfection by electroporation with high-voltage and short-pulse currents provides high-level and long-lasting gene expression. *Hum. Gene Ther.* **11,** 909–916.

18. Gossen, M. and Bujard, H. (1992) Tight control of gene expression in mammalian cells by tetracycline responsive promoters. *Proc. Natl. Acad. Sci. USA* **89,** 5547–5551.

19. Rivera, V. M., Clackson, T., Natesan, S., et al. (1996) A humanized system for pharmacological control of gene expression. *Nat. Med.* **2,** 1028–1032.

20. No, D., Yao, T. P., and Evans, R. M. (1996) Ecdysone-inducible gene expression in mammalian cells and transgenic mice. *Proc. Natl. Acad. Sci. USA* **93,** 3346–3351.

21. Wang, Y., O'Malley, B. W. Jr., Tsai, S. Y., and O'Malley, B. W. (1994) A regulatory system for use in gene transfer. *Proc. Natl. Acad. Sci. USA* **16,** 8180–8184.

22. *http://www.geneswitch.com/*

23. Rendahl, K. G., Leff, S. E., Otten, G. R., et al. (1998) Regulation of gene expression in vivo following transduction by two separate rAAV vectors. *Nat. Biotechnol.* **16,** 757–761.

24. Bohl, D., Salvetti, A., Moullier, P., and Heard, J. M. (1988). Control of erythropoietin delivery by doxocycline in mice after intramuscular injection of adeno-associated vector. *Blood* **92,** 1512–1517.

25. Ye, X., Rivera, V. M., Zoltick, P., et al. (1999) Regulated delivery of therapeutic proteins after in vivo somatic cell gene transfer. *Science* **283,** 88–91.

26. Hoppe, U. C., Marban, E., and Johns, D. C. (2000) Adenovirus-mediated inducible gene expression in vivo by a hybrid ecdysone receptor. *Mol. Ther.* **1,** 159–164

27. Burcin, M. M., Schiedner, G., Kochanek, S., Tsai, S. Y., and O'Malley, B. W. (1999) Adenovirus-mediated regulable target gene expression in vivo. *Proc. Natl. Acad. Sci. USA* **96,** 355–360.

28. Oligino, T., Poliani, P. L., Wang, Y., et al. (1998) Drug inducible transgene expression in brain using a herpes simplex virus vector. *Gene Ther.* **5,** 491–496.

29. Abruzzese, R. V., Godin, D., Burcin, M., et al. (1999) Ligand-dependent regulation of plasmid-based transgene expression in vivo. *Hum. Gene Ther.* **10,** 1499–1507.

30. Abruzzese, R. V., Godin, D., Mehta, V., et al. (2000) Ligand-dependent regulation of vascular endothelial growth factor and erythropoietin expression by a plasmid-based autoinducible GeneSwitch system. *Mol. Ther.* **2,** 276–287.

31. Cheung, J. and Smith, D. F. (2000) Molecular chaperone interactions with steroid receptors: an update. *Mol. Endocrinol.* **14,** 939–946.
32. Spitz, I. M., Bardin, C. W., Benton, L., and Robbins, A. (1998) Early pregnancy termination with mifepristone and misoprostol in the United States. *N. Engl. J. Med.* **338,** 1241–1247.
33. *http://www.fda.gov/cber/ptc/plasmid.txt* Points to consider on plasmid DNA vaccines for preventive infectious disease indications, Food and Drug Administration, docket no. 96N-0400, December 1996.

10

Photochemical Transfection

Light-Induced, Site-Directed Gene Delivery

Lina Prasmickaite, Anders Høgset, and Kristian Berg

1. Introduction

Most nonviral gene transfection vectors deliver transfecting DNA into cells through the endocytic pathway *(1,2)*. Poor escape from endocytic vesicles in many cases constitutes a major barrier for delivery of a functional gene, since the endocytosed transfecting DNA is unable to reach the cytosol and be further transported to the nucleus, but rather is trapped in endocytic vesicles and finally degraded in lysosomes *(3)*. Therefore, the development of endosome-disruptive strategies is of great importance for the further progress of gene transfection. We have developed a new technology, termed photochemical internalization (PCI), to achieve light-inducible permeabilization of endocytic vesicles *(4–8)*. The technology is based on photochemical reactions initiated by photosensitizers localized in endocytic vesicles and inducing rupture of these vesicles upon light exposure *(4)*. This leads to the release of endocytosed macromolecules such as transfecting DNA from endocytic vesicles into the cytosol (**Fig. 1**). As a light-dependent treatment, PCI-mediated transfection (photochemical transfection) allows the possibility of directing the gene delivery to a desired site, e.g., achieving tumor-specific expression of a therapeutic gene in gene therapy in vivo.

1.1 Photosensitizers and Photochemical Reactions

1.1.1. General Principles

Photosensitizers (S) are compounds that upon absorption of light become excited to an excited state (S*), which initiates further photochemical reac-

From: *Methods in Molecular Medicine, Vol. 69, Gene Therapy Protocols, 2nd Ed.*
Edited by: J. R. Morgan © Humana Press Inc., Totowa, NJ

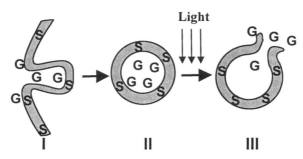

Fig. 1. Principle of photochemical internalization (PCI). I, endocytosis of the photosensitizer (S) and the transfecting gene (G); II, localization of the photosensitizer and the transgene in the same endocytic vesicles; III, rupture of endosomal membrane upon light exposure and subsequent release of the transfecting gene into the cytosol.

tions. The photochemical reactions may proceed via a type I process, with exchange of electrons between the photosensitizer and biomolecules, or a type II process, inducing formation of singlet oxygen ($^1O_2$). It is acknowledged that the photochemical reactions induced by the photosensitizers used in photodynamic therapy (PDT) and described here occur via type II reactions. Singlet oxygen is a highly reactive form of oxygen formed after interaction between the excited state of the photosensitizer (S*) and triplet ground-state molecular oxygen ($^3O_2$) *(9,10)*. Schematically:

$$S + light \rightarrow S^* \diagdown \begin{array}{l} ^3O_2 \\ ^1O_2 \rightarrow oxidative\ reactions \end{array}$$

Singlet oxygen can further oxidize various biomolecules such as unsaturated fatty acids, some amino acids, and guanine, inducing damage into various cellular structures *(11)*. However, singlet oxygen has a very short lifetime (<0.1 µs) and a short range of action (10–20 nm) *(12)*; therefore only targets close to the generated singlet oxygen will be photooxidized and damaged upon light exposure, leaving distant molecules intact. This fact is very important for the PCI technology, since the macromolecules to be released from the endocytic vesicles upon the photochemical treatment should stay intact and must maintain their biologic function.

1.1.2. Intracellular Localization of the Photosensitizers

Different photosensitizers, depending on their physicochemical properties, enter and localize inside the cell differently. Photosensitizers that enter the cell via endocytosis (like photosensitizers that aggregate) and hydrophilic photo-

TPPS$_{2a}$ **AlPcS$_{2a}$**

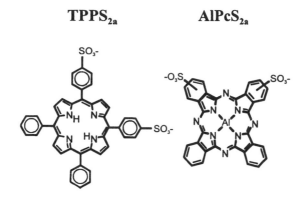

Fig. 2. Chemical structure of TPPS$_{2a}$ and AlPcS$_{2a}$.

sensitizers such as di- and tetrasulfonated tetraphenyl porphines (TPPS$_n$) *(13,14)* and aluminum phthalocyanines (AlPcS$_n$) *(15–17)* tend to accumulate in endocytic vesicles. Exposure to light activates such photosensitizers leading to the photochemical damage of endocytic vesicles and subsequent leakage of the vesicular content into the cytosol *(13,18)*. However, we have shown that not all photosensitizers that localize in endocytic vesicles are equally efficient in photochemical transfection. The most potent photosensitizers for photochemical transfection found so far seem to be TPPS$_{2a}$ and AlPcS$_{2a}$ (Prasmickaite et al., submitted). TPPS$_{2a}$ and AlPcS$_{2a}$ have two sulfonate groups on adjacent phenyl or phthalate rings (**Fig. 2**), making the photosensitizer molecules amphiphilic; therefore, they primarily localize not in the matrix but in the membranes of endocytic vesicles *(13,19)*. Upon light exposure, such photosensitizers destroy mainly vesicular membranes, whereas the content of the organelles (e.g., transfecting DNA) remains less affected.

1.1.3. Photochemically Induced Cytotoxicity and Photodynamic Therapy of Cancer

It should be mentioned that photochemical treatment in general induces cytotoxic effects and reduces cell survival. This fact has been successfully applied for therapeutic purposes in PDT, a cancer treatment modality in which the light exposure leads to photosensitizer-induced cytotoxic effects that kill the cancer cells *(20,21)*. In vivo many photosensitizers preferentially accumulate in tumor tissue compared with surrounding normal tissue *(20)*. The increased tumor uptake and retention of the photosensitizers seems to be related to certain properties of tumor tissue such as leaky vasculature, poor lymphatic drainage, the uptake by tumor-infiltrating macrophages of aggregated photosensitizers, and low pH in tumors *(20,21)*. A more specific uptake via the low-density lipoprotein (LDL) pathway might also play a role, since malignant cells, expressing elevated levels of LDL receptors, may take up lipophilic photosensitizers that bind to the LDL in the bloodstream *(22)*.

The level of cytotoxicity depends on the light dose and the properties of the photosensitizer, such as its intracellular localization. Thus, for example, lyso-somal rupture mediated by light-activated lysosomally localized photosensi-tizers seems not to be lethal *per se*, i.e., cells can survive a partial release of their lysosomal content induced by photochemical treatment *(13,18)*. This fact is important for photochemical transfection, since photochemically treated transfected cells still must be able to express a transgene.

1.2. Potential of the PCI Method for In Vivo Gene Therapy: Advantages and Disadvantages

So far photochemical transfection has been applied only for transfection of cells in culture; however, the advantages that this method could potentially provide make it particularly interesting for gene therapy in vivo.

The dependence of photochemical transfection on light treatment allows the possibility of directing the expression of a therapeutic gene to a specific loca-tion in the body, since only areas exposed to light will exhibit the increased transfection. For many kinds of therapeutic genes, like "suicide genes" or genes stimulating antitumor immune response (e.g., cytokine genes), the site-spe-cific expression achievable by photochemical transfection can be very advan-tageous, because it strongly reduces the risk of unwanted effects of the therapeutic gene outside the disease area.

Limited light penetration in tissues is a disadvantage of the method for in vivo use, since cells distant from the light source will not be affected by the treatment. Based on experience from PDT efficient light penetration, exploit-able for PCI, up to 1 cm could be expected *(23)*. However, this can also be considered an advantage since the limited light penetration means that the increased gene expression can be effectively confined to specific locations in the body. Moreover, illumination by interstitial fibers or via internal cavities or blood vessels by means of optical fiber devices allows cellular targets to be reached at many different locations in the body, e.g., in the lungs, pancreas, gastrointestinal tract, or brain *(20)*.

A potential drawback for both in vitro and in vivo applications of photo-chemical transfection is that the photochemical treatment induces cytotoxicity. We have previously shown that maximum transfection in surviving cells is achieved at photochemical doses killing about 50% of all cells *(5)*. In several clinical situations this cell killing cannot be tolerated; however, in many cases, e.g., for cancer therapy, it represents no problem or may even be beneficial. Also, in many gene therapy approaches immunologic responses or "bystander effects" *(24–26)* will play a major role, possibly making it unnecessary with very high transfection efficiency to achieve the desired therapeutic response. Furthermore, peptide sequences that can confer transport of therapeutic gene products into neighboring nontransfected cells have recently been described *(27)*. In many of these settings it may be more important to have a strict specificity in gene delivery rather than to achieve maximally high transfection efficiency, potentially making the specificity obtainable with photochemical transfection very valuable. It should also be noted that the technology is still far from optimal, and it is conceivable that under certain conditions a high efficiency of transfection could be achieved while maintaining cell viability.

In summary, compared with other gene delivery methods, photochemical transfection has several potential advantages: 1) it is a site-specific method, i.e., functional genes will be delivered and expressed only in areas that are exposed to light, reducing the risk of side effects; 2) the size of the nucleic acids to be delivered is not restricted; thus it should be functional with very large DNA molecules, as well as with DNA- or RNA-based oligonucleotides; and 3) the method can probably be combined with most other means for generating site and tissue specificity.

1.2.1. Photochemical Internalization In Vivo

We have already demonstrated the in vivo potential of the PCI technology in an animal model, by using the method to deliver the plant toxin gelonin to subcutaneous tumors (Selbo et al., submitted). Normally the toxicity of gelonin is limited because it is incapable of escaping from endocytic vesicles; therefore, if administrated alone, gelonin shows no effect on tumor growth. However, the cytotoxic potential of gelonin was increased tremendously by the photochemical treatment, which induces photochemical internalization of gelonin, resulting in complete tumor regression in 67% of treated mice. The conditions for administration of the photosensitizer and light treatment established in this study could also be used for the in vivo photochemical internalization of genes, since generally the optimal conditions for the photochemical treatment are largely independent of the molecules to be internalized.

2. Materials

2.1. Cell Culture

1. Adherent cells to be transfected, cultured according to recommendations. Nonadherent cells may also be used but have not so far been evaluated for photochemical transfection.
2. Cell growth medium supplemented with 10% fetal calf serum (FCS), 100 U/mL penicillin, 100 μg/mL streptomycin, and 2 mM glutamine (all Bio Whittaker,

Walkersville, MD). Store at 4°C. (We routinely use RPMI-1640 medium; however, other culture media can be used if recommended for the cells to be transfected.)

2.2. Stock Solution of the Photosensitizer AlPcS$_{2a}$

1. The photosensitizer aluminium phthalocyanine with two sulfonate groups on adjacent phthalate rings (AlPcS$_{2a}$; Porphyrin Products, Logan, UT).
2. 0.1 M NaOH.
3. Phosphate-buffered saline (PBS): 0.2 g/L KCl, 0.2 g/L KH$_2$PO$_4$, 8.0 g/L NaCl, 1.15 g/L Na$_2$HPO$_4$. Sterilize by filtration and store at 4°C.

2.3. Preparation of DNA/poly-L-Lysine Complex

1. Plasmid DNA, stock solution at 5 mg/mL in TE buffer (10 mM Tris-HCl, pH 7.5, 1 mM EDTA). Sterilize by filtration and store at –20°C.
2. Poly-L-lysine hydrobromide (MW 20,700; Sigma, St. Louis, MO). To make stock solution (1 mg/mL), dissolve 1 mg poly-L-lysine in 1 mL distilled water. Sterilize by filtration and store at 4°C.
3. Distilled water. Autoclave and store at 4°C.
4. Sterile polypropylene microcentrifuge tubes.

2.4. Photochemical Transfection

1. 6-well cell culture plates (cat. no. 3506, Costar),
2. Sterile tubes.
3. Light source for excitation of AlPcS$_{2a}$ at 670 nm (*see* **Note 1**).

3. Methods

The following protocol is for photochemical transfection of adherent cells in a 6-well culture plate (well diameter 3.5 cm). Amounts and volumes presented below are calculated for one such well but can be adjusted accordingly to any other cell culture plate, cell number, and volume.

3.1. Preliminary Experiments to Find the Light Dose

The light dose to be used for induction of photochemical transfection has to be found in advance. Thus cell survival as a function of light dose has to be measured individually for every cell line and for every light source; the light dose killing about 50% of the cells is recommended as a starting point (*see* **Note 1**). Cell survival can be measured by one of the common cell survival tests such as the 3-(4,5-dimethylthiazol-2-yl)-2,5-diphenyltetrazolium bromide) (MTT) test, clonogenic analysis, protein synthesis, or another test established in an individual laboratory.

3.2. Preparation of Adherent Cells for Photochemical Transfection

The cells are seeded out 1 day before the experiment. In a 6-well culture plate, seed cells in 2 mL/well of growth medium containing 10% FCS. The number of cells seeded out per well depends on the cell size and cell growth, however, the cells are usually 50–70% from confluency at the start of the incubation with the DNA complex. (Higher and lower cell densities may also be used; therefore adjust the seeding density for every cell line; *see* **Note 2**).

3.3. Preparation of AlPcS$_{2a}$ solutions

Work in subdued light (*see* **Note 3**).

3.3.1. Preparation of Stock Solution (5 mg/mL)

1. Dissolve 5 mg of AlPcS$_{2a}$ in a small volume (approx. 0.1–0.2 ml) of 0.1 M NaOH.
2. Dilute with PBS to a final volume of 1 mL. (The photosensitizers usually dissolve well this way. If higher concentrations are needed or complete solubilization of the photosensitizer is difficult, we recommend sonicator bath treatment of the solution.)
3. Sterilize by filtration, and store at –20°C in small aliquots for up to 6 months. Can be used several times after thawing and freezing (*see* **Note 3**).

3.3.2. Preparation of Working Solution (20 µg/mL)

Working solution should be freshly made before application to the cells. Add 4 µL of AlPcS$_{2a}$ stock solution (5 mg/mL) to 1 mL of growth medium containing 10% FCS.

3.4. Preparation of the Plasmid DNA/Poly-L-Lysine Complex

The complex should be freshly prepared before application to the cells (*see* **Note 4**).

We routinely make the complex with a charge ratio of 1.7 (*see* **Note 5**); therefore, the amounts presented below correspond to a complex with a charge ratio of 1.7.

1. Prepare plasmid DNA solution in a separate sterile microcentrifuge tube: 5 µg of plasmid DNA (we use 1 µL of DNA stock solution, 5 mg/mL) diluted to 75 µL in sterile water. Gently mix by pipetting the solution several times. Do not vortex.
2. Prepare poly-L-lysine solution in a separate sterile microcentrifuge tube: 5.33 µg of poly-L-lysine (i.e., 5.33 µL of stock solution, 1 mg/mL) diluted to 75 µL in sterile water. Gently mix by pipetting the solution several times. Do not vortex.
3. Slowly transfer the poly-L-lysine solution (*see* **step** 2) into the microcentrifuge tube containing the DNA solution (*see* **step 1**). The final volume of the mixture is 150 µL. Gently mix by pipetting the mixture several times. Do not vortex.

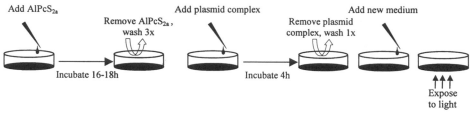

Fig. 3. Experimental scheme.

4. Leave the mixture on the bench at room temperature for 30 min to allow formation of the DNA/poly-L-lysine complex.
5. Transfer the whole DNA/poly-L-lysine mixture into another sterile tube containing growth medium to bring the final volume to 1 mL. Mix carefully by pipetting
   several times. The solution is now ready for application to the cells (*see* **Note 6**).

3.5. Photochemical Transfection

All the procedures starting from here should be carried out in subdued light
(*see* **Note 7**). A simplified experimental scheme is presented in **Fig. 3**.

1. Remove the growth medium from cells seeded out 1 day before the experiment
   (*see* **Subheading 3.2.**) and add 1 mL of growth medium containing 20 µg/mL
   $AlPcS_{2a}$ (*see* **Subheading 3.3.2.**).
2. Incubate overnight, i.e., for 16–18 h at 37°C in a CO_2 incubator.
3. Remove the medium containing $AlPcS_{2a}$ and wash the monolayer three times
   with growth medium.
4. Add growth medium containing freshly prepared plasmid DNA/poly-L-lysine
   mixture (*see* **Subheading 3.4.**).
5. Incubate at 37°C in a CO_2 incubator for the desired time. We usually incubate for
   4–6 h (*see* **Note 8**).
6. Remove the medium containing the DNA/poly-L-lysine complex and wash the
   cells once with growth medium.
7. Add 2 mL of growth medium.
8. Expose the cells to light. The optimal light exposure time depends on the cell line
   used; the light source and should therefore be determined individually (*see* **Note 1**).
9. Grow the cells further in the dark for 2 days (*see* **Note 9**) and analyze for transgene
   expression (*see* **Note 10**).

4. Notes

1. The light source used in our laboratory so far was homemade many years ago,
   and this complicates the reconstruction of an identical light source. However,
   any light source with suitable characteristics (see below) could be used. For every
   cell line to be transfected the treatment should be calibrated, since the light delivered from different lamps will be different and also the optimal light dose will

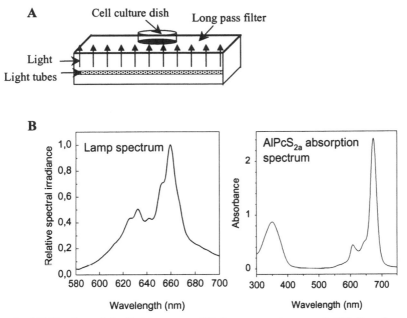

Fig. 4. **(A)** Design of the light source. **(B)** Lamp spectrum and AlPcS$_{2a}$ absorption spectrum.

vary between different cell lines. We routinely do calibration by measuring cell viability using one of the common cell survival tests (depending on the cell line), such as the MTT test, clonogenic assay, or measurement of protein synthesis. Generally a light dose killing about 50% of the cells would be a good starting point for photochemical transfection. We usually use light doses that give from 0 to 50% cell killing. In designing the light source the following points should be taken into consideration:

a. The light source should deliver light of a wavelength suitable for excitation of AlPcS$_{2a}$, which has an absorbtion peak at 670 nm (**Fig. 4B**). It is also an advantage to avoid too much UV light, since UV light may induce cytotoxic effects. However, AlPcS$_{2a}$ absorbs UVA light, and the UVA light doses needed for inducing photochemical reactions by means of AlPcS$_{2a}$ will usually be too low to induce cytotoxic reactions alone. If necessary, a filter excluding UV light may be employed. Furthermore, infrared emission from the lamp may lead to hyperthermia. It is therefore necessary initially to test the irradiation from the lamp for cytotoxic effects on the cells. We recommend that the light doses employed should be at least 3 times less than what causes toxic effects. If light-induced toxicity is experienced, we recommend the use of a fan or another cooling device. So far we have used a bench with four light tubes (model TL 20W/09, Philips) and a long-pass filter, with a cutoff at 550–600 nm (**Fig. 4A**). The light intensity, reaching the cells placed above the light tubes on the filter, is 13.5 W/m^2.

b. The light source should generate a homogeneous light field in the area where the cells are illuminated.

c. Although light penetrates the plastic dish before reaching the cell monolayer (**Fig. 4A**), the type of plastic dish does not seem to influence the results.

2. It should be noted that photochemical treatment is toxic for the cells, whereas only surviving cells can express the photochemically transfected gene. Since maximum transfection efficiency in the surviving cells is achieved at photochemical doses killing about 50% of all cells (i.e., at D_{50}), it should be taken into consideration that under the conditions giving the maximum transfection efficiency 50% of all cells will be lost. This might be important when deciding on the amount of cells to start the experiment with in order to have enough cells at the end, for example, for measuring the expressed product of the transgene.

3. $AlPcS_{2a}$ is a relatively photostable photosensitizer; however, we recommend protection of the photosensitizer solutions from light to avoid possible light-inducible damage to the photosensitizer itself. The photosensitizers may also aggregate after long-time storage or repeated freezing and thawing. Aggregation will reduce the efficacy of the photosensitizer and should be avoided. Usually the color of the stock solution of $AlPcS_{2a}$ is dark blue, and change of the color is usually a bad sign indicating aggregation or decomposition of the photosensitizer. In this case the quality of the stock solution should be checked by measuring the phototoxic effect on cells using cell viability tests.

4. Other nonviral cationic transfection vectors than polylysine could also be tested for application in photochemical transfection. We have tried the cationic polypeptide polyarginine, the cationic polymer polyethylenimine (PEI), different cationic lipids like dioleoyl-trimethyl ammonium propane (DOTAP), lipofectin, and others. In general, transfection mediated by polycationic vectors that did not show high efficiency alone in our cells (like polyarginine or PEI) was stimulated by light treatment. The effect of light on cationic lipid-mediated transfection was cell line dependent; transfection was stimulated in some cell lines and inhibited in other. Therefore readers are encouraged to try different transfection vectors applying the same main principles of photochemical transfection.

5. We routinely use DNA/poly-L-lysine complexes with a charge ratio of 1.7 (*see* **Subheading 3.4.**). However, charge ratios in the range 1.0–2.5 gave similar efficiency of photochemical transfection *(5)*; therefore the readers might try other charge ratios as well. (The charge ratio is the number of positive charges provided by the amino groups of polylysine divided by the negative charges provided by the phosphate groups of DNA. DNA at 1 μg has 3.03 nmol of phosphate groups, which are neutralized by the addition of 0.63 μg polylysine carrying 3.03 nmol of amino groups, so that a charge ratio of 1.0 is obtained.)

6. The presence of FCS in the media during transfection with polylysine complexes increases the transfection efficiency by about twofold *(5)*, but the method also works in serum-free media. This fact can be important when using cationic lipids as transfection vectors, where serum-free medium is usually recommended.

7. It is important to work in subdued light after the step in which the photosensitizer

AlPcS$_{2a}$ is added to the cells, to avoid uncontrollable activation of the photosensitizer and to protect the cells from undesirable photochemical damage. Therefore all procedures should be carried on in a room with a subdued light, and cell culture plates should be protected from direct light (e.g., may be packed in aluminum foil).

8. To reduce damage to the plasma membrane, it is not recommended to expose the cells to light until at least 4 h after the removal of AlPcS$_{2a}$. Therefore, we routinely incubate the cells with the DNA complex in AlPcS$_{2a}$-free medium for 4–6 h (*see* **Subheading 3.5., step 5**) to ensure that most of the AlPcS$_{2a}$ bound to the plasma membrane is washed out or internalized into the cell before irradiation. This ensures that the light exposure does not induce extensive photochemical damage to the plasma membrane, which is lethal for cells, and that the main effect will be due to induced rupture of endocytic vesicles, where AlPcS$_{2a}$ localizes. However, shorter incubation with the DNA complex (e.g., a 0.5–1-h pulse) is also possible, but in this case the cells should be chased first in photosensitizer-free medium before the incubation with the DNA complex, so that the total incubation time in AlPcS$_{2a}$-free medium before irradiation is not shorter than 4 h. It should be noted that the time needed to remove the bulk of photosensitizer from the plasma membrane varies between cell lines, but usually 4 h of incubation in photosensitizer-free medium has been found to induce efficient photochemical internalization of macromolecules.

9. It should be noted that photochemical treatment might temporarily inhibit transcription and translation *(28,29)*; thus it may delay transgene expression. This should be kept in mind for deciding when after the treatment to measure the expression. Generally, we analyze transgene expression 1–3 days after light treatment, but this will depend on the transgene used, the cell line, the promoter employed, and so on.

10. In the present text we have described the use of AlPcS$_{2a}$, the photosensitizer with which we have the most experience in photochemical transfection. However, some other photosensitizers may also be used, e.g., we have shown that photochemical transfection works equally well with TPPS$_{2a}$, whereas TPPS$_4$ and Photofrin are less efficient (Prasmickaite et al., submitted). If other photosensitizers are to be used it may, however, be necessary to use other light sources than for AlPcS$_{2a}$, to match the spectral characteristics of the different photosensitizers.

References

1. Zabner, J., Fasbender, A. J., Moninger, T., Poellinger, K. A., and Welsh, M. J. (1995) Cellular and molecular barriers to gene transfer by a cationic lipid. *J. Biol. Chem.* **270**, 18997–19007.
2. Friend, D. S., Papahadjopoulos, D., and Debs, R. J. (1996) Endocytosis and intracellular processing accompanying transfection mediated by cationic liposomes. *Biochim. Biophys. Acta* **1278**, 41–50.
3. Luo, D. and Saltzman, W. M. (2000) Synthetic DNA delivery systems. *Nat. Biotechnol.* **18**, 33–37.

4. Berg, K., Selbo, P. K., Prasmickaite, L., et al. (1999) Photochemical internalization: a novel technology for delivery of macromolecules into cytosol. *Cancer Res.* **59**, 1180–1183.

5. Høgset, A., Prasmickaite, L., Tjelle, T. E., and Berg, K. (2000) Photochemical transfection: a new technology for light-induced, site-directed gene delivery. *Hum. Gene Ther.* **11**, 869–880.

6. Prasmickaite, L., Høgset, A., Tjelle, T. E., et al. (2000) Role of endosomes in gene transfection mediated by photochemical internalisation (PCI). *J. Gene Med.* **2**, 477–488.

7. Selbo, P. K., Sivam, G., Fodstad, O., Sandvig, K., and Berg, K. (2000) Photochemical internalisation increases the cytotoxic effect of the immunotoxin MOC31-gelonin. *Int. J. Cancer* **87**, 853–859.

8. Selbo, P. K., Sandvig, K., Kirveliene, V., and Berg, K. (2000) Release of gelonin from endosomes and lysosomes to cytosol by photochemical internalization. *Biochim. Biophys. Acta* **1475**, 307–313.

9. Moan, J. and Sommer, S. (1985) Oxygen dependence of the photosensitizing effect of hematoporphyrin derivative in NHIK 3025 cells. *Cancer Res.* **45**, 1608–1610.

10. Weishaupt, K. R., Gomer, C. J., and Dougherty, T. J. (1976) Identification of singlet oxygen as the cytotoxic agent in photoinactivation of a murine tumor. *Cancer Res.* **36**, 2326–2329.

11. Jori, G. and Spikes, J. D. (1984) Photobiochemistry of porphyrins, in *Topics in Photomedicine* (Smith, K. C., ed.), Plenum, New York, pp.183–318.

12. Moan, J. and Berg, K. (1991) The photodegradation of porphyrins in cells can be used to estimate the lifetime of singlet oxygen. *Photochem. Photobiol.* **53**, 549–553.

13. Berg, K. and Moan, J. (1994) Lysosomes as photochemical targets. *Int. J. Cancer* **59**, 814–822.

14. Berg, K., Western, A., Bommer, J. C., and Moan, J. (1990) Intracellular localization of sulfonated meso- tetraphenylporphines in a human carcinoma cell line. *Photochem. Photobiol.* **52**, 481–487.

15. Peng, Q., Nesland, J. M., Madslien, K., Danielsen, H. E., and Moan, J. (1996) Subcellular photodynamic action sites of sulfonated aluminium phthalocyanines in a human melanoma cell line. *SPIE* **2628**, 102–113.

16. Moan, J., Berg, K., Kvam, E., et al. (1989) Intracellular localization of photosensitizers. *Ciba Found. Symp.* **146**, 95–107.

17. Moan, J., Berg, K., Bommer, J. C., and Western, A. (1992) Action spectra of phthalocyanines with respect to photosensitization of cells. *Photochem. Photobiol.* **56**, 171–175.

18. Moan, J., Berg, K., Anholt, H., and Madslien, K. (1994) Sulfonated aluminium phthalocyanines as sensitizers for photochemotherapy. Effects of small light doses on localization, dye fluorescence and photosensitivity in V79 cells. *Int. J. Cancer* **58**, 865–870.

19. Maman, N., Dhami, S., Phillips, D., and Brault, D. (1999) Kinetic and equilib-

rium studies of incorporation of di-sulfonated aluminum phthalocyanine into unilamellar vesicles. *Biochim. Biophys. Acta* **1420,** 168–178.

20. Pass, H. I. (1993) Photodynamic therapy in oncology: mechanisms and clinical use. *J. Natl. Cancer Inst.* **85,** 443–456.
21. Henderson, B. W. and Dougherty, T. J. (1992) How does photodynamic therapy work? *Photochem. Photobiol.* **55,** 145–157.
22. Maziere, J. C., Morliere, P., and Santus, R. (1991) The role of the low density lipoprotein receptor pathway in the delivery of lipophilic photosensitizers in the photodynamic therapy of tumours. *J. Photochem. Photobiol.* **8,** 351–360.
23. Wan, S., Parrish, J. A., Anderson, R. R., and Madden, M. (1981) Transmittance of nonionizing radiation in human tissues. *Photochem. Photobiol.* **34,** 679–681.
24. Bateman, A., Bullough, F., Murphy, S., et al. (2000) Fusogenic membrane glycoproteins as a novel class of genes for the local and immune-mediated control of tumor growth. *Cancer Res.* **60,** 1492–1497.
25. Kianmanesh, A. R., Perrin, H., Panis, Y., et al. (1997) A "distant" bystander effect of suicide gene therapy: regression of nontransduced tumors together with a distant transduced tumor. *Hum. Gene Ther.* **8,** 1807–1814.
26. Dilber, M. S. and Smith, C. I. (1997) Suicide genes and bystander killing: local and distant effects. *Gene Ther.* **4,** 273–274.
27. Phelan, A., Elliott, G., and O'Hare, P. (1998) Intercellular delivery of functional p53 by the herpesvirus protein VP22. *Nat. Biotechnol.* **16,** 440–443.
28. Moan, J., McGhie, J., and Jacobsen, P. B. (1983) Photodynamic effects on cells in vitro exposed to hematoporphyrin derivative and light. *Photochem. Photobiol.* **37,** 599–604.
29. Davies, C. L., Ranheim, T., Malik, Z., Rofstad, E. K., Moan, J., and Lindmo, T. (1988) Relationship between changes in antigen expression and protein synthesis in human melanoma cells after hyperthermia and photodynamic treatment. *Br. J. Cancer* **58,** 306–313.

11

Direct Gene Transfer and Vaccination Via Skin Transfection Using a Gene Gun

Chia-Feng Kuo, Jeng Hwan Wang, and Ning-Sun Yang

1. Introduction

Gene gun technology provides a useful means for direct transfer of DNA or RNA constructs that can result in transgenic protein expression from gene expression vectors *(1–8)*. The system has been applied to a broad spectrum of experimental studies on transgenic research, gene therapy approaches, and genetic vaccinations (*see* **Note 1**). The technology was first reported by Yang et al. in 1990 *(1)* for in vivo and in vitro gene transfer into mammalian somatic tissues and was later extended to various ex vivo gene transfer systems, including excised tissue explants or clumps, organic tissues placed in culture vessels, and their derived primary cultures *(9; see* **Note 2**).

During the past decade, the progressive development of gene gun devices and optimization of delivery parameters have created new methodologies for biologic research, such as assay of in vivo promoter activity in mammalian tissues *(2; see* **Note 3**), genetic immunization or nucleic acid vaccination *(10–16)*, transfection of cDNA expression libraries and profiling/elucidation of corresponding antibody species and immune protection against known or unknown pathogenic antigens *(17,18)*, cancer gene therapy and cancer vaccines *(15,19–24)*, skin wound-healing studies, and transgenic research into various leukocytes and hematopietic progenitor cells *(25,26; see* **Notes 4–7**). It is important to note that both humoral and cell-mediated immune responses, as well as Th1 and Th2 T-cell activity, can be effectively elicited by using gene gun-mediated vaccination of skin tissue in both small and large animals *(27)*, and in humans *(28,29; see* **Note 6**) An update, upgrade, and confirmation of the art of the

From: *Methods in Molecular Medicine, Vol. 69, Gene Therapy Protocols, 2nd Ed.*
Edited by: J. R. Morgan © Humana Press Inc., Totowa, NJ

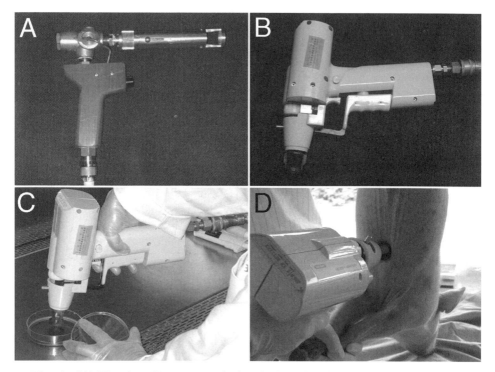

Fig. 1. (A) The Accell gene gun device designed and created by Dr. D. McCabe and associates of Agracetus, Inc. (Middleton, WI) and distributed as a proprietary instrument by PowderJect Vaccine, Inc. (Madison, WI). **(B)** The Helios gene gun device, developed as a modification of the Accell device, is now produced commercially by Bio-Rad. **(C)** In vitro gene delivery into attached monolayer cells using the Helios gene gun. **(D)** In vivo gene transfer into pig skin tissue using the Helios gene gun.

particle bombardment method for gene transfer are therefore desirable and are reported on here.

A number of gene gun devices were developed in the early stages of this technology *(7)*. Then the Acell gene gun (a hand-held, helium gas pressure-propelled gene delivery instrument designed by Dr. D. McCabe) was distributed under proprietary agreement by Aqracetus, Inc. and later by PowderJect Vaccine, Inc. (Madison, WI) **(Fig. 1A)**. The Helios gene gun **(Fig. 1B–D)** (Bio-Rad, Hercules, CA) became available to public users in late 1997 and is still the only standardized and routinely applicable gene gun device that is currently available commercially. A helium gas shockwave is used to propel the high-density DNA-coated gold particles to a very high velocity, efficiently enough to penetrate the cell membranes of targeted cells/tissues, resulting in direct physical delivery of plasmid DNA or RNA at high copy numbers and dry form. In vivo gene transfer has been used most successfully for skin gene transfer. The operational procedure related to this method of gene transfer is reasonably simple and reliable.

2. Materials
2.1. Instrumentation

1. The Dermal PowderJect-XR helium pulse gun (trademark of PowderJect Vaccine), formerly known as the Acell, gene gun, is now commercially available from Bio-Rad as the Helios gene gun. Instructions for the mechanical operation and/or manipulation of the gene gun and associated devices are given in the manufacturer's manual.
2. Compressed helium of grade 4.5 or higher.
3. Hearing protection device.

2.2. Elemental Gold Particle Preparations

Microscopic gold particles can be purchased from Degussa (South Plainfield, NJ). Gold particles of 2–3 μm were found to be the best for in vivo gene transfection into the skin *(8)*, 0.7–1-μm gold beads for in vitro or ex vivo transfection of various leukocytes, lymphocytes, and other small cells in suspension, and 1–2-μm gold beads for adhesive monolayer cells or cell aggregates and tissue clumps in culture (*see* **Notes 1** and **2**). It is important that these gold particles be obtained as elemental gold, not as gold salt or colloidal gold. If necessary, the gold particles can be washed and cleaned by rinsing them in distilled water, 70% ethanol, and 100% ethanol in sequence prior to use, and they can also be sterilized in phenol or $CHCl_3$ if necessary. It is important to examine each newly purchased gold particle preparation microscopically, making sure that the lot, particle size, and form are correct and appropriate as desired for the test systems.

2.3. DNA Vectors

A clean plasmid DNA dissolved in TE buffer (10 m*M* Tris-HCl, pH 7.0, 1 m*M* EDTA) or distilled water should be used for coating particles. Cocktails of different DNA vector systems or preparations can be mixed in desired molar ratios in aqueous solution and then effectively loaded onto gold particles as follows (*see* **Note 8**).

For exploratory gene transfer experiments, convenient and sensitive reporter gene systems that have low endogenous activity background (e.g., Luciferase) are recommended for verification of transient gene expression systems.

2.4. Coating of DNA onto Gold Particles and Gene Transfer

1. 0.05 M spermidine or polyethyleneglycol (PEG) in H_2O. Use fresh spermidine made weekly from a free-base solution (Sigma, St. Louis, MO).
2. 1 mg/mL polyvinylpyrrolidone (PVP; Sigma).
3. 2.5 M $CaCl_2$ in H_2O.
4. 100% ETOH kept at –20°C.

2.5. Treatment and Care of Skin

1. Oster electric hair clippers with no. 40 blade (Fisher, Pittsburgh, PA).
2. Tegaderm adhesive (3M, St. Paul, MN) or Scotch tape (3M).

2.6. Tissue Extraction Buffers and Enzyme Assay Systems

2.6.1. Luciferase Assay

1. Extraction buffer: 100 µL of cold 10% Triton X-100 in 9.9 mL of cold phosphate-buffered saline (PBS).
2. Luciferase assay substrate (Promega, Madison, WI), dissolved in Luciferase assay buffer at a concentration of 13.33 mg/mL.
3. Luminometer (Lumat LB 9507, Berthold Systems, Pittsburg, PA).

2.6.2. X-Gal Assay

1. 5-Bromo-4-chloro-3-indolyl-8-D-galactoside (X-gal) substrate, dissolved in dimethyl formamide at a concentration of 40 mg/mL.
2. Buffer solution: 44 mM HEPES, 15 mM NaCl, 1.3 mM $MgCl_2$, 3 mM K^+ ferricyanide, 3 mM K^+ ferrocyanide, pH 7.4.
3. X-gal buffer: Add X-gal substrate to buffer solution, making a final solution of 1 mg/mL.

2.6.3. hAAT Assay

1. Antibody dilution buffer: 5% fetal bovine serum (FBS) in PBS.
2. First antibody: Goat anti-human α_1-antitrypsin (Sigma) diluted in 50 mM carbonate buffer at a concentration of 3 µg/mL.
3. Second antibody: Rabbit anti-human α_1-antitrypsin (hAAT)(Sigma) at least 1000× diluted in dilution buffer.
4. Third antibody: Goat anti-rabbit IgG (H+L), peroxidase-conjugated (Pierce, Rockford, IL) at least 1000× diluted in dilution buffer.
5. Substrate solution: Add 10 µL H_2O_2 to 1 mL 3,3',5,5'-tetramethylbenzidine (TMB; Pierce) or 2,2'-azinobis(3-ethylbenzothiazoline-6-sulfonic acid)-diammonium salt (ABTS; Pierce).
6. Blocking solution: 0.5% bovine serum albumin (BSA) and 0.01% sodium azide in PBS.
7. Wash buffer: 0.05% Tween-20 in PBS, prepared freshly.

2.6.4. Cytokines and Other Transgenic Protein Products Assays

General extraction buffer: 9.5 mL PBS (0.2 g potassium chloride, 0.2 g potassium phosphate monobasic, 8.0 g sodium chloride, and 1.15 g sodium phosphate dibasic in 1 L distilled H_2O), 2.4 mg serine protease inhibitor Pefabloc® SC (4-[2aminoethyl]-benzenesulfonylfluoride hydrochloride; Roche), 0.5% Triton X-100, pH 7.2.

3. Methods

3.1. Coating DNA onto Gold Particles

1. Prepare DNA solution at a concentration of approx 1 μg/μL and store at 4°C.
2. Decide how many cartridges ("bullets") will be needed. One cartridge of the Helios gene gun device (conferring one transfection) contains approx 0.5 mg of gold particles.
3. Assuming that 40 cartridges will be needed, weight 21 g of gold particles into a 1.5-mL microcentrifuge tube.
4. Add 250 μL of 0.05 *M* spermidine or PEG to the tube with the gold particles (e.g., use 200–300 μL for 20–50 mg and 400–450 μL for 120 mg).
5. Vortex and sonicate for 3–5 s to break up gold clumps.
6. Add DNA at 2.5 DNA/gold loading rate (2.5 μg DNA/mg gold particle) to 21 mg of gold.
7. Add 250 μL of 2.5 *M* $CaCl_2$ (or the same volume as spermidine) dropwise while vortexing the tube at a low speed.
8. Incubate at room temperature for 10 min. While waiting, prepare a culture tube with 3 mL of 100% ETOH.
9. Microcentrifuge for 3–5 s. Remove and discard the supernatant. Break up the pellet by flicking the tube. Add 0.5 mL of cold 100% ETOH dropwise while gently vortexing the tube and then 0.5 mL more of 100% ETOH. Mix by inversion. Wash the pellet with cold ethanol 3 times.
10. Transfer particles into the culture tube containing 3 mL of 100% ETOH (7 mg gold particle/mL ETOH). Sonicate to disperse particles. Particle suspensions can be used immediately or stored at 4 or –20°C under stringent desiccation for 2–3 h.

3.2. Preparation of DNA Cartridges

1. Sonicate DNA-coated gold particle suspension for 2–3 s, vortex, add PVP at 0.01 mg/mL, sonicate again, and immediately load the suspension into the Tefzel tubing following the manufacturer's instruction (Bio-Rad).
2. Discard unevenly coated ends or portion of the tubing and cut the tubing into 0.5-inch pieces. Cartridges may be stored desiccated in a tightly sealed container at 4°C for several weeks.

3.3. Animal Care and Skin Treatment

1. Small experimental animals (including mouse, rat, hamster, and rabbit) usually do not need to be anesthetized for particle-mediated gene transfer into the skin. However, larger animals such as dogs and pigs in general need to be anesthetized before treatment by using ketamine (10 mg/kg body weight for intramuscular injection or 5 mg/kg for intravenous injection) or pentothal (12.5 mg/kg body weight for intramuscular injection or 5 mg/kg for intravenous injection) under the guidance of a consulting veterinarian as required by the Animal Welfare Act and administered by the institution's Animal Care and Use Committee.
2. Animal hair in the target area is removed with clippers and shearing blades. If a depilatory is used to remove stubble, animals should be anesthetized.

3.4. Epidermis and Dermis Gene Transfer (see Notes 1,5, and 6)

1. Allow the container with the DNA-gold cartridges to reach room temperature before opening.
2. Attach the regulator to the compressed helium tank. Connect the feed hose to the regulator and gene gun, and then plug in the device.
3. Load the prepared cartridges into the 12-chamber cartridge holder, following the manufacturer's instructions.
4. Put on a hearing protection device. Insert the loaded cartridge holder into the barrel of the gene gun device. Open the helium tank valve and the regulator valve.
5. Adjust the discharge pressure to the desired setting (usually 250–500 psi).
6. Restrain a mouse or other testing small animal in a hand or with a steady setup. Hold the nozzle of the device against the target skin area and discharge the device. If several skin transfections are required, turn the cartridge holder clockwise and discharge at another target skin area.
7. If desired, apply a semiocclusive skin dressing such as Tegaderm (3M) or Opsite (Smith and Nephew, Hull, UK) to the transfected epidermis.
8. To transfect dermal tissues, anesthetize a mouse, make an incision, and dissociate the full-thickness skin tissue of the target size (approx 3.2 cm^2) from the facies and muscle tissue using standard surgical procedures and tools. Flip the skin flap over, exposing the dermal tissue, and transfect the fibroblasts, muscle cells, and other stromal cell types. Moisten dermal tissues with sterile saline before closing. Close incision with sutures or wound clips.
9. Collect skin samples from target sites at a designated time point for transgene activities or sequences as described in **Subheading 3.5.**

3.5. Reporter Gene-Expression Assay (see Note 4)

3.5.1. Tissue Extraction

1. Collect transfected skin tissues from test animals by excising a small portion of the targeted skin area. A piece as small as 1–2 mm^2 can be enough for Luc or X-gal reporter gene expression assays. The transgenic expression products may also be stripped off the target sites of the skin with Scotch tape or duct tape.

2. Drop the freshly excised skin piece or tape into a tube with the appropriate buffer (0.2–0.5 mL) and keep on ice if performing the assay right away; otherwise freeze at –20°C.
3. Cut with scissors, grind, or homogenize the frozen skin or tape immediately before carrying out the assay. The samples should be kept on ice.
4. Sonicate and then centrifuge the sample at a high speed to separate the tissue or tape from the lysate. Crude tissues extracts are used for various reporter gene assays.

3.5.2. Luciferase Assay

1. After lysate has been collected, prepare dilutions in PBS if necessary.
2. Add 100 μL Luc assay substrate and 20 μL extraction buffer to the tube, and vortex.
3. Read relative light units in a luminometer and run a standard curve to quantify the results.

3.5.3. X-Gal Staining

1. Perform whole-mount tissue staining by placing the excised skin target into X-gal buffer. For best results, glue the skin tissue onto the surface of a 35-mm dish and keep the skin stretched out for better examination. The tissue can also be fixed in methanol/acetone (1:1) for 10 min and then stained with X-gal buffer.
2. Section the tissue into approx 10-μM sections by cryostat microtome or paraffin sectioning. Place the slide in a cold 1.5% glutaraldehyde solution for 10 minutes, and wash in cold PBS 5X for 5 min. Then stain the tissue slides with X-gal buffer for 1 hr. Avoid prolonged staining, because it can often result in a nonspecific greenish blue background that can develop in hair follicles of certain skin tissues.

3.5.4. hAAT Assay

1. Add 100 μL of skin tissue extract or tape-stripped cell extract to a 96-well plate precoated with anti-hAAT antibody.
2. Incubate at 37°C for 1 h. Add 90 μL of the second antibody (rabbit anti-hAAT, at least 1000× diluted), and incubate again for 1 h.
3. Wash 3–5 times with wash solution (300 μL/well).
4. Add 85 μL of the third antibody (anti-rabbit, at least 1000× dilution) and incubate for another 1 h. Wash 3–5 times.
5. Add 80 μL substrate solution. Allow color formation at room temperature for 1/2–1 h.
6. Measure transgenic hAAT levels by an enzyme-linked immunosorbent assay (ELISA) reader.

3.5.5. Cytokine Assay

ELISA tests are usually used for cytokine quantification. They may be purchased as a kit or antibody pair samples and used in a sandwich-style assay. The ELISAs can be run using skin crude tissue extract (with a general extraction buffer) or tape-stripped cell extract.

4. Notes

1. The gene gun method for gene transfer can be applied to a broad range of tissues and cell types; a key feature of this technology is its applicability to in vivo gene transfer to various organs, especially the epidermal skin cell layers *(1–4)*. The latter case thus permits a powerful nucleic acid- or gene-based vaccine strategy.

2. Two key technical advantages were observed for the particle-mediated gene delivery method: 1) a very wide DNA range (1 ng–10 µg/dose/transfection site) can be delivered into a 3–5-cm^2 surface area of targeted tissues, resulting in different efficacies of transgene expression, depending on target cells, tissues, or organ types as well as the in vitro, ex vitro, or in vivo experimental conditions; and 2) there is little or no restriction on the size or form of testing DNA, at least at a molecular size ≤40 kb as double-stranded DNA *(1–6,8)*.

3. The gene gun technique has been shown to provide an excellent experimental system for in vivo and in vitro assays of promoter strength, targeting various mammalian somatic tissues *(2,5)*.

4. Various transgenic proteins expressed by reporter transgenes (e.g., Luciferase and β-Gal), candidate therapeutic genes (e.g., glanulocyte macrophage-colony-stimulating factor or interleukin-12) or vaccine/antigen genes (e.g., influenza, foot and mouth disease, hepatitis B) have been readily detected at 0.6–2 ng for cytokines and luciferases, 0.2–5 µg or higher for relatively stable proteins such as human growth hormone or human α_1 anti-trypsin, and 10–200 pg for specific viral protein antigens (N.-S. Yang et al., unpublished data).

5. The high accessibility of the skin as an exposed tissue and the organization of the epidermis make the skin an excellent target for gene gun-mediated gene transfer, not only for genetic vaccination against infectious diseases, but potentially also for serving as a transgenic bioreactor for gene therapy approaches, including cytokine gene therapy for cancer and DNA cancer vaccines, using either in vivo or ex vivo gene delivery strategies *(9,30)*.

6. It is estimated that epidermal Langerhan cells and dendritic cells make up to 5% of the skin tissues, and it is believed that such cells are responsible for presenting the immunogens produced as transgenic proteins or peptides via skin transfection using a gene gun. This experimental system, applicable to small and large experimental animals and apparently also in humans (e.g., with U.S. Food and Drug Administration/National Institutes of Health authorized clinical trial studies) *(28, 29)*, may serve in the future as a highly valuable system for transgenic, cell biologic, and molecular studies of cellular and humoral immunities.

7. Gene gun technology has also been applied in clinical trials of human gene therapy as a cancer vaccine, as a hepatitis B DNA vaccine, and for HIV gene therapy approaches *(17,18,32,33)*.

8. Other than DNA and RNA *(8,20)*, peptide nucleic acid (PNA) *(31)* may also be effectively employed as a vector for gene-based vaccination using gene gun delivery.

References

1. Yang, N.-S, Burkholder, J., Roberts, B., Martinell, B., and McCabe, D. (1990) In vivo and in vitro gene transfer to mammalian somatic cells by particle bombardment. *Proc. Natl. Acad. Sci. USA* **87**, 9568–9572.
2. Cheng, L., Ziegelhoffer, P., and Yang, N.-S. (1993) In vivo promoter activity and transgenic expression in mammalian somatic tissues evaluated by using particle bombardment. *Proc. Natl. Acad. Sci. USA* **90**, 4454–4459.
3. Jiao, S., Cheng, L., Wolff, J., and Yang, N.-S. (1993) Particle bombardment-mediated gene transfer and expression in rat brain tissues. *Bio/Technology* **11**, 497–502.
4. Burkholder, J. K., Decker, J., and Yang, N.-S. (1993) Transgene expression in lymphocyte and macrophage primary cultures after particle bombardment. *J. Immunol. Methods* **165**, 149–156.
5. Thompson, T. A., Gould, M. N., Burkholder, J. K., and Yang, N.-S. (1993) Transient promoter activity in primary rat mammary epithelial cells evaluated using particle bombardment gene transfer. *In Vitro Cell Dev. Biol.* **29A**, 165–170.
6. Christou, P. (1994) Application to plants, in *Particle Bombardment Technology for Gene Transfer* (Yang, N.-S. and Christou, P., eds.), Oxford University Press, New York, pp. 71–99.
7. Yang, N.-S. and Ziegelhoffer, P. (1994) The particle bombardment system for mammalian gene transfer, in *Particle Bombardment Technology for Gene Transfer* (Yang, N.-S. and Christou, P., eds.), Oxford University Press, New York, pp. 117–141.
8. Qiu, P., Ziegelhoffer, P., Sun, J., and Yang, N.-S. (1996) Gene gun delivery of mRNA in situ result in efficient transgene expression and immunization. *Gene Ther.* **3**, 262–268.
9. Guo, Z., Yang, N.-S., Jiao, S., et al. (1996) Efficient and sustained transgene expression in mature rat oligodendrocytes in primary culture. *J. Neurol. Res.* **43**, 32–41.
10. Irvine, K. R., Rao, J. B., Rosenberg, S. A., and Restifo, N. P. (1996) Cytokine enhancement of DNA immunization leads to effective treatment of established pulmonary metastases. *J. Immunol.* **156**, 238–245.
11. Ross, H. M., Weber, L. W., Wang, S., et al. (1997) Priming for T-cell-mediated rejection of established tumors by cutaneous DNA immunization. *Clin. Cancer Res.* **3**, 2191–2196.
12. Feltquate, D. M., Heaney, S., Webster, R. D., and Robinson, H. L. (1997) Different T helper cell types and antibody isotypes generated by saline and gene gun DNA immunization. *J. Immunol.* **158**, 2278–2284.
13. Condon, C., Watkins, S. C., Celluzzi, C. M., et al. (1996) DNA-based immunization by in vivo transfection of dendritic cells. *Nature Med.* **2**, 1122–1128.
14. Iwasaki, A., Torres, C. A. T., Ohashi, P. S., Robinson, H. L., and Barber, B. H. (1997) The dominant role of bone marrow-derived cells in CTL induction following plasmid DNA immunization at different sites. *J. Immunol.* **159**, 11–14.

15. Mahvi, D. M., Sheehy, M. J., and Yang, N.-S. (1997) DNA cancer vaccines: a gene gun approach. *Immunol. Cell Biol.* **75,** 456–460.

16. Conry, R. M., Widera, G., LoBuglio, A. F., et al. (1996) Selected strategies to augment polynucleotide immunization. *Gene Ther.* **3,** 67–74.

17. Mahvi, D. M., Burkholder, J. K., Turner, J., et al. (1996) Particle-mediated gene transfer of granulocyte-macrophage colony-stimulating factor cDNA to tumor cells: implications for a clinically relevant tumor vaccine. *Hum. Gene Ther.* **7,** 1535–1543.

18. Tan, J., Yang, N.-S., Turner, J. G., et al. (1999) Interleukin-12 cDNA skin transfection potentiates human papillomavirus E6 DNA vaccine-induced antitumor immune response. *Cancer Gene Ther.* **6,** 331–339.

19. Hogge, G. S., Burkholder, J. C., Albertini, M. R., et al. (1998) Development of human granulocyte-macrophage colony-stimulating factor-transfected tumor cell vaccines for the treatment of spontaneous canine cancer. *Hum. Gene Ther.* **9,** 1851–1861.

20. Beardsley T. (1999) Innovative immunity. *Sci. Am.* **280,** 42–44.

21. Weber, S. M., Shi, F., Heise, C., Warner, T., and Mahvi, D. M. (1999) IL-12 gene transfer results in CD 8-dependent regression of murine CT26 liver tumors. *Ann. of Surg. Oncol.* **6,** 186–194.

22. Rakhmilevich, A. L., Janssen, K., Turner, J., Culp, J., and Yang, N.-S. (1997) Cytokine gene therapy of cancer using gene gun technology: superior antitumor activity of IL-12. *Hum. Gene Ther.* **8,** 1303–1311.

23. Rakhmilevich, A. L., Turner, J., Ford, M. J., et al. (1996) Gene gun-mediated skin transfection with interleukin 12 gene results in regression of established primary and metastatic murine tumors. *Proc. Natl. Acad. Sci. USA* **93,** 6291–6296.

24. Sun, W. H., Burkholder, J. K., Sun, J., Culp, J., Turner, J., and Lu, X. G. (1995) In vivo cytokine gene transfer by gene gun suppresses tumor growth in mice. *Proc. Natl. Acad. Sci. USA* **92,** 2889–2893.

25. Andree, C., Swan, W. F., Page, C. P., Macklin, M. D., Slama, J., Hatzis, D., and Eriksson, E. (1994) In vivo transfer and expression of an EGF gene accelerates wound repair. *Proc. Natl. Acad. Sci. USA* **91,** 12188–12192.

26. Ye, Z.-Q., Qiu, P., Burkholder, J. K., et al. (1998) Cytokine transgene expression and promoter usage in primary $CD34^+$ cells using particle-mediated gene delivery. *Hum. Gene Ther.* **9,** 2197–2205.

27. Fuller, D. H., Corb, M. M., Barnett, S., Steimer, K., and Haynes, J. R. (1997) Enhancement of immunodeficiency virus-specific immune responses in DNA-immunized rhesus macaques. *Vaccine* **15,** 924–926.

28. Fuller, D. H., et al. (2000) Oral presentation. *ASGT Meeting*, Denver, CO.

29. Swain, W. F., et al. (2000) Oral presentation. *2000 International Symposium on DNA Vaccine and Gene Therapy Technology*, Taipei, Taiwan.

30. Sun, W. H., Burkholder, J. K., Sun, J., et al. (1995) In vivo cytokine gene transfer by gene gun reduces tumor growth in mice. *Proc. Natl. Acad. Sci. USA* **92,** 2889–2893.

31. Felgner, P. L., et al. (2000) Oral presentation. *2000 International Symposium on DNA Vaccine and Gene Therapy Technology,* Taipei, Taiwan.

32. Woffendin, C., Yang, Z.-Y., Udaykumar, et al. (1994) No viral and viral delivery of a human immunodeficiency virus protective gene into primary human T cells. *Proc. Natl. Acad. Sci. USA* **91,** 11581–11585.
33. Hogge, G. S., Burkholder, J. K., Culp, J., et al. (1999) Preclinical development of human granulocyte-macrophage colony-stimulating factor-transfected melanoma cell vaccine using established canine cell lines and normal dogs. *Cancer Gene Ther.* **6,** 26–36.

12

Preparation of Pseudotyped Retroviral Vector

Hong Yu

1. Introduction

Retroviral vectors derived from murine leukemia retrovirus (MuLV) have been widely used for efficient gene transfer to achieve long-term expression of a chosen therapeutic gene in mammLian cells *(1)*. Disadvantages of this vector are the instability and low viral titers generated from packaging cells, low efficiency of gene transfer into human cells, especially *in vivo*, and the requirement for dividing cells. Some authors have attempted to increase the transduction efficiency by using strategies like low-spced centrifugation of viral supernatant with cells, multiple viral exposures *(2)*, or increasing viral titers by ultracentrifugation *(3)*; they were able to produce an average transduction efficiency of 10–60%. However, all such improvements in transduction efficiency require additional procedures, which are practically inefficient.

The envelope G glycoprotein from the vesicular stomatitis virus (VSV-G) has been used to construct a pseudotyped MuLV with significant improvement in stability and transduction efficiency *(4,5)*. VSV-G pseudotyped MuLV (VSV-G/MuLV) can be concentrated to titers exceeding 10^9 colony-forming units (cfu)/mL through ultracentrifugation with minimal loss of infectivity *(6)*. This pseudotyped vector has a much broader host range than the vectors with the conventional amphotropic Env and has been successfully used to transfer genes into human peripheral blood lymphocytes *(7,8)*, leukocytes *(9)*, hepatocytes *(10)*, and vascular tissues *(11)*. This chapter reviews the principles and procedures involved in generating such pseudotyped retroviral vectors.

From: *Methods in Molecular Medicine, Vol. 69, Gene Therapy Protocols, 2nd Ed.*
Edited by: J. R. Morgan © Humana Press Inc., Totowa, NJ

1.1. MuLV Pseudotyped with VSV-G Protein

Retrovirus contains two single-stranded RNA molecules associated with *gag* proteins and *pol* protein in a core structure. An estimated 100–300 *env* proteins protrude from the lipid bilayer of the particles. The packaging cells are genetically engineered to supply the *env*, *pol*, and *gag* sequences that encode the structural proteins necessary for the formation of the viral particle. Recombinant retroviral vectors are usually produced from a packaging cell line transfected with a retroviral vector DNA that is transcribed inside the cell; the resulting RNA is recognized by the structural proteins and packaged into retroviral particles that bud off from the plasma membrane. More details on retroviral vectors can be obtained from earlier chapters of this book.

The MuLV virus has a low transduction efficiency and is unstable. The envelope protein, a key component for efficient transduction, is composed of two subunits, a soluble unit and a transmembrane unit. The Env proteins interact with protein receptors on the cell surface to initiate internalization of the virus. The soluble unit of the envelope is easily shed from the viral particle, resulting in inactive Env protein. The labile structural characteristics of the envelope protein is thought to be a main reason for the unstable retrovirus. Therefore, modification of envelope components might result in a more stable particle.

VSV-G, an envelope protein from VSV, is a single chain protein. Emi et al. *(12)* demonstrated that the VSV-G protein could completely replace the env protein of MuLV and produce an infectious MuLV-based viral vector called VSV-G pseudotyped MuLV. VSV-G has been demonstrated to interact with phospholipid components of the plasma membrane *(13–15)*. Since virus entry seems not to be dependent on the presence of specific protein receptors, VSV-G/MuLV has a broader host range than that of a traditional retrovirus.

1.2. Packaging Cell for VSV-G Pseudotyped MuLV

VSV-G pseudotyped MuLV was originally produced from transient transfection of the 293T cell line with three plasmids: gag-pol expression plasmid, VSV-G expression plasmid, and retroviral vector plasmid (**Fig. 1**). Human 293T, a subline of Ad5-transformed embryonic kidney cell line 293, is chosen because it is more transfectable than the NIH 3T3 cell, from which most retroviral producer cell lines have been derived. In addition, the 293T cell line contains the simian virus 40 (SV40) large tumor antigen, whose expression may increase the replication of vectors containing the SV40 origin of replication.

Because VSV-G protein is toxic to the cell, stable expression of the protein was not achieved, and a packaging cell line had not been available until an inducible promoter was applied. The stable packaging cell lines for production

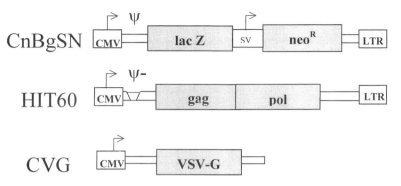

Fig. 1. Schematic representation of plasmids used for the production of viral vectors. CnBgSN is a viral vector plasmid carrying the *lacZ* reporter gene and the *neo*ʳ drug-resistant gene. HIT60 is a plasmid carrying *gag* and *pol* viral genes. CVG is the plasmid for VSV-G envelope expression.

of VSV-G/MuLV have been developed by applying a tetracycline-modulated promoter *(16,17)* or a recombination required system *(18)* for the controlled expression of VSV-G to minimize its associated toxicity.

In the packaging cell lines, the *gag-pol* genes are constantly expressed. In the tetracycline-regulatable promoter system, expression of the VSV-G gene is inhibited in the presence of tetracycline. Remove of tetracycline from the culture medium allows high expression of VSV-G and production of infectious viral particles. In the recombination-required system, the expression of VSV-G is completely silent because the RNA transcript terminates before the VSV-G coding sequence. By introducing Cre recombinase via adenovirus vector, the RNA transcript terminates are excised through a site-specific recombination, and VSV-G can be transcribed *(18)*. Here we have only used the tetracycline-modulated packaging cell line to generate stable producer cells for the production of VSV-G pseudotyped MuLV vectors.

1.3. Transduction Efficiency

Transduction efficiency of VSV-G/MuLV ranges from 70 to 95%, which is much higher than that of the tranditional amphotropic MuLV vector (10–50%). At the same multiplicity of infection (MOI), the transduction efficiency of VSV-G/MuLV remained about four-fold higher than that of Ampho/MuLV *(11)*. This high transduction efficiency could be a result of several factors: more stable VSV-G glycoprotein, and more receptors available for virus binding, which potentiates binding efficiency of the VSV-G pseudotyped virus to the cells.

By using the VSV-G pseudotyped vector, we may omit the step of selection of the transduced cells because most cells will be transduced with the vector.

This can significantly reduce the amount of time required for application of gene therapy.

1.4. Pseudotransduction

VSV-G/MuLV-mediated gene transfer has been reported to cause pseudotransduction (protein transfer) *(7,10)*. However, we have demonstrated via four different experiments that there was no protein transfer via the viral particles *(11)*. First, cotransfection of viral vector DNA with a nonviral vector DNA carrying the *lacZ* gene was performed to generate a viral vector. The proteins (ß-Gal) expressed from the nonviral vector in the producer cells could not be transferred to recipient cells via the viral particles generated from the cotransfection. This demonstrated that no protein transfer occurred via the viral vector.

Second, transduction was correlated with viral reverse transcriptase activity. When the activity of reverse transcriptase was blocked during transduction, targeted protein expression was not detected in the recipient cells, thereby indicating that the transduction was a result of reverse transcription from RNA into DNA.

Third, gene transfer using the MuLV vector requires cells to be dividing *(19)*, whereas protein transfer is independent of cell proliferation. When the proliferation of recipient cells was arrested with γ-irradiation, no cells were able to be transduced. The correlation of transduction and cell proliferation further supports the conclusion that no protein transfer occurs with the VSV-G pseudotyped vector.

Fourth, gene expression resulting from the transduction was stable. For pseudotransduction, the amount of transferred protein would be limited and its protein activity transient. Thus, after several passages, the population of cells retaining the transferred protein would decrease. On the other hand, the expression of the protein resulting from the transferred gene would be stable and the population of the cells possessing the gene would not decrease with passage. High protein expression of the transduced gene was detected following nine passages after transduction, indicating that the protein detected in the transduced cells truly arose from gene expression.

The viral supernatant reported with pseudotransduction was processed through ultracentrifugation, whereas we used the supernatant directly collected from cell cultures without concentration. Perhaps ultracentrifugation altered the pseudotyped vector, leading to pseudotransduction.

2. Materials
2.1. Cell Culture

1. Human 293 cells. The Humna 293T/17 cell line can be obtained from the American Type Culture Collection (CRL11268) *(20)*. Cells are maintained in Dulbecco's modified Eagle's medium (DMEM; Gibco BRL, Gaithersburg, MD) supplemented with 10% fetal bovine serum (FBS) and 2 mM glutamine (Gibco BRL).
2. The 293/GPG cell line, a VSV-G/MuLV packaging cell line *(17)*, is maintained in the DMEM medium described above with the addition of 1 μg/mL tetracycline (Sigma), 2 mg/mL puromycin (Sigma), 0.3 mg/mL G418 (Gibco BRL), and 1 mM sodium pyruvate (Gibco BRL).
3. VSV-G/MuLV producer cells are also maintained in the same medium as the packaging cells with tetracycline. To produce virus, the cells are cultured in the production medium, which is the same as the maintenance medium but without tetracycline, puromycin, or G418. All cells were maintained in a humidified 37°C incubator with 5% CO_2.
4. Phosphate-buffered saline (PBS): 0.9% NaCl, 0.0144% KH_2PO_4, 0.0795% $K_2HPO_4 \cdot 7H_2O$ (all w/v), pH 7.2
5. Trypsin-EDTA: 0.05% trypsin, 0.53 mM EDTA·4Na (Gibco BRL), stored at 4°C.
6. Tissue culture dishes (100-mm diameter).

2.2. Plasmids

1. The retroviral vector pCnBgSN *(21)* contains a *lacZ* gene encoding nuclcar-localized ß-Gal and a neomycin resistance gene (*neo*^{r}) encoding for neomycin phosphotransferase. The *lacZ* expression is driven by the hybrid 5' cytomegalovirus (CMV) long terminal repeat (LTR) promoter, and the *neo*^{r} is controlled by a Simian virus 40 (SV40) promoter (**Fig. 1**).
2. Plasmid pHIT60 is a Moloney (Mo) MuLV Gag- and Pol-expressing plasmid, whose expression is driven by a CMV promoter *(22)*. Use only JM109 cells to grow this plasmid.
3. Plasmid pCVG is a CMV-driven VSV-G expression vector *(11)*.
4. All plasmid DNA can be purified from *E. coli* using a Qiagen (Valencia, CA) Plasmid Kit.

2.3. DNA Transfection

1. 2× HBS: 50 mM HEPES, pH 7.1, 280 mM NaCl, 1.5 mM Na_2HPO_4, 10 mM KCl.
2. 2 M $CaCl_2$.
3. Sodium butyrate, 0.5 M, filtered and stored at –20°C.
4. Sterile distilled water.
5. Sterile 5-mL conical tubes and 1-mL pipets.

2.5. Virus Transduction

1. Polybrene (hexadimethrine bromide, H-9268; Sigma) stock solution (8 mg/mL).
2. G418 (neomycin phosphotransferase or Geneticin; GIBCO/BRL) stock solution (50 mg/mL).
3. X-Gal staining solution: prepare X-Gal in stock (40 mg/mL, Gibco BRL), and dilute just before use 1:40 with X-Gal staining buffer: 2 mM MgCl$_2$, 5 mM K$_3$Fe(CN)$_6$, and 5 mM K$_4$Fe(CN)$_6$ in PBS.
4. 10% formaldehyde.

3. Methods
3.1. Generation of VSV-G/MuLV from a Transient Transfection

VSV-G/MuLV vectors can be generated from a transient three-plasmid transfection system *(22)*. 293T/17 cells are transfected by calcium phosphate precipitation with plasmids pCnBgSN (or any viral vector you have), pHIT60 (for gal-pol), and pCVG (for VSV-G). The viral titers range from 10^6 to 10^7 cfu/mL (*see* **Note 1**).

1. Day 1. Prepare DNA and cells. For transfection in one 100-mm plate, 10 µg of each of the three plasmids is needed: pCnBgSN, pHIT60, and pCVG. Calculate the volume and pipet all required volume into a sterile microfuge tube. Add H$_2$O to make the total volume 100 µL. Add 1/10 volume (10 µL) of 3 M NaAc, 2.6 times volume (260 µL) of absolute ethanol. Place at –20°C overnight.
2. Trypsinize the 293T/17 cells (80% confluent) on a 100-mm tissue culture dish and pass them in 1:4 dilution. Culture overnight (15–18 h) at 7°C.
3. Day 2. Cells should be about 60% confluent. Change medium on the plate with fresh 10 mL medium and keep culturing. Perform transfection within 4 h after the medium change.
4. About 2 h after medium change, centrifuge DNA at 12,000g for 20 min at 4°C in a microfuge. The DNA pellet may be washed with 70% ethanol. Remove the supernatant, and air-dry the pellet for about 10 min inside a tissue culture hood. The pellet should appear clear at this point.
5. Add 438 µL of H$_2$O to dissolve the DNA pellet completely. Transfer the solution to a 5-mL sterile polypropylene tube.
6. Add 62 µL of 2 M CaCl$_2$ to the tube containing the DNA. This should be done slowly drop by drop. Do not mix. Leave it in the hood for 5 min.
7. Using a sterile 1-mL plastic pipet, take up 500 µL of 2× HBS. Insert the pipet into the DNA/CaCl$_2$ mixture until the tip is just above the bottom of the tube but not touching it. Very gently, add the HBS to the bottom, and then blow 30 bubbles. Do not mix. Cap the tube and leave it in the hood for 30 min for precipitate to form. A fine milky precipitate should be seen.
8. Using a 1-mL pipet, gently pipet up and down the DNA precipitate a few times. Add the precipitate to the 293T cell plate drop by drop while rocking the plates. Incubate at 37°C overnight.

9. Day 3, morning. Wash the cells with 5-mL prewarmed PBS to remove the precipitate. Culture the cells in 6 mL fresh medium containing 10 mM sodium butyrate. In the evening, after 8–12 h of incubation with sodium butyrate, replace the supernatant with 6 mL fresh DMEM medium and incubate the cells overnight.

10. Day 4, morning. Collect the supernatant, filter with a 0.45-μm filter, and aliquot. The cell plate can be continually cultured with 6 mL fresh medium and the supernatant harvested after about 24 h on day 4. The supernatants can be used for various assays or stored at –70°C. The supernatant should have a virus content of about 10^6–10^7 cfu/mL (*see* **Note 2**).

3.2. Generation of VSV-G/MuLV Producer Cell Line

The viral supernatants from the transient transfection are used to transduce the packaging cell line 293/GPG to generate the VSV-G/MuLV producer cell line. Because 293/GPG already has the G418-resistant marker gene, G418 selection cannot be used to select transduced cells. However, since the transduction efficiency of the VSV-G/MLV vector is >90%, the pools of the transduced 293/GPG, without selection, can be used directly as stable producer cells.

1. Day 1. Grow 293/GPG cells on 100-mm tissue culture dishes in DMEM medium with tetracycline, until 80% confluent. Trypsinize the cells and pass them in a 1:5 dilution. Culture overnight (15–18 h) at 37°C.
2. Day 2. Cells should be about 40–50% confluent. Thaw VSV-G/MuLV supernatant from transient transfection in 37°C water bath (requires 3 mL/100-mm plate), and add Polybrene (1:1000) to a final concentration of 8 μg/mL. Remove medium in the cell culture plate. Add the 3 mL supernatant to the plate and culture for 2 h at 37°C.
3. Add 10 mL DMEM with tetracycline to the virus-cell mixture plate. Culture overnight.
4. Day 3. Replace the medium with 10 mL fresh medium with tetracycline. Culture for 2–3 days until confluent.
5. Split the cells into five plates. Freeze the cells for later use or change medium into viral production medium to produce virus.

3.3. Generation of VSV-G/MuLV from a Producer Cell Line

1. Grow the producer cells (293/GPG/CnBgSN) to 80% confluent in maintenance medium.
2. Remove the medium and wash the cells with PBS once. Replace the medium with 6 mL production medium (without tetracycline), and culture overnight.
3. Collect the supernatant at 24 h, and replace with fresh 6 mL production medium. This can be repeated 3–4 times until most of the cells are detached and float up. This first collection has low virus titer (10^4–10^5 cfu/mL) and may be discarded.

Filter the following collections through a 0.45-μm filter and aliquot. The supernatant should have a virus content of about 10^6–10^7 cfu/mL.

3.4. Concentration of VSV-G/MuLV

1. Centrifuge the viral supernatant (100–500 mL) at 13,800g in a Beckman rotor JA-14 for 12 h.
2. The pellet is resuspended in <1 mL TNE buffer (0.01 M Tris-HCl, pH 7.2, 0.1 M NaCl, 0.001 EDTA), which will result in a 100–1000-fold concentration. This virus can be used directly.
3. The concentrated virus can be further purified by centrifugation through a 10–60% linear sucrose gradient in a Beckman SW-40 rotor at 30,000 rpm, for 2 h at 20°C.
4. The virus forms a white, milky band viewable with a black background, which is recovered using a syringe and punching through the centrifuge tube.
5. The recovered virus (1–2 mL) is suspended in 10 mL TNE buffer. The solution is pelleted by centrifugation at 39,800g in a Beckman rotor (JA-17) for 2 h at 4°C. The pellet is resuspended in TNE buffer at 1/1000 of the original supernatant volume.

3.5. Virus Titer and Transduction Effeciency Assay

The viral titer is analyzed by either G418-resistant colony formation or ß-Gal activity assay, which generate similar results. The transduction efficiency is measured by comparing ß-Gal-positive and -negative cells among transduced cells.

1. Day 1. Plate NIH 3T3 cells (or your target cells for the transduction efficiency assay) in 30-mm-diameter wells of a 6-well plate (5×10^4 cells/well).
2. Day 2. Mix viral supernatant with Polybrene (final concentration 8 μg/mL). For titration, do a series of 1:10 dilutions (up to 10^{-6}) of the viral supernatant with DMEM medium containing Polybrene.
3. Remove culture medium on cell plate. Add 1 mL viral supernatant to a well. For titration, use 10^{-1}–10^{-6} dilutions into six corresponding wells. For transduction efficiency, use the undiluted supernatant. Culture for 2 h at 37°C, followed by addition of 2 mL of fresh medium.
4. Day 3. Change the medium with 3 mL DMEM medium containing 0.8 mg/mL G418 for G418 selection assay or without G418 for ß-Gal assay.
5. For X-Gal staining, culture the cells for another 48 h, and then fix them with 0.5% glutaraldehyde in PBS followed by three washes with PBS for 10 min each time at room temperature. Then stain the cells for ß-Gal activity by incubating overnight at 37°C with 2 mL X-Gal staining solution. A cluster of blue cells is counted as one colony under an inverted microscope.
6. For G418 selection, culture the cells with medium containing 0.8 mg/mL G418 for 7 days. After 7 days, remove the media and fix the cells with 10% formalde-

hyde followed by a 3 times PBS wash. Stain the cells on the plate with 1 mL of 1% methylene blue (Sigma) in methanol for 5 min and wash with tap water. Count the resistant colonies, in blue. Calculate the titer (cfu/mL) as numbers of colonies/(dilution factor·virus volume added).
7. For transduction efficiency assay, stain the cells with X-Gal as described above. Calculate transduction efficiency by the ratio of blue cells (transduced) to the total cells contained in 10 randomly picked high-power fields (200×).

4. Notes

1. To check the transfection efficiency, the 293 cells can be fixed and X-Gal stained to measure the percentage of blue cells after collection of supernatant is finished.
2. If the transduction efficiency does not reach higher than 85% during production of a stable producer cell line, the titer of virus from the stable producer cells thus generated could be low. In this situation, a single colony should be isolated. Dilute the pool of the transduced cells to 1 cell/mL, plate 0.25 mL of the cell suspension onto a 96-well plate, and culture it until the single cell in some wells grows into a single colony. Pipet up and down the medium to dissociate the cells, pass them onto 24-well plate, and then duplicate the pass onto two 6-well plates. In one 6-well plate, the cells will be cultured in virus production medium and the supernatant will be collected and the titer checked. The plate will be discarded after viral production. The cells that generate the highest titer in the corresponding well of the other plate will be passed to a 100-mm plate and used to produce viral supernatant in large quantity.

References

1. Miller, A. D., Miller, D. G., Garcia, J. V., and Lynch, C. M. (1993) Use of retroviral vectors for gene transfer and expression. *Methods Enzymol.* **217,** 581–599.
2. Inaba, M., Toninelli, E., Vanmeter, G., Bender, J. R., and Conte, M. S. (1998) Retroviral gene transfer: effects on endothelial cell phenotype. *J. Surg. Res.* **78,** 31–36.
3. Zelenock, J. A., Welling, T. H., Sarkar, R., Gordon, D. G., and Messina, L. M. (1997) Improved retroviral transduction efficiency of vascular cells in vitro and in vivo during clinically relevant incubation periods using centrifugation to increase viral titers. *J Vasc. Surg.* **26,** 119–127.
4. Friedmann, T. and Yee, J. K. (1995) Pseudotyped retroviral vectors for studies of human gene therapy. *Nat Med.* **1,** 275–277.
5. Schnell, M. J., Buonocore, L., Kretzschmar, E., Johnson, E., and Rose, J. K. (1996) Foreign glycoproteins expressed from recombinant vesicular stomatitis viruses are incorporated efficiently into virus particles. *Proc. Natl. Acad. Sci. USA* **93,** 11359–11365.
6. Burns, J. C., Friedmann, T., Driever, W., Burrascano, M., and Yee, J. K. (1993) Vesicular stomatitis virus G glycoprotein pseudotyped retroviral vectors: concen-

tration to very high titer and efficient gene transfer into mammalian and nonmammalian cells [see comments]. *Proc. Natl. Acad. Sci. USA* **90,** 8033–8037.

7. Gallardo, H. F., Tan, C., Ory, D., and Sadelain, M. (1997) Recombinant retroviruses pseudotyped with the vesicular stomatitis virus G glycoprotein mediate both stable gene transfer and pseudotransduction in human peripheral blood lymphocytes. *Blood* **90,** 952–957.

8. An, D. S., Koyanagi, Y., Zhao, J. Q., et al. (1997) High-efficiency transduction of human lymphoid progenitor cells and expression in differentiated T cells. *J. Virol.* **71,** 1397–1404.

9. Sharma, S., Cantwell, M., Kipps, T. J., and Friedmann, T. (1996) Efficient infection of a human T-cell line and of human primary peripheral blood leukocytes with a pseudotyped retrovirus vector. *Proc. Natl. Acad. Sci. USA* **93,** 11842–11847.

10. Liu, M. L., Winther, B. L., and Kay, M. A. (1996) Pseudotransduction of hepatocytes by using concentrated pseudotyped vesicular stomatitis virus G glycoprotein (VSV-G)-Moloney murine leukemia virus-derived retrovirus vectors: comparison of VSV-G and amphotropic vectors for hepatic gene transfer. *J. Virol.* **70,** 2497–2502.

11. Yu, H., Eton, D., Wang, Y., et al. (1999) High efficiency in vitro gene transfer into vascular tissues using a pseudotyped retroviral vector without pseudotransduction. *Gene Ther.* **6,** 1876–1883.

12. Emi, N., Friedmann, T., and Yee, J. K. (1991) Pseudotype formation of murine leukemia virus with the G protein of vesicular stomatitis virus. *J. Virol.* **65,** 1202–1207.

13. Schlegel, R., Tralka, T. S., Willingham, M. C., and Pastan, I. (1983) Inhibition of VSV binding and infectivity by phosphatidylserine: is phosphatidylserine a VSV-binding site? *Cell* **32,** 639–646.

14. Mastromarino, P., Conti, C., Goldoni, P., Hauttecoeur, B., and Orsi, N. (1987) Characterization of membrane components of the erythrocyte involved in vesicular stomatitis virus attachment and fusion at acidic pH. *J. Gen. Virol.* **68,** 2359–2369.

15. Conti, C., Mastromarino, P., Ciuffarella, M. G., and Orsi, N. (1988) Characterization of rat brain cellular membrane components acting as receptors for vesicular stomatitis virus. Brief report. *Arch. Virol.* **99,** 261–269.

16. Chen, S. T., Iida, A., Guo, L., Friedmann, T., and Yee, J. K. (1996) Generation of packaging cell lines for pseudotyped retroviral vectors of the G protein of vesicular stomatitis virus by using a modified tetracycline inducible system. *Proc. Natl. Acad. Sci. USA* **93,** 10057–10062.

17. Ory, D. S., Neugeboren, B. A., and Mulligan, R. C. (1996) A stable human-derived packaging cell line for production of high titer retrovirus/vesicular stomatitis virus G pseudotypes. *Proc. Natl. Acad. Sci. USA* **93,** 11400–11406.

18. Arai, T., Matsumoto, K., Saitoh, K., et al. (1998) A new system for stringent, high-titer vesicular stomatitis virus G protein-pseudotyped retrovirus vector induction by introduction of Cre recombinase into stable prepackaging cell lines. *J. Virol.* **72,** 1115–1121.

19. Miller, D. G., Adam, M. A., and Miller, A. D. (1990) Gene transfer by retrovirus vectors occurs only in cells that are actively replicating at the time of infection [published erratum appears in Mol Cell Biol (1992) **12,** 433]. *Mol. Cell. Biol.* **10,** 4239–4242.

20. Pear, W. S., Nolan, G. P., Scott, M. L., and Baltimore, D. (1993) Production of high-titer helper-free retroviruses by transient transfection. *Proc. Natl. Acad. Sci. USA* **90,** 8392–8396.

21. Han, J. Y., Cannon, P. M., Lai, K. M., Zhao, Y., Eiden, M. V., and Anderson, W. F. (1997) Identification of envelope protein residues required for the expended host range of 10A1 murine leukemia birus. *J. Virol.* **71,** 8103–8108.

22. Soneoka, Y., Cannon, P. M., Ramsdale, E. E., et al. (1995) A transient three-plasmid expression system for the production of high titer retroviral vectors. *Nucleic Acids Res.* **23,** 628–633.

13

Quantitative Measurement of the Concentration of Active Recombinant Retrovirus

Stelios T. Andreadis and Jeffrey R. Morgan

1. Introduction

Recombinant retroviruses are among several virus vectors currently being tested in clinical trials for purposes of gene therapy. As the number of clinical studies increases, accurate quantitation of retrovirus stocks and comparisons between different laboratories and clinical trials will become increasingly important. However, work from our laboratory as well as others has shown that the most commonly used quantitative retrovirus measures (i.e., titer and multiplicity of infection [MOI]) cannot be used to make accurate comparisons (1–3).

The activity of a stock of recombinant retrovirus is typically quantitated by measuring titer, the number of gene transfer events per unit volume of retrovirus solution. To determine titer, the virus stock is first serially diluted and then used to transduce target cells. The titer, expressed as colony-forming units/mL (CFU/mL), is the number of colonies of transduced cells multiplied by the dilution factor and divided by the volume of retrovirus applied to the target cells. Visualization and quantitation of transduced cells is achieved by use of retroviral vectors that encode reporter genes, such as the *lacZ* gene, or antibiotic-resistance genes, such as the *neo* gene.

Numerous factors can influence the value of titer, including the time of exposure of cells to the virus, the number and type of target cells, the volume of the virus-containing medium, and the half-life of retrovirus (4–7). Moreover, critical physiochemical properties of the virus particles such as diffusivity and half-life have been largely ignored. To date, none of these parameters of the titer assay have been standardized, which makes quantitative comparisons difficult.

From: *Methods in Molecular Medicine, Vol. 69, Gene Therapy Protocols, 2nd Ed.*
Edited by: J. R. Morgan © Humana Press Inc., Totowa, NJ

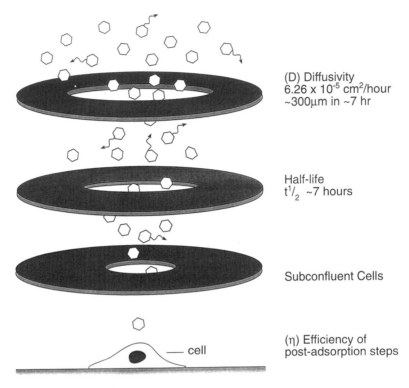

(D) Diffusivity
6.26 x 10^{-5} cm^2/hour
~300µm in ~7 hr

Half-life
t$^{1}/_{2}$ ~7 hours

Subconfluent Cells

—— cell

(η) Efficiency of
post-adsorption steps

Fig. 1. Most retrovirus particles are not able to infect a cell successfully. The barriers to infection (diffusion, half-life, and sparse cell coverage) are illustrated.

Thus, titer is not an absolute measure of the concentration of active retrovirus, rather it reflects the number of gene transfer events under a specific set of conditions.

1.1. Concentration of Active Retrovirus at the Start of Infection (C$_{vo}$)

Towards the goal of a more reliable quantitative measure of the activity of a stock of retrovirus, we have developed a method that helps to standardize the most important parameters of the titer assay and that takes into account the critical physiochemical properties of virus particles and cells *(8)*. From this method, we obtain a value for the concentration of biologically active retrovirus at the start of the infection (C$_{vo}$). C$_{vo}$ is a more reliable measure of the activity of a stock of retrovirus because it is not subject to the variables inherent in the titer assay. In this chapter, we present a method to determine C$_{vo}$ based on data obtained from a quantitative titer assay performed on a highly infectable reference cell type (NIH 3T3). This method and the mathematical analysis used to calculate C_{vo} takes into account and helps to standardize the major factors that influence measurements of titer.

The value of C_{vo} is higher than that of titer because C_{vo} is the concentration of active retroviruses in the viral stock at the start of the infection. Since retrovirus infection is limited by the relatively slow diffusion and rapid decay of the virus particles, only a fraction of the total number of active viruses in a stock is able to infect a cell successfully, and the titer assay scores only successful gene transfer events. Thus, most retrovirus particles, although active at the start of infection, are not able to infect a cell successfully due to the "barriers" of diffusion, virus half-life, and sparse coverage of cells on the tissue culture dish. Conceptually, these barriers are illustrated in **Fig. 1**. Therefore, for every successful gene transfer event seen in the titer assay, there must have been many more retrovirus particles that were active at the start of infection but were unable to infect a cell successfully (i.e., titer $<C_{vo}$; *see* **Note 1**).

We have derived equations that can be used to compute C_{vo} *(8)*. These equations also take into account the number of target cells as well as their size. These parameters are important because standard retroviral transductions are performed on subconfluent cultures to allow for cell division, a requirement for successful integration of the recombinant provirus. (Lentivirus vectors may not require cell division.) In this surface topography, viruses diffuse (and decay) and then adsorb on adsorbent patches (target cells) sparsely distributed on a nonabsorbent surface. We modified the mathematical analysis initially proposed by Soup and Szabo *(9)* for particle diffusion to a similar surface and derived an equation to calculate C_{vo}, the starting concentration of active retrovirus *(8)*:

$$C_{vo} = \frac{cfu}{\eta 4 a_c D N_{co} I} \tag{1}$$

where cfu is the number of colony-forming units as determined from a quantitative titer assay performed on a reference cell type (*see* **Note 2**), η is the efficiency of all postadsorption steps, a_c is the average radius of the target cells (cm), D is the diffusion coefficient of the virus (6.264×10^{-5} cm^2/h), N_{co} is the total number of cells at the start of transduction, and I is the integral (described below).

Since transduction is a multistep process (adsorption through gene expression) and not all steps proceed with 100% efficiency *(10)*, we have introduced the term η into **Eq. 1** to account for this fact. As discussed below, we have chosen NIH 3T3 cells as a reference cell type and for this cell type we have set the value of $\eta = 1$. However, it should be realized that even for highly infectable NIH 3T3 cells, transduction does not proceed with 100% efficiency ($\eta < 1$).

Table 1
Numerical Values of the Integral I for a Virus with a Half-Life of 7 Hours.[a]

Time (hr)	$$I = \int_0^t h(a_c,t)e^{-k_{dv}t}dt$$								
	$a_c = 7\mu m$	$a_c = 9\mu m$	$a_c = 10\mu m$	$a_c = 11\mu m$	$a_c = 13\mu m$	$a_c = 15\mu m$	$a_c = 17\mu m$	$a_c = 19\mu m$	$a_c = 21\mu m$
0.5	0.533	0.546	0.553	0.559	0.573	0.586	0.600	0.613	0.627
1.0	1.015	1.033	1.042	1.051	1.070	1.088	1.107	1.125	1.144
1.5	1.469	1.491	1.502	1.513	1.535	1.557	1.579	1.602	1.624
2.0	1.900	1.925	1.937	1.949	1.974	1.999	2.025	2.050	2.075
2.5	2.309	2.336	2.349	2.363	2.390	2.418	2.446	2.473	2.501
3.0	2.697	2.726	2.741	2.755	2.785	2.814	2.844	2.874	2.904
3.5	3.066	3.097	3.112	3.128	3.159	3.191	3.222	3.254	3.286
4.0	3.416	3.449	3.465	3.482	3.515	3.548	3.581	3.614	3.648
4.5	3.750	3.784	3.801	3.818	3.852	3.887	3.922	3.956	3.991
5.0	4.067	4.102	4.120	4.138	4.173	4.209	4.245	4.281	4.317
5.5	4.368	4.405	4.423	4.441	4.478	4.515	4.553	4.590	4.627
6.0	4.655	4.692	4.711	4.730	4.768	4.806	4.845	4.883	4.922
6.5	4.927	4.966	4.986	5.005	5.044	5.083	5.122	5.162	5.201
7.0	5.187	5.226	5.246	5.266	5.306	5.346	5.386	5.427	5.467
7.5	5.434	5.474	5.494	5.515	5.555	5.596	5.637	5.678	5.719
8.0	5.668	5.710	5.730	5.751	5.792	5.834	5.876	5.918	5.960

[a]Values of I are shown for target cells of various sizes (radius, a_c) and times of transduction. These numerical values can be used directly in **Eqs. 1** and **3**.

Therefore, values of C_{vo} from **Eq. 1** are a lower bound of the concentration of active retrovirus in the starting stock.

The integral I of **Eq. 1**, is a function of time, the diffusion coefficient, the half-life of the virus, and the radius of the target cells and is described by the following equation:

$$I = \int_0^t h(a_c, t)e^{-k_{dv}t}dt \qquad [2]$$

The function $h(a_c, t)$ was derived by Shoup and Szabo (9) and reflects the flux of particles on each cell at time t, over the steady-state particle flux. For convenience, we have evaluated the integral, I, for different times of infection of target cells with various sizes (radius, a_c). These numerical values of I are provided in **Table 1**.

1.2. Adsorbed Active Viruses per Cell (AVC): An Alternative to MOI

MOI is defined as the number of infectious viruses per cell and is typically used to predict the extent to which a cell population is transduced. MOI is calculated by multiplying titer (cfu/mL) by the volume of added virus and dividing this by the number of target cells at the start of transduction. However, since its calculation is based on titer, MOI suffers from the same lack of standardization and inaccuracy. Recent experimental results have shown that increasing the number of retroviral particles per target cell, either by increasing the volume of the retrovirus solution *(2,8,11)* or by decreasing the number of target cells *(2,10)*, did not result in increased levels of transduction, thus demonstrating the limitation of MOI. Similarly, the transduction efficiency did not correlate with MOI when nonadherent hematopoietic cells were transduced with recombinant lentiviruses *(3)*.

As an alternative, we propose AVC, the number of active retrovirus particles that will adsorb per cell during a given adsorption time *(8)*. AVC is a more reliable predictor as it is based on the concentration of active retrovirus (C_{vo}), which accounts for and thus helps standardize the physical parameters that govern transduction. AVC is calculated from the following equation:

$$AVC = \eta 4 a_c D C_{vo} I \qquad [3]$$

where, η is the efficiency of the postadsorption steps (for NIH 3T3 cells, which we use as a reference, we set $\eta = 1$; other cell types may vary), a_c is the average radius of the target cells (in cm), D is the diffusion coefficient of the virus $(6.264 \times 10^{-5} \text{ cm}^2/\text{h})$, C_{vo} is the initial concentration of active retrovirus at the start of the infection (**Eq. 1**), and the integral I is a function of time, the diffusion coefficient, the half-life of the virus, and the radius of the target cells.

Although we have used this analysis for recombinant retroviruses, the same concepts may also apply to other virus-based gene transfer vehicles such as recombinant lentiviruses, adenoviruses, or adenoassociated viruses. However, differences in key parameters such as diffusion coefficients, virus half-life, and the density of the target cells must be carefully considered before application of these equations.

1.3. Efficiency of Postadsorption Steps (η)

Transduction efficiency and cfu can vary depending on the target cell type. If two titer assays are run under identical conditions (same virus stock, time of infection, volume of virus, number of target cells), but two different cell types of the same size are used, experimentally the number of cfu may be significantly different. Since the same virus stock is used in this scenario, values of

C_{vo} cannot be different, rather the discrepancy is due to differences in the infectability or η: the efficiency of transduction of different cell types. Efficiencies of any number of postadsorption steps (internalization, reverse transcription, and so on) required for a successful gene transfer event may vary between cell types.

We have set the efficiency of the postadsorption steps of our reference cell type (NIH 3T3) equal to one (η = 1). NIH 3T3 cells were chosen to be the reference cell type because of their widespread use, ease of cultivation, infectability by ecotropic as well as amphotropic viruses and relatively high transduction efficiency. We have derived an equation that can be used to compute η for other cell types relative to the reference cell type. We transduce the reference cell type (NIH 3T3) and the new cell type with the same viral stock, and we use the following equation to calculate the relative transduction efficiency, η_x, of the unknown cell type:

$$\eta_x = \frac{cfu_x / N_{co,x}}{cfu_{3T3} / N_{co,3T3}} \left(\frac{a_{c,3T3}}{a_{c,x}} \right) \left(\frac{I_{3T3}}{I_x} \right) \qquad [4]$$

where cfu are the number of colony-forming units on both cell types, N_{co} is the total number of cells at the start of transduction for both cell types, a_c is the average radius of both cell types, and the integral I, which is a function of time, the diffusion coefficient, the half-life of the virus and the radius of the target cells, is computed for both cell types.

When making this comparison, it is important that the values of cfu used be obtained from the portion of the curve where cfu is linearly proportional to cell number *(8)*. This linear relationship is not always the case at very low or very high cell densities, presumably because transduction by retroviruses is strongly influenced by cell proliferation *(12)*, and proliferation can be perturbed at these extremes in cell density. Moreover, the linear portion of the curve of cfu versus cell number may vary between different cell types *(8)*.

2. Materials

2.1. Cell Culture

1. NIH 3T3 cells. The cells are split 1/10 twice a week.
2. Phosphate- buffered saline (PBS; Gibco BRL, Gaithersburg, MD): 1.0% NaCl, 0.025% KCl, 0.14% Na_2HPO_4, 0.025% KH_2PO_4 (all w/v), pH 7.2; sterile.
3. Dulbecco's modified Eagle's medium (DMEM; Gibco BRL) with glutamine containing 100 U/mL penicillin and 100 µg/mL streptomycin.
4. Tissue culture dishes (6-well plates, 35 mm in diameter; Costar, Cambridge, MA).
5. Bovine calf serum (BCS; HyClone, Logan, UT).
6. Fetal bovine serum (FBS; HyClone).

7. Trypsin (0.25%) containing 0.5 mM EDTA (Gibco BRL).
8. Filters, 0.45-μm pore size (Gelman, Ann Arbor, MI).
9. Polybrene (Sigma, St. Louis, MO). A concentrated stock solution is prepared at 800 μg/mL in sterile ddH$_2$O and stored at 4°C.

2.2. Quantitative Titer Assay for a lacZ Retrovirus

1. PBS.
2. PBS, 1 mM MgCl$_2$.
3. 0.5% glutaraldehyde (Sigma) in PBS (*see* **Note 3**).
4. Potassium ferricyanide, K$_3$Fe(CN)$_6$ (Sigma).
5. Potassium ferrocyanide K$_4$Fe(CN)$_6$•3H$_2$O (Sigma).
6. KC solution: add 0.82 g of K$_3$Fe(CN)$_6$ and 1.05 g of K$_4$Fe(CN)$_6$•3H$_2$O in 25 mL of PBS. Store wrapped in foil at 4°C.
7. X-gal (5-bromo-4-chloro-3-indolyl-β-D-galactopyranoside; Sigma): prepare fresh at 40 mg/mL in dimethylsulfoxide (DMSO).
8. Reaction mixture (10 mL): add 0.5 mL of KC solution and 0.25 mL X-gal to 9.25 mL of PBS/1 mM MgCl$_2$.

2.3. Measurement of Cell Size

1. CellTracker™ Orange CMTMR (5-[and-6]-[((4 chloromethyl)benzoyl)amino] tetramethylrhodamine) (Molecular Probes, Eugene, OR). Prepare a stock solution to a final concentration of 10 mM in high-quality anhydrous DMSO and store in small aliquots at –20°C protected from light.
2. PBS.

3. Methods
3.1. Cell Culture

NIH 3T3 cells and *lacZ* virus-producing cell lines are cultured in DMEM with 10% BCS, 100 U of penicillin, and 100 μg/mL streptomycin, at 37°C with 10% CO$_2$.

3.1.1. Isolation of Diploid Human Foreskin Fibroblasts (HFFs)

1. After obtaining necessary institutional approvals, obtain specimens of fresh newborn human foreskins from the hospital.
2. Trim the tissue to remove fat and muscle layers underlying the dermis.
3. Rinse repeatedly (8 times) in sterile PBS.
4. Cut the tissue into small pieces (0.2 × 0.2 cm^2) and place on the tissue culture plate, dermal side contacting the tissue culture plate.
5. After allowing the skin pieces to dry onto the tissue culture plate for approximately 45 min, add 10 mL of DMEM containing 20% FBS, 100 U penicillin, and 100 μg/mL streptomycin. Place in an incubator at 37°C with 10% CO$_2$.
6. Gently change the medium every 3–4 days.
7. The cells are passed for the first time 7–10 days later, when they migrate out from

the dermis and form a confluent monolayer on the surface of the tissue culture plate.

8. Thereafter, pass cells every week when they reach confluence.

3.2. Harvesting of Recombinant Retrovirus

1. Grow the virus producer cells to confluence in a 10-cm tissue culture dish.
2. Remove old medium and add 10 mL of fresh medium. Return plate to incubator at 37°C with 10% CO_2.
3. Harvest the virus containing medium 24 h later.
4. Filter the virus-containing medium through 0.45-µm pore sized filters, aliquot, and store at –80°C until use.

3.3. Quantitative Assay for a Retrovirus Encoding lacZ

1. Prepare six serial dilutions (10-fold) of the *lacZ* virus stock in cell culture medium.
2. Add Polybrene to each dilution to a final concentration of 8 µg/mL.
3. Add 2 mL of each virus dilution to the target cells (NIH 3T3 cells), which were plated in 6-well plates the previous day, at 60,000 cells/well. Target cell numbers may vary depending on target cell type (*see* **Note 4**).
4. After 24 h, remove the virus-containing medium, add fresh medium, and allow the cells to grow for 48–72 h (*see* **Note 5**).
5. Use three replicate wells for cell counting at the time the virus is added.
6. After 48–72 h, remove medium from wells.
7. Wash cells once with 2 mL PBS.
8. Fix cells with 2 mL of PBS containing 0.5% glutaraldehyde for 10 min at room temperature (*see* **Note 3**).
9. Wash cells with 2mL PBS/1 mM $MgCl_2$.
10. Incubate cells in 2 mL of reaction mixture for 3–4 h at 37°C or until blue-stained cells are obvious to the naked eye.
11. Remove the reaction mixture, and wash cells with 2 mL PBS.
12. Remove the PBS, wash, and allow plates to air-dry.
13. Count clusters of *lacZ*⁺ cells (blue) with the aid of a dissecting microscope. Do not count individual cells (*see* **Note 6**).
14. Calculate cfu of the viral stock by multiplying the number of blue cell clusters times the dilution factor. Do not divide this number by the volume of virus added (*see* **Note 2**).

3.4. Measurements of Target Cell Surface Area by Fluorescence Microscopy

1. Plate 50,000 cells/well in 6-well plates, and allow them to attach and spread overnight.
2. The next day, stain cells with 10 µM of the cytoplasmic dye CellTracker Orange CMTMR for 60 min at 37°C. The CellTracker dye is diluted 1:1000 from the

stock solution (10 mM in DMSO) in cell culture medium. It labels viable cells for at least 24 h after loading and often through several cell divisions (*see* **Note 7**).

3. Wash cells twice with 2 mL PBS to remove the unbound dye and reduce the background.
4. View cells with a fluorescence microscope using standard filter sets (the Orange CMTMR dye absorbs at 540 nm and emits at 566 nm).
5. Select multiple areas (usually 20 or more) at random in six identical wells for image analysis (*see* **Notes 8** and **9**).
6. Calculate the radius, a_c, of each cell from the measurement of its area, A ($a_c = \sqrt{A / \pi}$). Calculate the average radius as the mean of the radii of at least 50–100 cells.

3.5. Calculation of C_{vo}

1. Use **Eq. 1** to calculate C_{vo}, the concentration of active retrovirus at the start of infection (*see* **Note 10**).
2. For cfu, use the value obtained from the quantitative assay for a retrovirus encoding *lacZ* on a reference cell type such as NIH 3T3 cells (*see* **Subheading 3.3.**) (*see* **Note 2**).
3. For a_c, the radius of the cells, use the average value obtained by fluorescence microscopy. The average radius of our NIH 3T3 cells is 7×10^{-4} cm.
4. For D, the diffusion coefficient, use the value of 6.264×10^{-5} cm^2/h for retroviruses *(8,13)*.
5. For N_{co}, use the number of cells counted at the start of infection (*see* **Note 11**).
6. For the integral I, use values obtained from **Table 1**. Values from **Table 1** depend on the time of infection and the radius of the target cells. For a 4-h transduction of NIH 3T3 cells (radius $a_c = 7$ μm), $I = 3.416$.

3.6. Calculation of AVC

1. Use **Eq. 3** to calculate AVC, the number of adsorbed active particles per target cell at a given time interval.
2. For a_c, the radius of the cells, use the average value obtained by fluorescence microscopy.
3. For D, the diffusion coefficient, use the value of 6.264×10^{-5} cm^2/h for retroviruses.
4. For C_{vo}, the concentration of active retrovirus at the start of the infection, use the value calculated from **Eq. 1**.
5. For example, if NIH 3T3 cells are infected for 4 h with a viral stock with 10^6 active particles/mL (i.e., $C_{vo} = 10^6$ active particles/mL) and the virus has a 7 h half-life, the expected number of adsorbed active particles will be AVC = 0.6 particles/cell.
6. Under the same conditions, AVC = 0.13 active particles/cell for HFFs because of differences in η: the efficiency of postadsorption steps between the reference cell type, NIH 3T3 cells and HFFs.

3.7. Calculation of η

1. Use **Eq. 3** to calculate η_x; the efficiency of postadsorption steps of a new cell type (HFFs) relative to a reference cell type (NIH 3T3).
2. For cfu_X, use the number of colony-forming units obtained from the quantitative assay for a retrovirus encoding *lacZ* (*see* **Subheading 3.3.**) performed on HFFs.
3. For CFU_{3T3}, use the number of colony-forming units obtained from the quantitative assay for a retrovirus encoding *lacZ* (**Subheading 3.3.**) performed on NIH 3T3 cells (the reference cell type).
4. Be sure both values of cfu are obtained from a portion of the curve where cfu is linearly proportional to cell number (*see* **Note 4**).
5. For $N_{co,X}$ and $N_{co,3T3}$, use the total number of cells at the start of transduction for HFFs and NIH 3T3 cells, respectively.
6. For $a_{c,X}$ and $a_{c,3T3}$, use the average values of the radii obtained by fluorescence microscopy measurements of HFFs and NIH 3T3 cells, respectively. The average radius of our NIH 3T3 cells is 7×10^{-4} cm; for human fibroblasts it is 10×10^{-4} cm.
7. For I_X and I_{3T3}, use the values of the integrals obtained from **Table 1**. Values from **Table 1** depend on the time of infection and the radius of the target cells. For a 4-h transduction of NIH 3T3 cells (radius $a_{c,3T3} = 7$ μm), $I_{3T3} = 3.416$, whereas for HFFs (radius $a_{c,X} = 10$ μm) $I_X = 3.465$.
8. In prior experiments, the η (similar to relative transduction efficiency [RTE]) of HFFs was significantly lower than that of NIH 3T3 cells ($RTE_{3T3} = 1$ vs. $RTE_H = 0.15$) *(8)*.

4. Notes

1. The fact that the values of C_{vo} are greater than titer indicates that a significantly larger fraction of retrovirus in a stock is active than has been previously thought. It has been widely estimated and quoted that only between 0.1 and 1% of the retrovirus in a stock is active; it was also assumed that these particles were inherently inactive at the start. C_{vo} shows that this is not accurate and that there are many more active retroviral particles. They are active, but limited by decay and diffusion. This may have important implications in the use of retroviruses for in vivo gene transfer.
2. In **Eqs. 1** and **3**, use cfu and not cfu/mL, as determined from a titer assay. In a typical titer assay, a viral stock is serially diluted and a volume of virus is used to transduce the target cells. The titer is computed by multiplying the number of cfu times the dilution factor and dividing this number by the volume of virus added to the target cells. However, because of the slow diffusion and fast decay of retroviral particles, only particles in a small volume (approx 300 μL) above the cells can contribute to infection *(8,10)*. For larger volumes, such as those usually employed in viral infections, the number of cfu does not depend on the volume of viral supernatant. Titer, if expressed as cfu/mL, can introduce a considerable error into **Eqs. 1** and **3**. For example, if a virus stock is diluted 10,000 fold and either 10 or 2 mL of this dilution is used to infect 10-cm dishes containing identical target

cells, the numbers of cfu on both plates will be similar *(2)*. If this number of cfu is 100, then the plate that received 10 mL would report a titer of 1×10^5 cfu/mL, whereas the plate that received 2 mL would report a titer of 5×10^5 cfu/mL, a 5-fold difference. Therefore it is important that values of cfu and not cfu/mL be used in **Eqs. 1** and **3**. By taking into account diffusion and virus half-life, **Eqs. 1** and **3** help to standardize the data obtained from a titer assay.

3. Glutaraldehyde is corrosive and toxic. It is imperative that you prepare solutions in a fume hood wearing gloves and safety glasses. When mixing with water, use glass containers and stir small amounts of glutaraldehyde into water slowly. *Do not add water to glutaraldehyde.* Use cold water to prevent excessive heat generation. Finally, avoid generating vapors and do not use it with incompatible materials such as strong acids, strong alkalines and amines.

4. For **Eq. 1** to be accurate, cfu must be linearly proportional to cell number. Our data show that this is true over a narrow range of cell numbers *(8)* and that this range can vary between different cell types. At high cell numbers, linearity breaks down and cfu declines, possibly due to a decrease in the rate of cell growth, which is known to influence transduction *(12,14)*. Therefore, for each type of target cell used, it is important to do a preliminary experiment that determines the dependence of the number of gene transfer events (cfu) on cell number. The number of target cells used in subsequent quantitative assays must be from the linear part of this curve.

5. Do not let the target cells grow beyond confluence. When overgrown, they may produce endogenous β-galactosidase, which introduces background noise into the assay.

6. In this assay, the virus stock is diluted to very low particle concentrations, so that there are relatively few gene transfer events. After infection, a transduced cell undergoes several divisions to give rise to a cluster *(2–16)* of transduced cells. Low virus particle numbers ensure that clusters are widely separated from one another and that each cluster represents a single gene transfer event.

7. The CellTracker probe come in blue, orange, green, and yellow-green fluorescence derivatives that can be visualized with the appropriate filter set. These dyes enter the cells by passive diffusion through the cell membrane; once in the cytoplasm, they undergo a glutathione S-transferase–mediated reaction, producing a cell-impermeant reaction product. They can be used with a variety of cell types, as most cells contain high levels of glutathione (up to 10 mM) and glutathione transferase. For more information, visit www.probes.com.

8. When cells are close to each other so that their borders cannot be easily distinguished, they should be excluded from the analysis.

9. In each field, select several cells at random and use computer software (e.g., Metamorph or NIH Image) to delineate the perimeter of each cell and compute cell area. For statistically sound results, the average cell surface area should be based on measurements of a significant number of cells (50–100).

10. When using **Eqs. 1** and **3**, the radius of the target cells must be in cm and the diffusion coefficient in cm^2/h. Then the concentration of active retroviral particles, C_{vo}, is expressed in particles/mL.

11. The cell number is determined at the start of infection and is assumed to remain relatively constant during infection. This is a reasonable assumption for short infection times (e.g., 2–4 h), but it may not hold for longer times, especially for fast growing cells. Therefore, short infection times are recommended.

References

1. Kahn, M. L., Lee, S. W., and Dichek, D. A. (1992) Optimization of retroviral vector-mediated gene transfer into endothelial cells in vitro. *Circ. Res.* **71**, 1508–1517.

2. Morgan, J. R., LeDoux, J. M., Snow, R. G., Tompkins, R. G., and Yarmush, M. L. (1995) Retrovirus infection: effect of time and target cell number. *J. Virol.* **69**, 6994–7000.

3. Haas, D. L., Case, S. S., Crooks, G. M., and Kohn, D. B. (2000) Critical factors influencing stable transduction of human CD34(+) cells with HIV-1-derived lentiviral vectors. *Mol. Ther.* **2**, 71–80.

4. Kotani, H., Newton, P.B.I., Zhang, S., et al. (1994) Improved methods of retroviral vector transduction and production for gene therapy. *Hum. Gene Ther.* **5**, 19–28.

5. Paul, R. W., Morris, D., Hess, B. W., Dunn, J., and Overell, R. W. (1993) Increased viral titer through concentration of viral harvests from retroviral packaging cell lines. *Hum. Gene Ther.* **4**, 609–615.

6. Chuck, A. S. and Palsson, B. O. (1996) Membrane adsorption characteristics determine the kinetics of flow-through transductions. *Biotech. Bioeng.* **51**, 260–270.

7. LeDoux, J. M., Davis, H. E., Yarmush, M. L., and Morgan, J. R. (1999) Kinetics of retrovirus production and decay. *Biotech. Bioeng.* **63**, 654–662.

8. Andreadis, S., Lavery, T., Davis, H. E., et al. (2000) Toward a more accurate quantitation of the activity of recombinant retroviruses: alternatives to titer and multiplicity of infection. *J. Virol.* **74**, 3431–3439.

9. Shoup, D. and Szabo, A. (1982) Chronoamperometric current at finite disk electrodes. *J. Electroanal. Chem.* **140**, 237–245.

10. Kahn, M. L., Lee, S. W., and Dichek, D. A. (1992) Optimization of retroviral vector-mediated gene transfer into endothelial cells in vitro. *Circ. Res.* **71**, 1508–1517.

11. Chuck, A. C., Clarke, M. F., and Palsson, B. O. (1996) Retroviral infection is limited by brownian motion. *Hum. Gene Ther.* **7**, 1527–1534.

12. Miller, D. G., Adam, M. A., and Miller, A. D. (1990) Gene transfer by retrovirus vectors occurs only in cells that are actively replicating at the time of infection. *Mol. Cell. Biol.* **10**, 4239–4242.

13. Salmeen, I., Rimai, L., Luftig, R. B., et al. (1976) Hydrodynamic diameters of murine mammary, rous sarcoma and feline leukemia RNA tumor viruses: studies by laser beat frequency light-scattering spectroscopy and electron microscopy. *J. Virol.* **17**, 584–596.

14. Andreadis, S., Brott, D. A., Fuller, A. O., and Palsson, B. O. (1997) Moloney murine leukemia virus-derived retroviral vectors decay intracellularly with a half-life in the range of 5.5 to 7.5 hours. *J. Virol.* **71**, 7541–7548.

14

Minigene-Containing Retroviral Vectors Using an Alphavirus/Retrovirus Hybrid Vector System

Production and Use

Jarmo Wahlfors and Richard A. Morgan

1. Introduction

Retroviral vectors are optimal gene transfer vehicles for gene therapy, since they integrate into the target cell genome and therefore can provide permanent expression of a therapeutic gene. This is a particularly desired feature in gene therapy of monogenic inherited diseases, in which the ultimate goal is a single gene transfer event that would lead to life-long correction of the genetic defect. Moreover, retroviral vectors are thoroughly characterized, and their extensive use in preclinical and clinical studies has verified their efficacy and safety.

1.1. Utility of Complex Expression Cassettes

Numerous studies have demonstrated that noncoding genomic sequences like introns and 3' untranslated regions can improve and stabilize the expression of a transgene cassette, especially in vivo. The effect has been shown with human b-globin *(1,2)*, rat growth hormone *(3,4)*, human purine nucleoside dephosphorylase *(5)*, human clotting factor VIII *(6)*, and especially human clotting factor IX *(7–10)*. However, the combination of complex genomic expression cassettes, minigenes, and retroviral vectors is not simple, since the synthesis of retroviral RNA takes place in the nucleus and the intervening sequences are spliced out. Furthermore, noncoding genomic sequences may contain cryptic transcription termination signals that reduce the amount of the full-length retrovirus RNA and result in low titers. The problem of incompat-

From: *Methods in Molecular Medicine, Vol. 69, Gene Therapy Protocols, 2nd Ed.*
Edited by: J. R. Morgan © Humana Press Inc., Totowa, NJ

ibility between retroviral vectors and minigenes can be solved by cloning the minigene in the antisense orientation between the vector long terminal repeats (LTRs), but this has turned out to be difficult, and extensive modifications to the cassette and the vector were necessary for sufficient vector titers and transgene expression after transduction *(11,12)*. Another way to solve this problem would be to use a strategy whereby retroviral RNA is synthesized outside the nucleus with the aid of an enzyme other than eukaryotic RNA polymerase II.

1.2. Recombinant Alphaviruses: Tools for Cytoplasmic Gene Expression

Alphaviruses, like Semliki Forest virus (SFV) and Sindbis virus, are single-stranded RNA viruses that replicate in the cytoplasm of their target cells *(13)*. Alphaviral replication is very efficient and is carried out by its own enzymes, which form a replicase complex in concert with cellular factors. The replication leads, in the first phase, to amplification of the viral genomic RNA through a negative strand intermediate. The second phase of replication involves formation of excessive amount of subgenomic (26S) RNA, using the negative strand RNA as template. In the wild-type virus, this subgenomic RNA codes for viral structural proteins. However, in recombinant alphaviral vectors, the structural genes have been replaced by a transgene. Recombinant alphaviruses are efficient tools for transient production of large amounts of protein in target cells, but they can also be used to produce retroviral vector RNA that will be packaged into virions in a similar manner as RNA that has been synthesized in the cell nucleus.

1.3. SemLiki Forest Virus in Cytoplasmic Retroviral Vector Production

We have shown that SFV vectors can be manipulated to carry retroviral vector cassettes (RVCs) that encode RNAs readily packageable into biologically active retroviral vector particles *(14)*. The process starts with in vitro production of two SFV RNA species, the first RNA encoding the RVC (SFV-RVC) and second RNA encoding the SFV structural proteins (SFV-helper2). When these RNAs are electroporated together into BHK cells, the cells produce large quantities of RVC-carrying SFV virions. The SFV-RVC virions can then be used to transduce retrovirus packaging cells, in this case PHOENIX cells, which are derived from the human kidney cell line 293T. When alphaviral replication takes place, it starts producing the subgenomic RNA, which in this case is the RVC RNA that is almost identical to the retroviral virion RNA. This RNA is then packaged into retroviral particles that are produced by the PHOENIX cells (**Fig. 1**). These recombinant retroviruses are biologically fully func-

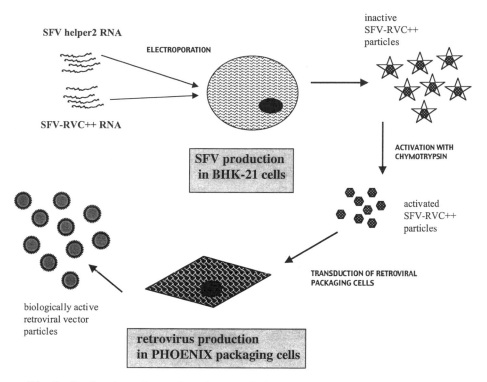

Fig. 1. Production of cytoplasmic retroviral vectors using an alphavirus/retrovirus hybrid system. RVC, retroviral vector cassette; SFV, Semliki Forest virus.

tional: they can enter their target cells through a receptor-mediated mechanism, undergo reverse transcription, and integrate into host cell's genome.

In addition to the method described here, there is an alternative method to produce retroviruses cytoplasmically with the aid of recombinant SFV system (*see* **Note 1**).

1.4. Minigenes in Cytoplasmically Produced Retroviral Vectors

Retroviral vectors produced with the SFV system are presumably independent of nuclear enzymatic activities of producer cells and therefore capable of carrying introns and other noncoding genomic sequences. We tested this hypothesis by cloning a human clotting factor IX minigene into SFV-RVC *(15)*. The resulting SFV vector, SFV-RVC++F9mg5'3', was used to transduce PHOENIX amphotropic packaging cells, and the supernatants were analyzed for the presence of biologically active recombinant retroviruses. The generated retroviral vector titers turned out to be comparable to the titers obtained earlier without the minigene, suggesting that the genomic sequences are not interfering with the cytoplasmic retroviral RNA synthesis. When the minigene-carrying retroviruses were used to transduce target cells (the human medulloblastoma cell line TE671) and the proviral structure in the cell population was analyzed, it was demonstrated that the intron was present in the integrated minigene. Furthermore, the minigene-harboring TE671 cell clones produced high amounts of human factor IX, the best clone producing > 1000 ng FIX/ million cells/24 h.

The capability of cytoplasmically produced retroviruses to carry minigenes and the advantage of the minigene vectors over the conventional, cDNA-carrying recombinant retroviruses has also been demonstrated by another group (*see* **Note 1**).

2. Materials
2.1. Cell Culture

1. BHK-21 cells (ATCC, Manassas, VA; cat. no. CCL-10). Grow in complete BHK medium with 5% fetal bovine serum (FBS; *see* **Note 2**) and split every 2–3 days 1:10.
2. PHOENIX retrovirus packaging cells with amphotropic envelope. These cells are provided by the ATCC (SD 3443), but permission from Dr. Garry Nolan is required. Find download-ready instructions and material transfer agreement forms at http://www.uib.no/mbi/nolan/NL-proto.htmL. Grow in Dulbecco's modified Eagle's medium (DMEM) with 10% FBS, and split 1:5 every 3-5 days. Avoid confluency and extended growth in culture, as the cells can lose their ability to produce retrovirus particles (*see* **Note 3**).
3. TE671 human medulloblastoma cells (ATCC, cat. no. CRL 8805). Grow in DMEM with 10% FBS and split every 2–3 days 1:10.
4. Phosphate-buffered saline (PBS), sterile: can be obtained from most cell culture media manufacturers.
5. Trypsin-EDTA (10× stock, Life Technologies, Gaithersburg, MD).
6. BHK complete medium. G-MEM, 2 mM glutamine, 20 mM HEPES buffer, and 10% tryptose phosphate broth. All components can be purchased from Life Technologies.
7. DMEM, supplemented with glutamine and high glucose (4.5 g/L); can contain antibiotics (penicillin + streptomycin or gentamicin).
8. FBS.
9. Tissue culture dishes: 10-cm plates, 3-cm plates, or 6-well plates.

2.2. Recombinant Semliki Forest Virus Plasmids

1. SFV Gene Expression System. This complete kit for production of SFV-based recombinant proteins or SFV virions can be obtained from Life Technologies (cat. no. 10180-016).
2. SFV vector containing the retrovirus vector cassette, RVC. The plasmid construct SFV-RVC++ *(14)* contains an SFV replicon that produces retroviral virion RNA from the 26S subgenomic promoter. The structure of the plasmid and the

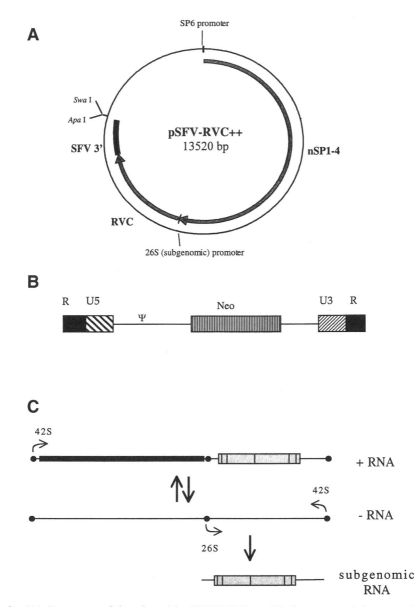

Fig. 2. **(A)** Structure of the plasmid pSFV-RVC++. **(B)** Structure of the retrovirus vector cassette (RVC). **(C)** Illustration of the SFV replication cycle that produces retroviral RNA.

RVC and an illustration of the SFV replication cycle that produces retroviral RNA are shown in **Fig. 2**.

3. Cloning of a transgene cassette/minigene into SFV-RVC++ is described in **Subheading 3.1.** The intermediate plasmids SFV1-RVC and pLRVCSpe++ are needed in this protocol. Requests for these reagents should be addressed to the corresponding author.

2.3. SFV-RVC RNA Synthesis

1. Restriction endonucleases for linearization of pSFV-RVC++ (*Apa*I or *Swa*I) and pSFV-helper2 (*Spe*I) plasmids.
2. Tris-EDTA–buffered phenol (pH 5.6) for extracting the linearized plasmid DNA (= template DNA).
3. Ethanol for precipitation.
4. RNAse-free H_2O for dissolving the template DNA.
5. SP6 mESSAGE mACHINE kit (Ambion, Austin, TX) for in vitro RNA synthesis. This kit contains all the reagents needed for the synthesis of 5' capped RNA molecules (*see* **Note 4** for an alternative way to synthesize RNA).

2.4. Generation of SFV-RVC Viruses

1. Culture of 50–75% confluent BHK-21 cells.
2. SFV-RVC++ and SFV-helper2 RNA.
3. Opti-MEM medium (Life Technologies) for suspending cells prior electroporation (*see* **Note 5**).
4. Electroporator, for example, ECM 600 from BTX (San Diego, CA).
5. Electroporation cuvets with 0.4-cm gap, for example, Gene Pulser Cuvette from Bio-Rad (Hercules, CA).

2.5. Titration of SFV-RVC Viruses

1. An SFV vector preparation with known titer (used as standard in the dot-blot-based titration of SFV-RVC++ vector stocks; *see* **Note 6**).
2. Dot-blot vacuum manifold (Schleicher & Schuell, Keene, NH).
3. Hybond N+ membrane (Amersham, Arlington Heights, IL).
4. 10× SSC: 1.5 *M* sodium chloride, 0.15 *M* sodium citrate.
5. UV-crosslinker (for example, Stratalinker by Stratagene, La Jolla, CA).
6. DNA probe for hybridization, for example a 2.6-kbp *Sac*II fragment from the plasmid pSFV1 (the plasmid is included in the SFV Gene Expression System kit).
7. Reagents for probe labeling, hybridization, and washes of the membrane. Standard methods of your laboratory can be used. Radioactive detection is preferred for better resolution and higher sensitivity.
8. Phosphorimager (for example, BAS1500 by Fuji Medical Systems) for quantitation of the dot intensities. Alternatively, scanned images of good-quality autoradiograms (X-ray films) can be used for quantitation.

2.6. Production and Characterization of Retroviruses

1. SFV-RVC++ vector stock with known titer.
2. Monolayers of 50–75% confluent PHOENIX ampho packaging cells for retrovirus production.
3. PBS with Mg^{2+} and Ca^{2+}.
4. Chymotrypsin A (Sigma; 10 mg/mL in PBS with Mg^{2+} and Ca^{2+}, stored as aliquots at –20°C) and 50 mM $CaCl_2$ solutions for SFV activation. Aprotinin–protease inhibitor (0.5 mg/mL; Sigma) for neutralizing chymotrypsin.
5. 0.45 μm syringe filters (Millipore).
6. Monolayers of TE-671 cells, 1×10^5 cells/6-well plate well, for retrovirus titration.
7. Polybrene 8 μg/mL (hexadimethrine bromide; Sigma), dissolved in water and sterile filtered.
8. G418 antibiotic (neomycin analog, Life Technologies) for selection of Neo^R clones.
9. Reagents for isolation of chromosomal DNA from retrovirus-transduced cells.
10. Probe for detecting proviral retrovirus sequences. For example, *Eco*RI – *Nco*I fragment from the retroviral plasmid pLN *(16)*, which hybridizes to the *Neo* gene.
11. Reagents for performing Southern blot analysis.

3. Methods

3.1. Construction of Semliki Forest Virus–Retrovirus Vector Cassette Hybrid Plasmids

The plasmid pSFV-RVC++ contains a retrovirus vector cassette with a Neo^R gene driven by the retroviral LTR promoter (**Fig. 2**). To clone an expression cassette/minigene into this vector is not straightforward and requires an intermediate cloning step. This is because the *Bcl*I site that can be used for minigene insertion is not unique in pSFV-RVC++ and the plasmid pLRVCSpe++, has to be used first. This plasmid is based on pLITMUS 38 (New England Biolabs, Beverly, MA), and the *Bcl*I enzyme cuts it only once. After inserting a minigene into pLRVCSpe++, an RVC-containing fragment can be cut out with *Spe*I and *Avr*II digestion and ligated with a 10-kbp *Spe*I fragment of the pSFV1-RVC plasmid to yield pSFV-RVC++ with your favorite minigene (*see* **Fig. 3** for the cloning strategy and **Note 7** for more information).

3.2. In Vitro Synthesis of SFV RNA

To produce recombinant SFV-RVC++ viruses, two SFV RNA species (SFV-RVC++ and SFV-helper2) need to be synthesized in vitro using the respective plasmids as templates. The RNA synthesis in this case is carried out using SP6 RNA polymerase; the resulting RNA is 5'-capped and thus is able to behave similarly to mRNA when inside the host cell. Since transcription by SP6 poly-

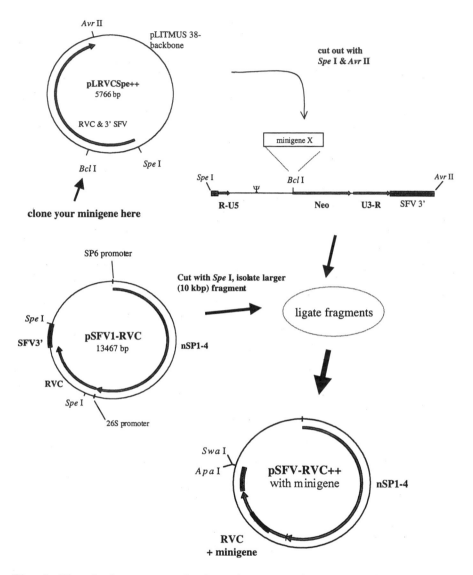

Fig. 3. The cloning strategy for inserting expression cassettes/minigenes into pSFV-RVC++.

merase is a runoff synthesis, the plasmid templates have to be linearized first for a proper 3' RNA termination.

1. Linearize at least 10 μg of the plasmids SFV-RVC++ (*Apa*I or *Swa*I) and SFV-helper2 (*Spe*I). Extract template DNAs using an equal volume of Tris-buffered phenol, precipitate DNA with ethanol, and wash at least twice with 70% ethanol. Dissolve in RNAse-free H_2O, aiming at 1 μg/μL concentration (*see* **Note 8**).

2. Perform in vitro transcription using the instructions from the SP6 mESSAGE mACHINE kit (Ambion). We routinely use 20 μL reaction volume, 1 μg linear template DNA and add 1 μL of the GTP solution (included in the kit; improves the yield with long templates). Incubate for 90–120 min at 37°C. Remove the template DNA by adding 1 μL of DNase I from the kit (2 U/μL) and incubate for an additional 15 min (*see* **Note 9**).

3. Precipitate RNA by adding 30 μL of nuclease-free H_2O and 25 μL lithium chloride precipitation solution from the kit. Chill overnight or at least 30 min at –20°C. Pellet RNA by spinning at 4°C for 15 min at maximum speed, wash the pellet twice with cold 70% ethanol, and air-dry briefly (*see* **Note 9**).

4. Dissolve the RNA pellet in 50 μL of nuclease-free H_2O. Keep on ice until electroporation or store at –20°C. Analyze concentration by spectrophotometer and quality by gel electrophoresis (*see* **Note 10**).

3.3. Generation and Characterization of SFV-RVC++ Virions

The RVC-containing SFV virions are produced in BHK-21 cells, using co-electroporation of vector RNA (SFV-RVC++) and helper RNA (SFV-helper2). The production is complete at 24 h after electroporation. The resulting SFV virions are conditionally inactive because of modification in their capsid protein and become activated upon chymotrypsin treatment *(17)*. Since the SFV-RVC++ particles do not express a marker gene, they can be titrated only indirectly. We use a dot-blot method that detects the number of SFV genomes in the supernatants *(18)*.

1. Grow BHK-21 cells to 50–70% confluency, trypsinize the cells, suspend the pellet in growth medium, and count the cells. Take 1.5×10^6 cells per electroporation reaction, wash twice with Opti-MEM, and suspend in Opti-MEM to yield 1.5×10^6 cells/550 μL.

2. Pipet 550 μL of the cell suspension to an electroporation cuvet (0.4-cm gap). Add 20 μg of SFV-helper2 RNA and 40–50 μg of SFV-RVC++ RNA (*see* **Note 11**), mix by flicking the cuvet and electroporate immediately.

3. Electroporation (*see* **Note 12**): we determined that three pulses of 7 ms each at 250 V gave optimal results with BHK-21 cells (when ECM 600 by BTX was used). Mix by flicking between pulses and plate onto a 3-cm plate with 2 mL of DMEM, 10% FBS. Incubate at 37°C for 24 h.

4. Collect SFV supernatants, filter slowly through a 0.45 μm syringe filter, divide into aliquots, and store at –80°C.

5. Determine the virus titer by a whole-virion dot-blot method. This method requires that an SFV virus stock with known titer be available. When the hybridization signal of the produced SFV-RVC++ stock is compared with the signal of the reference stock, the titer (number of RNA genomes in the preparation) can be estimated (*see* **Note 13**).

6. Make 10-fold dilution series of the SFV-RVC++ virus and the reference virus in 10× SSC and pipet the dilutions (180 μL each) onto Hybond N+ membrane with the aid of a dot-blot manifold. Wash the wells of the manifold 3 times with 10× SSC.

7. Air-dry and UV-crosslink the membrane. Prehybridize (for example, the following conditions can be used: incubate at 42°C for 1–3 h in solution containing 37.5% dextran sulfate, 7.5× sodium-saline-phosphate EDTA buffer [SSPE], 0.75% sodium dodecyl sulfate [SDS], 25% formamide, 5× Denhardt's solution, and 2 mg/mL denatured, sheared DNA).

8. Hybridize overnight at 42°C by adding a radiolabeled SFV probe to the prehybridization solution. We have used a 2.6-kbp *Sac*II fragment from the plasmid pSFV1. Wash the membrane using increasing stringency and expose to an autoradiogram reader screen or an X-ray film. Determine the signal intensities from SFV-RVC++ and the reference virus, using the dots of intermediate intensity, and calculate the titer.

3.4. Cytoplasmic Production of Retroviruses in PHOENIX Packaging Cell Lines

1. Grow amphotropic PHOENIX packaging cells to 80–90% confluency on 3- or 10-cm plates. Determine the approximate number of cells on the plate.

2. Activate enough SFV-RVC++ viruses for multiplicity of infection (MOI) 100: add 1/16 volume of chymotrypsin solution and 1/50 vol of 50 mM CaCl$_2$. Incubate for 15 min at room temperature and inactivate the protease by adding 1/4 vol of aprotinin. Store activated virus on ice (*see* **Note 14**).

3. Remove the growth medium from PHOENIX cells and wash them once with PBS containing Mg^{2+} and Ca^{2+}. Add activated virus in a volume of growth medium that barely covers the cells, and incubate for 90 min at 37°C. Remove the supernatant, wash the cells twice with PBS, and add an appropriate volume of growth medium (1 mL for a 3-cm plate, 5 mL for a 10-cm plate). Incubate for 16 h at 32°C, collect the retroviral supernatant, filter through a 0.45-μm syringe filter, divide into aliquots, and store at –80°C (*see* **Note 14**).

3.5. Analysis of the Produced Retroviral Vectors

Retroviral vectors that contain a functional *Neo* gene can be titrated based on the G418 resistance of transduced TE 671 cells (*see* **Note 15**). The same cells can be used to analyze the structure of integrated proviruses and the integrity of the minigene expression cassettes.

1. Titration: Plate 1×10^5 TE 671 cells/well of a 6-well plate (well diameter 3 cm) 24 before transduction (one plate per retrovirus stock).
2. Make a 1:10 dilution series of the virus in growth medium containing 8 µg/mL Polybrene; the final volume of each dilution is 2 mL. Aspirate the old growth medium from TE 671 cells and add virus-containing dilutions. Spin the plates at 32°C, 1,000g for 1 h, and incubate for 23 h at 32°C (*see* **Note 16**).
3. Replace the retrovirus-containing medium with 2 mL of growth medium with 0.8 mg/mL G418. Incubate for 7–14 days, changing the selection medium as necessary, until the majority of the cell layer is dead and separate colonies can be seen. Count the colonies from wells containing 10–100 colonies and calculate the titer.
4. Provirus analysis: Isolate and grow G418-resistant cell clones (*Neo* gene-containing viruses) or transduce TE 671 cells with as high an MOI of the virus as possible and grow the transduced population/isolate clones by limiting dilution.
5. Isolate chromosomal DNA from the provirus-harboring cells, using the standard method of your laboratory. Perform a Southern blot with appropriate enzymes that cut inside the provirus and reveal the minigene structure. Hybridize with a probe that recognizes either minigene or the retroviral sequences. The presence of an intron in the minigene can also be detected by a polymerase chain reaction method. Design primers that hybridize to exon sequences around the intron and detect the size of the PCR product (which is longer if the intron is present in the provirus).

4. Notes

1. Li and Garoff have *(19)* demonstrated an alternative way to use SFV vectors for cytoplasmic retrovirus production. In their system, an SFV vector carrying a retrovirus vector cassette almost identical to ours was used. However, in their protocol, the production was not carried out in retroviral packaging cells, but in BHK-21 cells. The coding sequences for retroviral virions were provided by two additional SFV vectors that encode a retroviral *Gag-Pol* gene cluster and a heterologous envelope, in their case from the murine amphotropic virus 4070A. When these three RNA species were electroporated together into BHK-21 cells, high titers of biologically active retroviruses were produced. Li and Garoff also showed that retroviruses are capable of carrying introns into target cells; the presence of these sequences increased the transgene expression significantly *(20)*. Interestingly, they did not use natural introns in their natural locations but still obtained enhanced gene expression, suggesting that the presence of any intervening sequence that interrupts the coding sequence will be beneficial.
2. BHK-21 cells grow well in most tissue culture mediums. However, only complete BHK medium with 5% FBS or DMEM with 10% FBS can be recommended, since they have been tested by us and shown to support SFV production.
3. PHOENIX packaging cells can loose their ability to support retrovirus production under some conditions (for example, cells are frequently grown to confluency or maintained in continuous culture for too long). Therefore, a stock of low-passage cells should be frozen and fresh cells should be taken into use frequently. If the

cells start deteriorating and backup cells are not available, a selection by drug resistance or surface marker is possible to a population with a packaging phenotype (for details, see http://www.uib.no/mbi/nolan/NL-helper.htmL).

4. RNA synthesis can also be performed without the kit. For instructions, see the manual of the SFV Gene Expression System.

5. Opti-MEM or DMEM without phenol red turned out to be the best media for suspending cells prior to electrophoresis. PBS also works fine, but care should be taken not to let the cells stay in warm buffer for more than 10 min (Cells start to loose their viability and capacity to produce SFV particles.)

6. SFV vectors with detectable marker genes like *lacZ* or *GFP* can be used as reference. These vector preparations are easy to titrate by transducing BHK-21 cells with a dilution series of the virus, followed by detection of blue cells after X-Gal staining (*lacZ*) or fluorescent cells in a flow cytometer (*GFP*). The plasmid pSFV3-lacZ is included in the SFV Gene Expression System and can be used to produce the reference SFV vector stock.

7. This cloning strategy saves the Neo^R gene but leaves it without a promoter. However, we have observed that the promoterless Neo^R gene can be used in G418 selection.

8. Phenol extraction, followed by ethanol precipitation, is the best way to purify long templates for RNA synthesis. Other methods (like GeneClean) may introduce breaks in the DNA and reduce the yield of full-length RNA. Linearize large amounts of plasmids (20–50 μg) and phenol-extract using volumes of at least 100 μL to obtain good yields. Verify the result by agarose gel electrophoresis before proceeding to RNA synthesis (>90% of the plasmid should be in linear form). This step should be performed in an RNase-free environment (preferably on a bench dedicated to RNA work, with clean pipets and other equipment).

9. Removal of the DNA template or precipitation of RNA may not be crucial steps, but they are recommended since we have used them in our experiments. The manual of the SFV Gene Expression System indicates that in vitro synthesized RNA is suitable for electroporation without any purification steps.

10. One reaction routinely yields 30–60 μg of purified RNA, and the typical A_{260}/A_{280} ratio is about 2.2. Lower yields may indicate a poor-quality DNA template or problems with the kit. (Always include a control reaction, using the template pTRI-Xef1 from the kit.) The quality of the RNA can be checked on regular 1.5% agarose gel, using TAE or TBE buffer. However, RNA is vulnerable in this kind of environment and degrades rapidly. Use high voltage and a short time to verify the overall integrity of the RNA; accurate size determination under these conditions is not possible. (Denaturing RNA gel with formaldehyde is required for more precise analysis of the produced RNA species.) Some templates are more difficult to transcribe, resulting in low proportions of the full-length product and varying number of shorter RNA species. These are likely to be caused by premature runoff of the SP6 polymerase at difficult secondary structures of the template DNA.

11. The amount of RNA for one electroporation is high, since the virus yields were directly proportional to the amount of SFV-RVC++ RNA (up to 50 μg). This is

probably because the RNA preparation consists of a mixture of different species and only up to 20% of it is full-length.

12. Electroporation conditions depend on the equipment in use. It is recommended that other electroporators than the one we used be optimized for maximal efficiency and virus production. The number of pulses given (under indicated conditions) can also vary; we have observed that BHK-21 cells can take up to 4 pulses, but in this study only 3 pulses were used.

13. The whole-virion dot-blot method is a rapid and easy way to get an estimate of the number of SFV-RVC++ virions in the preparation. There is no need for prior virion RNA purification, and different conditions for dot blotting and hybridization can be used. The sensitivity of this method is about 1000 SFV virions/mL, which is sufficient to give titers of any relevant virus preparation.

14. We used an SFV-RVC++ MOI of 100 on PHOENIX cells, although even higher virus amounts can yield better retrovirus production. When adjusting the volume of activated SFV-RVC++ for retrovirus packaging cell plates, be aware of the possibility that too small volumes can evaporate significantly during the 90-min incubation and result in low retrovirus yield (due to damaged producer cells). The growth temperature (32°C) and incubation time (16 h) of PHOENIX cells are optimized for maximal production and minimal degradation of amphotropic retroviruses.

15. Retroviral vectors with an inserted minigene (and promoterless *Neo* gene) cannot be titrated with the G418 resistance assay; use the dot-blot titration method described in **Subheading 3.3.** Alternatively, expression of a detectable transgene can be used for titer determination.

16. Spin transduction is not mandatory, but it increases the titers obtained. Also, incubation at 32°C stabilizes retroviral particles and yields a better transduction efficiency.

References

1. Grosveld, F., van Assendelft, G. B., Greaves, D. R., and Kollias, G. (1987) Position-independent, high-level expression of the human beta-globin gene in transgenic mice. *Cell* **51,** 975–985.

2. Buchman, A. R. and Berg, P. (1988) Comparison of intron-dependent and intron-independent gene expression. *Mol. Cell. Biol.* **8,** 4395–4405.

3. Brinster, R. L., Allen, J. M., Behringer, R. R., Gelinas, R. E., and Palmiter, R. D. (1988) Introns increase transcriptional efficiency in transgenic mice. *Proc. Natl. Acad. Sci. USA* **85,** 836–840.

4. Palmiter, R. D., Sandgren, E. P., Avarbock, M. R., Allen, D. D., and Brinster, R. L. (1991) Heterologous introns can enhance expression of transgenes in mice. *Proc. Natl. Acad. Sc.i USA* **88,** 478–482.

5. Jonsson, J., Foresman, M., Wilson, N., and McIvor, R. (1992) Intron requirement for expression of the human purine nucleoside phosphorylase gene. *Nucleic Acids Res.* **20,** 3191–3198.

6. Connelly, S., Gardner, J. M., McClelland, A., and Kaleko, M. (1996) High-level

tissue-specific expression of functional human factor VIII in mice. *Hum. Gene Ther.* **7,** 183–195.

7. Jallat, S., Perraud, F., Dalemans, W., et al. (1990) Characterization of recombinant human factor IX expressed in transgenic mice and in derived trans-immortalized hepatic cell lines. *EMBO J.* **9,** 3295–3301.

8. Kurachi, S., Hitomi, Y., Furukawa, M., and Kurachi, K. (1995) Role of intron I in expression of the human factor IX gene, *J. Biol. Chem.* **270,** 5276–5281.

9. Wang, J. M., Zheng, H., Sugahara, Y., et al. (1996) Construction of human factor IX expression vectors in retroviral vector frames optimized for muscle-cells. *Hum. Gene Ther.* **7,** 1743–1756.

10. Nakai, H., Herzog, R. W., Hagstrom, J. N., et al. (1998) Adeno-associated viral vector-mediated gene transfer of human blood coagulation factor IX into mouse liver. *Blood* **91,** 4600–4607.

11. Sadelain, M., Wang, C., Antoniou, M., Grosveld, F., and Mulligan, R. (1995) Generation of a high-titer retroviral vector capable of expressing high levels of the human beta-globin gene. *Proc. Natl. Acad. Sci. USA* **92,** 6728–6732.

12. Jonsson, J. J., Habel, D. E., and McIvor, R. S. (1995) Retrovirus-mediated transduction of an engineered intron-containing purine nucleoside phosphorylase gene. *Hum. Gene Ther.* **6,** 611–623.

13. Strauss, J. H. and Strauss, E. G. (1994) The alphaviruses: gene expression, replication, and evolution. *Microbiol. Rev.* **58,** 491–562.

14. Wahlfors, J. J., Xanthopoulos, K. G., and Morgan, R. A. (1997) SemLiki Forest virus-mediated production of retroviral vector RNA in retroviral packaging cells. *Hum. Gene Ther.* **8,** 2031–2041.

15. Wahlfors, J. J. and Morgan, R. A. (1999) Production of minigene-containing retroviral vectors using an alphavirus/retrovirus hybrid vector system. *Hum. Gene Ther.* **10,** 1197–1206.

16. Miller, A. and Rosman, G. (1989) Improved retroviral vectors for gene transfer and expression. *Biotechniques* **7,** 980–990.

17. Berglund, P., Sjoberg, M., Garoff, H., Atkins, G., Sheahan, B., and Liljestrom, P. (1993) SemLiki Forest virus expression system: production of conditionally infectious recombinant particles. *Biotechnology* **11,** 916–920.

18. Nelson, D. M., Wahlfors, J. J., Chen, L., Onodera, M., and Morgan, R. A. (1998) Characterization of diverse viral vector preparations, using a simple and rapid whole-virion dot-blot method. *Hum. Gene Ther.* **9,** 2401–2405.

19. Li, K. J. and Garoff, H. (1996) Production of infectious recombinant moloney murine leukemia-virus particles in BHK cells using Semliki Forest virus-derived RNA expression vectors. *Proc. Natl. Acad. Sci. USA* **93,** 11658–11663.

20. Li, K. J. and Garoff H. (1998) Packaging of intron-containing genes into retrovirus vectors by alphavirus vectors, *Proc. Natl. Acad. Sci.* USA **95,** 3650–3654.

15

Retrovirus-Mediated Gene Transfer to Human Hematopoietic Stem Cells

William P. Swaney, Enrico M. Novelli, Alfred B. Bahnson, and John A. Barranger

1. Introduction

The ability of hematopoietic stem cells (HSCs) to engraft in a recipient and establish long-term repopulation of the hematopoietic system makes them ideal targets for gene therapy vectors designed to correct inherited or acquired diseases affecting the hematopoietic and immune systems.

Monoclonal antibodies that bind to the CD34 antigen, a universally recognized marker for hematopoietic progenitor/stem cells (1–3), have been widely used to isolate HSCs from the hematopoietic tissues. Other cell surface antigens have also been targeted to subfractionate the heterogeneous CD34$^+$ cells into subpopulations with distinct phenotypes and HSC potential (4). In vitro and in vivo functional assays have defined a hierarchy of biologically different cells, from subpopulations enriched for early long-term repopulating, retransplantable HSCs, to those only including committed progenitor cells. The colony-forming cell (CFC) assay provides an estimate of myeloerythroid progenitor cell potential (5); the assays for long-term culture initiating cells (LTC-IC), blast colony-forming cells (CFC-blast), and cobblestone area-forming cells (CAFC) measure more primitive hematopoietic progenitors (6). On the other end of the spectrum, immunodeficient mouse models, such as the NOD/SCID assay, and other xenogeneic transplantation models like the preimmune fetal sheep allow measurement of the ability of cells to generate sustained, multilineage hematopoiesis after in vivo transplantation and therefore represent surrogate HSC assays (7). Animal assays are particularly valuable in gene therapy because they permit assessment of the efficiency of transduction and

From: *Methods in Molecular Medicine, Vol. 69, Gene Therapy Protocols, 2nd Ed.*
Edited by: J. R. Morgan © Humana Press Inc., Totowa, NJ

persistence of expression of transferred genes in human hematopoietic cells in vivo. Of particular interest is the observation that transduced human CD34$^+$ cells were able to repopulate the NOD/SCID strain *(8–12)*, thereby providing a model for gene transfer studies in patients.

Both in vitro *(13,14)* and in vivo assays *(15,16)* have indicated that the CD34$^+$ subpopulation that is negative for the surface antigen CD38 is enriched for primitive HSCs. These studies have shown that the cells responsible for long-term, retransplantable hematopoiesis in NOD/SCID mice and fetal sheep reside in the CD34$^+$ CD38$^-$ fraction. However, the definitive human HSC phenotype still remains elusive as additional putative markers for HSC such as AC133 *(17,18)* and KDR *(19)* are continuously discovered. Recent studies also demonstrate the presence in both primates and rodents of a rare CD34-Lin cell subset *(20)*, which could be a more immature precursor of CD34$^+$ cells and their subpopulations *(21)*. It will be necessary to elucidate fully the functional characteristics and biologic behavior of these populations before targeted, highly efficient gene therapy strategies for human HSCs can be implemented consistently.

A major impediment to retroviral transduction of HSCs has been their paucity in hematopoietic tissues. The CD34$^+$ cells account for only 1–4% of mononuclear cells in the bone marrow (BM), <1% in peripheral blood (PB), and <1% in cord blood (CB) *(22,23)*. CD34$^+$ enrichment with or without subsequent culture and expansion is therefore necessary for most gene therapy applications.

CD34$^+$ enrichment provides the advantage of greatly reducing the quantities of viral supernatants needed to achieve a useful multiplicity of infection (MOI), compared with requirements for nonenriched cells from these sources. For example, with a titer of 10^6 colony-forming units (cfu)/mL, full-strength supernatant will achieve an MOI of 10:1 when cells are suspended at the concentration of 10^5 cells/mL. An average marrow infusion of 2 × 10^8 nucleated cells for a 70-kg patient would require suspension of the cells in 140 L of supernatant to achieve this MOI. Enrichment reduces these requirements by a factor of 100, making the procedure practical. It also allows the use of fewer growth factors (GFS) and less dimethylsulfoxide (DMSO), thereby lowering the costs and discomfort of the procedure.

Ex vivo expansion of enriched HSC populations may be required to provide adequate cell numbers for the clinical application of stem cell therapy and to allow the use of low-abundance HSC sources such as CB. Expanded HSCs may provide higher engraftment levels of donor cells expressing the normal gene needed for correction of inherited enzyme deficiencies; expanded progenitor cells may achieve a faster, stronger support of hematopoietically compromised hosts. The ability to control HSC proliferation and self-renewal is also critical in the gene therapy arena, since successful gene transduction by retroviral vectors requires integration of the retroviral insert into cellular DNA, which in turn requires cellular proliferation *(24)*. The process by which target cells are primed for retroviral infection by stimulation in culture is called prestimulation. Ex vivo culture of HSC has utilized hematopoietic GFs *(10,25–27)*, with or without marrow stromal cells *(28,29)*. Interleukin 3 (IL-3), IL-6, and stem cell factor (SCF) have been extensively used for promoting retroviral transduction and survival of murine *(31–32)* and human *(33–36)* progenitor/stem cells. However, recent evidence suggests that the use of Flt-3 ligand (FL) and thrombopoietin (Tpo) may be associated with better preservation and in some cases limited expansion of progenitor/stem cell capacity in cultures of purified CD34$^+$ cells or CD34$^+$ subpopulations *(25,27,37–40)*. Newly developed lentiviral vectors may obviate the need to prestimulate target cell populations altogether, because of their ability to infect nondividing cells *(41,42)*.

Specific conditions for optimal retroviral transduction of CD34$^+$-enriched cells have not been conclusively determined. Reports and protocols differ with respect to many variables, e.g., type of retroviral vector employed, conditions of prestimulation (GFs used and timing), chemical (i.e., incubation with polybrene or over CH296 fibronectin fragment-coated plates) *(43)* and physical (i.e., centrifugal enhancement) *(44)* viral attachment factors used, and the presence or absence of stromal cells *(45,46)*. Using an optimized protocol based on coculture with the murine producer cells, we and others have demonstrated transduction of cells capable of long-term reconstitution of lethally irradiated murine primary recipients, which yielded progeny capable of reconstituting secondary lethally irradiated recipient mice *(47,48)*. Using an MFG-based vector, we have demonstrated expression in secondary recipients up to 12 months post transplantation *(47)*.

The studies of gene transduction and transplantation of genetically modified HSCs in humans have been less encouraging. This can be attributed to numerous factors, including the refractory status of the human HSC *(13)*, with consequently poor transduction efficiency, lack of sustained gene expression *(36)*, and poor engraftment following transplantation in non-myeloablated recipients *(45,46,49)*.

As of 1999, over 300 gene therapy clinical trials have been conducted in humans, for a wide range of diseases including AIDS (50), cancer *(51,52)*, and single gene disorders like adenosine deaminase (ADA) deficiency *(36,53)*, X-linked severe combined immunodeficiency (SCID) *(54)*, hemophilia B *(55)*, chronic granulomatous disease *(56)*, and Gaucher's disease *(45,57)*. In approximately half of these trials, a retroviral vector was used as the gene delivery vector. However, long-term clinical correction of the disease was only achieved in the trial of gene therapy for X-linked SCID by Cavazzana-Calvo et al. *(54)*.

This X-linked primary immunodeficiency disorder, caused by the γc cytokine receptor deficiency, represents an ideal target for gene therapy because the correction of the patients HSCs results in a selective advantage of their T-, natural killer (NK)-, and B-cell normal progeny. The selective expansion of the corrected cells may allow the limitation of poor transduction efficiency to be circumvented, waiveing the requirement for prior myeloablation of the recipient. In the trial by Cavazzana-Calvo et al. *(54)*, the BM CD34$^+$ cells from two patients were cultured in CH296 fibronectin fragment-coated, gas-permeable containers in serum-free medium supplemented with 4% fetal calf serum (FCS), SCF, IL-3, FL, and polyethylene-glycol-megakaryocyte differentiation factor for 4 days, with daily additions of γc-MFG-containing supernatant on days 2, 3, and 4. The transduction efficiency of CD34+ cells after infection was 20–40%. The infusion of $17–19 \times 10^6$ CD34$^+$ cells without myeloablation resulted in transgene-expressing T and NK cells at a 10-month follow-up in two patients, with T-, B-, and NK-cell counts and function comparable to those of age-matched controls. The authors speculate that improvements in the transduction method, i.e., use of the fibronectin fragment and a potent GF cocktail, may be responsible for the success of this trial compared with others conducted in ADA-deficient patients *(54)*.

In vivo selection of transduced cells may also be exploited in gene therapy strategies aiming at conferring multidrug resistance (MDR) to HSCs of patients undergoing "dose-intensive" chemotherapy for malignant tumors. Abonour et al. *(51)* showed that CD34$^+$ cells transduced with MDR engrafted to levels as high as 15% in the BM of patients with germ cell tumors, for more than a year. The circulating transgene-expressing cells decreased with time (and always remained below 1% of total PB leukocytes), but their number could be selectively expanded by the administration of etoposide. Similarly to Cavazzana-Calvo et al. *(54)*, the authors hypothesize that the transduction strategy employed, which consisted of two transduction steps on theCH296 fibronectin fragment after 2 days of prestimulation in SCF and IL-6 or granulocyte colony-stimulating factor (G-CSF), MGDF and SCF, was responsible for the relatively high transduction efficiency. However, the expression level was low and is unlikely to have provided clinical benefit.

In disorders in which the normal cells do not possess a selective advantage over the diseased counterpart, high-level engraftment of transgene-expressing HSCs in non-myeloablated patients has not been achieved. For example, in recent clinical trials for chronic granulomatous disease (CGD) *(56)* and Gaucher's disease *(45,57)*, gene marked cells could be detected in the pB of the patients, but at levels not sufficient to achieve consistent, if any, clinical efficacy.

In the clinical trial for the p47^{phox}-deficient CGD, PB CD34$^+$ cells cultured for 4 days in gas-permeable containers in serum-free medium supplemented with PIXY321 and G-CSF were transduced on days 2, 3, and 4 by centrifugal enhancement with a p47phox-encoding MFGS retroviral vector and transplanted at a dose of 0.1–4.7 Y 10^6/kg in five non-myeloablated patients. The peak correction of granulocytes in PB occurred 3–6 weeks after infusion and ranged from 0.004 to 0.05% of total PB granulocytes. In some patients, corrected cells could be detected for as long as 6 months after transplantation *(56)*. In another trial directed toward gp91phox-deficient CGD, which employed a similar design, slightly higher numbers of peak phox+ neutrophils were detected in the PB of two patients, ranging from 0.2 to 0.6% of total neutrophils *(58)*.

Three clinical trials of HSC gene transfer and autologous transplantation without myeloablation for type 1 Gaucher's disease have been performed. Two trials employed a protocol consisting of transduction on autologous BM stroma in the presence of IL-3, IL-6, and SCF (with or without IL-1), protamine-sulfate, and glucocerebrosidase (GC)-expressing viral supernatant for 3–5 days. In the first clinical trial, no gene-transduced cells engrafted in the patients *(59)*, whereas in the second a low level of corrected cells (0.02%) was detected in the PB, which did not result in GC expression and clinical benefit *(45)*. In the third clinical study, which was conducted by our group, we used a 1-day prestimulation of CD34+ cells in long-term BM culture medium containing IL-3, IL-6, SCF, and protamine sulfate, followed by two 2-h or one 4-h transduction with GC-expressing MFG viral supernatant, through the method of centrifugal enhancement. We provide the details of this gene therapy protocol below. Four patients received transplants of genetically modified cells. Transduction efficiency averaged 20%, and enzymatic activity of GC in transduced CD34+ cells increased 10-fold over baseline. The average transduction efficiency in HPP-CFC was 9% and for one of the transductions, 29% of the colonies were marked. Total PB lymphocytes and CD34+ cells harvested from the PB showed presence of the GC transgene by polymerase chain reaction (PCR) and increased GC enzymatic activity following transplantation. In one patient, the GC activity of total PBL rose to a level of as high as 80% of control, and the transgene was detected in the three different lineages of lymphocytes, polymorphonuclear leukocytes, and monocytes by fluorescence-activated cell sorting (FACS). These results permitted a gradual withdrawal of enzyme replacement therapy (ERT) over 9 months. During this time and for an additional 5 months, the enzymatic activity in PB lymphocytes remained substantially above deficient levels. Moreover, the dose reduction did not result in a decline in clinical status. However, over the next 12 months, clinical laboratory parameters slowly worsened, and ERT was reinstituted. In the other three patients, no signs of toxicity were observed. However, the engraftment of transduced CD34+ cells was not persistent, and the enzymatic activity was only marginally increased *(57)*. We conclude that CD34$^+$ cells from Gaucher's disease patients can be safely transduced with a retroviral vector and transplanted in non-myeloablated recipients. In one patient, the engraftment of transduced cells and the correction of the enzymatic deficit persisted for more than 2 years and were accompanied by an apparent clinical benefit. We hypothesize that improvements to the gene therapy protocol may lead to consistent, long-term efficacy.

Several strategies can be used to increase the efficiency of HSC gene therapy. First, improvements can be made to the retroviral vectors, by using oncoretrovirus-derived constructs with a better expression profile *(46)*, or by using non-oncoretroviral vectors like the HIV-derived vectors *(41)*. Second, in vitro conditions that would increase the cycling of HSC, leading to higher transduction efficiency, can be explored *(46)*. These include GF stimulation and the use of cyclins, cyclin-dependent kinases, and their inhibitors to act at the level of the intracellular regulators that control cell cycle *(60)*. Third, the BM microenvironment, which includes the stroma and molecules of the extracellular matrix, such as fibronectin, has been shown to enhance HSC transduction. It is therefore possible that once the biologic mechanisms underlying the interaction between the HSC and the microenvironment are elucidated, additional improvements will occur through the use of this strategy. Fourth, liposomal pretreatment of retroviral supernatants has been shown to enhance the effect of centrifugal enhancement on transduction efficiency of CD34$^+$ cells synergistically *(61)*. This procedure was readily transferable to a clinical setting (Swaney, personal communication) and thus is a candidate for large-scale clinical use in future trials.

2. Materials

2.1. Media

1. Human long-term bone marrow culture medium (LTBCM) is used for prestimulation and expansion of CD34$^+$ cells and for long-term cultures. It consists of Iscove's modified Dulbecco's medium (IMDM; Gibco-BRL, Gaithersburg, MD) containing 12.5% fetal bovine serum (FBS; Gibco-BRL), 12.5% horse serum (HS; Gibco-BRL), 2 mM L-glutamine, 1×10^{-6} M 2-mercaptoethanol, 1×10^{-6} M alphathioglycerol, and 1 µg/mL hydrocortisone.
2. Retroviral vector-packaging cells are grown in Dulbecco's modified Eagle medium (DMEM; Gibco-BRL) supplemented with 10% (v/v) calf serum (CS) and 2 mM L-glutamine.

2.2. Additional Reagents

1. Cytokines, IL-3, IL-6, and SCF are prepared as concentrated stock solutions in LTBMC. The 1000× solution contains 10 µg/mL of each cytokine.

2. Methylcellulose medium (without erythropoietin) is obtained ready to use from Stem Cell Technologies (Vancouver, BC).

2.3. Immunoaffinity Column and Anti-CD34 Antibody

The biotinylated anti-CD34 antibody, the avidin column, and necessary reagents (CEPRATE LC [CD34] Cell Separation System) are supplied by CellPro (Bothell, WA).

2.4. Supplies

1. Nunclon cell factories (Fisher cat. no. 12-565-39).
2. Teflon connector with O-ring (Fisher cat. no. 12-565-43).
3. Blue sealing caps (Fisher cat. no. 12-565-45).
4. Tyvek cover caps (Fisher cat. no. 12-565-44).
5. Gelman bacterial air vent (Fisher cat. no. 09-730-125).
6. Flex Boy 10-L collection bag (Stedim cat. no. 400-190).
7. Tubing, OD 3/8 inches ID 1/8 inches (Stedim cat. no. 200-004).
8. Connector MPC female (Stedim cat. no. 100-117).
9. 100-L or other custom size collection bag (Stedim cat. no 950-309).
10. Filter cartridges (0.45 μm).
11. 15-mL conical tubes.

3. Methods

3.1. Generation of Clinical Grade Amphotropic Vector Containing Supernatants

The production of amphotropic producer cell lines containing nonselectable markers has been previously described *(62)*. Briefly, amphotropic packaging cells are generated by infection with filtered supernatants from ecotropic producers, stably or transiently transfected with the vectors in plasmid form, and cloned by limiting dilution assay. Supernatants from these prospective clones are then screened for viral activity by infecting appropriate target cells and assaying for transgene expression by the relevant bioassay. High-titer clones can then be selected, preserved, and characterized; promising clones can be expanded into a master cell bank (MCB). Clinical grade retrovirus vector-containing supernatants (supes) are used for gene transfer into patient tissues. They are generated from viably thawed cells from an approved MCB and expanded in culture. Fresh medium is applied to and harvested from confluent monolayers of producer cells according to a predetermined schedule based on pilot studies. Individual supernatant collections harvested over the scheduled time are stored and then pooled into a single lot for filtering and distribution into final containers. A supernatant lot is considered to be a quantity of material that has been thoroughly mixed in a single vessel.

1. Clinical grade retroviral vector containing supernatants are produced from a certified MCB. Prior to release of a supernatant lot for clinical use, both the MCB and the supernatant must be fully tested and certified for safety according to current regulations and guidelines specified by the Center for Biologics Evaluation and Research (CBER). The tests currently include sterility, mycoplasma, the presence of adventitious viral agents, detection of endotoxin, general safety, the presence of replication-competent retrovirus (RCR) in cells and supernatant, and a characterization of the cells.

2. Clinical grade supernatant is produced in a biopharmaceutical manufacturing facility under aseptic processing conditions. One or more vials of cryopreserved cells are removed from the MCB, thawed, and diluted by dropwise addition of culture medium containing 10% CS (v/v). The cells are collected by centrifugation at $700g$ for 5 min at 4°C and the supernatant removed. The cells are resuspended, diluted with 10 mL of culture medium, and counted; the viability is then determined by Trypan blue dye exclusion.

3. Plate the cells in an appropriately sizes culture vessel and incubate at 37°C, 5% CO_2.

4. As the cells reach confluence, expand them as necessary to obtain a predetermined number of monolayer Nunclon cell factories. The final number of vessels will depend on the desired lot size and supernatant collection schedule.

5. When cells reach confluence, begin the scheduled collections by collecting the conditioned medium and refeeding with fresh complete medium daily. Under the hood, connect a collection bag (e.g., 10-L Flex boy) to the collection hose (*see* **Note 1**) assembly through the MPC connectors. Clamp off the unused tubing port and place the bag into a collection bin on the floor; keep the other end of the hose in the hood. Then place the Nunclon cell factories inside the hood, and remove the cap. Connect the collection hose is to the factory and then invert the factory. After the contents of the factory have completely emptied, add the appropriate volume of complete medium to the factory; then cap the factory, level it, and return it to the incubator. Repeat this process for all factories. (*See* **Note 2**)

6. Freeze and store the supernatants at –70°C.

7. Repeat **steps 5** and **6** until the predetermined volume of supernatant to be produced has been met.

8. At the completion of the collection schedule, trypsinize and pool the cells from all the production vessels. Count the cells and cryopreserve approximately 6% of the total at a concentration of 2×10^7 cells/vial (2 mL/vial). Current CBER guidelines call for cocultivation-RCR assay of viable postproduction cells at a level of 1% of the total cells or 1×10^8 cells, whichever is less. The excess cells provide for losses on thawing and for retention of backup samples.

9. Before final pooling, thoroughly clean and revalidate the production room, biologic safety cabinets, and all pertinent equipment.

10. Prior to filtration, and during the early, middle, and final stages of the filling, depending on the container volumes and numbers, set aside full or partially full

containers for safety testing as prescribed in current versions of *Points to Consider in Human Somatic Cell and Gene Therapy*. The list of samples to be tested, their volumes and containers, must be predetermined prior to initiation of the final processing.

11. Thaw and disinfect the outsides of the individual daily collections and place them in a convenient location near the biosafety hood. Open a collection bag of appropriate volume for the total pooled supernatant in the hood and then place it into the 30-gal. tank, keeping the collection end of the tubing inside the hood. Inside the hood, attach each individual daily collection bag to the pooling collection bag through the MPC couplers and allow its contents to drain into the pooling collection bag. Repeat this until all the individual daily collection bags have been emptied. Then clamp the tubing, and mix the contents thoroughly.

12. Set up a peristaltic pump and aliquot samples of the bulk unfiltered pool into the appropriate containers for safety testing. Test at least 5% of the total unfiltered bulk material for RCR and generally, save at least one backup set of samples for this purpose. If necessary, also save samples and backup samples for bulk sterility, mycoplasma, in vitro adventitious agents, from the unfiltered material.

13. Attach the filter to the end of the tubing and activate the pump to distribute filtered product into the final containers. Label each container with the supernatant lot number and a sequential number corresponding to the order of filling. Include additional information, e.g., the vector name, producer type, date, volume, and so on.

14. If the use of more than one filter cartridge is necessary, collect samples from the beginning and end of use of each filter to verify that equivalent titers are maintained throughout the life of the filter. For 21 CFR 610.12 sterility, surrogate final container samples may be prepared in the final containers by adding approximately 20 mL filtered product during the early, middle, and final stages of the filling process. Prepare one or more backup sets at this time, in case the sterility assay is invalidated and retesting is performed. Generally, 10% of final containers in the lot is submitted for sterility testing, up to a maximum of 20 samples.

15. Place the final containers into heat-sealable plastic pouches and seal them. Then freeze the final containers in a CO_2 cabinet using dry ice. Use >3 kg of dry ice/liter of supernatant. The amount of dry ice used is based on the amount of supernatant to be frozen and a heat of sublimation of 60 cal/g for dry ice.

16. After the supernatants are frozen to $<-20°C$, transfer the containers to $-80°C$ freezer(s) for long-term storage.

3.2. Enrichment of CD34+ Cells

Obtain autologous $CD34^+$ cells from patients who have been mobilized for 5 consecutive days with Neupogen (G-CSF) at 10 µg/kg. Subject the patients to two leukapheresis procedures on consecutive days, and enrich the $CD34^+$ cells using the Ceprate clinical system.

3.3. Prestimulation and Transduction of Human CD34+ Cells

The transduction protocol consists of two phases: an overnight 14–16 h prestimulation step in LTBMC medium containing the cytokines IL-3, IL-6, and SCF (10 ng/mL each) followed by centrifugation of the cells and vector-containing supernatants in blood bags for 4 h at 2400g at 4°C. In each experiment, nontransduced control cells are removed and cultured in parallel with the infected cells.

1. Culture the CD34+-enriched cells overnight for 14–16 h at 37°C, 5% CO_2 in LTBMC medium supplemented with IL-3, IL-6, and SCF (10 ng/mL each). After prestimulation, pool and count the cells and set aside samples for HPP-CFC and CFU-GM analysis and bulk culture. The centrifuge rotor can accommodate eight swinging buckets that can each hold three 100-mL blood bags. We have observed an inverse correlation between absolute cell number per surface area and volume of supernatant used (*see* **Note 3**). Therefore a maximum of 2.4×10^8 cells can be diluted with LTBMC to a final volume of 1200 mL and mixed with an equal volume of clinical grade supernatant.
2. Using a syringe, distribute 100-mL aliquots of the cell suspension/supernatant mixture into 100-mL blood bags.
3. Layer three blood bags in each swinging bucket, check for balance, and then place them into the centrifuge. Centrifuge the cells for 4 h at 2400g at 24°C.
4. After infection, empty the contents of the bag into a 200-mL conical tube. Then rinse each bag with 50 mL of LTBMC and also distribute the rinse into 200-mL conical tubes. Collect the cells by centrifugation for 10 min at 800g at 24°C and set aside the supernatant to be archived.
5. Resuspend the cells in 200 mL of LTBMC, collect and count them. Set aside samples of the infected cells for HPP-CFC and CFU-GM analysis and bulk culture.
6. Transport the cells to the bone marrow processing lab (BMP) or pharmacy using sterile technique. At the BMP, collect the cells by centrifugation at 800g at 4°C for 5 min. Remove the supernatant to a sterile container. Resuspend the cell pellet in 10 mL of Plasma-lyte containing 10 U/mL of heparin and a final concentration of 5% HSA (PL-Hep-HSA). Transfer the suspension to a 50-mL conical tube. Rinse the 200-mL bottle with an additional 10 mL of PL-Hep-HAS, and transfer the rinse to the 50-mL conical tube containing the cell suspension.
7. Pellet the cells again at 800g at 4°C for 5 min and remove the supernatant.
8. Resuspend the cell pellet in 10 mL of PL-Hep-HAS and repeat **step 7**.
9. Remove the supernatant and resuspend the pellet in 10 mL of PL-Hep-HSA. Count the cells.
10. Transfer the suspension to a 30 mL infusion syringe: rinse the conical tube with 10 mL of fresh PL-Hep-HSA and transfer the rinse to the syringe containing the cells. Infuse the cell suspension into the patient over a 1-min period (15 mL/min). Draw up 10 mL of PL-Hep-HSA into the same syringe, in order to rinse any

remaining cells from the syringe, and infuse them into the patient over a 1-min period (15 mL/min).
11. Archive aliquots of the supernatant from **step 6** for safety testing.
12. Plate cells for CFU-GM and HPP-CFC analysis according to the manufacturer's instructions. Perform transduction efficiency 12–14 d after infection by PCR amplification as previously described *(62)*. Expand the remaining bulk culture cells by culturing them in LTBMC containing IL-3, IL-6, and SCF or other cytokine combinations for 6 days. After culture, measure the specific enzymatic activity of the infected and noninfected cells by performing the appropriate bioassay.

4. Notes

1. A few days prior to the first scheduled collection, make enough collection hoses to collect for the anticipated number of collection days. In a biosafety cabinet, using sterile scissors, cut 4–5-ft sections of tubing. Place an MPC connector into one end of the tubing and a Teflon O-ring connector into the other. Place the assembly in a self-sealing autoclave pouch and sterilize.
2. Optimum conditions for collection of supernatant may vary between different packaging lines. With φ-CRIP producer lines, we have found that the highest titer supernatants are obtained after the cells reach confluence. As the dishes become confluent, feed the cells fresh medium daily to maintain a healthy monolayer, but do not split the cells. Instead, collect overnight supernatants from dishes (15 cm) fed with 30 mL of medium the previous afternoon, and collect 8-h supernatants during the afternoon from dishes fed with 10 mL of fresh medium in the morning.
3. We have observed an inverse correlation between cell number and transduction efficiency. When cells are transduced in a container with a fixed surface area and volume of supernatant, a 1 log increase in cell number generally results in a 30% reduction in transduction efficiency. Therefore, optimization of clinical transduction protocols requires a balance between lower cell numbers to increase TE and the logistics associated with infecting large amounts of cells.

5. References

1. Golde, D. W. (1991) The stem cell. *Sci. Am.* **265**, 86–93.
2. Civin, C. and Gore, S. (1993) Antigenic analysis of hematopoiesis: a review. *J. Hematother.* **2**, 137–144.
3. Molineux, G., Pojda, Z., Hampson, I., Lord, B., and Dexter, T. (1990) Transplantation potential of peripheral blood stem cells induced by granulocyte colony stimulating factor. *Blood* **76**, 2153–2158.
4. Novelli, E. M., Ramirez, M., and Civin, C. I. (1998) Biology of CD34+CD38- cells in lymphohematopoiesis. *Leuk. Lymphoma* **31**, 285–293.
5. Pluznik, D. H. and Sachs, L. (1966) The induction of clones of normal mast cells by a substance from conditioned medium. *Exp. Cell. Res.* **43**, 553–563.
6. Eaves, C. J., Cashman, J. D., and Eaves, A. C. (1991) Methodology of long-term culture of human hematopoietic cells. *J. Tissue Culture Methods* **13**, 55–62.

7. Orlic, D. and Bodine, D. M. (1994) What defines a pluripotent hematopoietic stem cell (PHSC): will the real PHSC please stand up! [Editorial]. *Blood* **84,** 3991–3994.
8. Hennemann, B., Oh, I. H., Chuo, J. Y., et al. (2000) Efficient retrovirus-mediated gene transfer to transplantable human bone marrow cells in the absence of fibronectin. *Blood* 96, 2432–2439.
9. Leung, W., Ramirez, M., Novelli, E. M., and Civin, C. I. (1998) In vivo engraftment potential of clinical hematopoietic grafts. *J. Invest. Med.* **46,** 303–311.
10. Novelli, E. M., Cheng, L., Yang, Y., et al. (1999) Ex vivo culture of cord blood CD34+ cells expands progenitor cell numbers, preserves engraftment capacity in nonobese diabetic/severe combined immunodeficient mice, and enhances retroviral transduction efficiency. *Hum. Gene Ther.* **10,** 2927–2940.
11. Schiedlmeier, B., Kuhlcke, K., Eckert, H. G., et al. (2000) Quantitative assessment of retroviral transfer of the human multidrug resistance 1 gene to human mobilized peripheral blood progenitor cells engrafted in nonobese diabetic/severe combined immunodeficient mice. *Blood* **95,** 1237–1248.
12. van der Loo, J. C. M., Hanenberg, H., Cooper, R. J., et al. (1998) Nonobese diabetic/severe combined immunodeficiency (NOD/SCID) mouse as a model system to study the engraftment and mobilization of human peripheral blood cells. *Blood* **92,** 2556–2570.
13. Hao, Q. L., Shah, A. J., Thiemann, F. T., Smogorzewska, E. M., and Crooks, G. M. (1995) A functional comparison of CD34 + CD38-cells in cord blood and bone marrow. *Blood* **86,** 3745–3753.
14. Hao, Q. L., Thiemann, F. T., Petersen, D., Smogorzewska, E. M., and Crooks, G. M. (1996) Extended long-term culture reveals a highly quiescent and primitive human hematopoietic progenitor population. *Blood* **88,** 3306–3313.
15. Civin, C.I., Almeida-Porada, G., Lee, M.J., et al. (1996) Sustained, retransplantable, multilineage engraftment of highly purified adult bone marrow stem cells in vivo. *Blood* **88,** 4102–4109.
16. Larochelle, A., Vormoor, J., Hanenberg, H., et al. (1996) Identification of primitive human hematopoietic cells capable of repopulating NOD/SCID mouse bone marrow: implications for gene therapy. *Nat. Med.* **2,** 1329–1337.
17. Gallacher, L., Murdoch, B., Wu, D. M., Karanu, F. N., Keeney, M., and Bhatia, M. (2000) Isolation and characterization of human CD34(-)Lin(-) and CD34(+)Lin(-) hematopoietic stem cells using cell surface markers AC133 and CD7. *Blood* **95,** 2813–2820.
18. Yin, A. H., Miraglia, S., Zanjani, E. D., Almeida-Porada, G., Ogawa, M., Leary, A. G., Olweus, J., Kearney, J., and Buck, D. W. (1997) AC133, a novel marker for human hematopoietic stem and progenitor cells. *Blood* **90,** 5002–5012.
19. Ziegler, B. L., Valtieri, M., Porada, G. A., De Maria, R., Muller, R., Masella, B., Gabbianelli, M., Casella, I., Pelosi, E., Bock, T., Zanjani, E. D., and Peschle, C. (1999) KDR receptor: a key marker defining hematopoietic stem cells. *Science* **285,** 1553–1558.
20. Bhatia, M., Bonnet, D., Murdoch, B., Gan, O. I., and Dick, J. E. (1998) A newly

discovered class of human hematopoietic cells with SCID-repopulating activity. *Nature Med.* **4**, 1038–1045.

21. Nakamura, Y., Ando, K., Chargui, J., Kawada, H., Sato, T., Tsuji, T., Hotta, T., and Kato, S. (1999) Ex vivo generation of CD34(+) cells from CD34(-) hematopoietic cells. *Blood* **94**, 4053–4059.

22. Bender, J., Unverzagt, K., Walker, D., Lee, W., Van Epps, D., Smith, D., Stewart, C. C., and To, L. B. (1991) Identification and comparison of CD34-positive cells and their subpopulations from normal peripheral blood and bone marrow using multicolor flow cytometry. *Blood* **77**, 2591–2596.

23. Nagler, A., Peacock, M., Tantoco, M., Lamons, D., Okarma, T. B., and Lamons, D. (1993) Separation of human progenitors from human umbilical cord blood. *J. Hematotherapy* **2**, 243–245.

24. Miller, D. G., Adam, M. A., Garcia, J. V., and Miller, A. D. (1990) Gene transfer by retrovirus vectors occurs only in cells that are actively replicating at the time of infection. *Mol. Cell Biol.* **10**, 4239–4242 .

25. Dorrell, C., Gan, O. I., Pereira, D. S., Hawley, R. G., and Dick, J. E. (2000) Expansion of human cord blood CD34(+)CD38(-) cells in ex vivo culture during retroviral transduction without a corresponding increase in SCID repopulating cell (SRC) frequency: dissociation of SRC phenotype and function. *Blood* **95**, 102–110.

26. Glimm, H. and Eaves, C. J. (1999) Direct evidence for multiple self-renewal divisions of human in vivo repopulating hematopoietic cells in short-term culture. *Blood* **94**, 2161–2168.

27. Kollet, O., Aviram, R., Chebath, J., et al. (1999) The soluble interleukin-6 (IL-6) receptor/IL-6 fusion protein enhances in vitro maintenance and proliferation of human CD34(+)CD38(-/low) cells capable of repopulating severe combined immunodeficiency mice. *Blood* **94**, 923–931.

28. Breems, D. A., Blokland, E. A. W., Siebel, K. E., et al. (1998) Stroma-contact prevents loss of hematopoietic stem cell quality during ex vivo expansion of CD34+ mobilized peripheral blood stem cells. *Blood* **91**, 111–117.

29. Shih, C. C., Hu, M. C., Hu, J., et al. (2000) A secreted and LIF-mediated stromal cell-derived activity that promotes ex vivo expansion of human hematopoietic stem cells. *Blood* **95**, 1957–1966.

30. Bodine, D., Karlsson, S., and Nienhuis, A. (1989) Combination of interleukins 3 and 6 preserves stem cell function in culture and enhances retrovirus-mediated gene transfer into hematopoietic stem cells. *Proc. Natl. Acad. Sci. USA* **86**, 8897–8901.

31. Bodine, D., Crosier, P., and Clark, S. (1991) Effects of hematopoietic growth factors on the survival of primitive stem cells in liquid suspension culture. *Blood* **78**, 914–920

32. Luskey, B. D., Rosenblatt, M., Zsebo, K., and Williams, D. A. (1992) Stem cell factor, interleukin-3, and interleukin-6 promote retroviral-mediated gene transfer into murine hematopoietic stem cells. *Blood* **80**, 396–402.

33. Cassel, A., Cottler-Fox, M., Doren, S., and Dunbar, C. E. (1993) Retroviral-medi-

ated gene transfer into CD34-enriched human peripheral blood stem cells. *Exp. Hematol.* **21,** 585–591.

34. Nolta, J. A. and Kohn, D. B. (1990) Comparison of the effects of growth factors on retroviral vector-mediated gene transfer and the proliferative status of human hematopoietic progenitor cells. *Hum. Gene Ther.* **1,** 257–268.

35. Nolta, J. A., Crooks, G. M., Overell, R. W., Williams, D. E., and Kohn, D. B. (1992) Retroviral vector-mediated gene transfer into primitive human hematopoietic progenitor cells: effects of mast cell growth factor (MGF) combined with other cytokines. *Exp. Hematol.* **20,** 1065–1071.

36. Kohn, D. B., Weinberg, K. I., Nolta, J. A., et al. (1995) Engraftment of gene-modified umbilical cord blood cells in neonates with adenosine deaminase deficiency. *Nature Med.* **1,** 1017–1023.

37. Conneally, E., Cashman, J., Petzer, A., and Eaves, C. (1997) Expansion in vitro of transplantable human cord blood stem cells demonstrated using a quantitative assay of their lympho-myeloid repopulating activity in nonobese diabetic-SCID/SCID mice. *Proc. Natl. Acad. Sci. USA* **94,** 9836–9841.

38. Dao, M. A., Hannum, C. H., Kohn, D. B., and Nolta, J. A. (1997) FLT3 ligand preserves the ability of human CD34+ progenitors to sustain long-term hematopoiesis in immune-deficient mice after ex-vivo retroviral-mediated transduction. *Blood* **89,** 446.

39. Piacibello, W., Sanavio, F., Garetto, L., et al. (1997) Extensive amplification and self-renewal of human primitive hematopoietic stem cell from cord blood. *Blood* **89,** 2644–2653.

40. Piacibello, W., Sanavio, F., Severino, A., et al. (1999) Engraftment in nonobese diabetic severe combined immunodeficient mice of human CD34(+) cord blood cells after ex vivo expansion: evidence for the amplification and self-renewal of repopulating stem cells. *Blood* **93,** 3736–3749.

41. Case, S. S., Price, M. A., Jordan, C. T., et al. (1999) Stable transduction of quiescent CD34(+)CD38(-) human hematopoietic cells by HIV-1-based lentiviral vectors. *Proc. Natl. Acad. Sci. USA* **96,** 2988–2993.

42. Miyoshi, H., Smith, K. A., Mosier, D. E., Verma, I. M., and Torbett, B. E. (1999) Transduction of human CD34+ cells that mediate long-term engraftment of NOD/SCID mice by HIV vectors. *Science* **283,** 683–686.

43. Hennemann, B., Chuo, J. Y., Schley, P. D., et al. (2000) High-efficiency retroviral transduction of mammalian cells on positively charged surfaces. *Hum. Gene Ther.* **11,** 43–51.

44. Bahnson, A. B., Dunigan, J. T., Baysal, B. E., et al. (1995) Centrifugal enhancement of retroviral mediated gene transfer. *J. Virol. Methods* **54,** 131–143.

45. Dunbar, C. E., Kohn, D. B., Schiffmann, R., et al. (1998) Retroviral transfer of the glucocerebrosidase gene into CD34+ cells from patients with Gaucher disease: in vivo detection of transduced cells without myeloablation. *Hum. Gene Ther.* **9,** 2629–2640.

46. Parkman, R., Weinberg, K., Crooks, G., Nolta, J., Kapoor, N., and Kohn, D. (2000) Gene therapy for adenosine deaminase deficiency. *Ann. Rev. Med.* **51,** 33–47.

47. Ohashi, T., Boggs, S., Robbins, P., et al. (1992) Efficient transfer and sustained high expression of the human glucocerebrosidase gene in mice and their functional macrophages following transplantation of bone marrow transduced by a retroviral vector. *Proc. Natl. Acad. Sci. USA* **89,** 11332–11336.

48. Lemishka, I. R., Raulet, D. H., and Mulligan, R. C. (1986) Developmental potential and dynamic behavior of hematopoietic stem cells. *Cell* **45,** 917–927.

49. Kohn, D. B., Bauer, G., Rice, C. R., et al. (1999) A clinical trial of retroviral-mediated transfer of a rev-responsive element decoy gene into CD34(+) cells from the bone marrow of human immunodeficiency virus-1-infected children. *Blood* **94,** 368–371.

50. Wong-Staal, F., Poeschla, E. M., and Looney, D. J. (1998) A controlled, phase 1 clinical trial to evaluate the safety and effects in HIV-1 infected humans of autologous lymphocytes transduced with a ribozyme that cleaves HIV-1 RNA. *Hum. Gene Ther.* **9,** 2407–2425.

51. Abonour, R., Williams, D. A., Einhorn, L., et al. (2000) Efficient retrovirus-mediated transfer of the multidrug resistance 1 gene into autologous human long-term repopulating hematopoietic stem cells. *Nat. Med.* **6,** 652–658.

52. Cowan, K. H., Moscow, J. A., Huang, H., et al. (1999) Paclitaxel chemotherapy after autologous stem-cell transplantation and engraftment of hematopoietic cells transduced with a retrovirus containing the multidrug resistance complementary DNA (MDR1) in metastatic breast cancer patients. *Clin. Cancer Res.* **5,** 1619–1628.

53. Blaese, R. M., Culver, K. W., Miller, A. D., et al. (1995) T lymphocyte-directed gene therapy for ADA-SCID: initial trial results after 4 years. *Science* **270,** 475–480.

54. Cavazzana-Calvo, M., Hacein-Bey, S., de Saint Basile, G., et al. (2000) Gene therapy of human severe combined immunodeficiency (SCID)-X1 disease. *Science* **288,** 669–672.

55. Kay, M. A., Manno, C. S., Ragni, M. V., et al. (2000) Evidence for gene transfer and expression of factor IX in haemophilia B patients treated with an AAV vector. *Nat. Genet.* **24,** 257–261.

56. Malech, H. L., Maples, P. B., Whiting-Theobald, N., et al. (1997) Prolonged production of NADPH oxidase-corrected granulocytes after gene therapy of chronic granulomatous disease. *Proc. Natl. Acad. Sci. USA* **94,** 12133–12138.

57. Novelli, E. M., Swaney, W. P., Kopp, D., et al. (2000) A phase I trial of retroviral-mediated gene therapy of Gaucher disease. *Blood* **96S.**

58. Malech, H. L. (2000) Use of serum-free medium with fibronectin fragment enhanced transduction in a system of gas permeable plastic containers to achieve high levels of retrovirus transduction at clinical scale. *Stem Cells* **18,** 155–156.

59. Schuening, F., Longo, W. L., Atkinson, M. E., et al. (1997) Retrovirus-mediated transfer of the cDNA for human glucocerebrosidase into peripheral blood repopulating cells of patients with Gaucher's disease. *Hum. Gene Ther.* **8,** 2143–2160.

60. Dao, M. A., Taylor, N., and Nolta, J. A. (1998) Reduction in levels of the cyclin-dependent kinase inhibitor p27(kip-1) coupled with transforming growth factor

beta neutralization induces cell-cycle entry and increases retroviral transduction of primitive human hematopoietic cells. *Proc. Natl. Acad. Sci. USA* **95,** 13006–13011.

61. Swaney, W. P., Sorgi, F. L., Bahnson, A. B., and Barranger, J. A. (1997) The effect of cationic liposome pretreatment and centrifugation on retrovirus-mediated gene transfer. *Gene Ther.* **4,** 1379–1386.

62. Bahnson, A. B., Nimgaonkar, M., Ball, E. D., and Barranger, J. A. (1997) Methods for retrovirus-mediated gene transfer to CD34+-enriched cells, in *Methods in Molecular Medicine, Gene Therapy Protocols*. Humana, Totowa, NJ, pp. 249-264.

16

Genetically Modified Skin Substitutes

Preparation and Use

Karen E. Hamoen, Gulsun Erdag, Jennifer L. Cusick, Hinne A. Rakhorst, and Jeffrey R. Morgan

1. Introduction

Skin loss due to burns or ulcers is a major medical problem and is the motivation for the development of skin substitutes and skin replacement technologies, many of which have had some success in the clinic *(1–4)*. Methods exist for the growth of large numbers of epidermal keratinocytes as well as dermal fibroblasts. These cells have been combined with various analogs of the dermis to fabricate composite skin substitutes. Different types of dermal matrices have been developed including a porous sponge of collagen and glycosaminoglycan *(5,6)* and a fibroblast contracted collagen lattice *(7,8)*. We and others have had success with a dermal analog of acellular human dermis that is deepithelialized and rendered completely acellular *(9–12)*. Acellular dermis is relatively nonimmunogenic and retains many of its structural elements after processing. Skin substitutes using acellular dermis can be formed in vitro and subsequently transplanted to athymic mice, generating a well-differentiated and fully pigmented epidermis with many characteristics of normal skin *(11,12)*.

Epidermal keratinocytes have been shown to be an attractive target for ex vivo gene therapy because the cells are easy to obtain by biopsy, grow rapidly under appropriate culture conditions, can be transplanted, and can be genetically modified by retroviral-mediated gene transfer *(13)*. Applications of genetically modified skin grafts include the systemic delivery of therapeutic proteins, such as clotting factor IX *(14)*; the local delivery of wound healing

From: *Methods in Molecular Medicine, Vol. 69, Gene Therapy Protocols, 2nd Ed.*
Edited by: J. R. Morgan © Humana Press Inc., Totowa, NJ

growth factors, such platelet-derived growth factor (PDGF), for the treatment of burns and ulcers *(15,16)*; and the correction of inherited defects of the skin, such as lamellar ichthyosis *(17)* or epidermolysis bullosa *(18,19)*.

There are some clear advantages to using recombinant retroviruses for the ex vivo genetic modification of keratinocytes versus other methods of gene transfer. The frequency of gene transfer by retroviruses is relatively high, and the transgene delivered by a recombinant retrovirus is stably integrated into the genome of the cell. Therefore, large numbers of keratinocytes in culture can be simultaneously genetically modified, and a high frequency of gene transfer increases the likelihood of modifying the epidermal stem cell. Moreover, since the introduced genes are stably integrated, the genetically modified cells can be further proliferated in vitro without loss of the introduced gene. Thus, for some applications in wound healing, this property can be used to prepare large cell banks of genetically modified cells.

In this chapter, we describe methods for the culture and ex vivo genetic modification of diploid human keratinocytes using retroviral-mediated gene transfer, the preparation of acellular dermis and the generation of composite skin substitutes of genetically modified keratinocytes seeded on acellular dermis, and the transplantation of these grafts to athymic mice.

2. Materials

2.1. Cell Culture and Genetic Modification of Keratinocytes

1. Human diploid keratinocytes established from neonatal foreskins and cultured using a feeder layer *(20)*.
2. Fibroblast feeder layer: 3T3-J2 mouse fibroblast cell line (originally provided by H. Green, Department of Physiology and Biophysics, Harvard Medical School, Boston, MA).
3. Cell lines producing recombinant retroviruses are constructed or are available from many academic sources.
4. Fibroblast tissue culture medium: Dulbecco's modified Eagle's medium (DMEM; high glucose, L-glutamine, 110 mg/L sodium pyruvate; Life Technologies, Gaithersburg, MD), bovine calf serum (BCS) 10% (HyClone, Logan, UT), penicillin/streptomycin (Life Technologies) 100 IU/mL/100 µg/mL.
5. Mitomycin C (Roche Biochemicals, Indianapolis, IN): 15 µg/mL in serum-free DMEM.
6. Keratinocyte tissue culture medium (KCM):
   a. DMEM/Ham's F12 medium (Life Technologies) 3:1.
   b. Fetal bovine serum (FBS) 10% (HyClone, Logan, UT).
   c. Penicillin/streptomycin 100 IU/mL/100 µg/mL.
   d. Adenine (6-aminopurine hydrochloride; Sigma, St. Louis, MO). Make up fresh at time medium is prepared. Prepare stock (50×) of 1.2 mg/mL in serum-free DMEM/Ham's F12 (3:1), adjust pH to 7.5 with 1 N NaOH; sterilize by

filtration using a 0.45-µm filter; add 2 mL of stock to 100 mL of KCM; final concentration: 1.8×10^{-4} M.

e. Cholera toxin (*Vibrio cholerae*, type Inaba 569 B; Calbiochem, La Jolla, CA): Prepare concentrated stock of 10^{-5} M in distilled water and store at 4°C; take 0.1 mL of concentrated stock and make up to 10 mL with DMEM (10% fetal calf serum [FCS]), sterilize by filtration using a 0.45-µm filter, aliquot in 1-mL portions (10^{-7} M), store at –20°C, and add 0.1 mL to 100 mL of KCM; final concentration: 10^{-10} M.

f. Epidermal growth factor (mouse; Collaborative Biomedical Products, Bedford, MA): Resuspend lyophilized material in distilled water to prepare stock of 10 µg/mL; sterilize by filtration using a 0.45-µm filter, aliquot in 1-mL portions, and store at –20°C; add 0.1 mL to 100 mL KCM with first medium change; final concentration: 10 ng/mL.

g. Hydrocortisone (chromatographic standard; Calbiochem): Prepare stock of 5 mg/mL in 95% ethanol and store at 4°C; take 0.4 mL of stock and make up to 10 mL with serum-free DMEM, sterilize by filtration using a 0.45-µm filter, and aliquot in 1-mL portions; add 0.2 mL to 100 mL of KCM; final concentration: 0.4 µg/mL.

h. Insulin (pork, 100 U/mL [3.8 mg/mL]; Novo Nordisk, Princeton, NJ): Add 0.13 mL of stock to 100 mL of KCM; final concentration: 5 µg/mL.

i. T/T3 stock (transferrin/triiodo-L-thyronine stock): T stock (transferrin, human, partially iron saturated; Boehringer) 5 mg/mL in phosphate-buffered saline (PBS); for T3 stock (3,3',5 triiodo-L-thyronine, sodium salt, Sigma) dissolve 13.6 mg in a minimum amount of 0.02 N NaOH and make volume up to 100 mL with distilled water, sterilize by filtration using a 0.45-µm filter, and store at –20°C (2×10^{-4} M); add 0.1 mL T3 stock to 9.9 mL of T stock, sterilize by filtration using a 0.45-µm filter, aliquot in 1-mL portions, and store at –20°C; add 0.1 mL of T/T3 stock to 100 mL of KCM; final concentration, transferrin: 5 µg/mL, triiodo-L-thyronine: 2×10^{-9} M.

7. PBS-EDTA solution: 5 mM EDTA (Life Technologies) in PBS; sterilize by filtration using a 0.2-µm filter.

8. Trypsin solution (trypsin 1-300; ICN Biochemicals, Costa Mesa, CA): D-dextrose 0.1% (w/v), trypsin 0.1% (w/v) in PBS, pH 7.5; sterilize by filtration using a 0.45-µm filter, store at –20°C, and avoid repeated thawing/freezing.

9. Trypsin/PBS-EDTA solution: Mix together trypsin solution and PBS-EDTA solution (1:1).

10. PBS: 138 mM NaCl, 2.7 mM KCl, 8.1 mM Na$_2$HPO$_4$, 1.5 mM KH$_2$PO$_4$, sterilize by filtration using a 0.2-µm filter.

2.2. Preparation of Acellular Dermis

1. Minipig (Minipig-YU Yucatan) (*see* **Note 1**).
2. Surgical prep: Sterile drapes, gowns, gloves, mask, 4 × 4 gauze, nylon gauze, 70% isopropyl alcohol, Betadine surgical scrub and prep solution, topical min-

eral oil (sterile), 0.9% sodium chloride irrigation solution, OR/scrub soap solution, OR scrub brush, disposable shaving razors, skin wax, clippers.

3. Sterile dermatome with blades (Padget or Zimmer).
4. Sterile disposable wide-mouth bottles (500 mL).
5. Cryopreservation solution (DMEM with 15% glycerol).
6. Surgical instruments: Sterile scissors, tissue forceps, #15 sterile disposable scalpels (Feather).
7. Sealable plastic or aluminum pouches, 6.5 × 8 inches (Kapak Seal Pak Pouches, Minneapolis, MN).
8. Liquid nitrogen.
9. Sterile PBS and DMEM (Life Technologies).
10. PBS with antibiotic cocktail: 300 µg/mL gentamycin (Sigma), 30 µg/mL ciprofloxacin (Bayer Corporation, West Haven CT), 7.5 µg/mL amphotericin B (Life Technologies), 300 IU/mL/300 µg/mL penicillin/streptomycin (Life Technologies) (*see* **Note 2**).

2.3. Preparation of Genetically Modified Skin Substitutes

1. 6-well tissue culture plates (Falcon).
2. Porous stainless steel squares 2 × 2 cm wide, 2–3 mm thick, and 100-µm pore size (Mott Corporation, Farmington, CT). Clean squares by soaking overnight in 10 N NaOH followed by extensive washing in distilled water (>1 h) and then autoclaving.
3. Surgical instruments: Two pairs of fine forceps (tweezers) without teeth.
4. Keratinocyte seeding medium *(22,23)*:
   a. DMEM/Ham's F12 medium (Life Technologies) (3:1).
   b. Fetal bovine serum 1% (HyClone).
   c. Penicillin/streptomycin 100 IU/mL/100 mg/mL (Life Technologies).
   d. Ascorbic acid (Sigma): This is light sensitive, so work quickly and cover the tubes and container with aluminum foil; add 1 g ascorbic acid to 20 mL sterile dH$_2$O and dissolve completely; sterilize by filtration using a 0.45-mm filter, aliquot in 1-mL portions, and store at –20°C; add 0.1 mL to 100 mL of seeding medium; final concentration 50 µg/mL.
   e. Final concentration 10^{-10} M cholera toxin (prepare stock as described under **Subheading 2.1.**).
   f. Final concentration 0.2 µg/mL hydrocortisone (prepare stock as described under **Subheading 2.1.**); add 0.1 mL to 100 mL of seeding medium.
   g. Final concentration 5 µg/mL insulin (prepare stock as described under **Subheading 2.1.**).
5. Keratinocyte priming medium: Equal to keratinocyte seeding medium supplemented with:
   a. Bovine serum albumin (Sigma): Add 1.63 g/L directly to medium after FBS, but before additional medium supplements; final concentration: 24 mM.
   b. L-serine (Sigma): Add 2.10 g of serine to 20 mL sterile dH$_2$O, dissolve completely, and sterilize by filtration through a 0.45-µm filter. Aliquot in 1-mL

portions and store at –20°C; add 0.1 mL per 100 mL of priming medium; final concentration: 1 m*M*.

   c. L-carnitine (Sigma): Add 49.4 mg of carnitine to 25 mL dH$_2$O, dissolve completely, and sterilize by filtration through a 0.45-mm filter. Aliquot in 1-mL portions and store at –20°C; add 0.1 mL per 100 mL of priming medium; final concentration: 10 µ*M*.

   d. Fatty acid cocktail. Oleic acid (Sigma): Add 790 µL to 10 mL 100% ethanol, aliquot in 1-mL portions, and store at –20°C. Linoleic acid (Sigma): Add 470 µL to 10 mL 100% ethanol, aliquot in 1-mL portions, and store at –20°C. Arachidonic acid (Sigma): Add 46.3 µL to 2 µL 100% ethanol; make this fatty acid fresh each time. Palmitic acid (Sigma): Add 638 mg to 10 mL 100% ethanol, aliquot in 1-mL portions, and store at –20°C. Warm 2 mL of each fatty acid in a water bath, add into a 15-mL tube, mix well, and aliquot in 400-µL portions in sterile Eppendorf tubes; store in upright manner to prevent leaking; add 40 µl per 100 mL of priming medium; final concentrations: 25 µM oleic acid, 15 µM linoleic acid, 7 µM arachidonic acid, 25 µM palmitic acid (**22**).

6 Air/liquid interface medium: Equal to serum-free keratinocyte priming medium supplemented with 1.0 ng/mL epidermal growth factor (Collaborative Biomedical Products). Prepare stock as described under **Subheading 2.1.** and add 10 µL/100 mL of air/liquid interface medium.

2.4. Transplantation of Genetically Modified Skin Substitutes

1. Surgical instruments: Small, sharp scissors, sterile disposable scalpels, two pairs of fine tissue forceps (tweezers) without teeth, small needle holder, 1-µL tuberculin syringes, razor blades, sterile cotton swabs.
2. 6-0 nylon sutures (Ethicon, Somerville, NJ) with a small (P-1) cutting needle.
3. Skin prep: 70% isopropyl alcohol, Betadine surgical prep solution, tincture of Benzoin skin adhesive, triple antibiotic ointment.
4. Skin dressing: Telfa no-stick gauze (Kendall, Mansfield, MA), Tegaderm® polyurethane occlusive dressing (3-M), 4.4 × 4.4-cm size, 3-M flexible Sports Band-Aid (must be at least 7/8–1 inch wide), 1-inch-wide elastic adhesive tape (Johnson and Johnson).
5. Anesthesia (2,2,2-tribromoethanol, Sigma: Prepare concentrated stock of 1.6 g/mL in 2methyl-2butanol and store at 4°C. Before use, prepare a working solution by diluting 12 µL (concentrated stock)/mL 0.9% NaCl at 40°C. Allow to slowly cool to room temperature before use. Sterilize by filtration using a 0.45-µm filter.
6. Skin substitutes in 6-well plates.
7. 37% formalin (Sigma): Dilute stock 1:10 (v/v) in PBS.
8. O.C.T. Compound (Sakura Finatek, Torrance, CA).
9. Dry ice (block form).
10. Camera for photo documentation (with macro lens).
11. Athymic nude mice, NIH Swiss nu, 6–8 weeks old, average weight 20–30 g (outbred, TAC:N;NIFS-nuDF; Taconic, Germantown, NY).
12. Dental wax (Byte ryte; Mizzy, Cherry Hill, NJ).

3. Methods

3.1. Cell Culture and Genetic Modification of Keratinocytes

1. For genetic modification, we prefer to use cultured primary keratinocytes isolated from neonatal foreskins and cultured using a feeder layer *(20,21)*.
2. Grow 3T3-J2 fibroblast feeder cells (used for the routine culture of keratinocytes) in 175-cm^2 flasks in fibroblast tissue culture medium in a 37°C incubator with 10% CO_2. Pass fibroblasts for a total of 12 times approximately twice a week, seeding 0.5×10^6 cells/flask at each passage.
3. To prepare a feeder layer for cocultivation with keratinocytes, plate 2×10^6 3T3-J2 cells in a 75-cm^2 flask in fibroblast tissue culture medium and incubate overnight in tissue culture medium at 37°C in 10% CO_2.
4. To limit the growth of the feeder layer cells, treat the 3T3-J2 cells with DMEM (no serum) supplemented with mitomycin C (15 µg/mL, in a 15 mL/75-cm^2 flask) for 2 h at 37°C in 10% CO_2.
5. After 2 h, remove the mitomycin C-containing medium and wash cells 3× with DMEM (no serum). Add 15 mL KCM (without epidermal growth factor [EGF]) and equilibrate flasks in humidified incubator (37°C, 10% CO2) for at least 2 h prior to adding keratinocytes.
6. For the genetic modification of human keratinocytes, use a fibroblast-based virus producer cell line that sheds the recombinant retrovirus of interest instead of the 3T3-J2 fibroblast feeder layer. Use the same cell density and mitomycin C treatment for the virus producer cell line. Prepare parallel plates of control unmodified (3T3-J2 fibroblast feeder layer) and genetically modified (virus producer cell line).
7. Prior to dissociation of the keratinocytes for passage onto the new 3T3-J2 feeder layer or the feeder layer of virus-producing cells, remove the old feeder layer by adding 10 mL of PBS-EDTA solution to the 75-cm^2 flask. Allow to sit for approximately 1 min and then shake flasks vigorously to dislodge the old feeder layer. Visualize that the feeder layer is detached under the microscope. Repeat if necessary (*see* **Note 3**).
8. Remove PBS-EDTA wash, gently rinse with a final PBS-EDTA wash, and remove.
9. Add 10 mL of prewarmed trypsin/PBS-EDTA solution, incubate at 37°C, 10% CO_2 for no longer than 8 min, and check periodically under microscope.
10. When cells are dissociated, add 10 mL of KCM (no EGF) to neutralize trypsin.
11. Pipet up and down about 5 times to break up cell clumps.
12. Spin down cells, resuspend and count.
13. Add 10^6 keratinocytes to each of the 75-cm^2 flasks that contain either a mitomycin C-treated 3T3-J2 fibroblast feeder layer (control unmodified) or a mitomycin C- treated virus producer cell line (genetically modified).
14. Next day, change the medium to KCM with EGF and feed every other day with this medium. After 4–6 days of cocultivation with either the 3T3-J2 fibroblast feeder layer or the virus producing cells, the keratinocytes have been genetically modified and are ready to be subcultured.

15. Dissociate keratinocytes by treatment with trypsin/PBS-EDTA (as described above), count, and pass cells to new dishes containing the normal 3T3-J2 fibroblast feeder layer to expand cell numbers for various purposes such as numerous analyses on the effects of genetic modification as well as cryopreservation of the cells. These keratinocytes can also be seeded on acellular dermis to prepare genetically modified skin substitutes for transplantation to athymic mice.

3.2. Preparation of Acellular Dermis

1. Acellular dermis for the preparation of genetically modified skin substitutes is prepared from split-thickness skin harvested from human cadavers or from pigs (*see* **Note 1**). What follows is the procedure for the preparation of acellular dermis from pig skin.
2. The person doing the prep and harvest does a full surgical scrub and then gowns in sterile attire including mask and cap.
3. Remove hair of prospective donor sites with clippers and disposable razors or strip using hair wax. Hair wax does not work on skin that has been shaved.
4. Wash the area to be harvested several times with soap solution followed by sterile saline. Prepping is performed in a circular fashion starting in the center and working out.
5. Apply Betadine surgical scrub to the skin and scrub the skin gently for several minutes with an OR scrub brush.
6. Wash the area with sterile saline and then paint 3 times using sterile gauze soaked in Betadine.
7. Allow the Betadine to dry (approx 5 min) and then wash the skin with sterile gauze soaked in 70% isopropyl alcohol prep solution.
8. Drape the area to be harvested and cover all unprepped areas of the body with sterile sheets or towels.
9. The person who prepped the area now regowns and regloves or a second person harvests the grafts.
10. Lock a fresh dermatome blade into the apparatus (Padget or Zimmer dermatome) and then screw the maximum width guard tightly down onto the blade. Set the skin harvest thickness at 0.015 inches.
11. Apply sterile mineral oil onto the donor area to help the dermatome move smoothly on the skin. Place the bevel of the dermatome on the donor site with the right hand. The angle of the dermatome to the skin is approx 45°. The left hand rests firmly on the skin behind the instrument and a second person makes countertraction on the other end of the skin. Depress the control button or lever and apply a steady and firm forward and downward pressure to begin the skin removal. To complete the cut, tilt the dermatome gradually with the bevel upward and release the control. Check skin thickness as you harvest because thickness may vary with applied pressure and site of harvest. Long harvests from the back of the animal are preferred.
12. Remove the skin from the dermatome with sterile forceps and place in a wide-

mouth 500-mL bottle containing 300 mL of sterile PBS two to three long (approx 40 cm) pieces of skin/bottle.

13. Repeat the harvesting of split-thickness skin from several new areas.
14. In a laminar flow hood, wash the skin in sterile PBS 3 times, transferring the skin each time to a new bottle with fresh PBS. All subsequent steps in handling the skin are performed in a laminar flow hood using sterile technique.
15. Place the skin in PBS with antibiotic cocktail and keep at 4°C for 1 day. The next day, transfer skin to a new bottle containing cryopreservation solution (300 mL) and return to the refrigerator for 2 h.
16. Remove the skin from the bottle and spread flatly onto sterile mesh gauze soaked in cryopreservation medium. Place the skin and gauze into sterile plastic or aluminum pouches and seal.
17. Freeze the pouches at –20°C for 24 h and transfer to –80°C for long-term storage.
18. To prepare acellular dermis, thaw each pouch of frozen skin by immersion in a tub of water at 15–20°C for about 10 min or until skin is pliable. Then refreeze the pouch rapidly in liquid nitrogen. This rapid freeze-thaw cycle devitalizes the cells and is repeated for a total of 3 times.
19. In a laminar flow hood using sterile technique, remove each piece of skin from its pouch and place in a sterile 500-mL bottle containing sterile PBS (300 mL). Wash the skin 3 times with sterile PBS and then place in a new bottle with PBS with antibiotic cocktail and refrigerate at 4°C for 1 week.
20. After 1 week at 4°C, replace the old PBS with antibiotic cocktail with fresh PBS with antibiotic cocktail and incubate the skin at 37°C for another week.
21. At the end of this week, place the skin in a sterile 15-cm dish and gently remove the epidermis by peeling or gentle scraping with forceps. Epidermis that remains adherent will detach spontaneously in the following weeks.
22. After removing most of the epidermis, return the skin to a new bottle, wash once with PBS, and then transfer to a new bottle with fresh PBS with antibiotic cocktail. Refrigerate at 4°C for 4 more weeks to remove any remaining cells (*see* **Note 4**).
23. After 4 weeks, the dermis is acellular and should be tested for sterility before use. Using sterile technique cut off a small piece, place in DMEM (no antibiotics) and incubate at 37°C for 48 h.
24. Prior to use, wash the acellular dermis 3 times with DMEM (no serum) to remove residual antibiotics (*see* **Note 5**).

3.3. Preparation of Genetically Modified Skin Substitutes

1. Use porous stainless steel squares (2×2 cm) as a bed for skin substitutes in both submerged and air/liquid cultures. Soak stainless steel squares in DMEM and then, using sterile forceps, place one in each well of a 6-well tissue culture dish (*see* **Note 6**).
2. Cut the acellular dermis into 1.25–1.5-cm^2 pieces using forceps, a #10 scalpel blade, and sterile technique. After removal of excess liquid, place each piece of dermis on a porous stainless steel square, papillary side up. The papillary side of

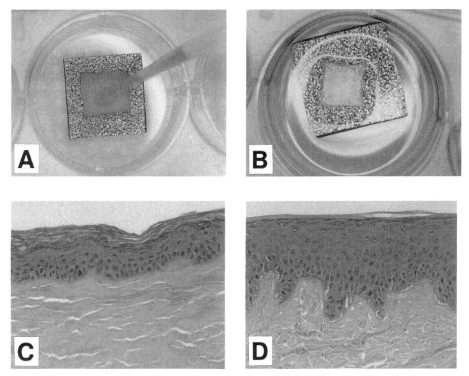

Fig. 1. (a) Skin substitutes are prepared by seeding human keratinocytes on acellular porcine dermis. Subsequently the skin substitute is cultured submerged for 3 days followed by 7 days at the air/liquid interface. **(b)** Macroscopically a sheen of epidermis on the surface of the dermis can be visualized. **(c)** Microscopically, keratinocytes form a stratified and differentiated epidermis. **(d)** Skin substitutes of keratinocytes genetically modified to secrete keratinocyte growth factor (KGF) form a thicker epidermis *(24)*.

the acellular dermis can be distinguished from the reticular side by a rougher feel and duller sheen. Incubate the dermis on the stainless steel square for 1/2 h at 37°C.

3. Trypsin treat control and genetically modified keratinocytes as described in **Subheading 3.1.**, resuspend in seeding medium, and seed onto the surface of each piece of acellular dermis (2.5×10^5 cells/ cm^2 in about 100 µL medium) (**Fig. 1**).

4. Incubate at 37°C for about 1/2 hour. The medium is absorbed by the dermis and porous stainless steel. Preliminary cell attachment occurs. Watch carefully and do not allow it to become too dry. Add 4 mL of seeding medium gently to the well so that the skin substitute is submerged under medium.

5. After 24 h, remove the seeding medium by aspiration. Be sure to place the tip of the aspirator onto the porous stainless steel square to remove spent medium inside

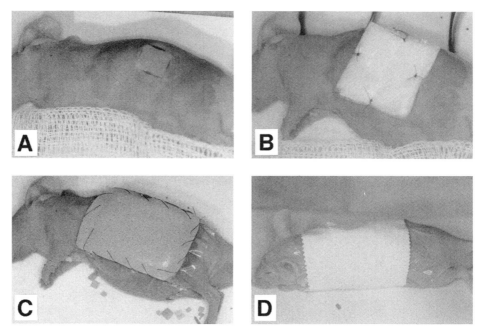

Fig. 2. Transplantation of skin substitutes. (**a**) Graft placed on a full-thickness defect on the dorsum of an athymic mouse. The graft is dressed first with Telfa and Tegaderm (**b**) and then a flexible Band-Aid (**c**) and then wrapped circumferentially with a tape (**d**).

the porous stainless steel. Add 4 mL of priming medium gently to the well so that the skin substitute is submerged under medium.

6. After 48 h, remove priming medium by aspiration. Add air/liquid medium until it is level with the top surface of the stainless steel. *Do not submerge the skin substitute.* Now the underside of the dermis is exposed to medium, and the keratinocytes on the surface are exposed to air (**Fig. 1**). Exposure to the air stimulates the keratinocytes to differentiate and form a stratified epidermis (**Fig. 1;** *see* **Note 6**).
7. Grow skin substitutes at air/liquid interface for a minimum of 7 days and no longer than 14 days. Feed with air/liquid medium every other day, being careful not to splash medium onto the surface of the skin substitute.

3.4. Transplantation of Genetically Modified Skin Substitutes

1. In a laminar flow hood, anesthetize athymic mice with an intraperitoneal injection of approximately 0.7–0.8 mL of 2,2,2-tribromoethanol (25-G, 3/4-inch needle, 30 μL working solution per gram of mouse).
2. Wash the left shoulder area of the mouse first with 70% alcohol and then with

Betadine. After the Betadine has dried, mark out a 1.25–1.5-cm^2 square on the left subscapular region of the mouse. (The square should be no larger than the graft to be transplanted.)

3. Use sharp straight scissors to create a full-thickness (including the panniculus carnosus) defect down to the fascia of the dorsal musculature and remove the entire piece of skin by scissors dissection. If bleeding is encountered, control it by 1–2 min of gentle pressure or cauterization (*see* **Note 7**).

4. Remove skin substitute from the stainless steel using a small flat spatula and transfer to the wound (**Fig. 2**).

5. Once on the wound, trim either the skin substitute or the wound edges to generate a precise fit of the graft to the defect. (A starting skin substitute slightly bigger than the defect is ideal.) Stitches in the corners can be used to anchor the graft to the wound, but have been unnecessary.

6. Cut two pieces of Telfa slightly larger than the wound. Apply triple antibiotic ointment to one piece of Telfa and apply this to the skin substitute. Apply another piece of Telfa on top of the first piece (to act as a compressive dressing), and secure both pieces with four 6-0 nylon sutures on each of the four sides (tacked to the surrounding mouse skin, not the graft). Be sure not to pull wound edges away from edges of the skin substitute.

7. Use a sterile cotton swab to apply tincture of Benzoin to the mouse skin on all sides of the wound. This will make the mouse skin sticky. Be sure not to get any on the skin substitute.

8. Apply a 4.4 × 4.4 cm piece of Tegaderm polyurethane occlusive dressing over the Telfa dressing. The Tegaderm, which is adhesive, should form a tight seal for several millimeters around the wound and adhere tightly to the mouse skin (aided by the Benzoin) (**Fig. 2**).

9. Using a tuberculin syringe, inject approximately 0.7 mL of keratinocyte culture medium (with EGF) through the Tegaderm into the Telfa pads to keep the grafts moist for the first few days.

10. Trim a 3-M flexible Sports Band-Aid so that only the pad and a thin (5-mm) adhesive strip around the pad remains. Place this Band-Aid over the Tegaderm on top of the wound. Use a 6-0 nylon suture to secure the Band-Aid to the surrounding mouse skin (Tegaderm included). This is a running stitch that travels around the entire circumference of the Band-Aid (**Fig. 2**).

11. Finally, wrap a long piece (approx. 30–40 cm) of 1-inch elastic tape circumferentially (3–4 times) around the mouse to cover the Band-Aid. Care must be taken not to wrap too closely to the mouse's hind legs, so as not to hinder the mouse's movement after surgery or too tightly, which would interfere with the mouse's breathing (**Fig. 2**).

12. If the whole procedure is performed correctly, the mouse retains its mobility and the dressing remains occlusive for at least 4–7 days (enough to allow epithelialization of the gaps between the mouse skin and the graft). The dressing should not be changed for at least 1 week and can, in fact, be left on indefinitely (*see* **Note 8**).

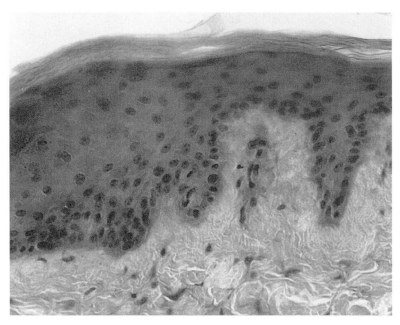

Fig. 3. Skin substitute in vivo. By 7 days the graft is well incorporated into the surrounding mouse skin, and histology reveals a fully stratified epidermis, an inter-digitating dermal-epidermal junction, and a revascularized dermal matrix that is well populated with host fibroblasts.

13. To harvest the graft, sacrifice the animal by CO_2 asphyxiation. Carefully remove dressing so as not to disturb graft. Photograph and trace the outline of the graft while it is still on the mouse.
14. Harvest the entire graft by making a full-thickness incision down to and through the dorsal thoracic musculature of the mouse with sharp scissors. The harvest should include the muscle underlying the graft, but not the mouse ribs. Be sure to include a segment of normal mouse skin around the edges of the graft.
15. Spread the dissected graft on a piece of dental wax (Byte ryte, Mizzy) and use a razor blade to divide the graft into two pieces. Place one piece immediately in a solution of 3.7% formalin and process for paraffin sectioning. Submerge the other piece in O.C.T. embedding medium for frozen tissue specimens. Snap-freeze this piece on a flat block of dry ice, label, and transfer to storage at −80°C. By one week, grafts have formed a stratified and differentiated epidermis and the dermis is repopulated with fibrovascular cells of the host (**Fig. 3**).

4. Notes

1. Young minipigs are preferred as a source of skin because they have less hair and the hair follicles have a relatively small diameter. These holes are present in the final acellular dermis and if they are too big they can present problems with seeded

cells on the surface. Acellular dermis prepared from pig split-thickness skin that is too thick may not transplant well but can be useful for in vitro analyses. Human acellular dermis is preferred for transplantation studies.

2. To prevent microbial growth, antibiotics should always be present in the storage solutions. According to our experience 3× higher doses of antibiotics are needed for porcine skin versus human skin.

3. Cells must be actively dividing in order to be genetically modified with recombinant retroviruses. It is imperative that optimal cell proliferation be maintained. Moreover, keratinocytes lose their growth potential after they reach confluence. Therefore, we recommend that the primary cultures of keratinocytes be passed onto the virus-producing feeder layer prior to reaching confluence. Methods to prepare primary cultures of keratinocytes from biopsies have been described previously *(21)*.

4. It cannot be overemphasized how important sterile technique is at all stages of the preparation of acellular dermis. Even the slightest contamination can blossom into fulminant bacterial overgrowth once the dermis is placed into culture medium.

5. When skin is prepared as described above, the dermis is rendered completely acellular, but it retains key architectural elements such as papillary projections and elastin fibers.

6. The stainless steel squares used for growth of skin substitutes at the air/liquid interface are highly porous and act like a sponge for culture medium. Their flat surface is superior to the meshed screens that we have used in the past. It is important to aspirate out the spent medium that is inside the stainless steel plate.

7. Many other researchers have described preservation of the panniculus carnosus to aid in survival of the transplanted skin graft. We do not find this to be necessary and have found that it serves to increase the length and difficulty of the procedure.

8. We do not use systemic antibiotics for the mice during or after the procedure. Good sterile technique results in few to no infections.

References

1. Gallico, G. G. D., O'Connor, N. E., Compton, C. C., Kehinde, O., and Green, H. (1984) Permanent coverage of large burn wounds with autologous cultured human epithelium. *N. Engl. J. Med.* **311**, 448–451.

2. Heimbach, D., Luterman, A., Burke, J., et al. (1988) Artificial dermis for major burns. A multi-center randomized clinical trial. *Ann. Surg.* **208**, 313–320.

3. Hansbrough, J. F., Boyce, S. T., Cooper, M. L., and Foreman, T. J. (1989) Burn wound closure with cultured autologous keratinocytes and fibroblasts attached to a collagen-glycosaminoglycan substrate. *JAMA* **262**, 2125–2130.

4. Falanga, V., Margolis, D., Alvarez, O., et al. (1998) Rapid healing of venous ulcers and lack of clinical rejection with an allogeneic cultured human skin equivalent. Human Skin Equivalent Investigators Group [see comments]. *Arch. Dermatol.* **134**, 293–300.

5. Yannas, I. V., Burke, J. F., Orgill, D. P., and Skrabut, E. M. (1982) Wound tissue can utilize a polymeric template to synthesize a functional extension of skin. *Science* **215**, 174–176.

6. Boyce, S. T., Christianson, D. J., and Hansbrough, J. F. (1988) Structure of a collagen-GAG dermal skin substitute optimized for cultured human epidermal keratinocytes. *J. Biomed. Mater. Res.* **22,** 939–957.

7. Bell, E., Ehrlich, H. P., Buttle, D. J., and Nakatsuji, T. (1981) Living tissue formed in vitro and accepted as skin-equivalent tissue of full thickness. *Science* **211,** 1052–1054.

8. Parenteau, N. L., Nolte, C. M., Bilbo, P., et al. (1991) Epidermis generated in vitro: practical considerations and applications. *J. Cell. Biochem.* **45,** 245–251.

9. Prunieras, M., Regnier, M., and Woodley, D. (1983) Methods for cultivation of keratinocytes with an air-liquid interface. *J. Invest. Dermatol.* **81,** 28s–33s.

10. Livesey, S. A., Herndon, D. N, Hollyoak, M. A., Atkinson, Y. H., and Nag, A. (1995) Transplanted acellular allograft dermal matrix. Potential as a template for the reconstruction of viable dermis. *Transplantation* **60,** 1–9.

11. Medalie, D. A., Eming, S. A., Tompkins, R. G., et al. (1996) Evaluation of human skin reconstituted from composite grafts of cultured keratinocytes and human acellular dermis transplanted to athymic mice. *J. Invest. Dermatol.* **107,** 121–127.

12. Medalie, D. A., Eming, S. A., Collins, M. E., et al. (1997) Differences in dermal analogs influence subsequent pigmentation, epidermal differentiation, basement membrane, and rete ridge formation of transplanted composite skin grafts. *Transplantation* **64,** 454–465.

13. Morgan, J. R., Barrandon, Y., Green, H., and Mulligan, R. C. (1987) Expression of an exogenous growth hormone gene by transplantable human epidermal cells. *Science* **237,** 1476–1479.

14. Gerrard, A. J., Hudson, D. L., Brownlee, G. G., and Watt, F. M. (1993) Towards gene therapy for haemophilia B using primary human keratinocytes. *Nat. Genet.* **3,** 180–183.

15. Eming, S. A., Medalie, D. A., Tompkins, R. G., Yarmush, M. L., and Morgan, J. R. (1998) Genetically modified human keratinocytes overexpressing PDGF-A enhance the performance of a composite skin graft. *Hum. Gene Ther.* **9,** 529–539.

16. Eming, S. A., Yarmush, M. L., Krueger, G. G., and Morgan, J. R. (1999) Regulation of the spatial organization of mesenchymal connective tissue: effects of cell-associated versus released isoforms of platelet- derived growth factor. *Am. J. Pathol.* **154,** 281–289.

17. Choate, K. A., Medalie, D. A., Morgan, J. R., and Khavari, P. A. (1996) Corrective gene transfer in the human skin disorder lamellar ichthyosis. *Nat. Med.* **2,** 1263–1267.

18. Vailly, J., Gagnoux-Palacios, L., Dell'Ambra, E., et al. (1998) Corrective gene transfer of keratinocytes from patients with junctional epidermolysis bullosa restores assembly of hemidesmosomes in reconstructed epithelia. *Gene Ther.* **5,** 1322–1332.

19. Seitz, C. S., Giudice, G. J., Balding, S. D., Marinkovich, M. P., and Khavari, P. A. (1999) BP180 gene delivery in junctional epidermolysis bullosa. *Gene Ther.* **6,** 42–47.

20. Rheinwald, J. G. and Green, H. (1975) Serial cultivation of strains of human epi-

dermal keratinocytes: the formation of keratinizing colonies from single cells. *Cell* **6,** 331–343.

21. Eming, S. A. and Morgan, J. R.(1997) Methods for the use of genetically modified keratinocytes in gene therapy, in *Methods in Molecular Medicine: Gene Therapy Protocols* (Robbins, P., ed.), Humana, Totowa, NJ, pp. 264–279.

22. Boyce, S. T. and Williams, M. L. (1993) Lipid supplemented medium induces lamellar bodies and precursors of barrier lipids in cultured analogues of human skin. *J. Invest. Dermatol.* **101,** 180–184.

23. Ponec, M., Weerheim, A., Kempenaar, J., et al. (1997) The formation of competent barrier lipids in reconstructed human epidermis requires the presence of vitamin C. *J. Invest. Dermatol.* **109,** 348–355.

24. Andreadis, S. T., Hamoen, K. E., Yarmush, M. L., and Morgan, J. R. Keratinocyte growth factor promotes hyperproliferation and delays differentiation of skin equivalents in vitro. *FASEB J.* **15,** 898–906.

17

Bioartificial Muscles in Gene Therapy

Courtney Powell, Janet Shansky, Michael Del Tatto, and Herman H. Vandenburgh

1. Introduction

Most recent gene therapy protocols describe in vivo delivery of foreign genes by means of injecting adenoviral, adeno-associated, or lentiviral particles *(1–4)*. Although these delivery systems show great promise, it is reasonable to believe that ex vivo gene therapy may move through clinical trials more swiftly due to the added safety factors associated with keeping all virus outside the patient. Ex vivo gene transfer into rapidly dividing cells by means of retroviral transduction or stable transfection has produced cells secreting a vast variety of growth factors, including insulin *(5)*, granulocyte colony-stimulating factor *(6)*, growth hormone *(7,8)*, dopamine *(9)*, and erythropoietin *(10)*. These cells can then be transplanted back into the body to provide a therapeutic protein treatment.

By combining ex vivo gene delivery and tissue engineering, our laboratory has developed an efficient method of transplanting ex vivo modified cells back into the body. We first genetically modify skeletal muscle cells and then engineer them into organized tissue constructs (bioartificial muscles [BAMs]) containing parallel arrays of postmitotic skeletal myofibers. The BAMs are then implanted in vivo as living protein factories capable of delivering predictable levels of protein depending on the number of BAMs implanted or the number of cells per BAM *(11)*. Although skeletal muscle cells are the primary focus of our studies, we also genetically modified and tissue engineered other cell types (fibroblasts, cardiomyocytes, osteoblasts) into organized constructs for potential use in gene therapy. This chapter describes the procedures we developed to

From: *Methods in Molecular Medicine, Vol. 69, Gene Therapy Protocols, 2nd Ed.*
Edited by: J. R. Morgan © Humana Press Inc., Totowa, NJ

form BAMs from primary adult mouse muscle cells and highlights some potential uses in the field of gene therapy. The protocols presented include isolation of primary mouse muscle cells, retroviral transduction, tissue engineering of BAMs, and subcutaneous implantation of BAMs.

1.1. Bioartificial Muscles for Gene Therapy

BAMs have been used in our laboratory for multiple purposes, including studies of muscle organogenesis (12), muscle atrophy (13), attenuation of muscle wasting (11), evaluation of bioreactor perfusion systems (14), and gene therapy/protein delivery (15,16). As a long-lived implantable living protein delivery device, BAMs are capable of producing and secreting high levels of recombinant proteins locally and into the systemic circulation (11,15). These devices have the potential to treat a wide variety of human disorders that are caused by alterations in protein turnover and are currently either being treated with daily injections of pharmaceutical agents or left untreated. Some potential applications are treating adult-onset growth hormone deficiency with recombinant human growth hormone (rhGH), cardiac failure with a combination of rhGH and recombinant human insulin-like growth factor I (17,18), limb ischemia with vascular endothelial growth factor, or diabetes with insulin. Skeletal myofibers have a high protein synthesis capacity and therefore have the ability to secrete a plethora of proteins limited mainly by vector development and integration of a foreign gene into the chromosomes. Our laboratory has formed BAMs expressing rhGH, recombinant human insulin-like growth factor I, erythropoietin, vascular endothelial growth factor, bone morphogenetic protein-6, and ß galactosidase.

1.2. Advantages of Bioartificial Muscles for Gene Therapy

As mentioned earlier, BAMs are implantable protein factories. Unlike many gene therapy applications, by using BAM technology the level of in vitro protein secretion can be monitored and used to predict the in vivo therapeutic level (11). This is important for reasons of safety, dosing, and predictability of side effects. Sustained protein delivery at low levels will have advantages over the bolus doses that result from daily injections (19). For example, our laboratory has shown that murine BAMs constituitively secreting low levels of rhGH and implanted subcutaneously into mice attenuate skeletal muscle wasting whereas daily injections of rhGH are ineffective (11). In addition, BAM therapy does not subject the patient to direct exposure to viral particles. This alleviates the possibility of infecting undesirable cells with circulating virus. Finally, the transduced cells are postmitotic and nonmigratory, so the genetically modified cells do not reenter the cell cycle and remain at the implant site.

Several limiting factors may be encountered when scaling up to treat human disorders based on rodent studies. One problem is the time needed for myoblast expansion. We can obtain autologous human muscle cells with ease using standard needle biopsy of the vastus lateralis. We then expand these cells to provide a sufficient number for gene therapy; however, the expansion process takes approximately 1 month *(16)*. BAM therapy therefore requires at least two physician visits, for muscle biopsy and for BAM implantation. Since the technology cannot be used to treat a disorder immediately, it is primarily aimed at treating chronic disorders such as muscle wasting, heart failure, and anemia. One pathway around this issue is the establishment of an allogeneic cell bank. Myoblasts from one patient might be expanded, transduced, and stored for the use in several, if not many, future patients. However, allogeneic cells would elicit an immune response in vivo and therefore the BAMs would need to be immunoprotected by way of a membranous structure. Fortunately, these obstacles can probably be overcome without major changes to the existing BAM technology.

2. Materials

2.1. Isolation of Primary Mouse Muscle Cells

1. Male C3HeB/FeJ mice, 4–6 weeks old (Jackson Laboratories, Bar Harbor, ME).
2. CO_2 chamber.
3. 70% ethanol.
4. Small bone scissors.
5. Small dissecting scissors.
6. Forceps, curved #7 (2 pair).
7. Phosphate-buffered saline (PBS).
8. Petri dishes (60-mm diameter, non-tissue culture-treated).
9. Fibroblast growth medium (FGM; Clonetics, cat. no. 3130).
10. Trypsin (Sigma, cat. no. T4549).
11. Type V collagenase (Sigma, cat. no. C9263).
12. Wheaton Double Sidearm Jacketed Cell Stirrer (Bellco Glass, cat. no. 1965-50050).
13. Fetal bovine serum (FBS; Sigma, cat. no. F2442).
14. Primary mouse myoblast growth medium (PMMGM): 20% (v/v) FBS, 1% (v/v) penicillin/streptomycin, and 1% (v/v) ITS+1 (Sigma, cat. no. I2521) in a 50:50 blend of FGM and Dulbecco's modified Eagle's medium (DMEM; Gibco, cat. no. 11995-040).
15. 15-mL centrifuge tubes.
16. 50-mL centrifuge tubes.
17. Magnetic stir plate.

2.2. General Cell Culture

1. Collagen-coated tissue culture dishes (100-mm diameter) (*see* **Note 1**).
2. PMMGM (*see* **Subheading 2.1., item 14**).
3. Primary adult mouse myoblasts. These cells are isolated using the protocol detailed below. The cells should not be allowed to grow beyond 70% confluence (*see* **Note 2**).
4. Earle's balanced salt solution, without Ca^{2+} and Mg^{2+} (EBSS).
5. Trypsin (Sigma, cat. no. T4549).
6. Versene (Gibco, cat. no. 15040-066).
7. 50-mL centrifuge tubes.

2.3. Retroviral Transduction of Primary Mouse Muscle Cells

Most applications for the use of bioartificial muscles for gene therapy require long-term secretion of a desired protein. For this reason, our laboratory primarily uses retroviral transduction as the means for stable gene insertion into the primary mouse muscle cells. However, individual labs may choose a method of gene insertion that best suits their needs. The materials listed below are those needed if one is beginning with a stock of frozen virus-containing-medium (VCM).

1. VCM (*see* **Note 3**).
2. Polybrene (hexadimethrine bromide; Sigma cat no. H9268).
3. PMMGM (*see* **Subheading 2.1., item 14**).
4. Collagen-coated 6-well plates (35-mm-diameter wells).
5. Collagen coated tissue culture dishes (100-mm diameter).
6. Parafilm.
7. Heated centrifuge (Jouan, model CT422).

2.4. Fabrication of Tissue Molds

1. Razor blades.
2. Medical grade silicone rubber tubing, 0.25 i.d. (Nalgene, cat. no. 8060-0060).
3. 0.01-inch-thick silicone rubber sheeting (Silicone Specialty Fabricators, Paso Robles, CA).
4. Nonadhesive backing Velcro (available at local fabric stores).
5. Clear RTV silicone adhesive (General Electric).
6. 6-well plates (35-mm-diameter wells).
7. Collagen spray.

2.5. Bioartificial Muscle Formation

1. Primary mouse myoblasts expressing a foreign gene.
2. Chilled sterile pipets and pipetor tips.
3. Type I rat tail collagen (Collaborative Biomedical Products, cat. no. 354236).
4. Matrigel™ (Collaborative Biomedical Products, cat. no. 354234).

5. Sterile 0.1 *N* NaOH.
6. Ice.
7. PMMGM (*see* **Subheading 2.1., item 14**).
8. Primary mouse myoblast fusion medium. This medium contains 10% (v/v) horse serum (HS), 1% (v/v) penicillin, in a 50:50 blend of FGM and DMEM.

2.6. Implantation of Bioartificial Muscles

1. Scalpel, no. 10 blade.
2. DMEM (Gibco, cat. no. 11995-040).
3. Bovine serum albumin (Sigma, cat. no. A4161).
4. C3HeB/FeJ mice.
5. Electric razor (Oster Professional Products, cat. no. 59-03).
6. Betadine.
7. Cyclosporin A.
8. Sterile curved forceps (2 pair).
9. Metofane™ (Schering-Plough Animal Health).
10. 70% isopropanol pads.
11. 50-mL conical tube.
12. Gauze sponges soaked in Metofane.
13. Fibrin glue (Tisseel, ImmunoAG).
14. 9-mm stainless steel wound clips.
15. Heating pad.
16. Glass dessicator.

3. Methods

This section details the methods for engineering and implanting bioartificial muscles (BAMs) from transduced primary mouse muscle cells. A schematic drawing outlining the complete procedure is shown in **Fig. 1**.

3.1. Isolation of Primary Adult Mouse Muscle Cells

Using this cell isolation protocol, we obtain a population of primary mouse muscle cells containing predominately myoblasts (80–95%). Although several myoblast isolation methods are detailed in the literature *(20,21)*, we have had the greatest success using the following procedure.

1. Prepare 25 mL of enzyme solution. Enzyme solution: 0.5 mL trypsin in 12 mL FGM; 6.25 mg type V collagenase dissolved in 12.5 mL FGM (filtered through 0.22-μm filter); combine diluted trypsin and collagenase solutions.
2. Sacrifice one mouse using CO_2. Move mouse to a sterile hood.
3. Spray hindlimbs with 70% (v/v) ethanol.
4. Remove hindlimbs from mouse by making a clean cut at each hip joint. Remove the skin from each leg and place the legs in a small Petri dish containing sterile PBS. Rinse in several dishes containing fresh PBS to rinse any remaining fur off the muscle.

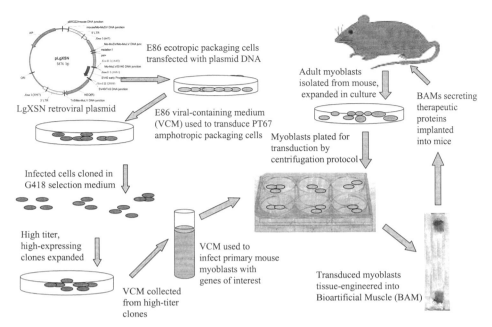

Fig. 1. This schematic illustrates the entire process for the use of bioartificial muscles (BAMs) for gene therapy. The procedure can take as few as 3 weeks from the time of cell isolation to BAM implantation.

5. Remove the muscle from the bone by pulling with a pair of sterile forceps. Dissecting scissors can also be used to aid this process. Place the muscle in clean, sterile PBS and mince into 1-mm pieces, as quickly as possible.
6. Transfer minced muscle pieces and PBS to a sterile 15-mL centrifuge tube.
7. Let muscle pieces settle out of the PBS. Carefully aspirate PBS from the tube without disturbing muscle pieces. Add 10 mL enzyme solution to the tube.
8. Transfer muscle pieces and enzyme solution to Wheaton Cell Stirrer.
9. Digest on a magnetic stir plate at low speed (the stir bar should rotate approx 60 rpm) for 10 min at 37°C. After digestion, remove supernatant to 15-mL tube and centrifuge at $3.5g$ for 2 min. (This slow speed will settle the large pieces of muscle without pelleting the cells.) Remove supernatant to a 50-mL tube and add 4 mL FBS to stop enzymatic action.
10. Add 10 mL fresh enzyme solution to the Cell Stirrer and repeat **step 9**.
11. Pool supernatants from both digestions and spin at $800g$ for 10 min at room temperature.
12. Aspirate supernatant and resuspend pellet in 1 mL PMMGM by carefully drawing up and down in a sterile 2-mL Pasteur pipet.
13. Count cells with a hemacytometer (*see* **Note 4**).
14. Plate cells into 6–8 collagen-coated 100-mm diameter tissue culture dishes, approximately 50,000 cells/100-mm dish, containing 10 mL PMMGM each (*see* **Note 2**).

15. Feed cells fresh PMMGM after 4–5 days. This extended incubation in the initial growth medium will allow additional cells to migrate out of any small muscle pieces that were transferred to the dishes. Individual cells may not be visible in dishes until the 3rd or 4th day in culture.

3.2. Retroviral Transduction of Primary Mouse Muscle Cells

As mentioned previously, our laboratory has had the greatest success using cells expressing a foreign protein after retroviral gene insertion. If you are using cells transfected/transduced using a different procedure, skip to **Subheading 3.3.** Our retroviral transduction protocol uses a modified centrifugation protocol described by Springer and Blau *(22)*. This requires a Jouan heated centrifuge with an H-1000 (B) rotor. Laboratories without this centrifuge and/or rotor may use an alternative retroviral transduction method (*see* **Note 5**).

1. When primary mouse muscle cells are 70–80% confluent in 100-mm dishes, subculture cells as follows:
   a. Aspirate medium from cells and rinse with sterile EBSS.
   b. Detach cells from 100-mm plates using 5 mL trypsin/versene (1:1 [v/v] trypsin [diluted 1:25, v/v] in EBSS): versene) per plate. Cell detachment should take approximately 3–5 min. Cells may be incubated at 37°C while in trypsin solution to enhance the process.
   c. Collect medium with suspended cells in a 50-mL tube; rinse plate with 5 mL PMMGM and add to the suspended cells in the 50-mL tube.
   d. Pellet cells by centrifugation at 800*g* for 5 min at room temperature.
2. Resuspend cells in 5 mL PMMGM and count cells using a hemacytometer.
3. Plate cells in collagen-coated 6-well plates at 40,000 cells/well in PMMGM.
4. Begin transduction 1 day after plating cells in 6-well plates.
   a. Quickly thaw VCM by placing in 37°C water bath. Aspirate the medium from each well and add 1.5 mL VCM containing 8 µg/mL Polybrene. Incubate in a 37°C, 5% CO_2 incubator for 15 min.
   b. Wrap each 6-well plate in Parafilm to prevent spills.
   c. Spin 6-well plates in Jouan CT422 centrifuge with the H-1000(B) rotor at 1100*g* for 30 min at 32°C.
   d. Remove VCM from wells, add 2 mL fresh PMMGM, and return plates to incubator.
5. Repeat **steps 4a.–4d.** above between 6 and 8 h later.
6. The following day, repeat **steps 4** and **5**.
7. Approximately 2 days after completing transduction (when cells are 70–80% confluent, approximately 300,000 cells/well) subculture the cells as detailed in **step 1** of this section and replate in collagen-coated 100-mm tissue culture dishes at 200,000 cells/dish. Cells may be expanded until the desired number of cells is obtained. At this point, cells may also be frozen in FBS containing 10% dimethylsulfoxide (DMSO) and stored in liquid nitrogen for future use.

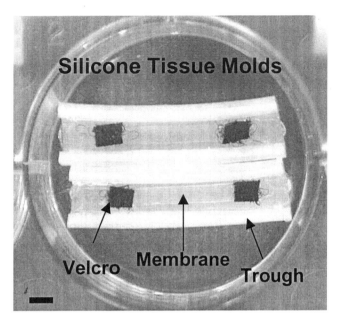

Fig. 2. BAMs are plated in silicone tissue molds. Refer to the Methods section for details. Scale bar = 4 mm.

3.3. Construction of Silicone Tissue Molds

1. Rinse silicone tubing in distilled water. Wipe dry with lint-free paper.
2. Cut tubing at 30-mm intervals. Cut these pieces lengthwise, removing the top one-third of the tubing. The remaining piece is called a trough through the remainder of this chapter and will hold the tissue engineered BAM (**Fig. 2**).
3. Cut silicone rubber sheeting into 30-mm-long by 3-mm-wide pieces.
4. Cut Velcro (loops) into 3 × 4-mm pieces.
5. Using RTV adhesive, glue Velcro pieces to the ends of the silicone rubber sheeting leaving 15 mm between inside edges of the two Velcro pieces. When dry, glue sheeting to the bottom of the trough; apply minimal glue to the membrane at the ends only, as the BAM/silicone rubber sheet combination will be removed from the trough for implantation.
6. Glue two silicone tissue molds into the bottom of each well of a 6-well plate.
7. After allowing the RTV adhesive to dry and cure for at least 24 h, rinse the 6-well dishes containing the silicone tissue molds with distilled water 2 times quickly and 1 time for 30 min on a rotating orbital shaker. Incubate with distilled water overnight at 37°C in a humidified incubator.
8. Allow dishes and molds to dry completely at room temperature. Spray with collagen (*see* **Note 1**) and sterilize using ethylene oxide. Sterile silicone tissue molds can be stored at room temperature for up to several months if the sterilization bags are securely sealed.

3.4. Formation and In Vitro Maintenance of Bioartificial Muscles

1. Thaw Matrigel on ice. This will take 2–4 h, depending on the amount of Matrigel in the bottle.
2. Subculture transduced primary mouse muscle cells as described in **Subheading 3.2. (step 1)**.
3. Place a medium aliquot containing the necessary number of cells (2×10^6 cells/BAM) on ice while preparing matrix.
4. Calculate required volume of each solution needed (total casting volume is 400 µl/BAM).
   a. Volume of Matrigel is total casting volume $\times$ 0.14.
   b. Volume of collagen solution is total casting volume $\times$ 0.86. This solution is 1.6 mg/mL rat tail type I collagen in PMMGM, neutralized with 0.1 N NaOH to pH 6.8–7.2 using 10% (v/v) of the stock collagen volume to neutralize.
5. Make collagen solution by combining rat tail type I collagen, NaOH, and PMMGM. Make at least 10% more than needed, as some solution is lost in mixing. Use chilled pipets and pipetor tips for collagen solution. Keep solution on ice.
6. Pellet cells by centrifugation at 800g for 5 min. Resuspend cell pellet in collagen solution prepared in **step 5** above. Use a chilled pipet and keep solution on ice.
7. Using a chilled pipet, add the appropriate volume of Matrigel. Mix well by gently drawing up and down in a chilled pipet. Keep solution on ice.
8. Cast 400 µl cell-gel solution into each tissue mold using a chilled pipetor tip. Spread the cell-gel solution between the Velcro attachment sites and carefully integrate into the Velcro.
9. Let the plates sit at room temperature in the sterile laminar flow hood for 5–10 min after casting; carefully transfer the plates to a humidified 37°C incubator without disturbing.
10. Two to 4 hours later, gently add 6 mL of PMMGM to each well.
11. Three days after casting, switch the BAMs to fusion medium for 4–5 days, changing medium every 2–3 days. After 4–5 days in fusion medium, switch BAMs back to PMMGM for the remainder of the in vitro experiment, changing the medium every 2 days (*see* **Note 6**).

3.5. Implantation of Bioartificial Muscles

1. Remove the necessary number of BAMs from the silicone tissue molds by cutting the silicone rubber sheeting away from the trough and lifting the BAM by the membrane/Velcro end piece (*see* **Note 7**).
2. Rinse the BAMs in new 6-well plates (2 BAMs/well) with each well containing 6 mL DMEM plus 0.125% (w/v) BSA for 60 min before implantation.
3. Anesthetize mouse in a glass dessicator containing Metofane soaked gauze pads for 5–10 min. To ensure that the mouse stays under anesthesia during the course of surgery, use a 50-mL conical tube filled with Metofane soaked gauze pads as a nose cone. Metofane should be used under a fume hood or with a vapor recovery system.

4. Shave the mouse's back with an electric razor and sterilize the area by wiping first with Betadine and then with 70% isopropanol.

5. Make a subcutaneous incision with a no. 10 scalpel, approximately 2 cm in length, along the midline of the back.

6. Reflect the skin on one side of the incision by lifting it with sterile forceps and gently tease the skin from the underlying fascia to make a "pocket." Insert 2–4 BAMs (*see* **Note 8**) longitudinally into the "pocket," placing the membrane side of the BAM against the fascia. Apply 150 µl fibrin sealant to each of the Velcro ends and under the membrane using the manufacturer's two-barrel mixing syringe (*see* **Note 9**).

7. Close the wound with four 9-mm stainless steel wound clips and return the mouse to its cage. Place the cage on a heating pad until the mouse regains consciousness.

8. Measure foreign serum protein levels by obtaining blood samples via tail bleeds or other approved blood collection methods and assaying the serum using the appropriate radioimmunoassay or enzyme-linked immunosorbent assay (ELISA) kit. Serum proteins can be assayed from 1 day post surgery to more than 6 months.

4. Notes

1. Collagen coating of tissue culture surfaces is necessary for attachment and expansion of primary mouse myoblasts. Our laboratory sprays all surfaces with 1 mg/mL rat tail type I collagen in 1% acetic acid. Using an airbrush, we apply 6 even coats of collagen to the surface, allowing each coat to dry before applying the next coat. An alternative collagen coating method is solution coating. For this method, dilute 1 mg/mL rat tail type I collagen in DMEM 1:100 (v/v). Cover the surface of each plate with the collagen solution and incubate for 1 h at 37°C. Remove solution and let the plates air-dry *(20)*.

2. Primary mouse muscle cells need to be subcultured before they become confluent or they will begin to fuse spontaneously even in growth medium. When expanding the cells after the initial plating, the cells should be plated at no less than 200,000 cells/100-mm dish. Cells plated at lower density grow in clusters and tend to fuse at an earlier time point. In addition, we have seen some variability between different primary mouse muscle cell preparations. Cells from some preparations simply do not grow, for unknown reasons. Other cultures are eventually taken over by nonmyoblast cell types. If either of these events occur, it is best to begin with another cell preparation. In a good culture, the percentage of myoblasts will increase with time. Our laboratory has expanded myoblast preparations for more than 15 passages.

3. The VCM used for retroviral transduction is very important. Our laboratory collects VCM from high-titer, high-secreting, clonally derived, replication defective retroviral packaging cells. We collect VCM from these packaging cells when they become confluent and continue for a total of 3 days. To increase transduction efficiency, concentration of viral particles may be necessary. We follow the method of Bowles et al. *(23)* to concentrate the VCM 10 times. This has resulted

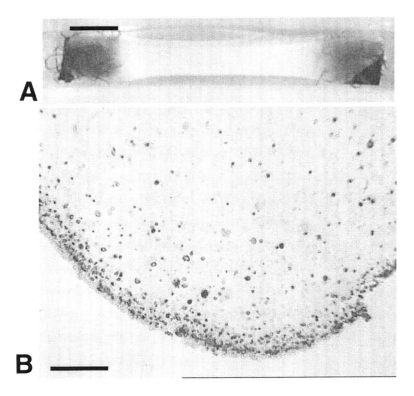

Fig. 3. (A) Unstained BAM, 2 weeks post casting. When the cell/gel mixture dehydrates (2–6 h after casting), the BAMs pull off the membrane and are held in place only at the Velcro end attachment sites. This results in passive internal longitudinal tensions on the order of 100 mg and causes cell alignment parallel to the long axis of the BAM. **(B)** Tropomyosin-stained cross-section. Muscle fibers concentrate around the perimeter of the BAM with a low density of fibers scattered throughout the interior. Scale bar = 4 mm in (A) and approximately 100 μm in (B).

     in an approximately 3 times higher titer (measured using 3T3 cells) and a 5 times increase in protein secretion from the packaging cell line. VCM can be stored in 15–50-mL aliquots at −80°C prior to use after snap freezing on dry ice.
4. The cell aliquot contains isolated cells (mainly myoblasts and fibroblasts), red blood cells, debris, and muscle chunks. The desired cells (the myoblasts and fibroblasts) are distinguishable in the hemacytometer from the red blood cells because of their round shape and larger size. Expect a cell yield of approximately 250,000–500,000 cells per mouse.
5. We have tried several methods of retroviral transduction. The centrifugation method yields the highest transduction efficiency and protein expression, but incubating the cells in VCM containing 8 μg/mL Polybrene for 3 days with daily changes to fresh VCM *(6)* has also proved successful. Each laboratory should

experiment to find a transduction method that works best within that laboratory setting.

6. BAMs contain parallel arrays of postmitotic muscle fibers by 10–12 days after casting. These fibers can be visualized by staining with antisarcomeric tropomyosin or antimyosin heavy chain antibodies (Sigma) using standard immunohistochemistry *(16)*. These fibers are highly concentrated around the perimeter of the BAM, with some fibers scattered throughout the center (**Fig. 3**).

7. BAMs are typically implanted 10–16 days after formation. Prior to implantation, we determine the BAM protein secretion level by collecting conditioned medium over a 4–24-h period and using the appropriate radioimmunoassay or ELISA assay kit.

8. Implant as many BAMs as necessary to get the desired level of in vivo protein delivery. Our laboratory has implanted as many as 4 BAMs in one mouse. Various other implantation sites are currently being investigated.

9. Fibrin glue is used to tack the BAMs down. This helps keep the BAMs under tension and in place and decreases abrasion of the skin by the Velcro.

Acknowledgments

This work was supported by grants from NASA and the NIH.

References

1. Li, S. and Huang, L. (2000) Nonviral gene therapy: promises and challenges. *Gene Ther.* **7,** 31–34.
2. Lien, Y. H. and Lai, L. W. (1999) Entering the gene therapy era. *J. Formos. Med. Assoc.* **98,** 718–721.
3. Snyder, R. O. (1999) Adeno-associated virus-mediated gene delivery. *J. Gene Med.* **1,** 166–175.
4. Buchschacher, G. L. and Wong-Staal, F. (2000) Development of lentiviral vectors for gene therapy for human diseases. *Blood* **95,** 2499–2504.
5. Gros, L., Riu, E., Montoliu, L., Ontiveros, M., Lebrigand, L., and Bosch, F. (1999) Insulin production by engineered muscle cells. *Hum. Gene Ther.* **10,** 1207–1217.
6. Bonham, L., Palmer, T., and Miller, A. D. (1996) Prolonged expression of therapeutic levels of human granulocyte colony-stimulating factor in rats following gene transfer to skeletal muscle. *Hum. Gene Ther.* **7,** 1423–1429.
7. Barr, E. and Leiden, J. M. (1991) Systemic delivery of recombinant proteins by genetically modified myoblasts. *Science* **254,** 1507–1509.
8. Dhawan, J., Pan, L. C., Pavlath, G. K., et al. (1991) Systemic delivery of human growth hormone by injection of genetically engineered myoblasts. *Science* **254,** 1509–1512.
9. Cao, L., Zhao, Y. C., Jiang, Z. H., et al. (2000) Long-term phenotypic correction of rodent hemiparkisonism by gene therapy using genetically modified myoblasts. *Gene Ther.* **7,** 445–449.
10. Hamamori, Y., Samal, B., Tian, J., and Kedes, L. (1994) Persistent erythropoiesis by myoblast transfer of erythropoietin cDNA. *Hum. Gene Ther.* **5,** 1349–1356.

11. Vandenburgh, H., Del Tatto, M., Shansky, J., et al. (1998) Attenuation of skeletal muscle wasting with recombinant human growth hormone secreted from a tissue-engineered bioartificial muscle. *Hum. Gene Ther.* **9,** 2555–2564.

12. Vandenburgh, H. H., Swasdison, S., and Karlisch, P. (1991) Computer aided mechanogenesis of skeletal muscle organs from single cells in vitro. *FASEB J.* **5,** 2860–2867.

13. Vandenburgh, H., Chromiak, J., Shansky, J., Del Tatto, M., and LeMaire, J. (1999) Space travel directly induces skeletal muscle atrophy. *FASEB J.* **13,** 1031–1038.

14. Chromiak, J., Shansky, J., Perrone, C. E., and Vandenburgh, H. H. (1998) Bioreactor perfusion system for the long term maintenance of tissue-engineered skeletal muscle organoids. *In Vitro Cell Dev. Biol.* **34,** 694–703.

15. Vandenburgh, H. H., Del Tatto, M., Shansky, J., et al. (1996) Tissue engineered skeletal muscle organoids for reversible gene therapy. *Hum. Gene Ther.* **7,** 2195–2200.

16. Powell, C., Shansky, J., Del Tatto, M., et al. (1999) Tissue-engineered human bioartificial muscles expressing a foreign recombinant protein for gene therapy. *Hum. Gene Ther.* **10,** 565–577.

17. Stromer, H., Cittadini, A., Douglas, P. S., and Morgan, J. P. (1996) Exogenously administered growth hormone and insulin-like growth factor-1 alter intracellular Ca^{2+} handling and enhance cardiac performance. *Circ. Res.* **79,** 227–236.

18. Lombardi, G., Colao, A., Ferone, D., et al. (1997) Effect of growth hormone on cardiac function. *Horm. Res.* **48,** 38–42.

19. Sun, Y. N., Lee, H. J., Almon, R. R., and Jusko, W. J. (1999) A pharmacokinetic pharmacodynamic model for recombinant human growth hormone effects on induction of insulin-like growth factor I in monkeys. *J. Pharmacol. Exp. Ther.* **289,** 1523–1532.

20. Naffakh, N., Pinset, C., Montarras, D., et al. (1996) Long-term secretion of therapeutic proteins from genetically modified skeletal muscles. *Hum. Gene Ther.* **7,** 11–21.

21. Lui, C., Dunigan, J. T., Watkins, S. C., Bahnson, A. B., and Barranger, J. A. (1998) Long-term expression, systemic delivery, and macrophage uptake of recombinant human glucocerebrosidase in mice transplanted with genetically modified primary myoblasts. *Hum. Gene Ther.* **9,** 2375–2384.

22. Springer, M. L. and Blau, H. M. (1997) High-efficiency retroviral infection of primary myoblasts. *Somat. Cell Mol. Genet.* **23,** 203–209.

23. Bowles, N. E., Eisensmith, R. C., Mohuiddin, R., and Pyron, M. (1996) A simple and efficient method for the concentration and purification of recombinant retrovirus for increased hepatocyte transduction in vivo. *Hum. Gene Ther.* **7,** 1735–1742.

18

Cytokine Gene-Modified Cell-Based Cancer Vaccines

R. Todd Reilly, Jean-Pascal H. Machiels, Leisha A. Emens, and Elizabeth M. Jaffee

1. Introduction

Antitumor immunity was first suggested in animals that reject tumor challenge after immunization with autologous inactivated tumor cells. Later, the discovery of tumor antigens recognized by T-cells strongly reinforced the concept that the tumor can be targeted by the immune system. In 1991, Boon and colleagues described the first human tumor antigen, *MAGE-1*, that is expressed in 50–60% of melanomas *(1)*. The identification of T-cell-dependent tumor antigens (*MAGE* family, *BAGE, GAGE, HER2/neu, p53, MART-1, tyrosinase, HPV,* and others) has opened the route of antigen-specific immunotherapy strategies *(2,3)*. Despite these important advances in tumor immunology, most tumor antigens are still unknown. Until more common tumor-specific antigens have been identified and their prevalence and relevance have been evaluated, the tumor cell itself remains one of the most convenient sources of antigens. Preclinical studies have shown that immunization with modified inactivated tumor cells can generate systemic antitumor immunity in vivo *(4)*. Currently, many clinical studies are investigating the safety and efficacy of autologous and allogeneic whole cell-based cancer vaccines *(5)*.

This chapter gives a brief overview of the most recent technical advances in the field of cytokine gene-modified cell-based cancer vaccines. The difficulties in translating this approach into the clinic and ways to overcome such difficulties are examined. Finally, a protocol to transfer cytokine genes retrovirally into tumor cells is provided and extensively discussed.

From: *Methods in Molecular Medicine, Vol. 69, Gene Therapy Protocols, 2nd Ed.*
Edited by: J. R. Morgan © Humana Press Inc., Totowa, NJ

1.1. The Role of T-cells in the Antitumor Immune Response

The rationale for genetically modified tumor vaccines for the treatment of established cancer depends on the existence of antigens within the tumors that can be recognized by the host immune response. It is the tremendous diversity of the T- and B-cell receptors that endows the immune system with the ability to distinguish fine antigenic differences among cells. Preclinical studies have yet to demonstrate a major role for B-cell antitumor responses following vaccination for the treatment of solid tumors. However, preclinical studies have demonstrated that activation of both CD4+ and CD8+ T-cells is critical for generating potent antitumor immune responses capable of eradicating preexisting burdens of tumor *(4)*.

For the generation of an antitumor immune response, at least two criteria must be fulfilled. First, tumor antigens must be available. Several techniques for the direct isolation of tumor antigens have confirmed the existence of tumor antigens that can be recognized by T-cells *(1–3)*. Currently, melanoma is the only human tumor for which a series of T-cell antigens has been identified. These antigens fall into several major categories including reactivated embryonic antigens and overexpressed differentiation antigens. The fact that some of these antigens are differentially expressed relative to normal cells as opposed to tumor-specific implies that a T-cell repertoire specific for these antigens must be available in the host and capable of activation by the immunizing cells (i.e., not irreversibly tolerized).

Second, T-cells must be appropriately activated to respond to the antigens presented on tumors. Even if tumor-specific T-cell precursors exist, tumor recognition and eradication is not guaranteed. T-cell activation is promoted when antigen is presented to the T-cell in the presence of inflammatory cytokines such as granulocyte-macrophage colony-stimulating factor (GM-CSF), resulting in the simultaneous expression of costimulatory molecules such as B7. Typically, these inflammatory cytokines recruit professional antigen-presenting cells (APCs), which provide adequate costimulation. Engagement of the T-cell receptor by antigen without the second signal that is provided by costimulatory molecules typically results in ignorance, anergy (active suppression of T-cell function), or apoptotic death of the antigen-specific T-cell. As tumors accumulate neoantigens during the process of tumorigenesis, the absence of associated inflammatory processes at these early stages probably produces a circumstance in which these neoantigens are viewed by the immune system as self-antigens, thereby inducing tolerance. Many mechanisms of immunune tolerance to tumor antigens have been described; a few of them are thymic or peripheral deletion of high-affinity T-cells specific for tumor antigens, secretion by the tumor of immunosuppressive factors such as transforming growth factor-ß (TGF-ß), ignorance, suppression, and immune deviation *(6,7)*.

Consequently, when tumor cells are used as cancer vaccines, they are frequently altered to increase their immunogenicity. Historically, such approaches have included administration with the vaccine of nonspecific immunologic adjuvants such bacille Calmette-Guérin (BCG) or *Corynebacterium parvum (8)*; or modification of the tumor cells by irradiation *(4)*, by nonpathogenic virus *(9)*, or by genetic manipulations *(4)*. More recently, gene transfer of genes encoding MHC molecules, costimulatory molecules (i.e., B7) and cytokines has shown preliminary encouraging results in animal models *(4,10,11)*. As shown in **Fig. 1**, there are two ways to modify tumor cells genetically to become immunogenic. Genes that encodes cytokine are commonly introduced into autologous or allogeneic tumor cells.

1.2. Genes and Principles of Gene Transfer to Tumor Cells

Gene transfer into tumor cells can be accomplished by physical or biologic methods. Physical methods include calcium phosphate transfection *(12)*, microinjection *(13)*, electroporation *(14)*, lipid-DNA complexes *(15)*, naked DNA *(16)*, DNA ligands *(17)*, and partical bombardment *(18)*. However, stable gene expression, which requires integration of the transfected DNA into a high proportion of the transfected DNA into the host's genome, is a rare event (around 1%) using physical methods. In vitro selection of those cells that have successfully incorporated the transferred DNA can be accomplished when DNA that encodes for a selectable marker (i.e., neomycin resistance) is cotransfected with the DNA of interest. However, although selection may increase the number of cells expressing the transgene to nearly 100%, it required 2–3 weeks of cell culture, which increases the theoretical risk of losing antigen expression by the tumor cell, thereby potentially decreasing vaccine efficacy.

Viral vectors for the introduction of cytokine genes offer the advantage of higher transduction efficiency. Retroviruses have been the most commonly employed vectors to transduce tumor cells because they can infect most mammalian cells and integrate into the host genome, which is a critical requirement for efficient gene transfer and expression in a stable fashion *(19,20)*. Moreover, second and third generations of retroviral vectors that have taken advantage of modifications in the intronic gag sequence as well as positioning of the start sites of the inserted gene allow efficient gene transfer into primary tumor explants in the absence of selection *(21,22)*. These vectors have the potential for being free of helper virus and are therefore extremely safe.

However, there are several potential disadvantages of using retroviral vectors:

1. The most significant risk, although never described in any clinical trials, is helper virus production. Apparition of replication-competent retroviruses (RCR) should be monitored closely since one study showed that injection of RCR in immuno-suppressed monkeys may lead to lymphomas.

2. Entry into the target cell is dependent on the surface expression of appropriate receptors for retroviral envelope proteins. Because the identity of most retroviral receptors is still unknown, it is difficult to determine whether a particular host cell expresses these receptors at a sufficient level for transduction by a given retroviruses. One way that is being explored to overcome this problem is the use of modified retroviral packaging cell lines (i.e, pseudotyped vectors), which contain alternative envelope proteins that can bind to better characterized and more abundant cell surface receptors *(23)*.

3. Due to the biology of the retrovirus, integration of the viral transgene into the host cell's genome requires a cell division. Infection of nondividing cells by retroviruses is usually very low. However, newer generations of retroviral systems are under development to overcome this problem. One such example, lentiviruses *(24)*, is described elsewhere in this book.

4. It is theoretically possible that random integration of the transgene by the retrovirus may lead to insertional mutagenesis.

5. Viral particles are not stable and are rapidly destroyed by the human complement system.

6. Transgenes larger than 9 kb can not be introduced easily into a retroviral vector.

In addition to retroviruses, adenoviral vectors are particularly appealing because they can transfer genes into nondividing cells *(25)*, they can accommodate larger transgenes (up to approx 20 kb), and they offer the particular advantage of higher titers, which results in higher gene transfer efficiency. The major limitation of adenoviral vectors for many gene therapy approaches is that because expression does not integrate stably into the genome of the transduced cell, it is lost after 1 to 4 weeks. This is not a problem, however, in the setting of gene transfer into tumor cells since the vaccine cells are usually irradiated and thus are eliminated within 1 week of injection. Prior exposure by the host to adenovirus can generate host immunity, which can be an obstacle to repeated adenoviral gene transfer. Recently, a clinical trial convincingly demonstrated that antitumor responses could be safely obtained in children with advanced neuroblastoma by adenovector-mediated transfer of the interleukin-2 (IL-2) gene *(26)*.

1.3. Murine Gene-Transduced Tumor Vaccines

There has been great interest in the study of immune responses generated by tumor cells genetically engineered to secrete various cytokines. This does not involve inducing the expression of any foreign genes in tumor cells but rather

Table 1
Animal Studies of Genetically Modified Cytokine-Secreting Tumor Vaccines

Cytokine	Local rejection after injection of a live vaccine	After irradiation, Systemic immunity after injection of a live vaccine	increased immunity compared with irradiated wild type-tumor cell
Il-2	+	+	–
Il-3	+	NT	NT
Il-4	+	+/–	–
Il-6	+	+	+
Il-7	+	+	+
IL-12	+	+	+
IFN-γ	+	+/–	–
TNF-α	+	–	–
GM-CSF	–	+	++
G-CSF	+	–	–

seeks to alter the immunologic environment of the tumor cell locally so as to either enhance antigen presentation of tumor antigens to the immune system or to enhance the activation of tumor-specific lymphocytes. One of the most important concepts underlying the use of cytokine gene-transduced tumor cells is that cytokine is produced at very high concentrations local to the tumor, but systemic concentrations generally remain quite low. This paracrine physiology much more closely mimics the natural biology of cytokine action than does the systemic administration of recombinant cytokines.

Many cytokine genes have been introduced into tumor cells with varying effects on both tumorigenicity and immunogenicity. Given the number of studies done to date with cytokine-transduced tumor cells, it is not surprising that variable results have been seen when different tumor systems are analyzed. Additional variables to the cytokines employed include cell dose, level of cytokine expression, location of immunization and challenge sites, and vaccination schedule. The following cytokine genes have been investigated: IL-1, -2, -3, -4, -5, -6, -7, -8, -9, -10, and -12, interferon-α (IFN-α), IFN-ß, IFNγ, tumor necrosis factor-α (TNF-α), tumor growth factor-ß (TGF-ß), M-GSF, G-CSF, GM-CSF, and Flt3 ligand (**Table 1**). In some murine models (see review in **ref. *27***), live tumor cells transduced with IL-2, -4, -6, -7, or IL-12, IFN-γ, G-CSF, or TNF-α generate a local tumor specific immune response after animal inoculation and are rejected or grow more slowly than wild-type tumor cells. In addition, vaccinated animals may develop a T-cell-dependent systemic

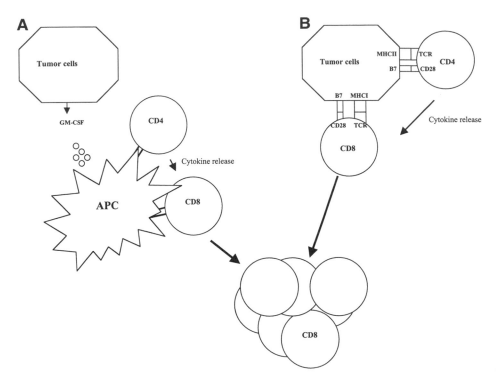

Fig 1. There are two ways to modify tumor cells genetically to increase their immunogenicity. (**A**) Transfer of gene-encoding cytokines, i.e., GM-CSF. Tumor cells secreting GM-CSF attract APCs that process tumor antigens and present them to CD4 and CD8 (cross-priming). (**B**) Transfer of gene-encoding MHC and/or costimulatory molecules (i.e., B7). Tumor cells become better APCs and stimulate CD4 and/or CD8 directly.

immunity conferring a protective response against secondary tumor challenge *(28–36)*. In contrast, in most tumor models, GM-CSF-transduced live tumor cells are not rejected in vivo and kill most of the recipient mice (presumably as a consequence of secretion to a toxic level of GM-CSF by the proliferative tumor cells) *(4)*. However, when GM-CSF-transduced tumor cells are irradiated, they are able to generate a potent antitumoral systemic immunity *(4)*.

Given the large number of potential cytokine genes and the technical difficulties in transducing human tumor cells to make vaccines, it is critical that they be compared for efficacy. Because most tumors show significant immunogenicity when simply irradiated, identification of genes that truly enhance a tumor's immunogenicity above that of irradiated wild-type cells is important. The first study that directly compared multiple cytokine and other genes in murine tumor models used a highly transmissible, defective retroviral vector

(4). This study demonstrated that, in a number of poorly and moderately immunogenic tumors, including a murine renal cell carcinoma, immunization with GM-CSF-transduced tumors produced the greatest degree of systemic immunity, which was enhanced relative to irradiated nontransduced tumors. It is noteworthy that in this study, the levels of antitumor immunity provided by live cytokine-transduced tumor cells (IL-2 and -4, TNF-α, IFN-γ) could also be achieved through the use of irradiated cells alone. To date, compared with more than 20 cytokines now screened in the B16 murine melanoma tumor vaccine model, GM-CSF is still the most potent. Moreover, tumor cells genetically altered to express GM-CSF can cure mice with preestablished small burdens of tumor and therefore represent a model for the minimal residual disease state common in patients with cancer *(4)*.

Analysis of the mechanism by which antitumor immunity is generated has demonstrated that GM-CSF is able to recruit antigen-presenting cells (APCs) and that immunity depends on CD4+ and CD8+ T-cells *(4,37)* (**Fig. 1**). APCs first take up and process tumor antigens and subsequently prime the T-cell arm of the immune response. Analysis of the effector phase of tumor rejection indicates a far broader role for CD4+ T-cells in orchestrating the host response to tumor *(38)*. This form of immunization leads to the simultaneous induction of Th1 and Th2 responses, both of which are required for maximal systemic antitumor immunity. Cytokines produced by these CD4+ T-cells activate eosinophils as well as macrophages that produce superoxide and nitric oxide. Both of these cell types then collaborate within the site of tumor challenge and contribute to its destruction.

In addition to GM-CSF, IL-4, -6, -7, and -12-transduced tumor cells have also been demonstrated to augment T-cell immunity compared with irradiated nontransduced tumor cells *(39,40)*. In particular, IL-12 appears to be very promising and has been shown to be more effective when used in combination with the transduction of tumor cells with B7.1 *(41)*.

1.4. Autologous Human Gene-Transduced Tumor Vaccines

Animal studies strongly support the investigation of cell-based cancer vaccines in clinical trials. Several phase I clinical trials for malignant melanoma, renal cell carcinoma, neuroblastoma, pancreatic adenocarcinoma, prostate cancer, sarcoma, or small cell lung carcinoma have already been completed or are currently activated *(5)*. Among the most frequently investigated cytokines in humand are IL-2 (neuroblastoma, melanoma, and small cell-carcinoma), GM-CSF (melanoma, pancreatic adenocarcinoma, prostate cancer, and renal cell carcinoma), IL-12 (melanoma), IL-7 (melanoma), and IFN-γ (melanoma). Phase I clinical trials have shown that the use of cytokine-modified autologous tumor vaccines is extremely safe *(26,42–48)*. The only side effects reported are grade I/II local skin reaction at the vaccine site, mild fever, and myalgia. Although in one study three patients presented with vitiligo after vaccination with IL-2-expressing autologous melanoma cells *(42)*, no major autoimmunity symptoms have been reported. In a few studies, observation of delayed-type hypersensitivity (DTH) reactions to autologous tumor cells provided evidence of immune priming *(42,44,45,47)*. Sometimes DTH reactions were correlated with tumor regression *(44,45)*.

We performed a randomized clinical trial comparing the administration of escalating doses of lethally irradiated autologous tumor cells either nontransduced or transduced with the human GM-CSF gene *(45)*. That study showed a fourfold magnitude difference in the total cutaneous reaction at the autologous tumor DTH test sites of patients who received the GM-CSF vaccine compared with nontransduced irradiated vaccine. DTH reactions against autologous tumor cells were increased with increasing vaccine dose administration, suggesting a dose effect. A patient receiving the GM-CSF treatment vaccine demonstrated a >80% regression of pulmonary metastases. However, one-fourth of the patients who underwent nephrectomy did not received vaccination because enough vaccine cells could not be prepared. The maximal toxic dose and dose with maximal bioactivity could not be determined in this trial because it was not technically feasible to obtain enough vaccine cells, illustrating a first limit of this vaccine approach.

More recently, Dranoff and collaborators *(44)* conducted a clinical trial with irradiated autologous melanoma cells engineered to secrete human melanoma GM-CSF. Interestingly, they reported that, in 11 of 16 patients, metastatic lesions biopsied after vaccination were infiltrated with T-lymphocytes and plasma cells and showed >80% of tumor destruction. Autologous nonirradiated tumor cells transduced with adenovirus IL-2 induced antitumor responses in children with neuroblastoma *(26)*. In this study, 5 of 10 patients had tumor responses, and 4 patients also demonstrated cytotoxic activity against tumor cells, suggesting a correlation between this test and clinical outcome. Vaccination with IL-12 and IL-7 gene-modified autologous melanoma cells induced immunologic activation demonstrated by an increased number of tumor-reactive proliferative and cytolytic cells in the peripheral blood of a few patients *(46,47)*. Infiltration of metastases by CD4+ and CD8+ T-cells was also observed after vaccination with Il-12-transduced autologous melanoma cells *(47)*. In another study, antimelanoma immunoglobulin G antibody responses were correlated with tumor regression after vaccination with IFN-γ gene-modified autologous melanoma tumor cells *(43)*. These early studies suggest that potent systemic antitumor immunity can be generated in vivo by this approach and therefore strongly support the continued clinical development of cytokine-secreting whole tumor cell vaccines.

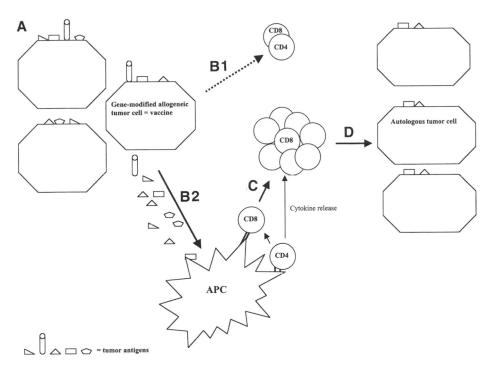

Fig 2. Rationale for using allogeneic tumor cell vaccine. **(A)** Most of the human antigens are shared by at least 50% of patients. **(B1)** Allogeneic tumor cells are poor APCS (absence of costimulatory molecules or/and compatible MHC). **(B2)** Secretion by allogeneic vaccine of engineered cytokines (i.e., GM-CSF) attracting APCs and release of tumor antigens. **(D)** APCs process the antigen and present it to CD4+ and CD8+ by a mechanism called cross-priming. **(E)** Activated CD8+ kill the autologou tumor cells.

1.5. Allogeneic Gene-Transduced Tumor Vaccines

The aforementioned autologous secreting vaccine trials have demonstrated highly encouraging results. However, there are significant difficulties with routinely expanding autologous tumor cells to the numbers required for adequate vaccination level. Autologous cell lines will need to be established from each patient to be treated, as a means of providing tumor antigens at the site of activation of the immune system. Even when subselection of continuously growing tumor lines can be achieved, there is a high likelihood that, after extended passage, the antigenic composition will change significantly, relative to the original primary tumor from which the tumor line originated. Another major problem is the variable and often lengthy time required for autologous vaccine preparation (approximately 4–12 weeks). During that time, disease may progress, and some patients may become ineligible for treatment *(42,44)*. These technical limitations make the use of autologous tumor cell vaccines for the treatment of most human cancers impractical.

The use of allogeneic cell lines that can be easily expanded in vitro may overcome these problems. Allogeneic vaccines can be generated from cell lines selected to provide predefined tumor antigens and their preparation can be standardized, decreasing the cost and the risk of this strategy. Two recent advances provide the immunologic rationale for an allogeneic vaccine approach (**Fig. 2**). First, studies evaluating human melanoma antigens have demonstrated that most of the human melanoma antigens identified are shared among at least 50% of patients, regardless of whether or not they share the same human leukocyte antigen type *(1–3,49,50)*. Second, studies have shown that the professional APCs of the host, rather than the vaccinating tumor cells themselves, are responsible for priming CD4+ and CD8+ T-cells *(51,52)* (**Figs. 1** and **2**). These data imply that the vaccinating cells used as the source of tumor antigens do not have to be MHC compatible with the host for successful priming of an antitumor immune response. Thus these studies suggest that relevant tumor antigens can be delivered by an allogeneic tumor for priming an antitumor immune response against the host tumor.

A few clinical trials using allogeneic melanoma cells (HLA-A2 matched or not) transduced with IL-2 have already been published *(53–55)*. Like autologous cytokine-based cancer vaccines, no major adverse events have been reported. Weak clinical responses or disease stabilization have been sparsely reported *(54,55)*. T-cell responses against antigens of autologous untransduced tumor have been observed in vitro post vaccination. These preliminary results, although promising, need to be confirmed in additional clinical trials. Clinical studies are currently ongoing with IL-2 (neuroblastoma, renal cell carcinoma, and melanoma), GM-CSF (pancreatic and prostate cancer), and Il-4 (melanoma) gene-transduced allogeneic tumor cells *(5,56)*.

Another way to overcome the technical limitations of autologous tumor cell vaccines is to use cytokine-expressing bystander cells such as fibroblasts. These cytokine-transduced fibroblasts are mixed with autologous tumor cells. This mixture is used as the vaccine. The feasibility and safety of this strategy have already been demonstrated for the treatment of tumors in which adequate numbers of tumor cells are available from resected tumor specimens without requiring in vitro cell expansion *(57,58)*.

2. Materials

1. Retroviral vector producer lines: The MGF retroviral producer cell lines were obtained from R.C. Mulligan (Chidren's Hospital, Harvard Medical School, Boston, MA).

a. Both the amphotropic and ecotropic retroviral producer cell lines, CRIP and CRE, respectively, are grown in Dulbecco's modified Eagle's medium (DMEM) with high glucose (4500 g/L), supplemented with 10% bovine calf serum, penicillin (100 U/mL final concentration), streptomycin (100 µg/mL final concentration), L-glutamine (2 mM final concentration), and gentamycin (50 µg/mL final concentration), at 37°C.

b. 0.25%/0.1% trypsin/EDTA for passaging the cell lines.

2. Tumor cell lines: All tumor cell lines to be transduced should be maintained in their optimal growth media before and after performing the transduction procedure to enhance the proliferative capacity of the cell population.

3. Retroviral gene transfer to tumor cells: Tumor cells and retroviral supernatant that has been prepared as described in **Subheading 3**.

a. In addition, DEAE-dextran 10 mg/mL stock solution prepared by dissolving 1 g into 100 mL of the producer line growth media (DMEM + 10% calf serum) and filtered through a 0.45-µm filter. Store in sterile 5-mL aliquots at 4°C for up to 6 months.

b. Sterile tumor growth media and sterile 1× phosphate-buffered saline (PBS).

4. Immunization of mice: Freshly thawed vaccine tumor cells (tumor cells transduced with the cytokine gene of interest and producing the adequate level of cytokine).

a. Mice between 6 and 12 weeks of age.

b. Sterile Hank's balanced salt solution (HBSS; pH 7.1–7.3) stored at 4°C.

c. 1-mL syringes.

3. Methods

3.1. Maintenance of Retroviral Vector Producer Lines

1. Grow the retroviral producer line in culture to confluency in large flasks ($\geq$162-cm^2). These cells grow in an adherent monolayer (*see* **Note 1**).

2. When the cells have reached confluency, remove the supernatant, and incubate the cells with enough trypsin-EDTA to cover the bottom of the flask (usually 2–3 mL) at 37°C until the cells become nonadherent (usually 1–2 min) (*see* **Note 2**).

3. Quench the trypsin with at least 3–4 vol of the growth media containing calf serum (*see* **Note 3**).

4. Centrifuge for 10 min at 1430g (1500 rpm) and 4°C.

5. Remove the supernatant and count the cells.

6. Replate the cells at about a 1:10 dilution of the number of cells in the confluent flask (approx 2×10^6 cells/162-cm^2 flask)

7. Split the cell lines 1:10 every 3–4 days or when each flask reaches confluency (*see* Notes 4 and 5).

3.2. Preparation of Retroviral Supernatant

1. Two days prior to transduction, trypsinize the producer cells, wash them once, and replate them at a density of 2×10^6 cells/100-mm culture dish (*see* **Notes 6–11**).

2. One day prior to transduction, remove the media, and add 10 mL of fresh media to the cells.
3. On the day of transduction, collect a 24-h retroviral supernatant and filter through a 0.45-μm filter to remove contaminating retroviral producer cells (*see* **Note 12**).

3.3. Maintenance of Tumor Cells Lines

1. Grow the tumor cells in culture in optimal tumor growth media to confluency (*see* **Note 13**).
2. On the day prior to transduction, replate the cells at a density of $2–5 \times 10^5$ cells/ 75-cm^2 culture dish (*see* **Note 14**).

3.4. Performing Retroviral Gene Transfer to Tumor Cells

1. Incubate the freshly collected retroviral supernatant with 10 μg/mL final concentration of the transduction enhancer DEAE-dextran for approximatively 10 min at room temperature, so that the retrovirus will bind to the enhancer prior to exposure to the tumor cells (*see* **Notes 15 and 16**).
2. Remove the growth media from the tumor cells and replace it with 10 mL of retroviral supernatant containing the enhancer.
3. Incubate the cells at 37°C for 24 h (*see* **Notes 17 and 18**).
4. Following incubation of the tumor cells with the retroviral supernatant, remove the supernatant and wash the cells twice with sterile PBS to rinse away residual retroviral supernatant (*see* **Notes 19–24**).
5. Add 10 mL of tumor growth media and allow the cells to grow for 48 h.

3.5. Testing for Cytokine Gene Product

1. At 48 h following transduction, remove the growth media and add 10 mL of fresh tumor growth media.
2. Collect a 24-h supernatant for evaluation of cytokine secretion. To do this, remove the supernatant, centrifuge or filter through a 0.45-μm filter to remove the cells, and aliquot the supernatant into three 1-mL, sterile aliquots that can be stored frozen at –70°C until the time of testing for cytokine secretion. Three separate aliquots should be stored so that repeat testing can be performed without multiple freeze/thaws (which might reduce the concentration of the gene product).
3. Following collection of the cell supernatant, take up the cells and count them. Record the total number of cells that contributed to the production of the cytokine over 24 h. This number will be used to calculate the concentration of cytokine secretion per given number of tumor cells after the concentration of cytokine in the supernatant is determined (*see* **Notes 25–29**).

3.6. Testing for In Vivo Efficacy

1. Grow the cytokine-transduced tumor cells (vaccine cells) and the parental tumor cell line in optimal tumor growth media to confluency.
2. On the day of vaccination, trypsinize the vaccine tumor cells, wash them twice in

HBSS, count them, and resuspend the cells in HBSS at the adequate concentration. (We propose 10^6 in 100 µL.)

3. Irradiate the vaccine cells (*see* **Note 35**).

4. Vaccinate the mice (6–12 weeks old). For the initial experiments, although this may vary with the tumor system used and should be optimized for each model (*see* **Notes 30–38**), we propose 10^6 vaccine cells in 100 µL of HBSS injected subcutaneously close to a lymph node area (i.e., left hindlimb). Maintain the mice in an animal facility under supervised care (sterile water and food) (*see* **Notes 30–35**).

5. Challenge the mice with untransduced parental tumor cells (in 100 µL of HBSS) 14 days after the first vaccination at a site distant from the vaccine cells (i.e., right hindlimb) (*see* **Notes 36–38**).

6. Follow the mice for occurrence of tumors (>3 mm) and tumor growth. (Measure with a caliper in two or three dimensions.)

4. Notes

1. Producer lines derived from the NIH 3T3 fibroblast cell line grow in a adherent monolayer. They do best when they are plated at a threshold density of about one-tenth the flask's total cell capacity. Plating the cells at a lower density may result in loss of the cell line.

2. Exposure to trypsin results in rapid release of the producer cells from the tissue culture flask. Be aware that overexposure to trypsin will result in significant cell death.

3. The MGF producer line grows in media supplemented with bovine calf serum. Substitution of fetal bovine serum may result in a change in growth kinetics and viral particle production. The growth requirements recommended by the laboratory in which the producer line originated should always be used to culture the producer line being employed.

4. It is advantageous to expand enough of the producer cells initially to allow for freezing a large stock of aliquots for two reasons. First, prolonged passage in culture may increase the possibility of recombination events within the producer cells that may result in the production of helper virus. Second, there is the theoretical concern that prolonged passage in culture may result in a decrease in the population of producer cells capable of efficiently producing the viral particles.

5. The MGF producer cell lines freeze well in 90% calf serum plus 10% DMSO. Recovery of viable producer cells will be severely compromised if these cells are frozen in other types of serum. Each producer line should be frozen in the same type of serum that is used for in vitro growth. These cells can be stored long term in liquid nitrogen.

6. Retroviruses are difficult to titer because they do not form plaques. It is therefore recommended that every producer line be titered for transduction efficiency using an easily transducible cell line. NIH 3T3 cells are a good choice for comparison with other murine cell lines. For the transduction of human primary cultures, a human cell line may be a more appropriate cell line for comparison. Titering can

be accomplished by using the transduction procedure described in **Subheadings 3.1.–3.5.**, and by performing serial two- to fivefold dilutions of the retroviral supernatant prior to exposure of the retrovirus to the tumor cell line. Dilutions of the retroviral supernatant should be made with the cell line's growth media for best results. Most supernatants are optimal either undiluted, or between a 1:2 and 1:10 dilution.

7. Before assuming that insufficient transduction rates are due to low-titer supernatants as a result of a poor supernatant collection, it is important to determine first whether the producer cells themselves are still capable of generating large quantities of retroviral particles. Because the producer cells themselves also express the gene encoded by the retroviral vector, an easy way to evaluate the producer line for production of the vector is to assay the cells for expression of the gene product. However, in vitro loss of high titer producer lines owing to long-term culture can easily be avoided by routinely thawing a new aliquot of producer cells every 3–4 weeks.

8. If expression is at the expected level, then the problem is more likely to be caused by a low-titer retroviral supernatant resulting from suboptimal supernatant collection. There are two major causes of low-titer retroviral supernatants: inadequate retroviral supernatant collection resulting from insufficient numbers of producer cells or overgrowth of producer cells; and suboptimal growth conditions for retroviral supernatant collection. Potential problems include a bad lot of calf serum, use of the wrong media and supplements, and inadequate CO_2 concentration during incubation.

9. A study performed to evaluate the improvement of retroviral vector production observed that the growth of 21/22 producer lines at 32°C for up to 2 weeks after the cells reached 100% confluence increased vector titers *(59)*. Growth of the producer lines at 32°C is thought to increase the stability of the viral particles. In addition, improved vector production may be a result of the decreased metabolism of the producer cells at this lower temperature.

10. For optimal transduction efficiencies, freshly collected retroviral supernatants should be used. Although it is possible to store the supernatants at 4°C for several days and to freeze these supernatants at −70°C for several weeks, the efficiency of transduction may decrease by as much as 50% following thaw of the supernatant.

11. Producer lines must be frozen in the same type of seram used for in vitro growth unless otherwise advised. Other serum may not support the growth of these cells well and may result in significant cell death during freezing and storage.

12. Occasionally freshly collected retroviral supernatant contains high concentrations of retroviruses, and gene transfer efficiency is low because of other contaminating toxins within the supernatant. Therefore, performing a titer by using serial dilution of the retroviral supernatant is recommended. The titer can be performed on NIH 3T3 cells. Dilution of supernatant can be made using the NIH 3T3 cell growth media.

13. Most proliferating cell lines can be transduced with a retroviral vector. However, the efficiency of transduction will depend on the percentage of cells that are ac-

tively proliferating at the time of exposure to the retrovirus, since integration into the host genome is required for stable expression of the transferred gene. Therefore, the growth conditions for each cell line being transduced should be optimized before attempting this transduction procedure. Most long-term cell lines already have defined growth conditions that support optimal growth. However, it is now possible to transduce many primary, short-term cancer cell lines, and conditions for optimizing their growth may already have been described.

14. The actual density of cells in the flask should be optimized for every tumor cell type, keeping in mind that the cells will need to be able to proliferate maximally for at least 48 h following transduction to allow for optimal integration and expression of the gene. For cells with a 48–72 h doubling time, adequate transduction can be achieved by plating the cells at a density that will result in approximately one-third confluency of the flask on the day of transduction.

15. Both a negative and a positive control group should be included in each transduction experiment to confirm that gene expression is the result of gene transfer. A good negative control is to incubate the flask of each cell type to be transduced with retroviral producer cell growth media containing the enhancing polymer alone. An adequate positive control would include the transduction of any easily transducible cell line with the same lot of retroviral supernatant used to transduce the test tumor cells.

16. Transduction efficiency can be enhanced by the addition of polymers to the retroviral supernatant just prior to exposure of the target cells to the retroviral vector. Enhanced gene transfer is thought to occur via a charge-mediated mechanism that affects virus binding to or penetration of the target cell. The polycations protamine, Polybrene, and DEAE-dextran are routinely used for this purpose *(4,60)*. In addition, liposome-forming compounds such as dioxeoyl-trimethyl ammonium propane (DOTAP; Boehringer Mannheim, Indianapolis, IN) have also been used successfully to enhance retroviral gene transfer and may be less toxic to the host cell than other enhancers. Liposome-forming agents probably enhance gene transfer into the host cell by first forming stable interactions with the virus, then adhering to the cell surface, and then fusing with the cell membrane and releasing the virus into the cytoplasm *(61)*. Because most enhancers are toxic to the cell lines at high concentrations, yet higher concentrations of polymers may be required for enhanced transduction efficiency to some cell lines, it is recommended that a titer of the enhancer be performed on each new batch of enhancers used, to determine the least toxic, most enhancing concentration of the polycation or lipid coumpound. **Table 2** illustrates recommended ranges of polycation and lipid reagent concentrations for the commonly employed transduction enhancers.

17. Longer incubation times will increase the number of proliferating cells that are exposed to the retroviral vector and therefore may increase the efficiency of transduction. Hardy tumor cell lines may tolerate the retroviral supernatant containing low concentrations of enhancer for 24–48 h without significant cell death. However, primary human tumor cultures may not tolerate a change in the growth media for more than several hours. Therefore, it is best to perform a pilot study evaluat-

Table 2
Commonly Employed Transduction Enhancing Reagents

Transduction enhancer	Target cell type (tumor cell lines)	Concentration range (µg/mL)	Incubation time[b] (h)
DEAE-dextran	Murine	5–10	24
(Sigma, St. Louis, MO)	Human	10–100	4–24
Polybrene	Murine	5–10	24
(Sigma or Aldrich, Milwaukee, WI)	Human	10–100	
Protamine sulfate	Murine	5–10	24
(Lilly, Indianapolis, IN)	Human	10–100	4–24
DOTAP	Murine	5–10	24
(Boehringer Mannheim, Indianapolis, IN	Human	10–100	4–24

[a] Final concentratiaon in retroval supernantant.

[b] Concentrations of transduction inhibitors can be toxic to cells with prolonged incubation times. We recommend a pretransduction feasibility study looking at toxicity of each enhancer over time.

ing the rate of tumor cell death over time under exposure to the retroviral supernatant containing the enhancer, to optimize the transduction procedure.

18. Recent evidence suggests that the efficiency of retroviral transduction can be improved by a 90-min centrifugation at $2000g$ (2500 rpm) and 32°C, prior to an overnight incubation (at 32°C) of the tumor cells with the retroviral supernatant. However, some tumor cells may not tolerate an overnight incubation at 32°C *(59)*.

19. It is often useful to perform the initial transduction studies on new tumor cell lines using the retroviral vector containing a marker gene (for example, the *lacZ* gene expresses the cytoplasmic enzyme ß-galactosidase, which will turn the cytoplasm of the transduced cell blue after exposure to the substrate Bluogal or X-Gal). Marker genes can be used quantitatively to determine the number of the tumor cells in the transduction population that are capable of expressing the transferred gene (the transduction efficiency of the vector for a particular tumor cell line).

20. It is not uncommon to have a high-titer retroviral supernatant. If this is the case, the supernatant can be diluted 1:5 or 1:10 (depending on titer) with target-cell media, prior to the transduction procedure, to decrease target-cell toxicity from the retroviral supernatant. As mentioned earlier, be aware that a dilution of a high-titer supernatant may be necessary because higher titer supernatants may contain inhibitors against successful retroviral transduction.

21. The procedure described in **Subheadings 3.1.–3.5.** can be used for transduction of adherent and nonadherent tumor cell lines and does not require any special modifications. For nonadherent cell lines, media changes require that the tumor cells be taken up with the supernatants and centrifuged to remove the supernatant. The cells are then resuspended in the fresh media and replated.

22. Freezing of large stocks of the transduced tumor cells is recommended to prevent loss of gene expression, as well as to prevent in vitro selection with loss of antigen expression. Transduced tumor cell lines can be frozen down and stored in liquid nitrogen long term without loss of gene expression. Controlled-rate freezing is recommended to prevent a significant decrease in viability following thawing. A cheap and efficient way to perform controlled-rate freezing is to immerse the freezing vial of cells in a propanol bath (Nalgene Cryo 1°C Freezing container) and to place the apparatus into a −70°C freezer overnight. This will freeze the cells at approximately 1°C/min. The cells can then be placed into liquid nitrogen for long-term storage.

23. Primary human tumor lines are more difficult to transduce than long-term established lines. However, with the increasing applications of gene therapy in clinics, there is an increasing need for improved methods of gene transfer to these cells. The most important criteria for efficient gene transfer to primary human tumor cultures are to optimize the growth conditions for maximal proliferation capacity. In addition, increasing the concentration of transduction-enhancing polymer may result in improved transduction efficiency. It is often beneficial to screen the different enhancing polymers initially for the upper limits of polymer concentration, and incubation time that each primary tumor cell line can tolerate before significant cell death is observed.

24. Recently, some evidence has indicated that modifying the packaging cell line to express alternative envelope proteins other than those of ecotropic or amphotropic cell surface receptors can enhance retroviral uptake and expression by human tumor cells. This possibility should be considered if gene expression cannot be improved with the aforementioned modifications.

25. Following transduction of tumor cells with the cytokine gene, the transduced cells should be evaluated for the total quantity of cytokine produced and for the quantity of cytokine that is bioactive. ELISA best determines the total quantity of cytokine produced. The ELISA kits are now commercially available for quantitation of most murine and human cytokines (Genzyme, Boston, MA; Endogen; R&D Systems). Although these kits are expensive, they usually have a sensitivity of 1–4 pg/mL and are specific for the cytokine being tested. Bioassays are also available for many murine and human cytokines. Although they are often not as sensitive or specific as ELISA, they provide important information concerning the function of the cytokine being secreted by the tumor cells. Cell line bioassays of common murine and human are listed in **Table 3**. For these assays, serial dilutions are made of the tumor cell supernatants collected as described in **Subheading 3.5**. Most of the bioassays rely on cell lines that are growth factor dependent. In these assays, the degree of proliferation of the cell line in the presence of the serially diluted cytokines is determined by [^{3}H]thymidine incorporation. A recombinant standard is also run along with the test samples to quantitate the cytokine accurately in the test samples. Because several cytokines may stimulate the same cell line, duplicate curves are often run for each sample, one curve in the presence of cytokine-blocking antibody, to evaluate the percent of prolifera-

Table 3
Common Bioassays Used to Quantitate Cytokine Production

Cytokine	Cell line for bioassay (ref.)
Human IL-2	CTLL-2 *(62)*
Murine IL-2	CTLL *(63)*
Human IL-3	TF-1 cells *(64)*
Murine IL-3	NFS 60 *(65)*
Human IL-4	PHA-activated peripheral blood monuclear cells *(65)*
Murine IL-4	CT4S or HT-2 cells *(62,66)*
Human IL-5	TF-1 cells *(63)*
Murine IL-5	TF-1 cells *(67)*
Human IL-6	T 1165.85.2.1 cells *(67)*
Murine IL-6	T 1165.85.2.1 cells *(67)*
Human IL-7	PHA-activated Peripheral blood monuclear cells *(66)*
Murine IL-7	PHA-activated peripheral blood monuclear cells *(66)*
Human IL-12	PHA-activated human lymphoblasts *(62)*
Murine IL-12	PHA-activated human lymphoblasts *(62)*
Human GM-CSF	TF-1 cells *(63)*
Murine GM-CSF	NFS 60 *(64)*
Human interferon-γ	Antiviral assay *(65)*
Murine interferon-γ	Antiviral assay *(65)*
Human TNF-α	Cytotoxic assay *(62)*
Murine TNF-α	Cytotoxic assay *(62)*

tion that is specifically the result of that cytokine. Exceptions are the TNF-α (cytotoxic) and INF-γ (antiviral) assays. The exact procedures for performing these assays can be found in the references listed in **Table 3**.

26. Genes encoding cytokines are currently one of the most commonly employed genetically altered tumor vaccine strategies in preclinical models and in clinical trials. However, other gene-modified vaccine strategies, including tumor cell surface expression of MHC class I and II molecules and costimulatory cell surface molecules (for example, B7), are also under investigation. Successful gene transfer of these gene products can be assayed using cell surface staining with a MAb specific for the gene product and analyzed by standard flow cytometric methods.

27. Evaluation of vector copy number should be considered, particularly in cases of suboptimal gene product expression, to determine whether the problem is at the level of transcription or owing to inadequate transduction. Vector copy number can be evaluated by Southern blot hybridization using standard procedures.

28. If the problem is owing to inadequate transduction and all the transduction condi-

tions are optimized, it is possible to improve on transduction efficiency significantly by subjecting the transduced cells to one or two or more rounds of transduction.

29. Untransduced unirradiated tumor cells (in vivo *and* in vitro) have been demonstrated to secrete varying quantities of many cytokines. Therefore, the untransduced tumor cells should also be evaluated for cytokine production as control cells.

30. The number of vaccine cells to inject to obtain adequate priming should be determined for each tumor model. Studies have shown that there is a minimal number of vaccine cells below which vaccine potency is significantly diminished *(68)*. In addition, increasing the number of vaccine cells above the minimal cell number that generates adequate immune priming did not significantly enhance the elicited antitumor immune response *(68)*.

31. The optimal route of administration of the vaccine cells may vary with the tumor model and should be determined in preliminary experiments. Immunization may be accomplished by one of several routes of vaccine administration, including subcutaneous, intradermal, intraperitoneal, intranodal, and intravenous. For treatment of solid tumor malignancies in the mouse, the subcutaneous space appears to provide an adequate area for the initial priming of immune responses, because this area has an adequate blood supply and is near the lower epidermis, where potential APCs reside. For the same reason the dermis is an area that should be considered in designing clinical trials, although in mice this area is too small for adequate vaccination access. Intravenous and intraperitoneal routes should be considered respectively for vaccination against hematologic malignancies and some solid tumors (i.e., ovarian cancer), which often metastasize to the peritoneum.

32. An evaluation of the optimal vaccine schedule, both in terms of the number of simultaneous vaccinations and the number and schedule of repeat vaccinations, should also be performed. The spatial distribution of the vaccine innoculum may have a significant impact on vaccine potency. For example, priming one lymph node region at a time may be less effective than priming multiple lymph node regions simultaneously. In addition, repeated weekly, biweekly, or monthly vaccinations may be more effective than a single vaccination *(68)*.

33. There appears to be a minimal cytokine production by vaccine cells required to achieve maximal antitumor immunity. For example, a minimal GM-CSF production of 36 ng/10^6 vaccine cells/24 h is necessary to generate an effective antitumor immune response in B16 melanoma tumor model. Dilution studies,in which the total number of cytokine-secreting tumor cells is serially decreased while the total number of antigen-expressing cells is kept constant (by dosing in wild-type irradiated tumor) can be performed to address this question *(68)*.

34. The cytokine-secreting vaccine should be compared with wild-type irradiated untransduced parental tumor cells, since many experimental animal tumors become immunogenic after irradiation other means of tumor cell inactivation. Consequently, three groups of mice should be considered: 1) mice vaccinated with the irradiated cytokine-secreting tumor cells; 2) mice vaccinated with the

irradiated parental tumor cells; and 3) control mice injected with HBSS. To allow statistical analysis, at least 10 mice should be vaccinated per group in models in which 100% of unvaccinated mice develop tumors *(68)*.

35. Each histologic tumor type will require different doses of γ-irradiation. It is therefore important to evaluate the optimal dose of irradiation that will result in loss of proliferation without loss of short-term cytokine secretion. In vitro studies from our laboratory have shown that lethally irradiated, cytokine-secreting murine and human tumor cell lines will continue to secrete adequate levels of cytokine for 7 days following irradiation, before the concentration of cytokines secreted begins to decline. Minimal cytokine production is usually detected 2 weeks following irradiation.

36. The number of tumor cells necessary for the challenge varies with the tumor cells and the model used. The optimal number of cells has to be previously determined in titration experiments. (Generally a tumor should appear between 1 and 2 weeks after the tumor challenge in the control group.) The route of injection varies with the tumor model used (intravenous for metastatic models or hematologic malignancies and subcutaneous for solid tumor).

37. The protection assay (challenge 14 days after the first vaccination) is the best method for the initial screening of immune response priming against a tumor by different cytokine-secreting tumor vaccines. In particular, this assay provides a sensitive method for studying the magnitude of priming that has resulted following vaccination. Because the challenge antigen load is given following vaccination and the volume of tumor can be controlled, even a low magnitude of priming can be detected, and small differences in priming between different vaccines can be observed. In contrast, cure experiments are best employed to study clinically relevant antitumor immune effects. In these studies, cytokine vaccines can be evaluated for their ability to impact on established cancer.

38. The safety issues concerning clinical trials are beyond the scope of this review (screening for helper virus contamination, reinjection of live vaccine cells, and so on). In the setting of autologous vaccine, each patient will require surgical resection of their primary tumor, followed by in vitro expansion and genetic modification. The best source of tumor antigens may be the primary tumor rather than a metastasis, because metastases probably develop from only one or a few clones of cells that arise from the primary tumor. Protocols to isolate tumor cells from primary tumor have been reported elsewhere *(45,69)*.

References

1. van der Bruggen, P., Traversari, C., Chomez, P., et al. (1991) A gene encoding an antigen recognized cytotoxic T lymphocytes on a human melanoma. *Science* **254,** 1643–1648.
2. Van den Eynde, B. J. and van der Bruggen, P. (1997) T-cell defined tumor antigens. *Curr. Opin. Immunol.* **9,** 684–693.
3. Rosenberg, S. A. (1999) A new era for cancer immunotherapy based on the genes that encode cancer antigens. *Immunity* **10,** 281–287.

4. Dranoff, G., Jaffee, E. M., Lazenby, A., et al. (1993) Vaccination with irradiated tumor cells engineered to secrete murine GM-CSF stimulates potent, specific, long lasting anti-tumor immunity. *Proc. Natl. Acad. Sci. USA* **90**, 3539–3543.
5. Human gene marker/Therapy clinical trial protocols. (1999) *Hum. Gene Ther.* **10**, 1043–1092.
6. Kruisbeek, A. M. and Amsen, D. (1996) Mechanisms underlying T-cell tolerance. *Curr. Opin. Immunol.* **8**, 815–821.
7. Sotomayor, E. M., Borrello, I., and Levistky, H. (1996) Tolerance and cancer. *Crit. Rev. Oncog.* **7**, 433–456.
8. Lipton, A., Harvey, H. A., Balch, C. M., et al. (1991) *Corynebacterium parvum* versus bacille Calmette-Guérin adjuvant immunotherapy of stage II malignant melanoma. *J. Clin. Oncol.* **9**, 1151–1156.
9. Schirrmacher, V., Ahlert, T., Probstle, T., et al. (1998) Immunization with virus-modified tumor cells. *Semin. Oncol.* **25**, 677–696.
10. Wallich, R., Bulbuc, N., Hammerling, G., et al. (1985) Abrogation of metastatic properties of tumor cells by de novo expression of H-2K antigens following H-2 gene transfection. *Nature* **315**, 301–305.
11. Pulaski, B. A. and Ostrand-Rosenberg, S. (1998) Reduction of established spontaneous mammary carcinoma metastases following immunotherapy with major histocompatibility complex class II and B7.1 cell-based tumor vaccines. *Cancer Res.* **58**, 1486–1493.
12. Chen, C. A. and Okayama, H. (1988) Calcium phosphate-mediated gene transfer: a highly efficient transfection system for stably transforming cells with plasmid DNA. *Biotechniques* **6**, 882–886.
13. Gordon, J. W. (1990) Micromanipulation of embryos and germs cells: an approach to gene therapy? *Am. J. Med. Genet.* **35**, 206–214.
14. Kubiniec, R. T., Liang, H., and Hui, S. W. (1990) effects of pulse length and pulse strength on transfection by electroporation. *Biotechniques* **8**, 16–20.
15. Hug, P. and Sleight, R. G. (1991) Liposomes for the transformation of eukaryotic cells. *Biochim. Biophys. Acta* **1097**, 1–17.
16. Wolff, J. A., Malone, R. W., Williams, P., et al. (1990) Direct gene transfer into mouse muscle in vivo. *Science* **247**, 1465–1468.
17. Wu, G. Y., Zhan, P., Sze, L. L., Rosenberg, A. R., and Wu, C. H. (1994) Incorporation of adenovirus into a ligand-based DNA carrier system results in retention of original receptor specificity and enhances targeted gene expression. *J. Biol. Chem.* **269**, 11542–11546
18. Jiao, S., Cheng, L., Wolff, J. A., and Yang, N.-S. (1993) Particle bombardment-mediated gene transfer and expression in rat brain tissues. *Biotechnology* **11**, 497–502.
19. Mulligan, R. C. (1991) Gene transfer and gene therapy. Principles, prospects, and perspective, in *Etiology of Human Diseases at the DNA Level* (Lindsten, J. and Pettersson, U., eds.), Raven, New York, pp. 143–181.
20. Danos, O. and Mulligan, R. C. (1988) Safe and efficient generation of recombinant retroviruses with amphotropic and ecotropic host ranges. *Proc. Natl. Acad. Sci. USA* **85**, 6460–6464.

21. Armentano, D., Sheau-Fung, Y., Kantoff, P., et al. (1987) Effect of internal viral sequences on the utility of retroviral vectors. *J. Virol.* **61**, 1647–1650.
22. Jaffee, E. M., Dranoff, G., Cohen L. W., et al. (1993) High efficiency gene transfer into primary human tumor explants without cell selection. *Cancer Res.* **53**, 2221–2226.
23. Kashara, N., Dozy, A. M., and Kan, Y. W. (1994) Tissue-specific targeting of retroviral vectors through ligand-receptor interactions. *Science* **266**, 1373–1376.
24. Naldini, L., Blomer, U., Gallay, P., et al. (1996) In vivo gene delivery and stable transduction of nondividing cells by a lentiviral vector. *Science* **272**, 263–267.
25. Quantin, B., Perricaudet L. D., Tajbakhsh, S., Mandel, and J. L. (1992) Adenovirus as an expression vector in muscle cells in vivo. *Proc. Natl. Acad. Sci. USA* **89**, 2581–2584.
26. Bowman, L., Grossmann, M., Rill, D., et al. (1998) Il-2 adenovector-transduced autologous tumor cells induce antitumor immune responses in patients with neuroblastoma. *Blood* **92**, 1941–1949.
27. Barth R. J. Jr. and Mule, J. J. (1996) Cytokine gene transfer into tumor cells: animal models, in *Gene Therapy in Cancer* (Brenner, M. K. and Moen, R. C., eds.), Marcel Dekker, New York, pp. 73–94.
28. Fearon, E. R., Pardoll, D. M., Itaya, T., et al. (1990) Interleukin-2 production by tumor cells bypasses T helper function in the generation of an antitumor response. *Cell* **60**, 397–403.
29. Golumbek, P. T., Lazenby, A. J., Levitsky, H. I., et al. (1991). Treatment of established renal cancer by tumor cells engineered to secrete interleukin-4. *Science* **254**, 713–716.
30. Gansbacher,B., Zier, K., Daniels, B., Cronin, K., Bannerji, R., and Gilboa, E. (1990) Interleukin 2 gene transfer into tumor cells abrogates tumorigenicity and induces protective immunity. *J. Exp. Med.* **172**, 1217–1224.
31. Tepper, R. I, Pattengale, P. K., and Leder, P. (1989) Murine interleukin-4 displays potent anti-tumor activity in vivo. *Cell* **57**, 503–512.
32. Hock, H., Dorsch, M., Diamantstein, T., and Blankenstein, T. (1991) Interleukin-7 induces CD4+ T-cell dependent tumor rejection. *J. Exp. Med.* **174**, 1291–1298.
33. Asher, A. L., Mule, J. J, Kasid, A., et al. (1991) Murine tumor cells transduced with the gene for tumor-necrosis factor-alpha. *J. Immunol.* **146**, 3227–3234.
34. Porgador, A., Tzehoval, E., Katz A., et al. (1992) Interleukin-6 gene transfection into lewis lung carcinoma tumor cells suppresses the malignant phenotype and confers immunotherapeutic competence against parental metastatic cells. *Cancer Res.* **52**, 3679–3686.
35. Colombo M. P., Ferrari, G., Stoppacciaro, A., et al. (1991) Granulocyte colony-stimulating factor gene transfer suppresses tumorigenicity of a murine adenocarcinoma in vivo. *J. Exp. Med.* **173**, 889–897.
36. Gansbacher, B., Bannerji, R., Daniels, B., et al. (1990) Retroviral vector-mediated gamma-interferon gene transfer into tumor cells generates potent and long lasting antitumor immunity. *Cancer Res.* **50**, 7820–7825.
37. Hung, K., Hayashi, R., Lafond-Walker, A., et al. (1998) The central role of CD4(+) T-cells in the antitumor immune response. *J. Exp. Med.* **188**, 2357–2368.

38. Inaba, K., Inaba, M., Romani, N., et al. (1992) Generation of large numbers of dendritic cells from mouse bone marrow cultures supplemented with granulocyte/macrophage colony-stimulating factor. *J. Exp. Med.* **176,** 1693–1702.

39. Hallez, S., Detremmerie, O., Giannouli, C., et al. (1999) Interleukin-12-secreting human papillomavirus type 16-transformed cells provide a potent cancer vaccine that generates E7-directed immunity. *Int. J. Cancer* **81,** 428–437.

40. Tepper, R. I. and Mule, J. J. (1994) Experimental and clinical studies of cytokine gene-modified tumor cells. *Hum. Gene Ther.* **5,** 153–164.

41. Zitvogel, L., Robbins, P. D., Storkus, W. J., et al. (1996) Interleukin-12 and B7.1 co-stimulation cooperate in the induction of effective antitumor immunity and therapy of established tumors. *Eur. J. Immunol.* **26,** 1335–1341.

42. Schreiber, S., Kampgen, E., Wagner, E., et al. (1999) Immunotherapy of metastatic malignant melanoma by a vaccine consisting of autologous interleukin 2-transfected cancer cells: outcome of a phase I study. *Hum. Gene Ther.* **10,** 983-993.

43. Abdel-Wahab, Z., Weltz, C., Hester, D., et al. (1997) A phase I clinical trial of immunotherapy with interferon-γ gene-modified autologous melanoma cells: monitoring the humoral response. *Cancer* **80,** 401–412.

44. Soiffer, R., Lynch, T., Mihm, M., et al. (1998) Vaccination with irradiated autologous melanoma cells engineered to secrete human granulocyte-macrophage colony-stimulating factor generates potent antitumor immunity in patients with metastatic melanoma. *Proc. Natl. Acad. Sci. USA* **95,** 13141–13146.

45. Simons, J. W., Jaffee, E. M., Weber, C. E., et al. (1997) Bioactivity of autologous irradiated renal cell carcinoma vaccines generated by ex vivo granulocyte-macrophage colony-stimulating factor gene transfer. *Cancer Res.* **57,** 1537–1546.

46. Moller, P., Sun, Y., Dorbic, T., et al. (1998) Vaccination with Il-7 gene-modified autologous melanoma cells can enhance the anti-melanoma lytic activity in peripheral blood of patients with a good clinical performance status: a clinical phase I study. *Br. J. Cancer* **77,** 1907–1916.

47. Sun, Y., Jurgovsky, K., Moller, P., et al. (1998) Vaccination with Il-12 gene-modified autologous melanoma cells: preclinical results and a first clinical phase I study. *Gene Ther.* **5,** 481–490.

48. Palmer, K., Moore, J., Everard, M, et al. (1999). Gene therapy with autologous, interleukin-2 secreting tumor cells in patients with malignant melanoma. *Hum. Gene Ther.* **10,** 1261–1268.

49. Bernhard, H., Karbach, J., Wolfel, T., et al. (1994) Cellular immune response to human renal-cell carcinomas: definition of a common antigen recognized by HLA-A2-restricted cytotoxic T-lymphocyte (CTL) clones. *Int. J. Cancer* **59,** 837–842.

50. Kawakami, Y., Eliyahu, S., Delgado, C. H., et al. (1994) Cloning of the gene coding for a shared human melanoma antigen recognized by autologous T-cells infiltrating into tumor. *Proc. Natl. Acad. Sci. USA* **91,** 3515–3519.

51. Huang, A. Y. C., Golumbek, P., Ahmadzadeh, M., et al. (1994) Role of bone marrow-derived cells in presenting MHC class I-restricted tumor antigens. *Science* **264,** 961–965.

52. Thomas, M. C., Greten, T. F., Pardoll, D. M, and Jaffee, E. M. (1998) Enhanced

tumor protection by granulocyte-macrophage colony-stimulating factor expression at the site of an allogeneic vaccine. *Hum. Gene Ther.* **9**, 835–843.

53. Arienti, F., Sule-Suso, J., Belli, F., et al. (1996) Limited antitumor T-cell response in melanoma patients vaccinated with interleukin-2 gene-transduced allogeneic melanoma cells. *Hum. Gene Ther.* **7**, 1955–1963.
54. Belli, F., Arienti, F., Sule-Suso, J., et al. (1997) Active immunization of metastatic melanoma patients with interleukin-2-transduced allogeneic melanoma cells: evaluation of efficacy and tolerability. *Cancer Immunol. Immunother.* **44**, 197–203.
55. Bowman, L. C., Grossmann, M., Rill, D., et al. (1998) Interleukin-2 gene-modified allogeneic tumor cells for treatment of relapsed neuroblastoma. *Hum. Gene Ther.* **10**, 1303–1311.
56. Jaffee, E. M., Abrams, R., Cameron, J., et al. (1998) A phase I clinical trial of lethally irradiated allogeneic pancreatic tumor cells transfected with the GM-CSF gene for the treatment of pancreatic adenocarcinoma. *Hum. Gene Ther.* **9**, 1951–1971.
57. Veelken, H., Mackensen, A., Lahn, M., et al. (1997) A phase-I clinical study of autologous tumor cells plus interleukin-2-gene-transfected allogeneic fibroblasts as a vaccine in patients with cancer. *Int. J. Cancer* **70**, 269–277
58. Mackensen, A., Veelken, H., Lahn, M., et al. (1997) Induction of tumor-specific cytotoxic T lymphocytes by immunization with autologous tumor cells and interleukin-2 gene transfected fibroblasts. *J. Mol. Med.* **75**, 290–296
59. Kotani, H., Newton, P. B., Zhang, S., et al. (1994) Improved methods of retroviral vector transduction and production for gene therapy. *Hum. Gene Ther.* **5**, 19–28.
60. Cornetta, K. and Anderson F. (1989) Protamine sulfate as an effective alternative to Polybrene in retroviral-mediated gene transfer. *J. Virol. Methods* **23**, 187–194.
61. Leventis, R. and Silvius, J. R. (1990) Interactions of mammalian cells with lipid dispersions containing novel metabolizable cationic amphiphiles. *Biochem. Biophys. Acta* **1023**,124–132.
62. Gearing, A. J. H. and Bird, C. B. (1987) Production and assay of Il-2, in *Lymphokines and Interferons, a Practical Approach* (Clemens, M. J., Morris, A. G., and Gearing, A. J. H, eds.), IRL, Washington, DC, pp. 291–301.
63. Coligan, J. E., Kruisbeck, A. M., Margulies, D. H., Shevach, E. M., and Strober, W. (1991) *Current Protocols in Immunology*, Greene and Wiley-Interscience, New York.
64. Kitanura, T., Tojo, A., Kuwaki, T., et al. (1989) Identification and analysis of human erythropoietin receptors on a factor-dependent cell line, TF-1. *Blood* **73**, 375–380.
65. Holmes, K. L., Palaszymski, E., and Fredrikson, T. (1985) Correlation of cell-surface phenotype with the establishment of interleukin3-dependent cell lines from wild-mouse murine leukemia with virus-induced neoplasms. *Proc. Natl. Acad. Sci. USA* **82**, 6687–6691.
66. Yokota, T., Otsuka, T., Mosmann, T., et al. (1986) Isolation and characterization of a human interleukin cDNA clone, homologous to mouse B cell stimulatory

factor 1, that expresses B cell stimulatory factor 1, that expresses B cell- and T-cell-stimulating activities. *Proc. Natl. Acad. Sci. USA* **83,** 5894–5898.

67. Nordan, R. P., Pumphrey, J. G., and Rudikoff, S. (1987) Purification and NH2-terminal sequence of a plasmocytoma growth factor derived from the murine macrophage cell line P388D1. *J. Immunol.* **193,** 813–817.

68. Jaffee, E. M., Thomas, M. C., Huang, A. Y., et al. (1996) Enhanced immune priming with spatial distribution of paracrine cytokine vaccines. *J. Immunother. Emphasis Tumor Immunol.* **19,** 176–183.

69. Jaffee, E. M., Schutte, M., Gossett, J., et al. (1998) Development and characterization of a cytokine-secreting pancreatic adenocarcinoma vaccine from primary tumors for use in clinical trials. *Cancer J. Sci. Am.* **4,** 194–203.

19

HIV-Based Vectors

Preparation and Use

Antonia Follenzi and Luigi Naldini

1. Introduction

The development of gene transfer vectors from lentiviruses, such as the human immunodeficiency virus 1 (HIV-1), has opened exciting perspectives for the genetic treatment of a wide array of inherited and acquired diseases, because of their ability to achieve the efficient delivery, integration, and long-term expression of transgenes into dividing and nondividing cells both in vitro and in vivo.

This chapter explores the various practical aspects of production and assay of lentiviral vector (LV), in addition to the structure and genetics of HIV-1 and the design and biosafety of LVs. More detailed and extensive reviews describing the biology of lentiviral vectors are also available *(1–4)*.

1.1. HIV-1 Structure and Genetics

HIV-1 is a member of one of the five major primate lineages of the lentivirus family of retroviruses. Lentiviruses are spherical particles of approximately 110 nm in diameter. They have lipid-enveloped particles comprising a homodimer of linear, positive-sense, single-stranded RNA genomes of about 9.7 kb. In the early stages of infection, the virion RNA genome is converted into double-stranded linear DNA by the process of reverse transcription. This linear viral DNA is integrated into the host cell genome to produce the provirus *(5,6)*.

Like retroviruses, HIVs and other lentiviruses have two long terminal repeats (LTR) at their ends. LTRs and neighboring sequences act in *cis* during expres-

From: *Methods in Molecular Medicine, Vol. 69, Gene Therapy Protocols, 2nd Ed.*
Edited by: J. R. Morgan © Humana Press Inc., Totowa, NJ

sion, packaging, reverse transcription, and integration of the genome. These sequences frame a colinear array of the *gag, pol,* and *env* genes. The *gag* gene encodes a polyprotein that forms the inner virion structures matrix, capsid, and nucleoprotein. The *pol* gene encodes the viral enzymes protease, reverse transcriptase, RNase H, and integrase. The *env* gene encodes the surface and transmembrane components of the envelope glycoprotein. Although the basic steps of the HIV-1 life cycle are the same as for other retroviruses, six virally encoded regulatory/accessory proteins (Tat, Rev, Vif, Vpr, Vpu, and Nef) that are not found in other classes of retroviruses impart novel levels of complexity to lentiviral replication *(7)*. The *tat* and *rev* proteins act at the transcriptional and posttranscriptional levels, respectively. The *tat* protein binds to a stem-loop structure (TAR) in the nascent HIV LTR RNA and tethers the cyclin T-CDK9 complex to polymerase II, leading to increased transcriptional elongation *(8,9)*. Rev binds to an RNA motif (Rev responsive element [RRE]), found in the envelope coding region of the HIV transcript, and bridges it to the nuclear export factor exportin 1 (CRM) *(10–12)*, promoting the cytoplasmic export of unspliced and singly spliced viral transcripts expressing the late viral proteins *(13)*.

The four accessory genes of HIV are *vif, vpr, vpu,* and *nef.* The Vif gene product acts during virion assembly to promote the early steps of infection, but its requirement is limited to virions produced from primary lymphocytes and macrophages, and not from most cell lines *(6,14)*. Moreover, Vif binds to HIV-1 RNA in the cytoplasm of virus-producing cells. Vif-RNA binding could be displaced by Gag-RNA binding, suggesting that Vif protein may mediate viral RNA interaction with HIV-1 Gag precursors *(15)*. The Vpu gene product stimulates the release of virions from the surface of infected cells *(16)*. Furthermore, Vpu binds CD4 in the endoplasmic reticulum and targets it for proteolysis by recruitment into the cytosolic ubiquitin-proteasome pathway *(17)*. The Nef gene product facilitates the routing of CD4 from the cell surface and Golgi apparatus to lysosomes, resulting in receptor degradation and preventing inappropriate interactions with Env, as for Vpu *(18)*; it also promotes the infectivity of HIV virions *(19)*, but the latter effect seems to be restricted to virions that penetrate cells by fusion at the plasma membrane *(20)*. It has recently been shown that HIV-1 Nef protein downregulates the cell surface expression of MHC-1 and probably thereby promotes immune evasion by HIV-1 *(21)*. The Vpr gene product induces growth arrest in G_2 phase *(22)*, when expression of the viral genome is most efficient *(23)*. Vpr can play opposite roles in the regulation of apoptosis, which may depend on the level of its intracellular expression at different stages of HIV-1 infection *(24–26)*. Vpr is incorporated in the virions and contributes to the nuclear import of the preintegration complex by interaction with importin-α as well as nucleoporins *(27–30)*. In spite of their nonessential role in vitro, the accessory genes of HIV are strictly conserved, and their gene products are essential virulence factors in vivo.

1.2. Development of Hybrid Lentiviral Vectors

LV are replication-defective, hybrid viral particles made by the core proteins and enzymes of a lentivirus and the envelope of a different virus, most often the vesicular stomatitis virus (VSV) *(31)*. The use of VSV.G yields higher vector titers and results in greater stability of the vector virus particles *(32)*. The general strategies employed in the design of LV involve segregation of *trans*-acting sequences that encode for viral proteins from *cis*-acting sequences involved in the transfer of the viral genome to target cells. The vector particles are assembled by viral proteins expressed in the producer cell from constructs stripped of the majority of viral *cis*-acting sequences. The viral *cis*-acting sequences are linked to the transgene and are introduced into the same cell. As the vector particle can only transfer the latter construct, the infection process is limited to a single round without further spreading.

The packaging functions for the LV are provided by at least two separate expression plasmids that use transcriptional signals different from those of the virus. A "core" packaging construct, derived from the HIV-1 proviral DNA, expresses the viral proteins but not the *env* gene that has been deleted. A separate construct expresses a heterologous envelope that is incorporated into the vector particles (pseudotyping) and allows entry into the target cells. A third construct, the transfer vector, expresses RNA that contains the viral *cis*-acting sequences required for packaging by the vector particles, reverse transcription, nuclear translocation, and integration in the target cells. It transfers an expression cassette for the transgene containing an internal promoter (**Fig. 1**).

1.3. Vector Packaging Systems

In the first generation of hybrid LV, the whole HIV genome under heterologous transcriptional control, lacking the packaging signal and deleted in the *env* gene, was employed as a packaging construct *(32)*. Advanced vector design was reached by deleting unnecessary genes (*Vif, Vpr, Vpu,* and *Nef*), resulting in the production of so-called multiply attenuated or "second generation" packaging systems without causing reduction in the transduction efficiency *(3,33–36)*. A further improvement in biosafety was achieved by the elimination of the *tat* and *rev* genes from the core packaging construct. The function of the *tat* gene could be replaced by strong costitutive promoters upstream of the vector transcriptional start site *(34,37)*, and *Rev* is provided in *trans* by a fourth construct driven by a Rous sarcoma virus promoter. These results achieved an

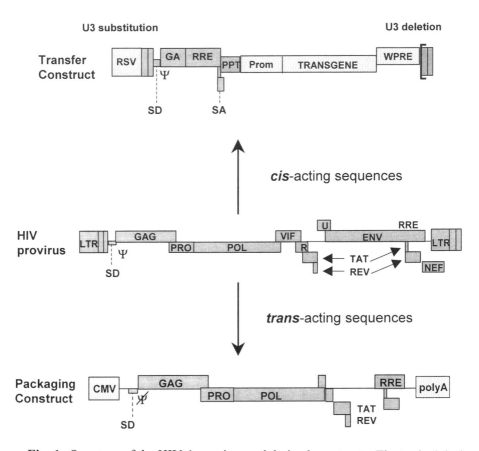

Fig. 1. Structure of the HIV-1 provirus and derived constructs. The typical design of lentiviral vectors is based on separation of the sequences acting in *cis* from the viral genes that express proteins working in *trans* during transduction and that must be expressed in the producer cells. All the *cis* functions that will allow efficient encapsidation, reverse transcription, and integration of the vector into the target cells are located in the transfer construct. The packaging construct expresses the indicated viral genes from the cytomegalovirus promoter (CMV) and polyadenylation sites (polyA). Abbreviations: LTR, long terminal repeats: the 5' LTR of the transfer construct is chimeric, with the enhancer-promoter of the RSV replacing the U3 region, and the 3' LTR has an almost complete deletion of the U3 region (sin-18); RSV, Rous sarcoma virus promoter; SD and SA, splice donor and acceptor sites; ψ, the packaging sequence; GA, 5' portion of the *gag* gene with truncated reading frame including extended packaging signal; RRE, Rev response element; cPPT, central polypurine tract; Prom, internal promoter; WPRE, posttranscriptional element from the genome of the woodchuck hepatitis virus.

important biosafety gain because any recombinant replication-competent virus (RCR) that could be generated during vector manufacturing would lack all factors essential for HIV-1 replication and virulence in vivo.

Using the new minimal packaging constructs, it is now possible to generate stable producer cell lines that allow better characterization and scale-up of vector manufacturing. A number of approaches are being explored toward realizing this goal, inluding the use of inducible expression systems *(38,39)*.

1.4. Transfer Vectors: Design and Performance

The transfer vector contains the complete HIV leader sequence and the 5' splice donor site, followed by part of the *gag* gene (extended packaging signal; with the *gag* reading frame closed by a synthetic stop codon) as well as *env* sequences containing the RRE and the splice acceptor sites of the third exon of the *tat* and *rev* genes. For the requirement of HIV-derived *cis*-acting sequences in the transfer vector, see also other works *(40–43)*. Considerable progress in biosafety was also achieved by the successful generation of self-inactivating LVs *(44–46)*. These vectors are produced by transfer constructs that carry an almost complete deletion in the U3 region of the HIV 3' LTR. The U3 region contains the viral enhancer and promoter. Transduction of vector deleted in the 3' U3 results in duplication of the deletion in the upstream LTR and the transcriptional inactivation of both LTRs. Remarkably, self-inactivating LVs could be produced with infectious titers and levels of transgene expression driven by an internal promoter similar to vectors carrying wild-type LTRs. Furthermore, the lack of promoter sequences in the LTR should improve the performance of an internal tissue-specific or regulated promoter.

Further elements were also incorporated in the expression cassette to improve its performance. A regulatory sequence from the genome of the woodchuck hepatitis virus (WPRE) was added to the 3' end of the transfer vector. This element acts at the posttranscriptional level, in all probability promoting the efficiency of polyadenylation of the nascent transcript, and increasing its total amount in the cells *(47)*. Addition of the WPRE to HIV-derived vectors improved substantially the level of transgene expression from several types of promoters in vitro and in vivo *(48,49)*.

A recently described modification of the transfer vector backbone restored infectivity of vector particles to the level of wild-type virus *(50,51)*. An additional copy of the polypurine tract, which primes transcription of the viral *plus* strand DNA, is present in the middle of the HIV genome, included in the *pol* gene. This sequence (cPPT) was shown to be required for efficient replication of HIV-1 by mutation analysis *(52–54)*. In the design of LV described above, the cPPT sequence was maintained only in the packaging construct, as part of the *pol* gene. The introduction of the cPPT sequence in the transfer vector backbone strongly increased the total amount of genome integrated into the DNA of target cells *(50,51)*.

1.5. Advantages and Limitations of Lentiviral Vectors for Gene Transfer and Gene Therapy

The ability of hybrid LV to infect nondividing or terminally differentiated cells, including neurons, retinal photoreceptor cells, several types of epithelial tissues, and pluripotent hematopoietic stem cells (for review, *see* **ref.** *3*), may expand the range of disease targets and facilitate the development of gene therapy approaches. In addition, LVs may increase the efficacy of gene transfer into target cells and decrease the need for much of the ex vivo manipulation required.

LVs offer significant advantages for optimizing the expression of a transgene. First, they allow for stable long-term expression. No evidence of significant transcriptional shutoff has yet been reported after lentiviral gene transfer into a wide spectrum of target tissues, at the longest times examined, often of several months *(55)*. The capacity to integrate up to 9 kb of foreign DNA into the host genome and the ability to carry regulatory elements (i.e., tissue-specific enhancers/promoters, splice sites, and so on) without interference from the viral genome allows for greater control of transferred gene expression *(56)*. The lack of preexisting immunity in the recipient may facilitate in vivo delivery by LVs, compared with vectors derived from viruses commonly infecting the host population.

Some of the concerns still associated with the current LV systems are the possibility of insertional mutagenesis by the vector as it integrates in the host genome and the possibility of contamination of vector batches by RCR. Regarding insertional mutagenesis by LVs, it is encouraging that this phenomenon has not been observed, even in patients with HIV infection who have high viral loads *(57)*. The use of self-inactivating vector diminishes the concern for oncogenesis by promoter insertion, and it alleviates significantly the risk of vector mobilization and recombination with the wild-type virus *(58)*.

The use of HIV-derived vectors systems clearly raises concern that a recombinant RCR could be generated during vector production even though no such a virus has been observed in the systems studied so far. However, the assays currently available for detection of such an RCR are of limited sensitivity and need to be further developed. One can predict that the only features of the parental virus shared by any of the eventual RCRs arising from late-generation systems would be those dependent on the *gag* and *pol* genes. Thus RCR monitoring could be performed by assays based on HIV-1 *gag* detection, such as p24 immunocapture assay, and *gag* RNA-polymerase chain reaction assay.

However, in vivo testing is required to score RCR growth on multiple target cells and to monitor pathogenic effects. Development of appropriate in vivo testing remains a major goal to advance LV applications. Given the strict species specificity of lentiviral pathogenesis, this testing should be performed initially in the natural host of the parental virus. For vectors derived from HIV-1 or HIV-2, this is a major challenge, and substitutes may be sought either among nonhuman primates or in chimeric mice harboring human hematolymphoid tissues *(59)*. Appropriate in vivo testing will also permit the monitoring of transmission of vector sequences to the germline, a risk factor difficult to evaluate but important for consideration in any protocol involving in vivo administration of an integrating vector that cannot be targeted to specific cells.

2. Materials
2.1. Cell Culture

1. Human 293T cells. 293T cells were originally called 293tsA1609ne *(60)* and were derived from 293 cells, a continuous human embryonic kidney cell line transformed by sheared type 5 Adenovirus DNA (ATCC, no. CRL-1573) *(61)*. These cells were subsequently transfected with the tsA 1609 mutant gene of SV40 large T-antigen and the *Neo^r* gene of *E. coli*, and clones were selected for resistance to neomycin. The cells are split 1:10 every 3 or 4 days (*see* **Note 1**).
2. Sterile phosphate-buffered saline (PBS): 1.0% NaCl, 0.025% KCl, 0.14% Na$_2$HPO$_4$, 0.025% KH$_2$PO$_4$ (all w/v), at pH 7.2.
3. Iscove's modified Dulbecco's medium (IMDM) containing 10% fetal bovine serum (FBS) and antibiotics (penicillin and streptomycin) (*see* **Note 2**).
4. Tissue culture dishes.
5. Trypsin (0.05%) and EDTA (4 mM, pH 7.3) in PBS.

2.2. Recombinant Lentiviral Vector Plasmids

It is important to use pure DNA for transfection. Plasmids should be prepared and purified by double CsCl gradient centrifugation or other commercially available column methods yielding endotoxin-free DNA (*see* **Note 3**).

2.3. Recombinant Lentiviral Vector Plasmids

1. Monolayers of 293Tcells at approximately 80% confluence.
2. 2× HBS: 281 mM NaCl, 100 mM HEPES, 1.5 mM Na$_2$HPO$_4$, pH 7.12, 0.22 μm filtered and stored in aliquots at –20°C (*see* **Note 4**).
3. 2.5 M CaCl$_2$: Tissue culture grade, 0.22 μm filtered and stored at –20°C.
4. 0.1× TE buffer, 0.22 μm filtered and stored at 4°C.
5. Double-processed tissue culture water H$_2$O (endotoxin-free).
6. Tissue culture dishes (100 × 20 mm, Falcon Code no. 353003).
7. 50-mL polypropylene tubes.

8. Recombinant lentiviral vector plasmids: CsCl-purified preparations.
9. Vortex.

2.4. Concentration of Lentiviral Vectors

1. Beckman Optima XL-100K Ultracentrifuge with swing-out rotor SW28.
2. Centrifuge tubes in polyallomer (25 × 89 mm; Beckman cat. no. 326823).
3. PBS/1% bovine serum albumin (BSA).

2.5. Titration of Lentiviral Vectors

1. HeLa cells (ATCC, cat. no. CCL2).
2. Six-well and 24-well tissue culture dishes.
3. Filtered sterile tips and automatic micropipets (1000, 200, 20, and 10 μL)
4. Lentiviral vector stock to be titered.
5. Polybrene stock 8 mg/mL in PBS (final concentration 8 μg/mL).
6. 5-mL fluoresence-activated cell sorting (FACS) tubes.
7. Fixing solution: 1% formaldehyde EM grade, 2% fetal bovine serum (FBS) in PBS.
8. FACS apparatus.
9. Software for analysis.

2.6. RCR Test

1. C8166 or SupT1, human T-lymphoblastoid cell lines, cultured in RPMI-1640, 10% FBS, maintained at 37°C with 5% CO_2 (58). Make sure the cells are in log growth phase before they are used.
2. 300 μl vector prep. (It is usually diluted 1:100.)

2.7. HIV-1 p24 Assay

1. HIV-1 p24 Core Profile ELISA Kit (NEN Live Sciences Products, cat. no. NEK060A).
2. Automated microplate reader.

3. Methods
3.1. Generation of Lentiviral Vectors

This section outlines the calcium phosphate method for cotransfecting 293T cells with four plasmids (third-generation core packaging plasmids: pMDLg/pRRE and pRSV.REV, self-inactivating (SIN) transfer vector plasmid pRRL-SIN-18PPT.CMV.EGFPWpre, and envelope plasmid pMD₂VSV.G. [32,37,45,50]) for rescue of the recombinant HIV genome with the gene of interest and packaging into viral particles. The cell line of choice is always 293T, because these cells are good recipients of DNA in DNA-mediated gene-transfer procedures, and the backbones of the vector constructs contain SV40 origin of replication (**Fig. 2**).

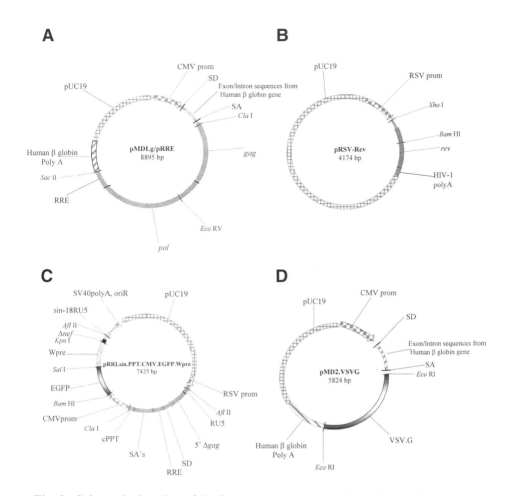

Fig. 2. Schematic drawing of the four constructs required to make third-generation HIV-1-derived lentiviral vectors. **(a)** The conditional packaging construct expressing the *gag* and *pol* genes from the CMV promoter. **(b)** A non-overlapping construct, RSV-Rev expressing the *rev* cDNA under the control of the Rous sarcoma virus (RSV) promoter. **(c)** The SIN transfer construct contains HIV-1 *cis*-acting sequences and an expression cassette for the transgene (enhanced green fluorescent protein [EGFP]) driven by the CMV promoter. **(d)** A fourth construct encodes a heterologous envelope to pseudotype the vector, here shown coding for protein G of the vesicular stomatitis virus Indiana serotype (VSV.G) under the control of the CMV promoter. Upon transfection of all four constructs into human 293T cells, high titers of replication-defective, self-inactivating vectors are produced. For other abbreviations, see **Fig. 1** legend.

We routinely use calcium-phosphate precipitation to obtain infectious particles. The reader is encouraged to evaluate other transfection techniques as well; good transfection efficiency is also reached by lipid-based methods.

3.1.1. Transfection of DNA Using Calcium-Phosphate Precipitation

1. Seed and incubate 5×10^6 293T cells in a 10-cm dish, approximately 24 h before transfection in IMDM, 10% FBS, penicillin (25 U/mL), streptomycin (25 U/mL).
2. Change medium 2 h before transfection.
3. The plasmid DNA mix is prepared by adding 3.5 µg ENV plasmid (pMD2-VSV-G), 5 µg core packaging plasmid (pMDLg/pRRE), 2.5 µg pRSV-REV, and 10 µg gene transfer plasmid (pRRLsin18.PPT.CMV.GFP.Wpre) together per dish. The plasmid solution is made up to a final volume of 450 µL with 0.1X TE/dpH$_2$O (2:1) in a 50-mL polypropylene tube. Finally, add 50 µL 2.5 M CaCl$_2$.
4. The precipitate is formed by dropwise addition of 500 µL of the 2× HBS solution to the 500 µL of DNA/TE/CaCl$_2$ mixture from **step 3** while vortexing at full speed. The precipitate should be added to the 293T cells immediately following addition of the 2× HBS (*see* **Note 5**).
5. The CaP-precipitated plasmid DNA should be allowed to stay on the cells for 14–16 h, after which the media should be replaced with fresh media for virus collection to begin. Discard medium as infectious waste.
6. Collect the cell supernatants at 24 and 48 h after changing the media.
7. Centrifuge at 300g for 5 min at room temperature (RT) and filter supernatant through a 0.22-µm pore nitrocellulose filter.
8. Use the conditioned supernatant as is or proceed to concentration.

3.2. Concentration of Lentiviral Vectors

1. 24 h after media replacement, collect the 293T cell supernatant, centrifuge at 300g for 5 min at RT, and filter supernatant through 0.22-µm pore nitrocellulose filter.
2. Concentrate the conditioned medium by ultracentrifugation at 50,000g (with an SW28 rotor in a Beckman ultracentrifuge Optima XL-100K) x 140 min at RT.
3. Discard the supernatant by decanting, and resuspend the pellets in a small volume (≥200 ml if only one centrifugation is performed) of PBS containing 1% BSA, pool in a small tube, and rotate on a wheel at RT for 1 h. If only one centrifugation is performed, go to **step 8**.
4. Store the concentrate vector from the first collection at 4°C until the end of the second collection and ultracentrifugation of the following day.
5. Repeat **steps 1–3** for the second collection.
6. Pool the vector suspension from the two collections, dilute in PBS, and concentrate again by ultracentrifugation.
7. Resuspend the final pellet in a very small volume (1/500 of the starting volume of medium) of sterile PBS/1% BSA. Resuspension of the second pellet requires prolonged incubation on a rotating wheel and pipeting at RT.
8. Split into small aliquots (20 µL), store at –80°C, and titer after freezing.

3.3. Titration of Lentiviral Vectors

1. Plate 5×10^4 HeLa cells/well a day before in a 6-well COSTAR plate. (The day after you will have approximately 10^5 cells/well.)
2. The following day prepare serial 10-fold dilutions of viral stocks. Take a 24-well plate and add 1.8 mL of medium (IMDM/10% FBS) to each well. Then add to the first well of the first row 200 μL of viral stock (10^{-1} dilution). After pipeting several times, change the tip, take 200 μL of the 10^{-1} dilution, and put it in the second well. Take 200 μL of 10^{-2} dilution, put it in the third well, and continue until you have 10^{-6} dilution.
3. If you are titrating concentrated vector stock, you should add 2 mL of medium to the first well of the dilution plate and add 2 μL of the concentrated vector preparation and continue with the serial dilutions as described above. (In this case you have dilutions from 10^{-3} to 10^{-8}.)
4. Take HeLa cells from the incubator and aspirate the media. Add 1 mL of medium with 2× Polybrene. (Use 1000× stock, 8 mg/mL.)
5. Add to wells 1 mL of serial dilutions changing tips each time or starting from the most diluted sample.
6. Incubate for 72 h at 37°C in a 5% CO_2 incubator.
7. Wash wells with 2 mL of PBS.
8. Add 200 μL of trypsin in PBS to each well. Wait until cells are detached from plates (5 min at 37°C).
9. Add 2 mL of PBS to each well and harvest cells in FACS tubes.
10. Centrifuge at 300g for 5 min at RT.
11. Aspirate the supernatant and add 1 mL of fixing solution to the pellet, and vortex the tubes. (Samples are stable at 4°C for a few days.)
12. Process the samples by FACS and calculate the titer (*see* **Note 6–8**).

3.4. RCR Test

1. Infect 2×10^4 SupT1 (or C8166) cells with 200 μL of the vector preparation to be tested diluted 1:100 in 1 mL medium (RPMI/10% FBS) with 2 μg/mL Polybrene, and incubate at 37°C for 16 h.
2. The following day wash cells 3 times with medium.
3. Resuspend the cells in 1 mL of fresh medium and incubate at 37°C, 5% CO_2.
4. Every 3–4 days, take about 0.5 mL of supernatant for HIV-1 p24 assay. Split the cells 1:5–1:7.
5. Collect the supernatant up to 25 days.
6. Measure p24 in the supernatant; usually do undiluted and 1:10 diluted samples.

3.5. HIV-1 p24 Assay

To perform the assay, follow the instructions provided by the manufacturer except for the sample preparation, for which you should:

1. Use the dilutions from the titration on HeLa cells.
2. If you have conditioned medium, use dilutions from 10^{-3} to 10^{-6}; and

3. If you have concentrated vector, use dilutions from 10^{-5} to 10^{-8}.

4. Notes

1. 293T cells should not be allowed to overgrow and should be used at low passage number (<p20) and kept as frozen stocks in liquid nitrogen.
2. Transfection efficiency for vector production is sensitive to *Mycoplasma* contamination. Check your cells often for this kind of microorganism. Keep your gold cell batch in the absence of antibiotics and add them only when you plate the 293T cells for transfection.
3. Transform all vector plasmids in TOP 10 *E. coli* cells. Grow bacteria in Luria-Bertani (LB) media in the presence of carbenicillin instead of ampicillin; the advantages of the first are that it is temperature resistant and the bacteria can be grown for >16 h.
4. After transfection, high-magnification microscopy of the 293T cells should reveal a very small granular precipitate of the CaPi-precipitated plasmid DNA, initially above the cell monolayer, and after incubation in the 37°C incubator overnight, on the bottom of the plate in the spaces between the cells.
5. Vector titer is calculated from the percent of HeLa cells transduced at low input of vector, as shown in the formula: % green fluorescence protein (GFP)$^+$/100 × n^o of cells infected × dilution factor = transduction units (TU)/mL. The range obtained is around 10^6–10^7 if you have titered a conditioned media and 10^9 TU/mL if you have a 500X concentrated vector preparation.
6. The infectivity of vector preparation is calculated by the ratio of transduction or infectious units (TU)/mL/ng/mL of p24 and should be >10^4.
7. The multiplicity of infection (MOI): TU/mL/n of cells to be infected for the following ranges: infection efficiency is primarily controlled by vector concentration (TU/mL), and MOI should be considered only for certain ranges of concentration of cells (>10^4 cells/mL) and vector (>10^6 TU/mL).

Acknowledgments

The financial support of Telethon–Italy (grant A.143), MURST (grant 9905313431), and the EU (grant QLK3-1999-00859) is gratefully acknowledged. A.F. is the recipient of an AIDS program fellowship from ISS, Italy.

References

1. Naldini, L. (1998) Lentiviruses as gene transfer agents for delivery to non-dividing cells *Curr. Opin. Biotechnol.* **9,** 457–463.
2. Federico, M. (1999) Lentiviruses as gene delivery vectors *Curr. Opin. Biotechnol.* **10,** 448–453.
3. Vigna, E. and Naldini, L. (2000) Lentiviral vectors: excellent tools for experimental gene transfer and promising candidates for gene therapy. *J. Gene Med.* **2,** 308–316.
4. Trono, D. (2000) Lentiviral vectors: turning a deadly foe into a therapeutic agent. *Gene Ther.* **7,** 20–23.

5. Coffin, J., Hughes, S. H., and Varmus, H. E. (2000) *Retroviruses*. Cold Spring Harbor Laboratory Press,Cold Spring Harbor, NY.1

6. Frankel, A. D. and Young, J. A. (1998) HIV-1: fifteen proteins and an RNA. *Annu. Rev. Biochem.* **67,** 1–25.

7. Emerman, M. and Malim, M. H. (1998) HIV-1 regulatory/accessory genes: keys to unraveling viral and host cell biology. *Science* **280,** 1880–1884.

8. Wei, P., Garber, M. E., Fang, S. M., Fischer, W. H., and Jones, K. A. (1998) A novel CDK9-associated C-type cyclin interacts directly with HIV-1 Tat and mediates its high-affinity, loop-specific binding to TAR RNA. *Cell* **92,** 451–462.

9. Bieniasz, P. D., Grdina, T. A., Bogerd, H. P., and Cullen, B. R. (1999) Recruitment of cyclin T1/P-TEFb to an HIV type 1 long terminal repeat promoter proximal RNA target is both necessary and sufficient for full activation of transcription. *Proc. Natl. Acad. Sci. USA* **96,** 7791–7796.

10. Neville, M., Stutz, F., Lee, L., Davis, L. I., and Rosbash, M. (1997) The importin-beta family member Crm1p bridges the interaction between Rev and the nuclear pore complex during nuclear export. *Curr. Biol.* **7,** 767–775.

11. Fornerod, M., Ohno, M., Yoshida, M., and Mattaj, I. W. (1997) CRM1 is an export receptor for leucine-rich nuclear export signals. *Cell* **90,** 1051–1060.

12. Stade, K., Ford, C. S., Guthrie, C., and Weis, K. (1997) Exportin 1 (Crm1p) is an essential nuclear export factor. *Cell* **90,** 1041–1050.

13. Pollard, V. W. and Malim, M. H. (1998) The HIV-1 Rev protein. *Annu. Rev. Microbiol.* **52,** 491–532.

14. von Schwedler, U., Song, J., Aiken, C., and Trono, D. (1993) Vif is crucial for human immunodeficiency virus type 1 proviral DNA synthesis in infected cells. *J. Virol.* **67,** 4945–4955.

15. Zhang, H., Pomerantz, R. J., Dornadula, G., and Sun, Y. (2000) Human immunodeficiency virus type 1 Vif protein is an integral component of an mRNP complex of viral RNA and could be involved in the viral RNA folding and packaging process. *J. Virol.* **74,** 8252–8261.

16. Gottlinger, H. G., Dorfman, T., Cohen, E. A., and Haseltine, W. A. (1993) Vpu protein of human immunodeficiency virus type 1 enhances the release of capsids produced by gag gene constructs of widely divergent retroviruses. *Proc. Natl. Acad. Sci. USA* **90,** 7381–7385.

17. Schubert, U., Anton, L. C., Bacik, I., et al. (1998) CD4 glycoprotein degradation induced by human immunodeficiency virus type 1 Vpu protein requires the function of proteasomes and the ubiquitin-conjugating pathway. *J. Virol.* **72,** 2280–2288.

18. Garcia, J. V. and Miller, A. D. (1991) Serine phosphorylation-independent downregulation of cell-surface CD4 by nef. *Nature* **350,** 508–511.

19. Miller, M. D., Warmerdam, M. T., Gaston, I., Greene, W. C., and Feinberg, M. B. (1994) The human immunodeficiency virus-1 nef gene product: a positive factor for viral infection and replication in primary lymphocytes and macrophages. *J. Exp. Med.* **179,** 101–113.

20. Aiken, C. (1997) Pseudotyping human immunodeficiency virus type 1 (HIV-1) by the glycoprotein of vesicular stomatitis virus targets HIV-1 entry to an

endocytic pathway and suppresses both the requirement for Nef and the sensitivity to cyclosporin A. *J. Virol.* **71**, 5871–5877.

21. Piguet, V., Wan, L., Borel, C., et al. (2000) HIV-1 Nef protein binds to the cellular protein PACS-1 to downregulate class I major histocompatibility complexes. *Nat. Cell Biol.* **2**, 163–167.

22. Bartz, S. R., Rogel, M. E., and Emerman, M. (1996) Human immunodeficiency virus type 1 cell cycle control: Vpr is cytostatic and mediates G2 accumulation by a mechanism which differs from DNA damage checkpoint control. *J. Virol.* **70**, 2324–2331.

23. Goh, W. C., Rogel, M. E., Kinsey, C. M., et al. (1998) HIV-1 Vpr increases viral expression by manipulation of the cell cycle: a mechanism for selection of Vpr in vivo. *Nat. Med.* **4**, 65–71.

24. Nishizawa, M., Kamata, M., Mojin, T., Nakai, Y., and Aida, Y. (2000) Induction of Apoptosis by the Vpr Protein of Human Immunodeficiency Virus Type 1 Occurs Independently Of G(2) Arrest of the Cell Cycle. *Virology* **276**, 16–26.

25. Patel, C. A., Mukhtar, M., and Pomerantz, R. J. (2000) Human immunodeficiency virus type 1 Vpr induces apoptosis in human neuronal cells. *J. Virol.* **74**, 9717–9726.

26. Conti, L., Matarrese, P., Varano, B., et al. (2000) Dual role of the HIV-1 vpr protein in the modulation of the apoptotic response of T cells. *J. Immunol.* **165**, 3293–3300.

27. Heinzinger, N. K., Bukinsky, M. I., Haggerty, S. A., et al. (1994) The Vpr protein of human immunodeficiency virus type 1 influences nuclear localization of viral nucleic acids in nondividing host cells. *Proc. Natl. Acad. Sci. USA* **91**, 7311–7315.

28. Popov, S., Rexach, M., Zybarth, G., et al. (1998) Viral protein R regulates nuclear import of the HIV-1 pre-integration complex. *EMBO J.* **17**, 909–917.

29. Gallay, P., Swingler, S., Aiken, C., and Trono, D. (1995) HIV-1 infection of nondividing cells: C-terminal tyrosine phosphorylation of the viral matrix protein is a key regulator. *Cell* **80**, 379–388.

30. Gallay, P., Swingler, S., Song, J., Bushman, F., and Trono, D. (1995) HIV nuclear import is governed by the phosphotyrosine-mediated binding of matrix to the core domain of integrase. *Cell* **83**, 569–576.

31. Burns, J. C., Friedmann, T., Driever, W., Burrascano, M., and Yee, J. K. (1993) Vesicular stomatitis virus G glycoprotein pseudotyped retroviral vectors: concentration to very high titer and efficient gene transfer into mammalian and nonmammalian cells. *Proc. Natl. Acad. Sci. USA* **90**, 8033–8037.

32. Naldini, L., Blomer, U., Gallay, P., et al. (1996) In vivo gene delivery and stable transduction of nondividing cells by a lentiviral vector. *Science* **272**, 263–267.

33. Gasmi, M., Glynn, J., Jin, M. J., et al. (1999) Requirements for efficient production and transduction of human immunodeficiency virus type 1-based vectors. *J. Virol.* **73**, 1828–1834.

34. Kim, V. N., Mitrophanous, K., Kingsman, S. M., and Kingsman, A. J. (1998)

Minimal requirement for a lentivirus vector based on human immunodeficiency virus type 1. *J. Virol.* **72,** 811–816.

35. Mochizuki, H., Schwartz, J. P., Tanaka, K., Brady, R. O., and Reiser, J. (1998) High-titer human immunodeficiency virus type 1-based vector systems for gene delivery into nondividing cells. *J. Virol.* **72,** 8873–8883.

36. Zufferey, R., Nagy, D., Mandel, R. J., Naldini, L., and Trono, D. (1997) Multiply attenuated lentiviral vector achieves efficient gene delivery in vivo. *Nat. Biotechnol.* **15,** 871–875.

37. Dull, T., Zufferey, R., Kelly, M., et al. (1998) A third-generation lentivirus vector with a conditional packaging system. *J. Virol.* **72,** 8463–8471.

38. Kafri, T., van Praag, H., Ouyang, L., Gage, F. H., and Verma, I. M. (1999) A packaging cell line for lentivirus vectors. *J. Virol.* **73,** 576–584.

39. Klages, N., Zufferey, R., and Trono, D. (2000) A stable system for the high-titer production of multiply attenuated lentiviral vectors. *Mol. Ther.* **2,** 170–176.

40. Parolin, C., Dorfman, T., Palu, G., Gottlinger, H., and Sodroski, J. (1994) Analysis in human immunodeficiency virus type 1 vectors of cis-acting sequences that affect gene transfer into human lymphocytes. *J. Virol.* **68,** 3888–3895.

41. Carroll, R., Lin, J. T., Dacquel, E. J., et al. (1994) A human immunodeficiency virus type 1 (HIV-1)-based retroviral vector system utilizing stable HIV-1 packaging cell lines. *J. Virol.* **68,** 6047–6051.

42. McBride, M. S., Schwartz, M. D., and Panganiban, A. T. (1997) Efficient encapsidation of human immunodeficiency virus type 1 vectors and further characterization of cis elements required for encapsidation. *J. Virol.* **71,** 4544–4554.

43. Berkowitz, R. D., Hammarskjold, M. L., Helga-Maria, C., Rekosh, D., and Goff, S. P. (1995) 5' Regions of HIV-1 RNAs are not sufficient for encapsidation: implications for the HIV-1 packaging signal. *Virology* **212,** 718–723.

44. Miyoshi, H., Blomer, U., Takahashi, M., Gage, F. H., and Verma, I. M. (1998) Development of a self-inactivating lentivirus vector. *J. Virol.* **72,** 8150–8157.

45. Zufferey, R., Dull, T., Mandel, R.J., et al. (1998) Self-inactivating lentivirus vector for safe and efficient in vivo gene delivery. *J. Virol.* **72,** 9873–9880.

46. Iwakuma, T., Cui, Y., and Chang, L. J. (1999) Self-inactivating lentiviral vectors with U3 and U5 modifications. *Virology* **261,** 120–132.

47. Donello, J. E., Loeb, J. E., and Hope, T. J. (1998) Woodchuck hepatitis virus contains a tripartite posttranscriptional regulatory element. *J. Virol.* **72,** 5085–5092.

48. Deglon, N., Tseng, J. L., Bensadoun, J. C., et al. (2000) Self-inactivating lentiviral vectors with enhanced transgene expression as potential gene transfer system in Parkinson's disease. *Hum. Gene Ther.* **11,** 179–190.

49. Zufferey, R., Donello, J. E., Trono, D., and Hope, T. J. (1999) Woodchuck hepatitis virus posttranscriptional regulatory element enhances expression of transgenes delivered by retroviral vectors. *J. Virol.* **73,** 2886–2892.

50. Follenzi, A., Ailles, L. E., Bakovic, S., Geuna, M., and Naldini, L. (2000) Gene transfer by lentiviral vectors is limited by nuclear translocation and rescued by HIV-1 pol sequences. *Nat. Genet.* **25,** 217–222.

51. Zennou, V., Petit, C., Guetard, D., et al. (2000) HIV-1 genome nuclear import is mediated by a central DNA flap. *Cell* **101,** 173–185.
52. Charneau, P., Alizon, M., and Clavel, F. (1992) A second origin of DNA plus-strand synthesis is required for optimal human immunodeficiency virus replication. *J. Virol.* **66,** 2814–2820.
53. Charneau, P., Mirambeau, G., Roux, P., et al. (1994) HIV-1 reverse transcription. A termination step at the center of the genome. *J. Mol. Biol.* **241,** 651–662.
54. Ilyinskii, P. O. and Desrosiers, R. C. (1998) Identification of a sequence element immediately upstream of the polypurine tract that is essential for replication of simian immunodeficiency virus. *EMBO J.* **17,** 3766–3774.
55. Blomer, U., Naldini, L., Kafri, T., et al. (1997) Highly efficient and sustained gene transfer in adult neurons with a lentivirus vector. *J. Virol.* **71,** 6641–6649.
56. May, C., Rivella, S., Callegari, J., et al. (2000) Therapeutic haemoglobin synthesis in beta-thalassaemic mice expressing lentivirus-encoded human beta-globin. *Nature* **406,** 82–86.
57. An, D. S., Wersto, R. P., Agricola, B. A., et al. (2000) Marking and gene expression by a lentivirus vector in transplanted human and nonhuman primate CD34(+) cells. *J. Virol.* **74,** 1286–1295.
58. Bukovsky, A. A., Song, J. P., and Naldini, L. (1999) Interaction of human immunodeficiency virus-derived vectors with wild-type virus in transduced cells. *J. Virol.* **73,** 7087–7092.
59. Bonyhadi, M. L. and Kaneshima, H. (1997) The SCID-hu mouse: an in vivo model for HIV-1 infection in humans. *Mol. Med. Today* **3,** 246–253.
60. DuBridge, R. B., Tang, P., Hsia, H. C., et al. (1987) Analysis of mutation in human cells by using an Epstein-Barr virus shuttle system. *Mol. Cell Biol.* **7,** 379–387.
61. Graham, F. L., Smiley, J., Russell, W. C., and Nairn, R. (1977) Characteristics of a human cell line transformed by DNA from human adenovirus type 5. *J. Gen. Virol.* **36,** 59–74.

20

Packaging Cell System for Lentivirus Vectors

Preparation and Use

Narasimhachar Srinivasakumar

1. Introduction

Packaging systems based on human immunodeficiency virus type 1 (HIV-1) can be used to transfer genes (both ex vivo and in vivo in experimental animals), with high efficiency into a wide variety of cell types including nondividing and terminally differentiated cells such as muscle and neuronal cells *(1–6)*. In this chapter, I begin with an overview of HIV-1 structure, genetics, and replication cycle, emphasizing those features essential for the creation of packaging systems based on HIV-1. Next, I describe the plasmid constructs for preparation of HIV-1 vector stocks, including some recent advances in the design of helper and gene transfer vectors. Finally, a detailed protocol for generation and testing of vectors is provided. Although HIV-1 is used as a prototype for other lentiviruses and there are many similarities between different viruses, there are also significant differences. The extensive knowledge base of HIV-1-based gene transfer systems can be readily adapted to develop gene transfer vectors based on other lentiviruses such as feline immunodeficiency virus (FIV; *see* Chapter 23). These other lentivirus vectors are likely to share many of the advantages of HIV-1 vectors. Because of space limitations, comparisons with other lentiviruses that highlight similarities or differences are not given. I recommend the excellent treatise on retroviruses *(6a)* edited by J.M. Coffin, S.H. Hughes, and H.E. Varmus (CSHL,1997) for this purpose and to support many of the statements made in this introduction.

From: *Methods in Molecular Medicine, Vol. 69, Gene Therapy Protocols, 2nd Ed.*
Edited by: J. R. Morgan © Humana Press Inc., Totowa, NJ

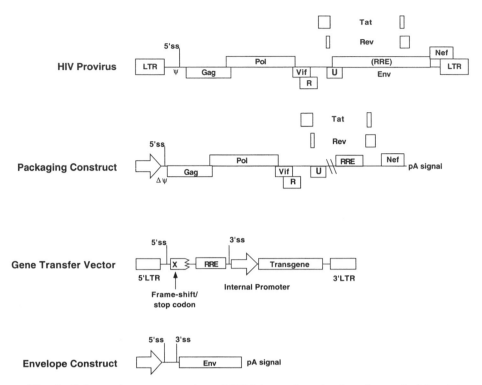

Fig. 1. Schematic representation of HIV-1 provirus (top) and a typical lentivirus packaging system consisting of a packaging plasmid, a gene transfer vector, and an envelope expression plasmid. Only relevant portions of the various constructs are shown. LTR, long terminal repeat; RRE, Rev responsive element.

1.1. HIV-1 Structure and Genetics

The RNA genome of HIV-1 is about 9.7 kb in size. The reverse-transcribed and integrated form of the HIV-1 genome is called the provirus. The organization of the HIV-1 provirus is shown in **Figure 1**. Like other retroviruses, the provirus contains the long terminal repeat (LTR) elements at both the 5' and 3' ends of the provirus. Each LTR contains U3 (unique 3), R (repeat), and U5 (unique 5) regions. The U3 contains promoter and enhancer elements, and the R region contains the sequence for *trans*-acting response (TAR) element and signals for polyadenylation. The 5' end of U3 and the 3' end of U5 also contain sequences for recognition by the viral integrase protein to bring about recombination between the virus cDNA and the host chromosomal DNA. Between the two LTRs are coding regions for viral structural and regulatory proteins. Simple retroviruses have coding sequences for only three genes: *gag*, *pol*, and *env*. Complex retroviruses such as HIV contain coding sequences for at least six additional proteins: two regulatory proteins (Tat and Rev) and four accessory proteins (Vif, Vpr, Vpu, and Nef). Transcription begins at the first nucleotide of R in the 5' LTR and terminates at the last nucleotide of 3' R. Termination may not be very efficient and may proceed through to cellular flanking sequences. The various structural and regulatory proteins are synthesized from either full-length (Gag and Gag-Pol), partially spliced (e.g., Env and Vpu), or fully spliced messages (e.g., Tat, Rev, and Nef).

1.2. Virion Structural Proteins

HIV-1, like other retroviruses, is an enveloped virus that contains two copies of a positive or message sense RNA genome within a cone-shaped core. Gag and Gag-Pol are polyproteins that contribute to most of the internal proteins of the virion. Upon proteolytic cleavage by the viral protease, the Gag polyprotein is split into the matrix (p17), capsid (p24), nucleocapsid (p7), and p6 protein. Upon cleavage, the Pol region of the Gag-Pol precursor yields protease, reverse transcriptase (RT), and integrase proteins (IN). The viral protease is released by autoproteolytic cleavage from the Gag-Pol precursor and then acts on the other cleavage sites in the Gag and Gag-Pol proteins. Except for the p17 or matrix protein that lines the inner aspect of the virus envelope, the other proteins conglomerate around the RNA genome to form the virion core. Vif, Vpr, and Nef are other proteins that are encapsulated in the virion. Although it was previously believed that Vif was packaged into virions by a nonspecific mechanism, recent evidence seems to indicate that Vif is incorporated into virions via its interaction with the genomic RNA *(6b)*.

The lipid envelope of the virion contains the transmembrane (TM) and surface (SU) glycoproteins gp41 and gp120, which are coded for in the *env* region. The SU glycoprotein is noncovalently attached to the TM protein and contains the binding sites for the CD4 receptor and the chemokine coreceptors such as CCR5 or CXCR4 depending on whether the virus is macrophage or T-cell tropic, respectively. The SU protein effects binding of the virion to target cells, whereas fusion of virion and target cell membrane is accomplished by the TM protein. The HIV-1 envelope glycoproteins can be replaced with envelope glycoproteins from other viruses such as Moloney murine leukemia virus (MMLV) or the vesicular stomatitis virus G (VSV-G) glycoprotein. This enables us to prepare virus stocks that can infect a wide variety of cells that do not bear either CD4 or the chemokine coreceptors.

1.3. Lentivirus Life Cycle

Following binding of the virus particle to the cell surface receptor(s), fusion of virion and plasma membrane of the target cell ensues. This results in deposition of the capsid together with the RNA genome into the cytoplasm of the target cell. The reverse transcriptase then converts the genomic RNA into a double-stranded cDNA by a complicated procedure that involves the use of a cellular t-RNA primer that is packaged together with the genomic RNA in the virion and multiple jumps from one end of the molecule to the other. The reverse transcription process recreates a functional promoter at the 5' end of the provirus *(6a)*. Following reverse transcription, the cDNA is transported through the nuclear pore into the nucleus. Recent studies have shown that during reverse transcription, a central DNA flap is created due to the presence of a central polypurine tract (cPPT) and central termination sequence (CTS) within the Pol coding sequence *(7,8)*. The creation of this unique triple-stranded DNA region appears to be important for the nuclear import of preintegration complex (PIC), in both dividing and nondividing cells *(7,9)*. Recent vector designs have, therefore, incorporated the *cis* signals (cPPT and CTS) for the creation of this central DNA flap and thereby allow more efficient nuclear import of vector genome.

The viral matrix (p17), integrase, and Vpr proteins have also been implicated in the nuclear import of PICs in nondividing cells *(10–14)*. Although both p17 and integrase have nuclear localization domains, there is some controversy regarding their roles in the nuclear import of PICs *(15–17)*. Once the cDNA is imported, the viral integrase functions to insert the cDNA into the host-cell chromosome at a site that appears to be chosen in a random fashion. Upon activation of the infected cell, full-length messages are transcribed from the viral LTR that undergoes splicing to yield messages that code for synthesis of Tat, Rev, and Nef proteins. Tat and Rev enter the nucleus to increase gene expression from the viral LTR and effect nucleocytoplasmic transport of full-length and partially spliced messages, respectively, as described below. The viral structural proteins, the full-length RNA genome, and the t-RNA primer come together at the plasma membrane of the cell decorated with the envelope glycoprotein to assemble into a virus particle.

1.4. HIV-1 Accessory and Regulatory Proteins

1.4.1. Tat

The Tat protein is a transcriptional activator that enhances transcription from the viral LTR promoter and also increases the processivity of the RNA PolII transcription machinery, ensuring abundant amounts of full-length genomic message *(18–20)*. The Tat protein achieves this by binding to the TAR element present at the 5' end of the newly synthesized RNA. Tat binds to a bulge in this stem loop of TAR and recruits cellular proteins to TAR (and thereby the viral LTR promoter). The host proteins recruited include the positive transcription elongation factor b (P-TEFb) and Tat-associated kinase complexes, which consist of, among other protein subunits, cyclin-T1 and the cyclin-dependent kinase CDK9. This complex phosphorylates the C-terminal domain of RNA polymerase II and increases its processivity. Tat has also been shown to recruit histone acetyltransferases (HAT) p300 and p300 CREB binding protein associated factor (p/CAF) to the viral promoter *(21–23)*. The HAT proteins appear to be essential for optimal transcriptional activation of the viral promoter in the context of integrated vector but not in transient transfection assays. Tat is required in HIV-1-based packaging systems to ensure that adequate amounts of the gene transfer vector RNA are synthesized for encapsidation into the virus particles *(24,25)*. Replacement of HIV enhancer and promoter elements within the U3 of the 5' LTR with elements from other promoters overcomes this requirement for Tat protein and allows one to create Tat-independent gene transfer vectors (*see* **Subheading 1.6.2.3**) *(25,26)*.

1.4.2. Rev

The Rev protein binds to a target sequence within the envelope coding region referred to as the Rev responsive element (RRE). Rev recruits the nuclear export factors Crm1 to RRE to bring about the transport of intron-containing messages (i.e., partially spliced and unspliced messages such as those coding for Gag and Pol proteins or the Env glycoprotein) into the cytoplasm *(27–29)*. Without Rev, there is, therefore, no synthesis of these essential structural proteins, resulting in a nonreplicating virus. Thus, Rev is an attractive target for control of HIV-1 replication. Rev and RRE are required for expression of HIV-1 subgenomic messages such as those used in the expression of viral Gag and Gag-Pol proteins in packaging or helper constructs *(24,30)*. The RRE is also part of the gene transfer vector in most of the HIV-1 vector constructs reported to date. The RRE sequence is required to ensure efficient transport of the vector RNA into the cytoplasm in the presence of Rev, which allows encapsidation of the RNA into the assembling virion *(24,31–33)*.

1.4.3. Nef

The Nef protein is synthesized early after infection together with Tat and Rev and has pleiotropic effects. Nef has been shown to downmodulate cell surface expression of CD4 by inducing endocytosis, resulting in the degradation of this molecule in lysosomes *(34–37)*. Since Nef is believed to an important contributor in the pathogenicity of AIDS by HIV *(38)*, it may be prudent to remove this from HIV-based packaging systems in spite of its known positive effect on increasing the efficiency of reverse transcription during viral entry *(39,40)*. Fortunately, this effect of Nef is restricted to certain types of envelope glycoproteins used for pseudotyping HIV vectors. For instance, Nef increases

infectivity of virions pseudotyped with HIV-1 or amphotropic MMLV envelopes but not those pseudotyped with VSV-G *(41,42)*. Nef can, therefore, be
eliminated from packaging systems that employ VSV-G to pseudotype vectors.

1.4.4. Vpr

Vpr is another protein with pleiotropic effects. It can cause cells to arrest in
the G2 phase of the cell cycle *(43–46)*. Cell-cycle arrest in G2 results in a
modest increase in transcriptional output from the viral LTR promoter *(46)*.
Vpr has also been shown to be essential for importation of PIC into the nucleus
of macrophages *(14,47–49)*. Vpr is packaged into the virion at a ratio of one
Vpr molecule to seven capsid protein (p24) monomers *(50)*. Vpr is recruited
into the virus particle through its interaction with the p6 region of the Gag
molecule *(51,52)*. When concentrated stocks of HIV-1 vector stocks are used
in gene transfer experiments, Vpr delivered by the virus particles can result in
apoptosis of target cells *(53–55)*. So the advantages of using Vpr for nuclear
import of PICs in some types of cells must be weighed against its propensity to
cause apoptosis.

1.4.5. Vif and Vpu

Vif has been shown to increase infectivity of virions produced in certain
human T-cell lines *(56,57)*. Vif appears to affect events after binding and
fusion, during either uncoating or RT *(58)*. Vpu has been shown to increase
virion export in human cells but not in cell lines of simian origin *(59)*. In spite
of these known functions of Vif and Vpu, it appears that one can safely eliminate these proteins for preparation of high-titer virus stocks when using 293T
cells (see below).

1.5. Constitutive Transport Element (CTE)

Simpler retroviruses, such as the Mason-Pfizer monkey virus (MPMV), a
type D retrovirus, utilize a structured RNA element that does not require the
coexpression of a viral protein to achieve the transport of unspliced or intron-
containing messages from the nucleus to the cytoplasm *(60,61)*. These elements have been dubbed constitutive transport elements (CTEs). The CTE has
been shown to substitute for Rev-RRE function not only in the context of the
HIV-1 provirus *(62,63)* but also in subgenomic constructs used for the expression of viral helper proteins or gene transfer vector RNAs *(24,32,33,64,65)*. In
fact, packaging systems that completely lack Rev have been described *(24,32)*,
although these systems seem to perform a little less efficiently than the classical Rev-RRE system for production of lentivirus vector stocks. Further refine-
ments in the use of these alternative transport elements from other viruses may

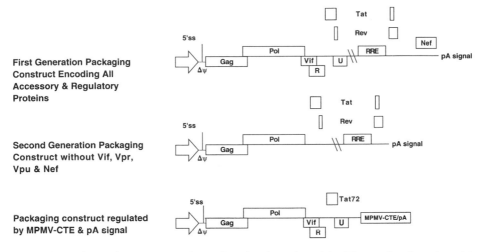

First Generation Packaging Construct Encoding All Accessory & Regulatory Proteins

Second Generation Packaging Construct without Vif, Vpr, Vpu & Nef

Packaging construct regulated by MPMV-CTE & pA signal

Fig. 2. Schematic representation of packaging plasmids. The packaging plasmid regulated by MPMV-CTE/pA signal (bottom) also expresses a functional single-exon or 72-amino acid Tat protein (Tat72) *(33)*. LTR, long terminal repeat; RRE, Rev responsive element; MPMV-CTE, Mason-Pfizer monkey virus/constitutive transport element.

allow enhanced titers and increased gene expression in transduced target cells. Rev-independent packaging systems may be useful for delivery of dominant-negative Rev genes into HIV-1 susceptible cells to effect a type of "intracellular immunization" against HIV-1 infection *(65)*.

1.6. Lentivirus Packaging Systems

A typical lentivirus packaging system (**Fig. 1**) consists of three components: 1) helper or packaging plasmids that encode for essential virus structural and regulatory proteins (with the exception of the envelope glycoprotein); 2) a plasmid that expresses an envelope glycoprotein from an unrelated virus such as the amphotropic MMLV or VSV-G glycoprotein; and 3) a gene transfer vector.

1.6.1. Packaging or Helper Constructs

1.6.1.1. FIRST GENERATION PACKAGING/HELPER PLASMID

Early versions of packaging plasmids (**Fig. 2**) encoded most viral structural and regulatory proteins with the exception of the envelope glycoprotein *(5,66–68)*. A deletion within the 5' untranslated region upstream of *gag* and within the encapsidation signal was engineered to reduce the efficiency of incorporation of the helper RNA and thereby prevent recombination between vector and helper RNA molecules. The viral LTR promoter was replaced with a heterologous promoter such as the cytomegalovirus immediate early promoter. The downstream or 3' LTR was replaced with a heterologous polyadenylation signal (e.g., from SV40 or the insulin gene or another gene). A deletion or mutation within Env ensured that no envelope glycoprotein was synthesized. The helper plasmid still expressed all other regulatory and accessory proteins of HIV-1.

1.6.1.2. SECOND-GENERATION PACKAGING PLASMID WITHOUT CODING REGIONS FOR VIF, VPR, OR VPU

More recent versions of packaging plasmids have further deletions and modifications to eliminate Vif, Vpr, and Vpu expression altogether (**Fig. 2**) or in various combinations *(1,5,6,26,33)*. These helper constructs thus lack many of the accessory proteins that may be involved in HIV-1 pathogenicity and may render the packaging system safer. The goal of creating these minimal constructs is to design helper constructs that express only those genes that are essential for gene transfer into a wide variety of cells at high efficiency.

1.6.1.3. THIRD-GENERATION PACKAGING PLASMID USING SEGREGATION OF GAG/GAG-POL AND REV FUNCTIONS

For expression of the HIV-1 Gag and Gag-Pol proteins, normally one needs to insert the RRE downstream of the coding sequence and also express viral Rev protein, which allows the *gag/pol* message to be transported into the cytoplasm for translation. Early versions of the packaging plasmids contained the RRE and also expressed Rev. To further enhance safety, packaging systems have been devised in which the HIV-1 *gag/pol* and *rev* sequences have been segregated into separate plasmids *(24,25)*. Tat has also been eliminated from this sytem by using Tat-independent gene transfer vectors (*see* **Subheading 1.6.2.3.**) Such a packaging system uses four plasmids for the creation of vector stocks instead of three. This increases the safety of the system by further decreasing the probability of recombination between the different helper plasmids and the gene transfer vector to recreate a replication-competent virus because multiple recombination steps need to occur to recreate a functional retrovirus.

1.6.1.4. RECENT MODIFICATIONS TO PACKAGING PLASMIDS

1.6.1.4.1. *Segregation of Gag and Pol Coding Regions.* A recent novel modification to the packaging system involves a clever approach toward separating Gag and Pol coding regions *(69)*. The Gag-Protease proteins are expressed using one construct. The Pol proteins (RT and IN) are expressed as a Vpr-RT-IN fusion protein. The Vpr-RT-IN fusion protein is drawn into the assembling virus particle through the interaction between the Vpr and the p6

5' U3 can be replaced with enhancers
from other promoters to make Tat-
independent vectors and to prevent
reconstitution of U3 in SIN vectors

Deletion in U3 increases
safety; improves gene
expression from
internal promoter

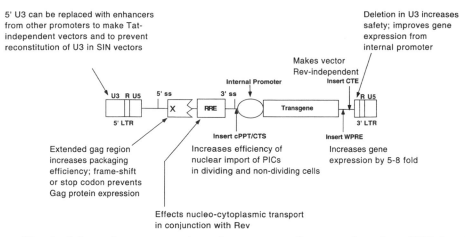

Makes vector
Rev-independent
Insert CTE

Internal Promoter

Extended gag region
increases packaging
efficiency; frame-shift
or stop codon prevents
Gag protein expression

Insert cPPT/CTS

Increases efficiency of
nuclear import of PICs
in dividing and non-dividing cells

Insert WPRE

Increases gene
expression by 5-8 fold

Effects nucleo-cytoplasmic transport
in conjunction with Rev

Fig. 3. Schematic representation of a gene transfer vector based on HIV-1 and modifications to improve safety and expression (see text for details). LTR, long terminal repeat; RRE, Rev responsive element; CTE, constitutive transport element; CPPT/CTS, central polypurine tract/central termination sequence; PIC, preintegration complex; SIN, self-inactivating; WPRE, woodchuck posttranslational regulatory element.

protein present in the C terminus of the Gag polyprotein. This approach appears to provide a higher margin of safety in lentivirus packaging system and considerably reduces the frequency of recombination between the packaging plasmids and the gene transfer vector.

1.6.1.4.2. Rev-Independent Packaging Plasmids. As alluded to earlier, the dependence of *gag/pol* expression on RRE and Rev can be eliminated by using CTEs from other viruses (**Fig. 2**) *(24,26,32,33,64,65,68)*. Studies have indicated that CTEs function better if they are positioned toward the 3' end of the expression cassette *(70,71)*. Also, one recent study showed that multiple copies of CTE may be better than single copies of the element for obtaining high levels of *gag/pol* expression *(70)*. Such constructs showed gene expression levels exceeding those based on Rev and RRE.

1.6.2. Gene Transfer Vectors

The ideal gene transfer vector will lack all viral protein coding sequences and have a capacity for large transgenes. It should possess, in addition to the transgene expression cassette, all *cis*-acting sequences for efficient packaging, nucleocytoplasmic transport, reverse transcription, and integration of the vector RNA. It is believed that HIV-1 vectors can deliver a transgene expression cassette of approximately 8 11 kb in length.

1.6.2.1. FIRST-GENERATION GENE TRANSFER VECTORS

The first-generation vectors (**Fig. 3**) carried a deletion between the middle of the *gag* coding region and well into the *env* coding region, thereby essentially eliminating all viral coding regions *(67)*. It may not be possible to eliminate *gag* sequences completely because the sequence at the beginning of the *gag* has been shown to contain elements that can increase packaging efficiency of the vector RNA *(72–77)*. A frame shift mutation or stop codon inserted within the *gag* reading frame can ensure premature termination of translation. The vectors also contain the RRE between the major 5' splice donor site and the 3' *tat/rev* splice acceptor site. The transgene expression cassette is usually positioned between the *Bam*HI site within the second coding exon of *rev* and the *Xho*I site within *nef*, effectively shutting out Nef expression. Polyadenylation of the RNA (including that derived from the internal promoter) occurs using signals present in the 3'U3 and R sequences within the LTR. The most common internal promoter enhancer elements used in HIV-1 vectors are those derived from the cytomegalovirus immediate early gene and the cellular phosphoglycerokinase promoter from mouse or human. It is believed that one may be able to replace these promoters with other tissue-specific or regulatable ones.

1.6.2.2. REV-INDEPENDENT HIV-1 VECTORS

The requirement of Rev for expression of vector RNA containing RRE can be eliminated by using the CTE downstream of the transgene expression cassette (**Fig. 3**) *(24,32,33,64)*. Vectors containing the CTE can be used to create a Rev-independent packaging system for delivery of transdominant negative Rev *(65)*.

1.6.2.3. TAT-INDEPENDENT HIV-1 VECTORS

The first-generation HIV-1 vectors required Tat to obtain optimal titers, since expression from the viral LTR promoter requires coexpression of Tat. To overcome this, hybrid promoters that use enhancer elements from other viruses, instead of those present in the U3 of HIV-1, have been created (**Fig. 3**) *(25,26)*. These vectors appear to be nearly as efficient in terms of vector titer as the original Tat-dependent vectors that contained the wild type HIV-1 LTR. Although there is a report showing that Tat, in addition to its effect on transcription from the viral LTR, can also effect the efficiency of reverse transcription *(78)*, studies using HIV-1 vectors have not revealed this requirement *(25,26,33)*.

1.6.2.4. MODIFICATIONS TO VECTOR TO IMPROVE SAFETY AND EFFICACY

Figure 3 shows various modifications to lentivirus vectors to improve efficacy and safety.

1.6.2.4.1. Self-Inactivating HIV-1 Vectors. It has been shown that most of the U3 region of the 3' LTR can be safely deleted without compromising vector titer *(79,80)*. Such a deletion ensures that transcription from the 5' LTR promoter is efficiently suppressed following reverse transcription and integration into the target cell chromosome. Another advantage of using vectors with deletions in the U3 region is the enhanced transgene expression noted from the internal promoters in such vectors *(79)*. This is probably because of a decrease in promoter competition between the viral LTR and the internal promoter.

1.6.2.4.2. Addition of Posttranscriptional Regulatory Elements from Woodchuck Hepatitis Virus. Hepatitis B viruses contain *cis*-acting elements that can substitute for Rev and RRE function in HIV-1 subgenomic constructs *(81–83)*. Experiments have revealed that addition of a 600 bp posttranscriptional regulatory element from woodchuck hepatitis virus (WPRE) in retroviral and lentiviral vectors downstream of the transgene can increase expression by five to eightfold **(Fig. 3)** *(84)*.

1.6.2.4.3. Addition of cPPT and CTS. Recent studies have demonstrated that a central DNA flap created during reverse transcription of lentivirus RNA is important in the nuclear importation of PICs *(7,9)* (*see* **Subheading 1.3.**). The *cis*-acting sequences required for the creation of this central DNA flap, the cPPT and CTS, within the *pol* coding region, are being increasingly utilized in lentivirus vectors to improve virus titer in both dividing and nondividing cells.

1.6.3. Envelope Glycoproteins Used for Pseudotyping Lentivirus Vectors

The most popular envelope glycoprotein that has been used for pseudotyping lentivirus vectors is the VSV-G glycoprotein, which is quite stable under high centrifugation forces. This allows vectors pseudotyped with VSV-G to be readily concentrated by ultracentrifugation. In contrast to VSV-G, the amphotropic MMLV envelope pseudotyped vectors cannot be concentrated by ultracentrifugation without loss of titer. To concentrate vectors pseudotyped with amphotropic envelope, one has to use alternative methods based on ultrafiltration using molecular weight cutoff filters (further described under **Subheading 3.2.2.**).

1.7. Packaging Systems

Depending on the RNA transport element being used, one can design three types of packaging systems for production of HIV-1 vector stocks (**Table 1**) *(33)*. The traditional or classical HIV-1 packaging system uses RRE and Rev for the expression of both helper and gene transfer vector RNAs. The second packaging system uses MPMV-CTE for expression of helper and gene transfer vector RNAs. The third type, called the combination or reciprocal packaging

Table 1
**Classification of Packaging Systems Based on RNA Transport Elements
Used for Expression of Helper and Gene Transfer Vector RNAs[a]**

Packaging system	RNA transport element in		Rev requirement
	Helper plasmid	Gene transfer vector	
RRE/Rev	RRE	RRE	Yes
CTE	CTE	CTE	No
Reciprocal-1	CTE	RRE	Yes
Reciprocal-2	RRE	CTE	Yes

[a] RRE, Rev responsive element; CTE, constitutive transport element.

system, utilizes Rev/RRE for expression of one component of the packaging system and CTE for the other component *(33,64)*. There are two regions of homology between the helper and gene transfer vector constructs. One is at the 5' end of the *gag* coding region, and the other is the shared RRE or CTE toward the 3' end of the helper construct and the same element present within the gene transfer vector. Using dissimilar transport elements at the 3' end of the helper plasmid and the gene transfer vector in the combination packaging system may render the system safer by reducing the chance of replication-competent retrovirus (RCR) formation due to a recombination event.

Taking this idea further, several investigators have resorted to decreasing the homology at the *gag* end of the helper and gene transfer vector constructs. One approach has been to "humanize" *gag* codons systematically in the helper plasmid *(85)*. This not only reduces the homology in the *gag* region, it also renders *gag* expression Rev independent. Elimination of homology at the *gag* end and at the 3' end by removal of RRE in the helper plasmid makes this packaging system one of the safest described to date. An alternate approach to the same goal has been to use a helper construct derived from simian immuno-deficiency virus (SIV) *(86)* based on the observation that SIV can cross-package HIV-1 RNA *(87)*. However such packaging systems are still in their infancy and need to be developed further.

1.8. Advantages and Disadvantages for Lentivirus Vectors

Because of their ability to integrate into the host chromosome, lentivirus vectors, like oncoretroviral vectors and unlike many other gene transfer vectors, can transduce cells permanently. Thus they have the capability to effect long-term gene expression. Lentiviruses also have the advantage that they can transduce terminally differentiated and growth-arrested cells efficiently.

Oncoretroviral vectors, on the other hand, can only transduce growing cells *(88–90)*. Because of the random nature of integration, lentiviruses have some of the same disadvantages as other retroviruses, such as the possibility of either inactivating cellular genes (e.g., tumor suppressor genes) or activating or causing overexpression of other genes (e.g., oncogenes). Inadvertent activation of cellular genes, including oncogenes, can be avoided by using later generations of self-inactivating vectors, which are devoid of promoter and enhancer elements in the viral 3' LTR (see above). It may be possible to overcome the drawback of random integration by redirecting PICs to specific regions of the host chromosome by fusing sequence-specific DNA binding domains to the viral integrase *(91,92)*.

Most investigators currently use a transient transfection approach to produce vector stocks. A drawback of this procedure is that there may be a significant degree of rearrangement of the input DNAs. This can lead to contamination of vector stocks with defective vector genomes. To produce "clean" vector stocks, it may be preferable to derive well-characterized packaging cell lines. The major hurdle for the creation of packaging cell lines appears to be the toxicity of certain viral proteins. For instance, Vpr causes cells to arrest in the G2 phase of the cell cycle *(43–45)* and therefore would not be conducive for producing cell lines that express high levels of this protein. Likewise, the viral protease has been shown to also cleave cellular proteins, which could interfere with the establishment of cell lines that stably express viral Gag-Pol proteins *(93,94)*. VSV-G protein expression is also toxic to cells. Although cell lines that constitutively express viral packaging proteins have been described *(24,95,96)*, more recent approaches toward the creation of lentivirus packaging cell lines resort to the use of inducible or regulatable promoters to overcome potential toxic effects of viral proteins *(66,97)*. The construction of packaging cell lines using regulated promoters is beyond the scope of this chapter and the interested reader is referred to recent articles *(66,97)* describing such cell lines.

2. Materials

2.1. Cell Culture

1. Adenovirus transformed human embryonic kidney cell line expressing SV40 T-antigen (293T cells). 293T cells can be obtained from American Type Culture Collection (Rockville, MD; ATCC cat. no. SD-3515).
2. Phosphate-buffered saline (PBS) without divalent cations.
3. Growth medium (DMEM/10% fetal bovine serum [FBS]). Dulbecco's modified Eagle's medium (DMEM) with 2 mm L-glutamine, 100 U/mL penicillin, 100 μg/mL streptomycin, and 10% heat-inactivated FBS (*see* **Note 1**).
4. Tissue culture flasks: 25 cm^2 or larger.

2.2. Calcium Phosphate-Mediated Transfection

1. Plasmid constructs: Production of vector stocks requires cotransfection of three or more plasmids, one or more packaging plasmids that provide helper function, a gene transfer vector, and an envelope glycoprotein-expressing plasmid. Various types of packaging constructs, gene transfer vectors and envelope expression constructs have been described above (*see* **Note 2**).
2. HEPES-buffered saline (HeBS), 2× (50 mM HEPES, 10 mM KCl, 12 mM dextrose, 280 mM NaCl, 1.5 mM Na$_2$HPO$_4$, pH 7.05). To make 500 mL of stock buffer, add 5.96 g HEPES, 0.37 g KCl, 1.08 g dextrose, 8.18 g NaCl, and 5 mL 150 mM Na$_2$HPO$_4$•7H$_2$O (20 g/500 mL of deionized distilled water). It is prudent to make several lots, adjusted to pH 7.05, since even slight differences in pH can have a dramatic effect on transfection efficiency. Filter through a 0.2-μM filter and store in aliquots at –20°C. Test aliquots from each lot for transfection efficiency using a green fluorescent protein (GFP) reporter plasmid. Keep the lot that gives the highest transfection efficiency. Before use, warm to room temperature and vortex to achieve uniform mixing. Prepare fresh lots every 6–12 months (*see* **Note 3**).
3. 2.5 M calcium chloride (CaCl$_2$). Store in aliquots at –20°C. Before use, warm to room temperature and vortex to achieve uniform mixing.

2.3. Concentration of Virus Stock

1. Ultracentrifuge tubes (Beckman). Place inverted in racks, wrap with aluminum foil, and autoclave at 120°C for 15 min.
2. Appropriate rotors and an ultracentrifuge.
3. Centrifugal ultrafiltration devices. 100,000-kDa molecular weight cutoff filters (e.g., Centricon Plus-20 or Centricon Plus-80 from Millipore or Macrosep centrifugal concentrators from Pall Gelman).
4. High-speed centrifuge.

2.4. Titration of Virus Stock

1. 6-well tissue culture plates.
2. PBS with calcium and magnesium.
3. Cell culture medium appropriate for cell line being used.
4. Polybrene (hexadimethrine bromide) or DEAE-dextran. Make 1 mg/mL stock in PBS. Filter using 0.45-μm filter and store at 4°C.
5. 4% paraformaldehyde. Add 4 g of paraformaldehyde to 85 mL of water. Warm to 60°C in a water bath. Add a few drops of 2 M NaOH. Return to water bath and mix contents every few minutes. When paraformaldehyde is completely dissolved, add 10 mL of 10× PBS. Adjust pH to 7.4–7.6. Filter through a 0.45 μm filter and store at 4°C. Use within 1 week of preparation.

2.5. Detection of Replication-Competent Retrovirus (RCR)

1. Materials for phlebotomy (tourniquet, sterile syringe with appropriate size needle, sterile gauze).

2. Heparin-containing tube (0.2 mL of 1000 U/mL) for collection of peripheral blood.
3. Histopaque.
4. Phytohemagglutinin (PHA; Sigma, cat. no. L8902).
5. Interleukin-2 (IL-2) (T-cell growth factor [TCGF]).
6. RPMI-10: RPMI-1640 with 10% heat-inactivated FBS, 100 U/mL penicillin, 100 μg/mL streptomycin, 0.05 μM 2-mercaptoethanol (2-ME).
7. HIV-p24 antigen detection kit (NEN-Dupont or Zeptometrix or other manufacturer).

3. Methods
3.1. Generation of Virus Stocks by Calcium Phosphate–Mediated Transient Transfection Method

293T cells can be transfected to high efficiency and are preferred for production of virus stocks by the calcium phosphate-mediated transient transfection method. The production of virus stocks requires the transfection of a helper plasmid(or plasmids) that expresses Gag, Gag-Pol, Tat, and Rev, a gene transfer vector encoding the transgene expression cassette, and an envelope glycoprotein expressing plasmid. Tat is required for expression of full-length gene transfer vector RNA, whereas Rev is required for the transport of this message into the cytoplasm for encapsidation by the Gag and Gag-Pol proteins into the virus particle. Packaging systems that do not require Tat or Rev protein for production of vector stocks have been described (*see* **Subheading 1.6.1.3.**) but are not in widespread use. It is important to use clean plasmid DNA for transfections (*see* **Note 2**). It is imperative that appropriate biosafety practices should be used while generating and using lentivirus vector stocks (*see* **Note 3**)

1. Seed 2.5×10^6 293T cells in each T25 flask in 4 mL of medium 1 day before transfection. This usually gives 60–80% confluency in 24 h.
2. The next day, make up DNAs in 450 μL of water in a sterile microcentrifuge tube (2-mL capacity). I typically use 3.75 μg of packaging plasmid, 0.2 μg of VSV-G-expressing plasmid, and 7.5 μg of gene transfer vector for transfection of cells in a T25 flask. Add 50 μL of 2.5 M CaCl$_2$. Mix well and add 500 μL of 2× HeBS, in drops, while bubbling air through the DNA/CaCl$_2$ mix. Bubbling can be carried out using a 1-mL sterile, cotton-plugged disposable plastic serologic pipet with the help of a pipet-aid. An EDP-Plus electronic pipet (Rainin) set in the "titrate" mode allows one to add the HeBS dropwise consistently between different samples. The procedure can be scaled up for larger size flasks (*see* **Note 5**). The amount of different DNAs to be used for transfection depends on many factors (*see* Note 6).
3. Add the DNA/CaCl$_2$ solution to the cells within 1–2 min of preparation.
4. Abundant fine granular precipitate is usually visible the next morning, both in the medium and around the cells. Replace with 4 mL fresh medium next morning.

Addition of chloroquine diphosphate or sodium butyrate can increase yields (*see* **Note 7**).

5. Harvest the medium at approx 60–72 h post-transfection by centrifugation at 1400g for 15 min at 4°C or by filtration through a 0.45-µm filter. One can get more mileage out of each transfection by harvesting and replacing medium every day for 3–4 days (*see* **Note 8**).

3.2. Concentration of Virus

The transient transfection method generally yields titers in the range of 10^5–10^6 infectious units/mL. To obtain higher titers, one can concentrate the virus-containing supernatant either by ultracentrifugation or by ultrafiltration using molecular weight cutoff centrifugation filters. Vectors pseudotyped with most envelope glycoproteins can be concentrated using ultrafiltration but this method suffers from some drawbacks (*see* **Note 9**). Ultracentrifugation is the usual method for concentration of VSV-G pseudotyped vectors.

3.2.1. Ultracentrifugation Method

1. Clarify virus-containing supernatant from transfected cell cultures by either centrifugation at 1400g for 15 min at 4°C or by filtration through a 0.45-µM filter.
2. Centrifuge at 100,000g for 2 h at 4°C using an SW-41 or SW28 (Beckman) rotor depending on the volume of supernatant being concentrated.
3. Carefully aspirate the supernatant and add 1/50th to 1/100th vol of DMEM/10% FBS. Vortex for 15–20 s and keep on ice. Repeat vortexing 3 or 4 times an hour for about 2 h.
4. Aliquot in 5–50-µL amounts and freeze at –80°C.

3.2.2. Concentration by Ultrafiltration Using Centrifugal Concentration Devices

See **Note 9**.

1. Clear supernatant to be concentrated by filtration through 0.45-µm filters or by centrifugation at 1,400g for 15 min at 4°C.
2. Add cleared supernatant to the centrifugation filter devices with 100-KDa molecular weight cutoff filters.
3. Spin at the manufacturer's recommended speed for a duration that yields the desired concentration.
4. Recover sample by simple decantation or by a reverse-spin procedure.
5. Aliquot and freeze down at –80°C.

3.3. Titration of Vectors

Titration to detect the number of infectious units/mL in the virus stock consists of testing serial dilutions of the stock on appropriate target cells. The method and time it takes to determine the titer depends on the particular marker gene encoded in the gene transfer vector. For example, vectors with GFP or luciferase marker can be harvested about 2–3 days postinfection. For vectors with drug-resistant markers (e.g., hygromycin phosphotransferase or neomycin phosphotransferase genes, it takes approximately 2 weeks for the drug-resistant colonies to form.

1. Subculture cells to be infected (target cells) into 6-well plates. Seed 2×10^5 cells/well 1 day prior to virus infection.
2. Make dilutions of virus stock in cell culture medium containing DEAE-dextran or Polybrene (8 µg/mL).
3. Rinse wells twice with PBS containing calcium and magnesium to remove any floating cells.
4. Add 1 mL of virus dilution from **step 2** to each well. Alternatively, if testing only small amounts of virus stock (e.g., from concentrates), add virus directly to well containing 1 mL of medium with an appropriate amount of DEAE-dextran or Polybrene.
4. Incubate plates at 37°C and 5% CO_2.
5. Next morning add 2 mL of complete medium to each well and continue incubation at 37°C and 5% CO_2 depending on the marker gene present in the vector.
6. For vectors encoding drug resistance markers (e.g., hygromycin or neomycin phosphotransferase genes), start selection 48 h postinfection by replacing medium with medium containing selection agent. Replace medium every 3–4 days with fresh medium containing selection agent.
8. Fix cells after 12–14 days with 0.5% crystal violet in 50% methanol when no live or adherent cells are seen in control (uninfected) wells and colonies are of sufficient size in test wells.
9. Enumerate colonies and estimate titer from amount or dilution used for infection.

3.3.1. Fixing Cells for GFP-Encoding Vectors

For GFP-encoding vectors, cells are typically harvested 48–72 h post-infection. Titer determination using GFP reporter may be complicated by pseudotransduction (*see* **Note 10**).

1. Rinse wells twice with PBS without calcium and magnesium (2 mL/wash).
2. Add 0.3 mL trypsin to each well.
3. Incubate for 5 min at 37°C.
4. Neutralize trypsin by adding 2 mL cell culture medium with serum.
5. Pipet up and down to break up clumps and transfer cells to a 15-mL conical tube.
6. Wash out wells with additional 2 mL of medium or PBS and add to previous harvest in the 15-mL tube.
7. Spin down cells at 200g at 4°C.
8. Wash cells once with 4 mL of PBS.

9. Remove PBS and dislodge the cells by tapping the tube or running the tube on top of a test tube rack.
10. Add 0.5–1 mL PBS with 4% paraformaldehyde (freshly prepared) dropwise while vortexing gently.
11. Estimate percent positive for GFP by flow cytometry. GFP released from transduced can adsorb nonspecifically onto the surface of nontransduced cells. However, the two populations can be usually distinguished by FACS analysis.

3.4. Assays for Replication Competent Retrovirus (RCR)

For obvious reasons, the chance of recreating a replication-competent HIV is impossible when using heterologous envelope constructs during transfection of 293T cells. There is, however, a possibility of formation of a novel RCR as a result of recombination among packaging plasmid, gene transfer vector, and a heterologous envelope-expressing construct. To detect RCR capable of growing in human peripheral blood mononuclear cells (PBMCs), an aliquot of concentrated virus stocks is used for infection of PBMCs (*see* **Subheading 3.4.1.**). The transduced cells are maintained in culture for at least 2 weeks and periodically monitored for the release of HIV-1 p24 in the culture supernatant.

3.4.1. Infection of PBMCs

1. Draw 15 mL of blood into a syringe containing 0.2 mL of heparin (1000 U/mL). Dilute to 30 mL with PBS.
2. Layer on 7.5 mL of Histopaque.
3. Centrifuge at 500g for 30 min at room temperature.
4. Remove top portion carefully with a 25-mL pipet.
5. Remove interphase containing PBMCs with a pipet and transfer to a new tube.
6. Dilute with 15 mL of PBS.
7. Spin down cells at 500g for 5 min.
8. Resuspend cells in 50 mL of PBS.
9. Spin cells down once again.
10. Repeat washes 2 more times (3 washes total). Count cells before final wash.
11. Resuspend cells to a concentration of 2×10^6 cells/mL in RPMI-10 with 0.05 μM 2-ME and and 2 μg of PHA/mL.
12. Incubate flask at 37°C, 5%CO_2 for 48 h.
13. Spin down all mononuclear cells from flask at 550g for 5 min.
14. Resuspend in RMPI-10 to give 4×10^6 cells/mL.
15. Distribute 0.5 mL of cell suspension in each well of a 24-well plate.
16. Add an aliquot of concentrated virus stock to each well.
17. Add Polybrene in PBS to each well to give 8 μg/mL final concentration.
18. Incubate cells at 37°C, 5% CO_2.
19. Next day, transfer mock and HIV-infected cultures to 15-mL screw-capped centrifuge tubes.
20. Spin down cells at 500g for 5 min.

21. Save supernatant for analysis of p24.
22. Resuspend cells in RPMI-10 with 5% IL-2 and return to original wells.
23. Repeat this procedure twice weekly for 14 days.
24. Test all saved supernatants for HIV-1 p24 antigen using a commercial kit per the manufacturer's instructions.

3.4.2. Marker Rescue/Marker Mobilization Assay

Gene transfer by retrovirus vectors to target cells is expected to be confined to one replicative cycle. That is, the supernatant from transduced cells should not contain vectors that can transfer the marker into naïve target cells—unless the initial stock used for production of virus stock contains RCR. The marker rescue/mobilization assays are devised to detect such RCRs. One can increase the sensitivity of detection of RCRs by using target cells that already harbor a vector that encodes a drug resistance marker.

1. Infect HeLa or other indicator cells with an aliquot of virus stock (approximately 10% of the volume used in the gene transfer experiment may be a suitable volume for detection of RCR).
2. The next day split the cells at a ratio of 1:50. Add Polybrene to 2 µg/mL in the medium to aid the spread of any RCR that may be present.
3. After 3–4 days, when 50% confluent, change to medium without Polybrene.
4. The following day harvest supernatant and clear by filtration through a 0.45-µm filter or by centrifugation at 1400g at 4°C for 15 min.
5. Concentrate the supernatant by using molecular weight cutoff filters or by ultracentrifugation (*see* **Subheading 3.2.1.**).
6. Infect fresh cells with unconcentrated and concentrated supernatant and assay for marker (either by starting drug selection or by assaying otherwise for the marker).

3.4.3. Assays for Partial Recombinants

The first step toward the creation of RCR probably involves a recombination event between the packaging plasmid and the gene transfer vector. Such a recombinant may be able to produce either Gag protein or one of the accessory or regulatory proteins such as Tat. A recombinant vector that contains a Tat coding sequence can be detected using cell lines that encode the β-galactosidase gene under control of the HIV-1 LTR (*see* **Note 11**). Gag production can be monitored by assaying supernatants of transduced cell cultures for HIV-1 p24 using commercial kits.

4. Notes

1. Maintain 293T cells in DMEM/10% FBS. Split cells at a ratio of approximately 1:10 or 1:20 every 3–4 days. For a T225 flask, seeding at 2.5×10^6 cells should yield a near confluent monolayer in about 4 days. The cells should not be allowed

to overgrow. The cells are very sensitive to drying and readily detach from the surface even during rinsing procedures prior to trypsinization. Freeze down cells when in a low passage number in growth medium containing 10% dimethyl sulfoxide at a concentration of 5×10^6 cells per vial. After about 20 passages, recover fresh cells from the frozen stock.

2. Plasmids used for transfection should be very clean. I use Qiagen columns for preparation of DNAs for transfection.

3. Make fresh 2× HeBS every 6–12 months and freeze in aliquots. If no precipitate is seen the day following the transfection, it is likely that the pH is not optimal. Make a fresh batch of buffer. The easiest way to determine transfection efficiency is by using a GFP-expressing reporter plasmid. Transfection efficiency should be >50% as judged by fluorescence microscopy.

4. Biosafety concerns. It is highly recommended that Centers for Disease Control (CDC)/National Institutes for Health (NIH) and institutional biosafety guidelines be followed for preparation and use of retroviral and lentivirus vectors. I currently use near BSL-3 practices in a BSL-2 laboratory. This involves, among other required safety practices, the disinfection of all materials that come in contact with virus with 10% bleach (made fresh daily) inside the biologic safety cabinet and bagging the potentially contaminated materials within the cabinet before bringing the material out for autoclaving and disposal. Details of how to set up a BSL-2 and BSL–3 laboratory should be available with your Institutional Biosafety Committee and are also available from the CDC and NIH (HHS Publication No. [CDC] 93-8395).

5. I have successfully scaled up the protocol for transfection in T225 flasks without a diminution of virus yield. For these larger volumes of DNA/CaCl$_2$ and HeBS solutions, use appropriately larger size tubes for making the calcium phosphate-DNA precipitates (e.g., 50-mL sterile screw-capped tubes).

6. The amount of various constructs to be used for transfection will obviously depend on expression levels obtained from each construct. This will vary from one construct to another depending on vector backbone and promoter/enhancer elements. Thus, for any new construct, the optimal amount to be used for transfection must be determined by a titration experiment to vary the different DNAs systematically.

7. Modest increases in virus yields can be achieved by using chloroquine diphosphate and/or sodium butyrate. For treatment with chloroquine, prior to transfection, remove medium and gently add fresh growth medium containing 25 μM of chloroquine. Replace medium after 10 h (7–12 h). For treatment with sodium butyrate, replace medium 1 day after transfection (approx 10 h later) with medium containing 10 mM sodium butyrate. Leave it on for 12 h and then change to medium without sodium butyrate. These modifications, either adding sodium butyrate alone or both chloroquine and sodium butyrate in the medium, in our hands, yielded about a twofold increase in virus titer.

8. Virus-containing supernatant can be harvested daily for 3 or 4 days starting 24 h after the first medium change. Clear each harvest by centrifugation or filtration through 0.45-μM filters and store on ice in a covered Styrofoam container at 4°C

until all the harvests have accumulated. The virus in the pooled supernatant can be concentrated by ultracentrifugation or by ultrafiltration.

9. Concentration of virus particles using molecular weight cutoff filters frequently concentrates inhibitors that reduce the titer of the preparation. The inhibitors are believed to consist of sulfated proteoglycans *(98,99)*. Attempts have been made to remove these inhibitors using enzymes such as chondroitinase ABC, resulting in modest improvements in titer *(98,100)*.

10. Pseudotransduction is a type of protein or DNA delivery in the absence of *bona fide* infection *(101–103)*. The marker protein or plasmid DNA is incorporated into the virion. Upon binding and fusion of the virus particles with target cell plasma membrane, the protein or DNA is then delivered to the cytoplasm. If sufficient quantity of protein or DNA is delivered, one may wrongly conclude that the target cells were expressing the protein as a result of authentic retrovirus infection process. Pseudotransduction has been observed with highly concentrated preparations of VSV-G pseudotyped vectors. To distinguish between authentic and pseudotransduction, one can use azidothymidine (AZT), an inhibitor of HIV-1 RT *(32)*. The treatment of target cells with AZT (10 μM/mL) will allow the virus particles to bind and fuse with the plasma membrane but should interfere with the reverse transcription of incoming virus RNA and thereby abort the infection process. Under these circumstances, pseudotransduction by protein or DNA delivery is still possible, but no true transduction can occur. As an alternative approach, investigators have used packaging plasmids that have a mutation in the viral integrase *(102)* to prevent integration of the reverse-transcribed genome into the host chromosome. In this situation, the GFP detected in the target cells may be attributable to transcription from unintegrated vector genome and not necessarily all from protein delivery. It has been observed that expression of GFP caused by pseudotransduction gradually decreases with time and is negligible after 5 days post infection. Thus it may be prudent to measure or enumerate GFP-expressing cells in the target cell population 5 or more days after infection.

11. To detect partial recombinants that express Tat, many investigators have used target cells containing the ß-galactosidase reporter gene under control of the HIV-1 LTR. HeLa-CD4-LTR-ß-Gal and MAGI-CCR5 are two such cell lines available from the NIH-AIDS Research and Reference Reagent Program. A detailed protocol for staining cells for detection of ß-galactosidase expression is provided with the cell lines.

References

1. Kafri, T., Blomer, U., Peterson, D. A., Gage, F. H., and Verma, I. M. (1997) Sustained expression of genes delivered directly into liver and muscle by lentiviral vectors. *Nat. Genet.* **17,** 314–317.
2. Miyoshi, H., Smith, K. A., Mosier, D. E., Verma, I. M., and Torbett, B. E. (1999) Transduction of human CD34+ cells that mediate long-term engraftment of NOD/SCID mice by HIV vectors. *Science* **283,** 682–686.

3. Miyoshi, H., Takahashi, M., Gage, F. H., and Verma, I. M. (1997) Stable and efficient gene transfer into the retina using an HIV-based lentiviral vector. *Proc. Natl. Acad. Sci. USA* **94,** 10319–10323.
4. Sutton, R. E., Wu, H. T., Rigg, R., Bohnlein, E., and Brown, P. O. (1998) Human immunodeficiency virus type 1 vectors efficiently transduce human hematopoietic stem cells. *J. Virol.* **72,** 5781–5788.
5. Mochizuki, H., Schwartz, J. P., Tanaka, K., Brady, R. O., and Reiser, J. (1998) High-titer human immunodeficiency virus type 1-based vector systems for gene delivery into nondividing cells. *J. Virol.* **72,** 8873–8883.
6. Zufferey, R., Nagy, D., Mandel, R. J., Naldini, L., and Trono, D. (1997) Multiply attenuated lentiviral vector achieves efficient gene delivery in vivo. *Nat Biotechnol* **15,** 871–875.
6a. Coffin, J. M., Hughes, S. H. and Varmus, H. E. (ed) (1997). *Retroviruses.* Cold Spring Harbor Laboratory Press, Plainview, NY.
6b. Khan, M. A., Aberham, C., Kao, S., et al. (2001). Human immunodeficiency virus type 1 Vif protein is packaged into the nucleoprotein complex through an interaction with viral genomic RNA. *J. Virol.* **75,** 7252–7265.
7. Zennou, V., Petit, C., Guetard, D., et al. (2000) HIV-1 genome nuclear import is mediated by a central DNA flap. *Cell* **101,** 173–185.
8. Charneau, P., Mirambeau, G., Roux, P., et al. (1994) HIV-1 reverse transcription. A termination step at the center of the genome. *J. Mol. Biol.* **241,** 651–662.
9. Follenzi, A., Ailles, L. E., Bakovic, S., Geuna, M., and Naldini, L. (2000) Gene transfer by lentiviral vectors is limited by nuclear translocation and rescued by HIV-1 pol sequences. *Nat. Genet.* **25,** 217–222.
10. Bukrinsky, M.I., Haggerty, S., Dempsey, M.P., et al. (1993) A nuclear localization signal within HIV-1 matrix protein that governs infection of non-dividing cells. *Nature* **365,** 666–669.
11. von Schwedler, U., Kornbluth, R. S., and Trono, D. (1994) The nuclear localization signal of the matrix protein of human immunodeficiency virus type 1 allows the establishment of infection in macrophages and quiescent T lymphocytes. *Proc. Natl. Acad. Sci. USA* **91,** 6992–6996.
12. Gallay, P., Swingler, S., Song, J., Bushman, F., and Trono, D. (1995) HIV nuclear import is governed by the phosphotyrosine-mediated binding of matrix to the core domain of integrase. *Cell* **83,** 569–576.
13. Gallay, P., Swingler, S., Aiken, C., and Trono, D. (1995) HIV-1 infection of non-dividing cells: C-terminal tyrosine phosphorylation of the viral matrix protein is a key regulator. *Cell* **80,** 379–388.
14. Heinzinger, N. K., Bukinsky, M. I., Haggerty, S. A., et al. (1994) The Vpr protein of human immunodeficiency virus type 1 influences nuclear localization of viral nucleic acids in nondividing host cells. *Proc. Natl. Acad. Sci. USA* **91,** 7311–7315.
15. Freed, E. O., Englund, G., and Martin, M. A. (1995) Role of the basic domain of human immunodeficiency virus type 1 matrix in macrophage infection. *J. Virol.* **69,** 3949–3954.

16. Freed, E. O. and Martin, M. A. (1994) HIV-1 infection of non-dividing cells (letter; comment). *Nature* **369,** 107–108.
17. Freed, E. O., Englund, G., Maldarelli, F., and Martin, M. A. (1997) Phosphorylation of residue 131 of HIV-1 matrix is not required for macrophage infection. *Cell* **88,** 171–173; discussion 173–174.
18. Laspia, M. F., Rice, A. P., and Mathews, M. B. (1989) HIV-1 Tat protein increases transcriptional initiation and stabilizes elongation. *Cell* **59,** 283–292.
19. Feinberg, M. B., Baltimore, D., and Frankel, A. D. (1991) The role of Tat in the human immunodeficiency virus life cycle indicates a primary effect on transcriptional elongation. *Proc. Natl. Acad. Sci. USA* **88,** 4045–4049.
20. Jones, K. A. (1997) Taking a new TAK on tat transactivation (comment). *Genes Dev.* **11,** 2593–2599.
21. Benkirane, M., Chun, R. F., Xiao, H., et al. (1998) Activation of integrated provirus requires histone acetyltransferase. p300 and P/CAF are coactivators for HIV-1 Tat. *J. Biol. Chem.* **273,** 24898–24905.
22. Marzio, G., Tyagi, M., Gutierrez, M. I., and Giacca, M. (1998) HIV-1 tat transactivator recruits p300 and CREB-binding protein histone acetyltransferases to the viral promoter. *Proc. Natl. Acad. Sci. USA* **95,** 13519–13524.
23. Yamamoto, T. and Horikoshi, M. (1997) Novel substrate specificity of the histone acetyltransferase activity of HIV-1-Tat interactive protein Tip60. *J. Biol. Chem.* **272,** 30595–30598.
24. Srinivasakumar, N., Chazal, N., Helga-Maria, C., et al. (1997) The effect of viral regulatory protein expression on gene delivery by human immunodeficiency virus type 1 vectors produced in stable packaging cell lines. *J. Virol.* **71,** 5841–5848.
25. Dull, T., Zufferey, R., Kelly, M., et al. (1998) A third-generation lentivirus vector with a conditional packaging system. *J. Virol.* **72,** 8463–8471.
26. Kim, V. N., Mitrophanous, K., Kingsman, S. M., and Kingsman, A. J. (1998) Minimal requirement for a lentivirus vector based on human immunodeficiency virus type 1. *J. Virol.* **72,** 811–816.
27. Hammarskjold, M. L., Heimer, J., Hammarskjold, B., et al. (1989) Regulation of human immunodeficiency virus env expression by the rev gene product. *J. Virol.* **63,** 1959–1966.
28. Malim, M. H., Hauber, J., Le, S. Y., Maizel, J. V., and Cullen, B. R. (1989) The HIV-1 rev trans-activator acts through a structured target sequence to activate nuclear export of unspliced viral mRNA. *Nature* **338,** 254–257.
29. Fornerod, M., Ohno, M., Yoshida, M., and Mattaj, I. W. (1997) CRM1 is an export receptor for leucine-rich nuclear export signals (see comments). *Cell* **90,** 1051–1060.
30. Smith, A. J., Cho, M. I., Hammarskjold, M. L., and Rekosh, D. (1990) Human immunodeficiency virus type 1 Pr55gag and Pr160gag-pol expressed from a simian virus 40 late replacement vector are efficiently processed and assembled into viruslike particles. *J. Virol.* **64,** 2743–2750.
31. Cui, Y., Iwakuma, T., and Chang, L. J. (1999) Contributions of viral splice sites and cis-regulatory elements to lentivirus vector function. *J. Virol.* **73,** 6171–6176.

32. Srinivasakumar, N. and Schuening, F. G. (2000) Novel Tat-encoding bicistronic human immunodeficiency virus type 1-based gene-transfer vectors for high-level transgene expression. *J. Virol.* **74,** 6659–6668.

33. Srinivasakumar, N. and Schuening, F. (1999) A lentivirus packaging system based on alternative RNA transport mechanisms to express helper and gene transfer vector RNAs and its use to study the requirement of accessory proteins for particle formation and gene delivery. *J. Virol.* **73,** 9589–9598.

34. Anderson, S. J., Lenburg, M., Landau, N. R., and Garcia, J. V. (1994) The cytoplasmic domain of CD4 is sufficient for its down-regulation from the cell surface by human immunodeficiency virus type 1 Nef [published erratum appears in *J. Virol.* 1994; 68: 4705]. *J. Virol.* **68,** 3092–3101.

35. Aiken, C., Konner, J., Landau, N. R., Lenburg, M. E., and Trono, D. (1994) Nef induces CD4 endocytosis: requirement for a critical dileucine motif in the membrane-proximal CD4 cytoplasmic domain. *Cell* **76,** 853–864.

36. Garcia, J. V. and Miller, A. D. (1992) Downregulation of cell surface CD4 by nef. *Res. Virol.* **143,** 52–55.

37. Garcia, J. V. and Miller, A. D. (1991) Serine phosphorylation-independent downregulation of cell-surface CD4 by nef. *Nature* **350,** 508–511.

38. Hanna, Z., Kay, D. G., Rebai, N., et al. (1998) Nef harbors a major determinant of pathogenicity for an AIDS-like disease induced by HIV-1 in transgenic mice. *Cell* **95,** 163–175.

39. Schwartz, O., Marechal, V., Danos, O., and Heard, J. M. (1995) Human immunodeficiency virus type 1 Nef increases the efficiency of reverse transcription in the infected cell. *J. Virol.* **69,** 4053–4059.

40. Aiken, C. and Trono, D. (1995) Nef stimulates human immunodeficiency virus type 1 proviral DNA synthesis. *J. Virol.* **69,** 5048–5056.

41. Luo, T., Douglas, J. L., Livingston, R. L., and Garcia, J. V. (1998) Infectivity enhancement by HIV-1 Nef is dependent on the pathway of virus entry: implications for HIV-based gene transfer systems. *Virology* **241,** 224–233.

42. Aiken, C. (1997) Pseudotyping human immunodeficiency virus type 1 (HIV-1) by the glycoprotein of vesicular stomatitis virus targets HIV-1 entry to an endocytic pathway and suppresses both the requirement for Nef and the sensitivity to cyclosporin A. *J. Virol.* **71,** 5871–5877.

43. Rogel, M. E., Wu, L. I., and Emerman, M. (1995) The human immunodeficiency virus type 1 vpr gene prevents cell proliferation during chronic infection. *J. Virol.* **69,** 882–888.

44. Bartz, S. R., Rogel, M. E., and Emerman, M. (1996) Human immunodeficiency virus type 1 cell cycle control: Vpr is cytostatic and mediates G2 accumulation by a mechanism which differs from DNA damage checkpoint control. *J. Virol.* **70,** 2324–2331.

45. Re, F., Braaten, D., Franke, E. K., and Luban, J. (1995) Human immunodeficiency virus type 1 Vpr arrests the cell cycle in G2 by inhibiting the activation of p34cdc2-cyclin B. *J. Virol.* **69,** 6859–6864.

46. Goh, W. C., Rogel, M. E., Kinsey, C. M., et al. (1998) HIV-1 Vpr increases viral

expression by manipulation of the cell cycle: a mechanism for selection of Vpr in vivo. *Nat. Med.* **4,** 65–71.

47. Popov, S., Rexach, M., Zybarth, G., et al. (1998) Viral protein R regulates nuclear import of the HIV-1 pre-integration complex. *EMBO J.* **17,** 909–917.

48. Popov, S., Rexach, M., Ratner, L., Blobel, G., and Bukrinsky, M. (1998) Viral protein R regulates docking of the HIV-1 preintegration complex to the nuclear pore complex. *J. Biol. Chem.* **273,** 13347–13352.

49. Vodicka, M. A., Koepp, D. M., Silver, P. A., and Emerman, M. (1998) HIV-1 Vpr interacts with the nuclear transport pathway to promote macrophage infection. *Genes Dev.* **12,** 175–185.

50. Muller, B., Tessmer, U., Schubert, U., and Krausslich, H. G. (2000) Human immunodeficiency virus type 1 vpr protein is incorporated into the virion in significantly smaller amounts than gag and is phosphorylated in infected cells (in process citation). *J. Virol.* **74,** 9727–9731.

51. Paxton, W., Connor, R. I., and Landau, N. R. (1993) Incorporation of Vpr into human immunodeficiency virus type 1 virions: requirement for the p6 region of gag and mutational analysis. *J. Virol.* **67,** 7229–7237.

52. Selig, L., Pages, J., Tanchou, V., et al. (1999) Interaction with the p6 domain of the gag precursor mediates incorporation into virions of vpr and vpx proteins from primate lentiviruses (in process citation). *J. Virol.* **73,** 592–600.

53. Stewart, S. A., Poon, B., Jowett, J. B., and Chen, I. S. (1997) Human immunodeficiency virus type 1 Vpr induces apoptosis following cell cycle arrest. *J. Virol.* **71,** 5579–5592.

54. Stewart, S. A., Poon, B., Jowett, J. B., Xie, Y., and Chen, I. S. (1999) Lentiviral delivery of HIV-1 Vpr protein induces apoptosis in transformed cells. *Proc. Natl Acad. Sci. USA* **96,** 12039–12043.

55. Stewart, S. A., Poon, B., Song, J. Y., and Chen, I. S. (2000) Human immunodeficiency virus type 1 vpr induces apoptosis through caspase activation. *J. Virol.* **74,** 3105–3111.

56. Madani, N. and Kabat, D. (2000) Cellular and viral specificities of human immunodeficiency virus type 1 vif protein. *J. Virol.* **74,** 5982–5987.

57. Simon, J. H., Miller, D. L., Fouchier, R. A., et al. (1998) The regulation of primate immunodeficiency virus infectivity by Vif is cell species restricted: a role for Vif in determining virus host range and cross-species transmission. *EMBO J.* **17,** 1259–1267.

58. von Schwedler, U., Song, J., Aiken, C., and Trono, D. (1993) Vif is crucial for human immunodeficiency virus type 1 proviral DNA synthesis in infected cells. *J. Virol.* **67,** 4945–4955.

59. Schwartz, M. D., Geraghty, R. J., and Panganiban, A. T. (1996) HIV-1 particle release mediated by Vpu is distinct from that mediated by p6. *Virology* **224,** 302–309.

60. Ernst, R. K., Bray, M., Rekosh, D., and Hammarskjold, M. L. (1997) Secondary structure and mutational analysis of the Mason-Pfizer monkey virus RNA constitutive transport element. *RNA* **3,** 210–222.

61. Ernst, R. K., Bray, M., Rekosh, D., and Hammarskjold, M. L. (1997) A structured

retroviral RNA element that mediates nucleocytoplasmic export of intron-containing RNA. *Mol. Cell Biol.* **17,** 135–144.

62. Zolotukhin, A. S., Valentin, A., Pavlakis, G. N., and Felber, B. K. (1994) Continuous propagation of RRE(-) and Rev(-)RRE(-) human immunodeficiency virus type 1 molecular clones containing a cis-acting element of simian retrovirus type 1 in human peripheral blood lymphocytes. *J. Virol.* **68,** 7944–7952.

63. Bray, M., Prasad, S., Dubay, J. W., et al. (1994) A small element from the Mason-Pfizer monkey virus genome makes human immunodeficiency virus type 1 expression and replication Rev- independent. *Proc. Natl. Acad. Sci. USA* **91,** 1256–1260.

64. Mautino, M. R., Keiser, N., and Morgan, R. A. (2000) Improved titers of HIV-based lentiviral vectors using the SRV-1 constitutive transport element. *Gene Ther,* **7,** 1421–1424.

65. Mautino, M. R., Ramsey, W. J., Reiser, J., and Morgan, R. A. (2000) Modified human immunodeficiency virus-based lentiviral vectors display decreased sensitivity to trans-dominant Rev. *Hum. Gene Ther.* **11,** 895–908.

66. Kaul, M., Yu, H., Ron, Y., and Dougherty, J. P. (1998) Regulated lentiviral packaging cell line devoid of most viral cis-acting sequences. *Virology* **249,** 167–174.

67. Naldini, L., Blomer, U., Gallay, P., et al. (1996) In vivo gene delivery and stable transduction of nondividing cells by a lentiviral vector. *Science* **272,** 263–267.

68. Gasmi, M., Glynn, J., Jin, M. J., et al. (1999) Requirements for efficient production and transduction of human immunodeficiency virus type 1-based vectors. *J. Virol.* **73,** 1828-34.

69. Wu, X., Wakefield, J. K., Liu, H., et al. (2000) Development of a novel trans-lentiviral vector that affords predictable safety. *Mol. Ther.* **2,** 47–55.

70. Wodrich, H., Schambach, A., and Krausslich, H. G. (2000) Multiple copies of the Mason-Pfizer monkey virus constitutive RNA transport element lead to enhanced HIV-1 Gag expression in a context-dependent manner. *Nucleic Acids Res.* **28,** 901–910.

71. Rizvi, T. A., Schmidt, R. D., and Lew, K. A. (1997) Mason-Pfizer monkey virus (MPMV) constitutive transport element (CTE) functions in a position-dependent manner. *Virology* **236,** 118–129.

72. Callahan, M. A., Handley, M. A., Lee, Y. H., et al. (1998) Functional interaction of human immunodeficiency virus type 1 Vpu and Gag with a novel member of the tetratricopeptide repeat protein family. *J. Virol.* **72,** 8461.

73. Clever, J., Sassetti, C., and Parslow, T. G. (1995) RNA secondary structure and binding sites for gag gene products in the 5' packaging signal of human immunodeficiency virus type 1. *J. Virol.* **69,** 2101–2109.

74. McBride, M. S. and Panganiban, A. T. (1996) The human immunodeficiency virus type 1 encapsidation site is a multipartite RNA element composed of functional hairpin structures [published erratum appears in *J. Virol.* 1997; 71: 858]. *J. Virol.* **70,** 2963–2973.

75. Parolin, P., Taddeo, B., Palu, G., and Sodroski, J. (1996) Use of cis- and trans-acting regulatory sequences to improve expression of human immunodeficiency virus vectors in human lymphocytes. *Virology* **222,** 415–422.

76. Parolin, C., Dorfman, T., Palu, G., Gottlinger, H., and Sodroski, J. (1994) Analysis in human immunodeficiency virus type 1 vectors of cis-acting sequences that affect gene transfer into human lymphocytes. *J. Virol.* **68,** 3888–3895.
77. Luban, J. and Goff, S. P. (1994) Mutational analysis of cis-acting packaging signals in human immunodeficiency virus type 1 RNA. *J. Virol.* **68,** 3784–3793.
78. Harrich, D., Ulich, C., Garcia-Martinez, L. F., and Gaynor, R. B. (1997) Tat is required for efficient HIV-1 reverse transcription. *EMBO J.* **16,** 1224–1235.
79. Zufferey, R., Dull, T., Mandel, R. J., et al. (1998) Self-inactivating lentivirus vector for safe and efficient in vivo gene delivery. *J. Virol.* **72,** 9873–9880.
80. Iwakuma, T., Cui, Y., and Chang, L. J. (1999) Self-inactivating lentiviral vectors with U3 and U5 modifications. *Virology* **261,** 120–132.
81. Donello, J. E., Beeche, A. A., Smith, G. J. 3rd, Lucero, G. R., and Hope, T. J. (1996) The hepatitis B virus posttranscriptional regulatory element is composed of two subelements. *J. Virol.* **70,** 4345–4351.
82. Huang, Z. M. and Yen, T. S. (1994) Hepatitis B virus RNA element that facilitates accumulation of surface gene transcripts in the cytoplasm. *J. Virol.* **68,** 3193–3199.
83. Huang, J. and Liang, T. J. (1993) A novel hepatitis B virus (HBV) genetic element with Rev response element-like properties that is essential for expression of HBV gene products. *Mol. Cell. Biol.* **13,** 7476–7486.
84. Zufferey, R., Donello, J. E., Trono, D., and Hope, T. J. (1999) Woodchuck hepatitis virus posttranscriptional regulatory element enhances expression of transgenes delivered by retroviral vectors. *J. Virol.* **73,** 2886–2892.
85. Kotsopoulou, E., Kim, V. N., Kingsman, A. J., Kingsman, S. M., and Mitrophanous, K. A. (2000) A Rev-independent human immunodeficiency virus type 1 (HIV-1)-based vector that exploits a codon-optimized HIV-1 gag-pol gene. *J. Virol.* **74,** 4839–4852.
86. White, S. M., Renda, M., Nam, N. Y., et al. (1999) Lentivirus vectors using human and simian immunodeficiency virus elements. *J. Virol.* **73,** 2832–2840.
87. Rizvi, T. A. and Panganiban, A. T. (1993) Simian immunodeficiency virus RNA is efficiently encapsidated by human immunodeficiency virus type 1 particles. *J. Virol.* **67,** 2681–2688.
88. Miller, D. G., Adam, M. A., and Miller, A. D. (1990) Gene transfer by retrovirus vectors occurs only in cells that are actively replicating at the time of infection [published erratum appears in Mol. Cell. Biol. 1992; 12:433]. *Mol. Cell. Biol.* **10,** 4239–4242.
89. Lewis, P., Hensel, M., and Emerman, M. (1992) Human immunodeficiency virus infection of cells arrested in the cell cycle. *EMBO J.* **11,** 3053–3058.
90. Lewis, P. F. and Emerman, M. (1994) Passage through mitosis is required for oncoretroviruses but not for the human immunodeficiency virus. *J. Virol.* **68,** 510–516.
91. Goulaouic, H. and Chow, S. A. (1996) Directed integration of viral DNA mediated by fusion proteins consisting of human immunodeficiency virus type 1 integrase and *Escherichia coli* LexA protein. *J. Virol.* **70,** 37–46.

92. Bushman, F. D. (1994) Tethering human immunodeficiency virus 1 integrase to a DNA site directs integration to nearby sequences. *Proc. Natl. Acad. Sc.i USA* **91,** 9233–9237.

93. Krausslich, H. G., Ochsenbauer, C., Traenckner, A. M., et al. (1993) Analysis of protein expression and virus-like particle formation in mammalian cell lines stably expressing HIV-1 gag and env gene products with or without active HIV proteinase. *Virology* **192,** 605–617.

94. Shoeman, R. L., Honer, B., Stoller, T. J., et al. (1990) Human immunodeficiency virus type 1 protease cleaves the intermediate filament proteins vimentin, desmin, and glial fibrillary acidic protein. *Proc. Nat.l Acad. Sci. USA* **87,** 6336–6340.

95. Corbeau, P., Kraus, G., and Wong-Staal, F. (1996) Efficient gene transfer by a human immunodeficiency virus type 1 (HIV-1)-derived vector utilizing a stable HIV packaging cell line. *Proc. Natl. Acad. Sci. USA* **93,** 14070–14075.

96. Corbeau, P., Kraus, G., and Wong-Staal, F. (1998) Transduction of human macrophages using a stable HIV-1/HIV-2-derived gene delivery system. *Gene Ther.* **5,** 99–104.

97. Kafri, T., van Praag, H., Ouyang, L., Gage, F. H., and Verma, I. M. (1999) A packaging cell line for lentivirus vectors. *J. Virol.* **73,** 576–584.

98. Le Doux, J. M., Morgan, J. R., Snow, R. G., and Yarmush, M. L. (1996) Proteoglycans secreted by packaging cell lines inhibit retrovirus infection. *J. Virol.* **70,** 6468–6473.

99. Le Doux, J. M., Morgan, J. R., and Yarmush, M. L. (1999) Differential inhibition of retrovirus transduction by proteoglycans and free glycosaminoglycans. *Biotechnol. Prog.* **15,** 397–406.

100. Le Doux, J. M., Morgan, J. R., and Yarmush, M. L. (1998) Removal of proteoglycans increases efficiency of retroviral gene transfer. *Biotechnol. Bioeng.* **58,** 23–34.

101. Liu, M. L., Winther, B. L., and Kay, M. A. (1996) Pseudotransduction of hepatocytes by using concentrated pseudotyped vesicular stomatitis virus G glycoprotein (VSV-G)-Moloney murine leukemia virus-derived retrovirus vectors: comparison of VSV-G and amphotropic vectors for hepatic gene transfer. *J. Virol.* **70,** 2497–2502.

102. Case, S. S., Price, M. A., Jordan, C. T., et al. (1999) Stable transduction of quiescent CD34(+)CD38(-) human hematopoietic cells by HIV-1-based lentiviral vectors. *Proc. Natl. Acad. Sci. USA* **96,** 2988–2993.

103. Gallardo, H. F., Tan, C., Ory, D., and Sadelain, M. (1997) Recombinant retroviruses pseudotyped with the vesicular stomatitis virus G glycoprotein mediate both stable gene transfer and pseudotransduction in human peripheral blood lymphocytes. *Blood* **90,** 952–957.

21

Lentiviral Vectors

Preparation and Use

Lung-Ji Chang and Anne-Kathrin Zaiss

1. Introduction

Lentiviruses, such as human immunodeficiency virus (HIV), feline immunodeficiency virus (FIV), and equine infectious anemia virus (EIAV), are members of the Retroviridae, viruses with enveloped capsids and a plus-stranded RNA genome. Like all retroviruses, the RNA genome of lentivirus is converted to DNA by reverse transcription after infection and is subsequently stably integrated into the host cell genome. However, unlike other retroviruses, which mainly infect dividing cells, lentiviruses can infect both dividing and nondividing cells *(1–3)*. Gene therapy viral vectors are produced by cotransfection of plasmids encoding viral envelopes, capsids, enzymes, and viral RNA. The specific tissue tropism of the lentiviral envelope restricts its use as a gene transfer vector. This limitation has been overcome by pseudotyping the viral envelope with vesicular stomatitis virus G (VSV-G) glycoprotein *(4,5)*. The aim of this chapter is to describe the general procedures developed for the preparation and use of an HIV type 1 (HIV-1)-based lentiviral vector system. Detailed descriptions of lentiviral molecular biology can be found in several recent reviews *(1,6,7)*.

1.1. Lentiviral Genome Structure and Vector Elements

Lentiviruses are classified as complex retroviruses because of the multiple regulatory steps involved in their life cycle. The lentivirus genome is a diploid, plus-stranded RNA that is reverse transcribed into DNA (called provirus) and

From: *Methods in Molecular Medicine, Vol. 69, Gene Therapy Protocols, 2nd Ed.*
Edited by: J. R. Morgan © Humana Press Inc., Totowa, NJ

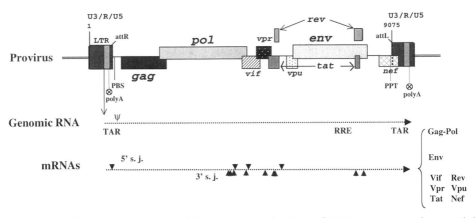

Fig. 1. Genomic structure of HIV-1. Organization of HIV-1 genes and essential genetic elements in the proviral DNA and viral RNA. The transcription of viral RNA starts from the 5' terminus of R. Viral proteins except for Gag-Pol are expressed from alternatively spliced viral RNAs using a list of 5' and 3' splice junctions (s.j.) as depicted by solid arrowheads.

integrated in the host cell chromosomes (**Fig. 1**). The proviral genome is comprised of three conserved genes termed *gag* (group-specific antigen), *pol* (polymerase, reverse transcriptase), and *env* (envelope), which are flanked by elements called long terminal repeats (LTRs). LTRs contain the left and right integration attachment sites (att) and are required for integration into the host genome. The LTRs also serve as enhancer-promoter sequences, controlling expression of the viral genes. The LTRs (U3-R-U5) consist of untranslated regions (U), which contain the transcription start site, a polyadenylation signal, and repeated sequences (R) important for reverse transcription. Other critical elements in the viral genome include the packaging sequence (psi, ψ), which interacts with capsid proteins during packaging and allows the viral RNA to be distinguished from other RNAs in the host cells, and the primer binding site (PBS) and polypurine tract (PPT) adjacent to the 5' and 3' LTR, respectively, both of which are essential for reverse transcription.

Once integrated, the proviral genome will synthesize a full-length viral transcript using host RNA polymerase. The full-length viral transcript encodes the viral Gag-Pol polyprotein that provides the capsid and enzymes for reverse transcription (reverse transcriptase) and integration (integrase). The synthesis and nuclear export of the full-length viral RNA require the interaction of viral regulatory proteins with different *cis*-acting sequences in the viral RNA. All lentiviral regulatory and accessory proteins are encoded by variously spliced viral mRNAs that are generated from this full-length viral RNA using different 5' and 3' splice site signals (**Fig. 1**). The two essential regulatory proteins, Tat

pHP (packaging helper construct)

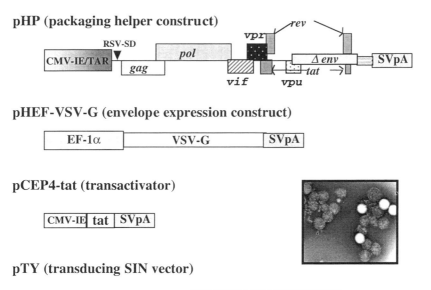

pHEF-VSV-G (envelope expression construct)

pCEP4-tat (transactivator)

pTY (transducing SIN vector)

Fig. 2. Plasmid constructs for the generation of HIV-1-derived lentiviral vectors. The HP/TV lentiviral vector system uses three basic vector constructs, pHP, pHEF-VSVG, and pTY in DNA cotransfections. An additional eukaryotic expression plasmid, pCEP4-tat, can be included that increases vector titer two to fivefold *(14)*. pHP expresses all viral proteins except for Env and Nef. pTY is a self-inactivating (SIN) transducing vector derived from pTV carrying necessary lentiviral packaging signals (ψ). Insert: Electron microscopic photograph of concentrated HP/TV vectors mixed with 142 nm latex beads (white particles).

and Rev, are required for efficient Gag-Pol synthesis. Tat interacts with a 59-nt RNA structure, TAR, at the 5' end of the viral RNA and promotes transcriptional elongation. Rev facilitates nuclear export of unspliced and singly spliced viral RNA by interacting with Rev responsive element (RRE) and other *cis*-regulatory sequences *(1)*.

Although the accessory genes of HIV-1 have been shown to be dispensable for lentiviral vector function *(8,9)*, their possible roles in transduction of different types of cells have not been extensively studied. For example, there is evidence that the *vpr* gene is needed for efficient transduction of resting T-cells *(10)*.

1.2. A Simple Lentiviral Vector System: Three Plasmids Plus a Human Cell Line

The vector system described here is based on HIV-1 (*see* Chapters 19 and 20). However, lentiviral vector systems based on simian immunodeficiency virus (SIV), FIV, and EIAV have also been established (*11–13*, and Chapters 23 and 24). The production of HIV-1-derived lentiviral vectors requires human cell lines because many of the viral regulatory functions are human cell dependent. Lentiviral vectors can be produced by cotransfection of the three different plasmid constructs coding for Gag-Pol, VSV-G envelope, and the vector genome (**Fig. 2**). An ideal vector system will have these components established separately without overlapping sequences. The helper construct of the lentiviral vector system, pHP, contains a chimeric promoter that drives expression of Gag-Pol. pHP has deletions in the 5' untranslated leader, *env*, *nef*, and the 3' LTR (*14*). Besides Gag-Pol, pHP also produces Tat, Rev, and other HIV-1 accessory proteins (Vif, Vpr, and Vpu) through RNA splicing. The complicated gene regulation scheme of lentivirus makes it difficult to express *gag-pol* independent of Tat, Rev, and *cis*-regulatory elements. Although Tat- and Rev-dependent lentiviral vectors have been generated, extensive testing of these vectors has not been reported. Several *cis*-acting sequences are required for *gag-pol* expression and are located across the entire viral genome (*15–18*). The deletion of these sequences from the helper vector construct diminishes *gag-pol* expression. In addition, constitutive expression of HIV-1 Gag-Pol and other viral proteins has been shown to be toxic to cells (*see* ref. *19* and Chang, unpublished data). The problems associated with the synthesis of lentiviral Gag-Pol have slowed the development of a stable high-titer vector system.

The transducing construct, pTV, which serves as a vehicle for foreign gene delivery, was derived from an LTR-modified recombinant HIV-1 plasmid (*20*). Lentiviral LTRs are transcriptionally silent without Tat. This means that the 5' LTR of pTV cannot be used as a functional promoter to drive foreign gene expression as commonly applied with the 5' LTR of MLV vectors. Nevertheless, to improve safety, efficacy, and viral titer, a self-inactivating (SIN) lentiviral vector (pTY) has been generated with extensive deletions in 3' U3 and U5, as well as insertion of a strong polyadenylation signal (*21*). In pTY, the gene of interest under control of its own enhancer/promoter should be inserted downstream of RRE, which is needed for efficient genomic RNA expression and packaging. The transducing constructs of lentiviruses require more than the conserved retroviral packaging signal (ψ) for efficient packaging. Sequences in TAR, *gag*, RRE, and *env* have all been reported to effect HIV-1 genome packaging (*22–25*).

In addition to pHP and pTY, a VSV-G expression construct, pHEF-VSVG, is needed for the synthesis of the viral envelope. The construct pHEF-VSVG uses a strong human elongation factor promoter to drive expression of VSV-G. Control of VSV-G levels may be necessary to reduce vector toxicity.

1.3. Safety Issues Concerning Lentiviral Vectors

Because lentiviruses are potentially lethal human pathogens, efforts must be made in the development of lentiviral vector systems to minimize the risk of replication-competent virus (RCV) production. This is best achieved by deleting or replacing viral sequences from the vector constructs. Sequences such as *env*, accessory genes, and most of the 3' U3 and U5 can be deleted without affecting vector function.

Additional safety steps should be taken by constantly monitoring RCV production. Since lentiviral transducing vectors do not carry any viral genes, the expression of *gag-pol* would signal RCV production. The most sensitive assay for RCV detection is the polymerase chain reaction (PCR). PCR is very sensitive, but the results are easily confounded by the presence of plasmid DNA. Another simple and sensitive method for detecting RCV is to use immunostaining to detect Gag expression. For example, HIV-1 Gag p24 expression can be detected at the single-cell level using anti-p24 antibodies *(14)*. Production of RCV of HIV-1-derived vectors can also be monitored by coculturing with cell lines that are sensitive to infection and spread of the virus. Alternatively, culture supernatant can be assayed for HIV-1 reverse transcriptase (RT) activity *(20)*. An active program of RCV monitoring should be established by all viral vector production facilities.

1.4. Applications of Lentiviral Vectors: Problems and Possible Solutions

1.4.1. Insertion and Expression of Foreign Genes

A foreign promoter/gene cassette of up to 8–9 kb can be inserted into the transducing vector pTY. Our recent studies suggest that the pTY transducing vector does not contain silencing or destabilizing elements, which are found in conventional retroviral vectors that affect long-term gene expression (*see* **refs. 26** and **27** and Zaiss and Chang, unpublished data). The expression of lentiviral transgenes has been shown to be sustained in vitro and in vivo for over 6 months *(14,28)*.

1.4.2. Titer, Stability, and Toxicity of Lentiviral Vectors

The key to production of large quantities of lentiviral vectors is to increase the per cell production level of the vectors. With optimal transfection conditions, up to 10^7 infectious units per mL, or approx 100 infectious units per cell can be produced in 24 h. For high-titer and scale-up production of lentiviral vectors, a high-titer producer cell line needs to be developed that will also reduce the risk of RCV production. A few lentiviral producer cell lines based on the tetracycline inducible system have been reported, but their stability and consistency of long-term, high-titer production await further evaluation *(29–31* and *see* Chapter 20).

VSV-G pseudotyped vectors are sensitive to human serum, which may present some problems in future systemic in vivo lentiviral vector delivery *(32,33)*. In addition to serum sensitivity, in vivo distribution of vector particles is also a rate-limiting step. These physical barriers may be overcome by modifications of lentiviral transduction protocols.

The toxicity of lentiviral Gag-Pol (HP), VSV-G (env), or other viral proteins may affect both vector production and its use (*see* **Subheading 1.4.3.**). This problem might be confounded by the high defective particle ratios with the HIV-1-derived vectors, which are approx 346:1 (particle: infectious unit) with unconcentrated vector stocks and approx 18:1 with the concentrated vector stocks *(33)*. Improvement of the defective particle to infectious unit ratio may increase the efficiency of vector transduction and reduce toxicity.

1.4.3. In Vitro and In Vivo Use of Lentiviral Vectors

In vivo lentiviral transduction of brain, eye, liver, and muscle tissues of mice or rats has been demonstrated *(34–37)*. Although VSV-G pseudotyped lentiviral vectors have expanded host cell tropism, efficient transduction of different tissue targets remains cell type and growth state dependent. Cells of different vertebrate species (including chicken, mouse, rat, monkey, and human) and of different tissue origins and growth states (including neural cells, pancreatic islets, retina, muscles, macrophages, dendritic cells, fibroblasts, endothelial cells, activated T-cells, embryonic stem cells, and different tumor cell lines) have been successfully transduced with lentiviral vectors *(11,14,36,38,39)*. However, efficient transduction of CD34 hematopoietic stem cells, nonproliferating hepatocytes, monocytes, resting T-cells, some primary tumor cultures, and mouse T-cells with the VSV-G pseudotyped lentiviral vectors has been limited or requires high multiplicity of infection (MOI) (*see* **refs. *11,40–43*** and Chang, unpublished data). Pseudotyping with chimeric or alternative viral envelopes may overcome the entry barriers. Efficiency of transduction may also be improved by treating the target cells with growth factors. For example, transduction of primary resting T-cells has been successfully demonstrated by supplementing with cytokine interleukin (IL)-2, IL-4, IL-7, or IL-15 in the culture *(41,44)*.

2. Materials

2.1. Cells and Culture Conditions

1. 293, 293T-cells (transformed human primary embryonic kidney cells) and TE671 cells (human rhabdomyosarcoma cells) can be obtained from American Type Culture Collection (ATCC), Rockville, MD.
2. Cells are maintained in Dulbecco's modified Eagle's medium (DMEM; Mediatech) supplemented with 10% fetal bovine serum (FBS; Gibco BRL) and 100 U/mL of penicillin-streptomycin (Gibco BRL).
3. Stock cells are maintained in T75 flasks and split 1/6 every 3–4 days. These cells should not be overgrown at any time.
4. 6-well tissue culture plates for transfection and 24-well plates for titration.

2.2. Plasmids

1. All plasmids used for transfection are purified by CsCl gradient centrifugation (*see* **Note 1**).
2. The two expression plasmids, pHP and pHEF-VSVG, which encode HIV-1 structural protein Gag-Pol and VSV-G envelope, respectively, provide necessary helper functions. The third plasmid, pTV or pTY (SIN vector), carries the transgene of interest (promoter plus the transgene open reading frame). A polylinker cloning vector, pTVlinker, or pTYlinker is available for easy insertion of foreign genes (*see* **Note 2**).
3. To increase vector titer (up to fivefold), a fourth plasmid, pCEP4Tat, which provides additional transactivation function, can be included in the DNA cotransfection.

2.3. DNA Transfection

It is important that all materials be sterile to prevent tissue culture contamination.

1. 6-well-tissue culture plates, T25 flasks, or other culture dishes.
2. 15-mL polystyrene or polycarbonate but not polypropylene tubes. Polypropylene reduces the efficiency of DNA transfection.
3. Sterile ddH$_2$O (*see* **Note 3**).
4. 2.5 M CaCl$_2$ (Mallinkrodt brand; *see* **Note 4**).
5. 2× BBS (BES-buffered solution): 50 mM N,N-bis(2-hydroxyethyl)-2 aminoethanesulfonic acid (BES; Calbiochem or Sigma), 280 mM NaCl, 1.5 mM Na$_2$HPO$_4$; adjust the pH with 1 N NaOH to precisely 6.95. Filter the 2× BBS solution with a 0.45-μm filter and freeze in aliquots at –20°C (*see* **Note 5**).

2.4. Viral Vector Harvesting and Concentration

1. Sarstedt 1.5-mL sterile screw-cap tubes and 15- and 50-mL conical tubes.
2. 0.45-μm, sterile-packed, low-protein-binding filters (Millex-HV, Millipore).
3. Polyethylene glycol 8000 (PEG 8000), 50% stock in ddH$_2$O, autoclaved.

4. NaCl, 5 *M* stock, autoclaved.
5. Microfuge (max. 20,000*g*) and bench top clinical centrifuge (max. 4100*g*).

2.5. Lentiviral Vector Titration for LacZ Reporter Gene Expression

1. 24-well tissue culture plates.
2. Polybrene (Sigma), 10 mg/mL stock (1000×), filter sterilized.
3. Phosphate-buffered saline (PBS).
4. Fixation solution: 1% formaldehyde (0.27 μL of 37.6% stock to make 10 mL), 0.2% glutaraldehyde (Sigma, 80 μL of 25% stock to make 10 mL) in PBS.
5. X-gal-staining solution: 10 mL solution of PBS containing 4 m*M* K-ferrocyanide (100 μL of 0.4 *M* stock), 4 m*M* K-ferricyanide (100 μL of 0.4 *M* stock), 2 m*M* MgCl$_2$ (20 μL of 1 *M*), and 0.4 mg/mL X-gal (200 μL of 20 mg/mL stock in dimethyl formamide). Stock ferrocyanide, ferricyanide, and X-gal should be kept frozen and away from light (*see* **Note 6**).

3. Methods

3.1. Generation of Recombinant Lentiviral Vectors

Viral vectors are generated by cotransfection of three or four plasmids into 293, 293T or TE671 cells (**Fig. 3**). TE671 cells are easier to handle because they attach to the culture dish better than 293 cells. All the following steps should be carried out in a biosafety hood. Lentiviral vectors should be handled using NIH BSL-2 safety guidelines. Extra caution should be taken if one is not familiar with the vector background and the monitoring procedures for RCV.

3.1.1. Cell Preparation

1. 17–20 h before transfection, seed each well of a 6-well plate with 7×10^5 TE671 cells in 2 mL DMEM. By the time of transfection, the culture should have reached 95–100% confluency. (Cells can also be split in T25 flasks with 5 mL complete DMEM; *see* **Note 7**).
2. The cells should be fed with fresh medium prior to adding the DNA. This can be done during DNA incubation time.

3.1.2. DNA Transfection Using Calcium Phosphate Precipitation Method

1. Wipe-clean a test tube rack and pipetmans with 70% ethanol. Prepare one polycarbonate tube per vector sample.
2. For each well of a 6-well plate, pipette sterile ddH$_2$O 90 μL and 2.5 *M* CaCl$_2$ 10 μL into one tube, and mix by vortexing (see examples in **Fig. 4**). Add the following amount of DNA (2 μg/μL DNA stocks) to each tube:15 μg pHP, 3 μg pHEF-VSV-G, 8 μg pTV (or pTY), and an optional 0.5 μg pCEP4Tat. Optional: a control

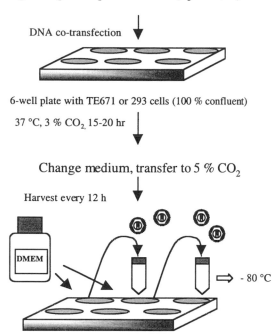

pHP + pTY + pHEFVSV-G (+pCEP4tat)

DNA co-transfection

6-well plate with TE671 or 293 cells (100 % confluent)

37 °C, 3 % CO_2, 15-20 hr

Change medium, transfer to 5 % CO_2

Harvest every 12 h

DMEM

⇒ - 80 °C

Fig. 3. A simple flow chart depicting lentiviral vector production. The HP/TV vector system uses three DNA constructs for cotransfection of TE671 cells in 6-well plates.

plasmid can be included in the DNA mixture for monitoring transfection efficiency, for example, a GFP expression plasmid if the pTV construct does not express GFP.

3. To the DNA solution for each well, add 100 μL of 2× BBS buffer at 2–3 drops each time with gentle shaking while adding the buffer. A fine precipitate will form immediately after mixing DNA with the BBS buffer (*see* **Note 8**).

4. Incubate at room temperature (RT) for 5–45 min depending on the turbidity (pale white suspension) of the DNA solution. If the DNA precipitates form quickly and the solution looks very turbid, allow the solution to sit at RT for only 5–10 min before adding it to the culture well.

5. During DNA-calcium phosphate incubation, replace the culture medium with 2 mL fresh medium (*see* **Note 9**).

6. Add the DNA precipitate (200 μL/well) dropwise while swirling the plate with the other hand. If DNA is added too fast, the strong local pH changes may result in reduced transfection efficiency.

7. Incubate the culture plates overnight at 37°C in a 3% CO_2 incubator, which further improves transfection efficiency.

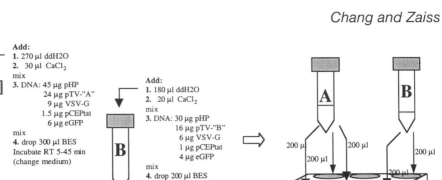

Fig. 4. Examples of producing two different lentiviral vectors. The diagram shows the production of two different lentiviral vectors, A and B, scaled for 3-well and 2-well transfections, respectively, using a 6-well tissue culture plate.

3.2. Viral Vector Harvesting and Concentration

3.2.1. Harvesting Virus

1. The next morning (15–20 h after DNA addition), remove the medium, wash the cells once with 1–2 mL medium if necessary, and add fresh medium (1–2 mL). At least 70–80% cells (up to 100%) should express the cotransfected green fluorescence protein (GFP), which can be monitored directly under an inverted fluorescent microscope.
2. The virus is secreted into the medium. From this time point on (15–20 h after DNA addition), virus can be harvested every 12 h for 3–4 days by collecting the cell culture medium. The highest titer falls in between the second and the fifth harvest (*see* **Note 10**).
3. For unconcentrated virus stocks, virus supernatant is filtered using a 0.45- or 0.2-μm low-protein-binding filter to remove cell debris. Virus aliquots should be stored at −80°C until use (extended exposure to RT or repeated freezing and thawing reduces the titer). The transfected cells normally produce lentiviral vectors with titers ranging from 10^5 to 10^6 transducing units per milliliter of culture medium.
4. For concentrated virus preparation, virus supernatant can be pooled and frozen at −80°C immediately after collection for later concentration use.

3.2.2. Lentiviral Vector Concentration by Microfuge Centrifugation

A small volume of lentiviral vectors can be concentrated 30–50-fold by a simple centrifugation protocol using a microcentrifuge.

1. Filtered virus supernatant is transferred as a 1-mL aliquot into a 1.5-mL sterile screw-cap tube with one side marked with a marker (Sarstedt) (*see* **Note 11**).
2. With the marked side facing out, spin the tube at 20,000*g* at 4°C in a microfuge for 2.5 h.

3. Carefully transfer the tubes to a biosafety hood and discard most of the supernatant by pipeting, but leave behind approx 20–30 µL vol.
4. Rigorously vortex each tube for 20 s and then vortex gently and continuously at 4°C in a mixer or shaker for 2–4 h or overnight.
5. The same vector sample should be pooled and realiquoted to avoid titer variations among different tubes. The virus aliquots should be stored at –80°C.

3.2.3. Concentration by PEG 8000 Precipitation

For a large volume of virus concentration, we recommend the PEG 8000 precipitation method. The protocol is quite gentle to lentivirus, and the recovery is close to 100%. The only drawback is the presence of a high concentration of PEG in the final virus pellet, which may affect subsequent operations. However, most tissue culture cells are not affected by the low volume of PEG in the concentrated virus stock.

1. Centrifuge the collected virus supernatant at 2500g for 10 min to remove dead cell debris, filter the supernatant through a 0.45- or 0.2-µm low-protein-binding filter into a 50-mL conical tube, and keep on ice.
2. To the virus supernatant, add 50% PEG 8000 to a final 5%, and leave on ice.
3. To the virus-PEG sample, add 5 M NaCl to a final 0.15 M, and mix by inverting the tube several times.
4. Incubate the virus and PEG mix on a rotating wheel at 4°C overnight.
5. After overnight incubation, centrifuge the virus-PEG mix in a bench top clinical centrifuge at 4100g for 10 min.
6. Discard the supernatant carefully so as not to disturb the white virus-PEG pellet.
7. The virus pellet can be resuspended in any desired buffer at low volume (usually at 1/100 of the original sample volume). The pellet can be easily dissolved in PBS or any buffer solution by gently pipeting up-and-down.
8. The virus stock can be stored at –80°C in small aliquots.

3.3. Titration of Lentiviral Vectors Carrying a lacZ Reporter Gene

The titration assay described below and in **Fig. 5** determines the number of infectious units per mL of virus carrying the *lacZ* reporter gene.

1. Split TE671 cells at 6×10^4 cells/well in a 24-well plate (approx 90% confluent). Allow cells to attach to the plate (2–4 h to overnight).
2. Mix different volumes of virus stock (usually 2–20 µL for virus with expected titer of 10^5–10^6/mL) with 200–300 µL growth media containing 4–8 µg/mL Polybrene.
3. Add the diluted virus to each well and incubate the plate at 37°C in a 5% CO_2 incubator overnight.
4. The next morning, add 0.5 mL growth media to each well and incubate the plate for another 24 h. (Incubation time can be up to 48 h from the time the virus was added.)

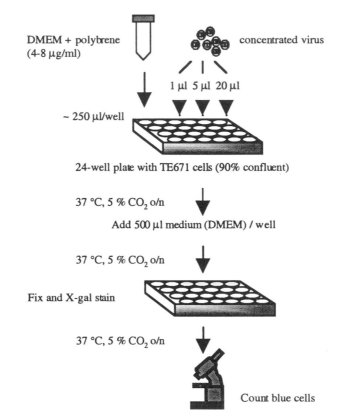

DMEM + polybrene
(4-8 μg/ml)

concentrated virus

1 μl 5 μl 20 μl

~ 250 μl/well

24-well plate with TE671 cells (90% confluent)

37 °C, 5 % CO_2 o/n

Add 500 μl medium (DMEM) / well

37 °C, 5 % CO_2 o/n

Fix and X-gal stain

37 °C, 5 % CO_2 o/n

Count blue cells

Fig. 5. Titration of lentiviral vectors carrying a *lacZ* reporter gene. The diagram illustrates titer determination by using different dilutions of a concentrated *lacZ* reporter vector stock on TE671 cells in a 24-well tissue culture plate.

5. For *lacZ* enzyme assay, wash cells twice with PBS, and fix at room temperature with the fixative for exactly 5 min (300 μL/well).
6. Wash cells 3 times with PBS, add 300 μL X-gal staining solution to each well, and incubate the plate at 37°C overnight (*see* **Note 6**).
7. Determine virus titer by counting blue cells directly using an inverted microscope.

4. Notes

1. CsCl banding is necessary *(45)*, because the quality of plasmid DNA is critical to transfection efficiency; purified DNA should be dissolved at 2 μg/μL in molecular biology grade ddH2O (BRL).
2. pHP, pTYlinker, pTYEFnlacZ, pTYEFeGFP, pHEF-VSVG, and pCEP4tat are all available from the NIH AIDS Research and Reference Reagent Program (http://www.aidsreagent.org/). pHP transformed bacteria should be grown at 30°C during plasmid amplification to prevent deletion. For the transducing vector con-

struct (pTY or pTV), an internal promoter is necessary, but a foreign polyadenylation signal should not be included.

3. The quality of water is critical to transfection. Molecular grade ddH_2O should be used for the preparation of all buffers, media, and DNA solutions.

4. Many other brands of $CaCl_2$ do not work.

5. The working stock of 2× BBS can be kept in room temperature, but the pH may change with time. Therefore, the DNA incubation time will need to be adjusted accordingly by monitoring the cloudiness of the DNA precipitates.

6. The pH of the X-gal solution for ß-galactosidase assay is critical if the cells have background enzyme activity, which can be prevented by adjusting the reaction pH using Tris-HCl buffer pH 8.0–8.5.

7. A consistent high titer is usually obtained from 6-well plate transfections.

8. The pH of BBS is critical to the transfection efficiency. If the DNA solution remains very clear or becomes very turbid immediately after adding 2× BBS, the transfection efficiency will be low. A slightly turbid mixture, which becomes more turbid with time, signals the success of DNA precipitation.

9. To ensure optimal pH, use a fresh bottle of medium (less than 1 month old) made with high quality ddH_2O as noted above (**Note 3**).

10. For harvest at 12-h periods, feed cells with a low volume of medium (1 mL/well of 6-well plates) to enrich virus concentration.

11. Mark one side of the tube and remove medium from the other side of the tube after centrifugation to avoid disturbing the virus pellet.

References

1. Tang, H., Kuhen, K. L., and Wong-Staal, F. (1999) Lentivirus replication and regulation. *Annu. Rev. Genet.* **33,** 133–170.

2. Verma, I. M. and Somia, N. (1997) Gene therapy—promises, problems and prospects. *Nature* **389,** 239–242.

3. Fouchier, R. A. and Malim, M. H. (1999) Nuclear import of human immunodeficiency virus type-1 preintegration complexes. *Adv. Virus Res.* **52,** 275–299.

4. Emi, N., Friedmann, T., and Yee, J. K. (1991) Pseudotype formation of murine leukemia virus with the G protein of vesicular stomatitis virus. *J. Virol.* **65,** 1202–1207.

5. Burns, J. C., Friedmann, T., Driever, W., Burrascano, M., and Yee J.-K. (1993) Vesicular stomatitis virus G glycoprotein pseudotyped retroviral vectors: concentration to a very high titer and efficient gene transfer into mammalian and nonmammalian cells. *Proc. Natl. Acad. Sci. USA* **90,** 8033–8037.

6. Coffin, J. M., Hughes, S. H., and Varmus, H. E. (1997) *Retroviruses.* Cold Spring Harbor Laboratory Press, Cold Spring Harbor, NY.

7. Emerman, M. and Malim, M. H. (1998) HIV-1 regulatory/accessory genes: keys to unraveling viral and host cell biology. *Science* **280,** 1880–1884.

8. Zufferey, R., Nagy, D., Mandel, R. J., Naldini, L., and Trono, D. (1997) Multiply attenuated lentiviral vector achieves efficient gene delivery in vivo. *Nat. Biotechnol.* **15,** 871–875.

9. Haas, D. L., Case, S. S., Crooks, G. M., and Kohn, D. B. (2000) Critical factors influencing stable transduction of human CD34(+) cells with HIV-1-derived lentiviral vectors. *Mol. Ther.* **2,** 71–80.

10. Chinnasamy, D., Chinnasamy, N., Enriquez, M. J., et al. (2000) Lentiviral-mediated gene transfer into human lymphocytes: role of HIV-1 accessory proteins. *Blood* 96, 1309–1316.

11. Poeschla, E. M., Wong-Staal, F., and Looney, D. J. (1998) Efficient transduction of nondividing human cells by feline immunodeficiency virus lentiviral vectors. *Nat. Med.* **4,** 354–357.

12. White, S. M., Renda, M., Nam, N. Y., et al. (1999) Lentivirus vectors using human and simian immunodeficiency virus elements. *J. Virol.* **73,** 2832–2840.

13. Olsen, J. C. (1998) Gene transfer vectors derived from equine infectious anemia virus. *Gene Ther.* **5,** 1481–1487.

14. Chang, L.-J., Urlacher, V., Iwakuma, T., Cui, Y., and Zucali, J. (1999) Efficacy and safety analyses of a recombinant human immunodeficiency virus type 1 derived vector system. *Gene Ther.* **6,** 715–728.

15. Olsen, H. S., Cochrane, A. W., and Rosen, C. (1992) Interaction of cellular factors with intragenic cis-acting repressive sequences within the HIV genome. *Virology* **191,** 709–715.

16. Amendt, B. A., Hesslein, D., Chang, L.-J., and Stoltzfus, C. M. (1994) Presence of negative and positive cis-acting RNA splicing elements within and flanking the first tat coding exon of human immunodeficiency virus type 1. *Mol. Cell. Biol.* **14,** 3960–3970.

17. Huffman, K. M. and Arrigo, S. J. (1997) Identification of cis-acting repressor activity within human immunodeficiency virus type 1 protease sequences. *Virology* **234,** 253–260.

18. Mikaelian, I., Krieg, M., Gait, M. J., and Karn, J. (1996) Interactions of INS (CRS) elements and the splicing machinery regulate the production of Rev-responsive mRNAs. *J. Mol. Biol.* **257,** 246–264.

19. Levy, J. A. (1993) Pathogenesis of human immunodefeciency virus infection. *Microb. Rev.* 57, 183–289.

20. Chang, L.-J., McNulty, E., and Martin, M. (1993) Human immunodeficiency viruses containing heterologous enhancer/promoters are replication competent and exhibit different lymphocyte tropisms. *J. Virol.* 67, 743–752.

21. Iwakuma, T., Cui, Y., and Chang, L. J. (1999) Self-inactivating lentiviral vectors with U3 and U5 modifications. *Virology* **261,** 120–132.

22. Berkowitz, R. D., Hammarskjold, M.-L., Helga-Maria, C., Rekosh, D., and Goff, S. P. (1995) 5' Regions of HIV-1 RNAs are not sufficient for encapsidation: implications for the HIV-1 packaging signal. *Virology* **212,** 718–723.

23. McBride, M. S., Schwartz, M. D., and Panganiban, A. T. (1997) Efficient encapsidation of human immunodeficiency virus type 1 vectors and further characterization of cis elements required for encapsidation. *J. Virol.* **71,** 4544–4554.

24. Cui, Y., Iwakuma, T., and Chang, L. J. (1999) Contributions of viral splice sites and cis-regulatory elements to lentivirus vector function. *J. Virol.* **73,** 6171–6176.

25. Clever, J. L., Taplitz, R. A., Lochrie, M. A., Polisky, B., and Parslow, T. G. (2000) A heterologous, high-affinity RNA ligand for human immunodeficiency virus Gag protein has RNA packaging activity. *J. Virol.* **74**, 541–546.
26. Kempler, G., Freitag, B., Berwin, B., Nanassy, O., and Barklis, E. (1993) Characterization of the Moloney murine leukemia virus stem cell-specific repressor binding site. *Virology* **193**, 690–699.
27. Challita, P. M., Skelton, D., El-Khoueiry, A., et al. (1995) Multiple modifications in cis elements of the long terminal repeat of retroviral vectors lead to increased expression and decreased DNA methylation in embryonic carcinoma cells. *J. Virol.* **69**, 748–755.
28. Blomer, U., Naldini, L., Kafri, T., et al. (1997) Highly efficient and sustained gene transfer in adult neurons with a lentivirus vector. *J. Virol.* **71**, 6641–6649.
29. Freundlieb, S., Schirra-Muller, C., and Bujard, H. (1999) A tetracycline controlled activation/repression system with increased potential for gene transfer into mammalian cells. *J. Gene Med.* **1**, 4–12.
30. Kafri, T., van Praag, H., Ouyang, L., Gage, F. H., and Verma, I. M. (1999) A packaging cell line for lentivirus vectors. *J. Virol.* **73**, 576–584.
31. Klages, N., Zufferey, R., and Trono, D. (2000) A stable system for the high-titer production of multiply attenuated lentiviral vectors. *Mol. Ther.* 2, 170–176.
32. Takeuchi, Y., Porter, C. D., Strahan, K. M., et al. (1996) Sensitization of cells and retroviruses to human serum by (alpha 1-3) galactosyltransferase. *Nature* **379**, 85–88.
33. Higashikawa, F., and Chang L.-J. (2000) Kinetic analyses of stability of retroviral and lentiviral vectors. *Virology* **280**, 124–131.
34. Miyoshi, H., Takahashi, M., Gage, F. H., and Verma, I. M.. 1997. Stable and efficient gene transfer into the retina using an HIV-based lentiviral vector. *Proc. Natl. Acad. Sci. USA* **94**, 10319–10323.
35. Kafri, T., Blomer, U., Peterson, D. A., Gage, F. H., and Verma, I. M. (1997) Sustained expression of genes delivered directly into liver and muscle by lentiviral vectors. *Nat. Genet.* **17**, 314–317.
36. Blomer, U., Naldini, L., Kafri, T., et al. (1997) Highly efficient and sustained gene transfer in adult neurons with a lentivirus vector. *J. Virol.* **71**, 6641–6649.
37. Naldini, L., Blomer, U., Gage, F. H., Trono, D., and Verma, I. M. (1996) Efficient transfer, integration, and sustained long-term expression of the transgene in adult rat brains injected with a lentiviral vector. *Proc. Natl. Acad. Sci. USA* **93**, 11382–11388.
38. Miyoshi, H., Takahashi, M., Gage, F. H., and Verma, I. M. (1997) Stable and efficient gene transfer into the retina using an HIV-based lentiviral vector. *Proc. Natl. Acad. Sci. USA* **94**, 10319–10323.
39. Kafri, T., Blomer, U., Peterson, D. A., Gage, F. H., and Verma, I. M. (1997) Sustained expression of genes delivered directly into liver and muscle by lentiviral vectors. *Nat. Genet.* **17**, 314–317.
40. Zack, J. A., Arrigo, S. J., Weitsman, S. R., et al. (1990) HIV-1 entry into quiescent primary lymphocytes: molecular analysis reveals a labile, latent viral structure. *Cell* **61**, 213–222.

41. Unutmaz, D., KewalRamani, V. N., Marmon, S., and Littman, D. R. (1999) Cytokine signals are sufficient for HIV-1 infection of resting human T lymphocytes. *J. Exp. Med.* **189,** 1735–1746.

42. Park, F., Ohashi, K., Chiu, W., Naldini, L., and Kay, M. A. (2000) Efficient lentiviral transduction of liver requires cell cycling in vivo. *Nat. Genet.* **24,** 49–52.

43. Case, S. S., Price, M. A., Jordan, C. T., et al. (1999) Stable transduction of quiescent CD34(+)CD38(-) human hematopoietic cells by HIV-1-based lentiviral vectors. *Proc. Natl. Acad. Sci. USA* **96,** 2988–2993.

44. Emerman, M. (2000) Learning from lentiviruses. *Nat. Genet.* **24,** 8–9.

45. Maniatis, T., Fritsch, E. F., and Sambrook, J. (1989) *Molecular Cloning: A Laboratory Manual.* Cold Spring Harbor Laboratory Press, Cold Spring Harbor, NY.

22

Simian Foamy Virus Vectors

Preparation and Use

Jeonghae Park and Ayalew Mergia

1. Introduction

The life cycle of retroviruses involves stable integration of viral genetic material into the host genome; expression of viral genes is, in part, regulated by host cell factors *(1)*. These features make retroviruses a widely used efficient means for introducing foreign DNA into the cell genome. Most retroviral vector systems used in clinical gene transfer are based on murine leukemia virus (MuLV) *(2,3)*. MuLV replication is cell cycle dependent, and these vectors are inadequate to transduce nondividing and/or terminally differentiated cells stably, which severely restricts their potential utility for clinical gene transfer *(4–6)*. In contrast, vectors based on the human immunodeficiency virus (HIV) can deliver genes into nondividing cells as efficiently as into proliferating cells *(7,8)*. HIV vectors, then, serve as vehicles to deliver a gene of interest into clinically important terminally differentiated, quiescent, and nondividing cells. The use of vectors derived from a pathogenic virus for human gene therapy, however, remains a major safety concern and health risk issue. Recently, it was shown that animal lentivirus vectors pseudotyped with vesicular stomatitis virus envelope glycoprotein G (VSV-G) can transduce human cells, overcoming the restricted host tropism *(9)*. These vectors deliver genes into nondividing cells as efficiently as HIV. However, the comparative safety and efficacy of vectors based on nonprimate lentiviruses in humans remain to be determined.

Foamy viruses have several inherent features that make them an ideal alternative vector system for human gene therapy. These viruses are distinct and

From: *Methods in Molecular Medicine, Vol. 69, Gene Therapy Protocols, 2nd Ed.*
Edited by: J. R. Morgan © Humana Press Inc., Totowa, NJ

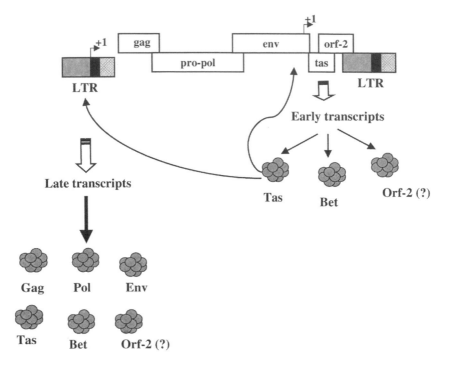

Fig. 1. Differential gene expression by SFV-1. Top: schematic representation of the genetic organization of SFV-1 provirus. +1 at the arrow represents transcription initiation sites by the internal promoter located at the 3' end of *env* and U3 promoter of the 5' LTR. The arrow indicates direction of transcription. Early transcripts are controlled by the internal promoter located at the end of *env* and late transcripts by the U3 promoter. Tas is required for expression of the late transcripts. It is not clear whether there is a protein product is encoded by the orf-2 region.

grouped into the *Spumavirus* genus of retroviruses. Foamy viruses are ubiquitous and are found in many mammalian species. The virus can be propagated efficiently in various cell types of several species *(10)*. Cultured epithelial and fibroblast cells as well as lymphoid and neurally originating cells support the growth of foamy viruses. Each foamy virus isolate can infect several mammalian species, and virus has been recovered from many organs in these animals *(11)*. To date, no disease has been attributed to foamy virus in naturally or experimentally infected animals *(12–15)*. Humans are not natural hosts of foamy virus infection and animal caretakers accidentally infected with simian foamy virus (SFV) remain healthy after more than 10 years of infection *(16,17)*. Importantly, foamy virus vectors can deliver genes efficiently into nondividing cells and human peripheral blood lymphocytes *(18)*. This chapter describes progress in vector development and practical aspects of foamy virus vector, focusing on the SFV-1 system.

1.1. Genomic Structure of Foamy Virus

Foamy viruses possess a complex genomic organization as well as a complex means of regulating gene expression. The genome of SFV-1 is 11,671 nucleotides long and contains the structural genes *gag, pro, pol*, and *env*, which encode core proteins, protease, reverse transcriptase, integrase, and envelope protein, respectively (**Fig. 1**) *(14,19)*. The long terminal repeat (LTR) of SFV-1 is 1621 nucleotides in length, which is unusually long for retroviruses. In addition to *gag, prt, pol, and env*, SFV-1 contains two open reading frames (ORFs) located at the 3' end of *env*. In contrast, the genome of the human foamy virus (HFV) contains three ORFs in the corresponding region *(20,21)*. Other primate foamy viruses, as well as bovine, feline, and equine isolates, have two ORFs similar to SFV-1 *(22–26)*. The first ORF encodes a transcriptional transactivator designated tas (transcriptional transactivator of spumavirus) that activates gene expression directed by the viral promoters *(22,27–32)*. It remains controversial whether there is a protein product encoded by the ORF-2 region *(33–37)*. A Bet protein, which is a product of a spliced message containing the first 88 amino acids of the Tas fused to the last 390 amino acids of ORF-2, was found to be highly expressed in cytoplasm of foamy virus-infected cells *(36,38–40)*. The function of bet has not been fully elucidated *(37,41–43)*.

1.2. Overview of Foamy Virus Life Cycle

Foamy viruses share many features with other retroviruses. These viruses also have distinct characteristics that make them unique among retroviruses. Viral gene expression is temporally regulated and controlled by different promoters like DNA viruses (**Fig. 1**). Similar to other retroviruses, the U3 domain of the LTR contains a promoter element. In addition, foamy viruses contain a second promoter located at the 3' end of *env* upstream from the *tas* gene *(39,44–46)*. Immediately after virus infection, the RNA genome is reverse transcribed into double-stranded DNA, which is then transported to the nucleus and integrates into the host genome. Viral gene expression commences with the internal promoter directing the expression of genes located at the end of *env* including *tas*. Following initial expression, Tas feedback on the internal promoter further amplifies the expression of regulatory genes *(39,45)*. Subsequently, Tas acts on the LTR to express the virion structural genes (*gag, prt, pol*, and *env*).

Newly synthesized viral genome and proteins assemble, and virus particles are formed by budding from the plasma membrane or mature inside the cell within vacuoles. Gag precursor protein is produced from unspliced genomic transcripts and Env from a singly spliced subgenomic message. Similar to retroviruses with a complex genomic organization, multiple splicing events are required to generate messages that encode the *tas*-ORF-2 regions of foamy viruses *(38)*. One feature common to all retroviruses that is not shared by the foamy viruses is that the Pol protein is not synthesized as a Gag-Pro-Pol precursor polyprotein. Instead, the Pro-Pol gene product of foamy viruses is translated from a spliced subgenomic message that lacks the Gag domain *(47,48)*. Translation of the protease-polymerase products is initiated from an ATG initiator codon located at the 5' end of the protease gene *(49,50)*. Distinct from other retroviruses, a second reverse transcription can occur late in the foamy virus replication cycle prior to virion release *(58,51,52)*. About 20% of foamy virus particles were shown to contain double-stranded DNA, suggesting that foamy viruses have a distinct replication pathway with features of both retroviruses and hepadnaviruses. DNA isolated from the virion was shown to be infectious implying that fully reverse-transcribed DNA can be incorporated into virus particles *(52)*. Mutational analysis in foamy viruses revealed that the *tas* gene is essential for viral replication, whereas ORF-2 is dispensable *(33,34,53)*. Therefore, in addition to the four basic genes (*gag*, *prt*, *pol*, and *env*) and *cis*-acting elements required for virus replication, the *tas* gene is critical for foamy virus vector construction, but the ORF-2 region is dispensable.

1.3. Development of Foamy Virus Vector

The first foamy virus vector was constructed by replacing a portion of the ORF-2 region with a foreign gene *(54)*. This vector is replication competent and is capable of transducing genes efficiently. Replication-defective foamy virus vectors have also been constructed by replacing the *env* gene and/or *tas* ORF-2 region with a marker gene *(53,55–60)*. For these vectors, the *env* and *tas* gene products were supplied in *trans* to express the vector genome and produce packaged vector particles that can infect target cells. Titers ranging from 10^3 to 10^5 can easily be obtained with these foamy virus vectors.

Traditional retroviral vectors with a minimal genome sequence contain the 3' and the 5' LTRs, the *cis* packaging sequences, and the polypurine tract, which is required for the second DNA strand synthesis during reverse transcription of the RNA genome *(2)*. Sequences important for packaging are usually located in the 5' leader region extending into the 5' end of the *gag* gene. The polypurine tract is mainly located upstream from the 3' LTR. A *cis*-acting element at the 3' end of the *pol* gene is critical for foamy vector-mediated gene transfer *(58–60)*. The requirement of a second *cis*-acting element located in the *pol* gene for foamy virus vector is unique among retroviruses. Therefore, foamy virus vector construction with minimal genome requires the 3' and the 5' LTRs, the 5' leader region extending to the 5' *gag*, the *cis*-acting element in the *pol* gene, and the polypurine tract. Since the LTR depends on Tas for gene expression, the *tas* gene can be included in the vector or supplied in *trans*.

Replacing the 5' U3 region of the LTR with the powerful cytomegalovirus (CMV) immediate early gene promoter can eliminate the need for Tas in foamy virus vector construction *(61–64)*. In fact, in a number of retroviral vectors production of high-titer helper-free particles can be obtained utilizing the CMV promoter to drive the expression of both the packaging components and the vector genome *(65–67)*. This system takes advantage of high-level transfectability and strong E1A-mediated stimulation of CMV-programmed transcription of the human embryonic kidney fibroblast-derived 293 cell line. The SFV-1 vector system, in which both the vector genome (pCV7-9) and the packaging plasmid (pCGPET) are placed under the control of CMV promoter, yields vector particles with titers up to 5×10^6/mL *(63)*. One potential problem with retroviral vectors is generating replication-competent genome as a result of recombination between the vector and the packaging system. A foamy virus vector system that lacks the *tas* gene is restricted to one round of replication even if recombination occurs between the vector and packaging constructs. This feature makes foamy virus an ideal safe vector system for gene therapy.

1.4. SFV-1 Vector System

Several versions of the SFV-1 vector containing the *lacZ* gene are available that effectively transduce a variety of cells from different species *(58,63)*. Vector pCV7-9 (**Fig. 2**) is commonly used to test the efficiency of the SFV-1 vector system. In pCV7-9, the vector genome is constitutively expressed from the CMV promoter. This vector contains SFV-1 sequences from the beginning of the R to the end of the *pol* gene, the PPT, and the 3' LTR. To monitor for the efficiency of transduction, a *lacZ* gene under the control of CMV promoter is placed between the *pol* gene and PPT. Packaging plasmid (pCGPET) for SFV-1 vector is constructed by placing the structural genes *gag*, *pol*, *env*, and *tas* downstream from the CMV promoter *(63)*. A splice donor of the 5' SFV-1 is provided between the CMV promoter and the *gag* gene in pCGPET. The SV40 polyA signal sequence is supplied at the 3' end. Recombination between pCGPET and pCV7-9 could result in the production of wild-type virus. To restrict replication to one cycle in case of recombination, a mutation is introduced in the pCGPET, creating a Tas-defective packaging plasmid (pCGPEΔT) (**Fig. 2**). A further modified SFV-1 packaging system is available in which the *gag-pol* and *env* viral coding regions are on separate expression plasmids *(63)*. Although this packaging system significantly reduces the chances of generating wild-type virus, there may be a decrease in vector production. Since Tas is critical for SFV-1 replication, the pCGPEΔT packaging system solves the prob-

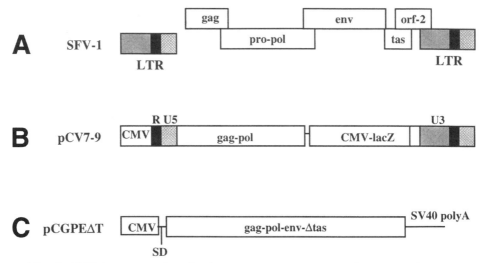

Fig. 2. SFV-1 vector and packaging constructs: SFV-1 provirus **(A)** and vector **(B)** and packaging **(C)** derivatives. CMV-lacZ is the coding sequence of the *lacZ* gene linked to the CMV immediate-early gene promoter. SD represents a synthetic DNA containing the 5' end of the SFV-1 splice donor cloned downstream from the CMV promoter.

lems of potential reduction in vector production caused by placing the structural genes in a separate plasmid and generating recombinant wild-type virus that spreads efficiently.

1.5. Packaging Cell Lines

Foamy viruses cause extensive cytopathology in cell culture. The cytopathic effect includes cell fusion, potentially mediated by the interaction of the *env* gene product and cell receptor. Establishing a packaging cell line for foamy virus vector in which sufficient amounts of viral proteins are synthesized for efficient vector production without toxicity to the cells can be a major obstacle in foamy virus vector development. Recently, two packaging cell lines were created for SFV-1 vector using the human kidney embryonic fibroblast-derived 293 cell line: helper DNA is placed under the control of either a constitutive CMV immediate early gene or inducible tetracycline (tetO) promoter for expression *(63)*. Transfection of pCV7-9 vector in the constitutive expressing and inducible cell lines produced SFV-1 vector particles with titers of 3.5×10^3 and 1.1×10^4 per mL, respectively. This level of vector production is several fold lower compared with transient vector expression when packaging and vector plasmids are cotransfected into 293T cells. Therefore, the 293T cell line will be an alternative for creating a packaging cell line that yields high-titer vector particles. The availability of stable packaging cell lines would allow a scale-up of production of vector stocks for gene therapy.

2. Materials

2.1. Cell Culture

1. Human embryonic kidney fibroblast-derived 293T cells. Cells are passaged by splitting 1:10 every 3–4 days. 293T cells can be obtained from American Type Culture Collection (Rockville, MD).
2. Fetal bovine serum (FBS).
3. Dubelco's modified Eagle's medium (DMEM).
4. Complete DMEM medium: DMEM with 10% FBS, 2 mM L-glutamine, 100 µg/mL, penicillin G and 100 µg/mL streptomycin.
5. Tissue culture flasks: 25 and 75cm^2.
6. Hanks' balanced salt solution (HBSS).
7. Trypsin-EDTA: 0.05% trypsin, 0.53 mM EDTA in HBSS.
8. A 37°C cell culture incubator under 5% CO_2 conditions.

2.2. Recombinant SFV-1 Plasmid Vector

To produce a pure plasmid of pCV7-9 vector for transfection, the DNA should prepared by CsCl gradient centrifugation. Alternatively, pure plasmid DNA can be prepared using a commercial Qiagen (Chatsworth, CA) plasmid preparation kit. It is always advisable to check purified plasmid by several restriction enzyme digestions to test for a possible deletion and/or rearrangements. A modified pCV7-9 vector is available with a unique *Sma*I site for cloning a foreign gene cassette upstream of the CMV-lacZ region.

2.3. Packaging Plasmid

Construction of the packaging plasmid pCGPEΔT is described in **Subheading 1.3.1**. Packaging plasmid DNA should also be prepared by CsCl gradient purification or a Qiagen plasmid preparation kit.

2.4. Generation of Recombinant Virus

1. 6-well cell culture plates.
2. Lipofectamine and Opti-MEM medium (Gibco-BRL).
3. Sodium butyrate (Sigma).
4. Recombinant SFV-1 vector and packaging plasmid (*see* **Subheadings 2.2.** and **2.3.**).
5. DMEM and complete DMEM medium (*see* **Subheading 2.1.**).
6. Phosphate-buffered saline (PBS): 1.0% NaCl, 0.025% KCl, 0.14% Na_2HPO_4, 0.025% KH_2PO_4.
7. Filters (0.45 µm; Millipore).
8. Disposable syringes (3 mL).

2.5. Titration of Recombinant Virus

1. 12-well tissue culture dishes.
2. Complete medium (*see* **Subheading 2.1.**).
3. PBS.
4. X-gal solution (1 mg/mL 5-bromo-4-chloro-3-indoyl-β-D-galactopyranoside in 2 mM MgCl$_2$, 5mM K$_4$Fe[CN]$_6$.3H$_2$O, and 5 mM K$_3$Fe(CN)$_6$ in 1× PBS).
5. 0.25% glutaraldehyde (Sigma).

2.6. Infection of Nonadherent Cells with SFV-1 Vector Particles

1. 12-well culture plates.
2. Microscope slides.
3. Cytofunnel (Shandon, Pittsburgh, PA).
4. Cytospin cytocentrifuge (Shandon, Pittsburgh, PA).
5. RPMI-1640 medium with 10% FBS.

3. Methods
3.1. Generation of Recombinant Virus

Recombinant virus can be produced by cotransfection of viral vector and packaging plasmid. We found that transfection in 293T cells yields the highest vector production. Vector particles can also be generated by transfection of the packaging cell line 293-2 or 293-24 *(63)* with viral vector plasmid. However, the vector production is significantly lower compared with that of particles harvested from 293T cells cotransfected with vector and packaging plasmids. The packaging cell lines 293-3 and 293-24 are derived from the 293 cell line. It is our experience that during transfection the 293T cells adhere better to the plate, whereas the 293 cells tend to lift from the surface of the culture dish; as a result, 293T cells yield several fold higher vector production. Therefore, we routinely prepare SFV-1 vector by cotransfecting pCV7-9 and pCPGEΔT into the 293T cell line

3.1.1. Transfection of DNA Using Lipofectamine

The transfection procedure provided here is slightly modified from the Lipofectamine-mediated transfection method suggested by the manufacturer (*see* **Note 1**). To enhance expression of both vector genome and viral structural proteins, cells can be treated with sodium butyrate after transfection (*see* **Note 2**).

1. In a 6-well plate, seed 4×10^5 cells (293T cells) per well in 2 mL of DMEM with 10% FBS (*see* **Note 3**).
2. Incubate the cells at 37°C under 5% CO$_2$ conditions until the cells are 80% confluent. This usually takes 24–48 h.

3. For each transfection, dilute 9 μL of Lipofectamine reagent in 50 μL of Opti-MEM.
4. Simultaneusly, for each transfection, dilute 1–3 μg (1.5 μg each of vector pCV7-9 and packaging pCGPEΔT) of DNA into 50 μL Opti-MEM and incubate for 30 min at room temperature. Peak activity should be at about 9–15 μL of Lipofectamine reagent for 3 μg DNA.
5. Allow diluted Lipofectamine and DNA to stand at room temperature for 30 min.
6. Mix the two solutions, vortex for 2 s at medium speed, and incubate for 30-45 min at room temperature to allow DNA-lipid complex formation. The solution may appear cloudy; however, this will not impede the transfection.
7. Wash the cells once with 2 mL of serum-free DMEM.
8. For each transfection, add 0.8 mL of serum-free DMEM to each tube containing the DNA-lipid complexes. Do not add antibacterial agents to media during transfection. Then slowly add the mix dropwise to each well of the plate containing the 293T cells.
9. Incubate the cells overnight (but not for a full 24 h) at 37°C in a 5% CO_2 incubator.
10. Replace media the next morning with 1–1.5 mL DMEM containing 10 m*M* sodium butyrate. Cells must be 80–90% confluent before adding sodium butyrate. Maintain in sodium butyrate for 8–12 h (*see* **Note 4**).
11. Wash cells with HBSS and replace media after 8–12 h with complete DMEM medium.
12. Maintain cells for 4–5 days prior to harvesting supernatant. Filter harvested vector particles through 0.45-μm filters and store at –80°C.
13. Transfection efficiency can be checked by staining the cells for β-galactosidase (β-Gal) expression as described in **Subheading 3.1.2.** after harvesting the supernatant. We find that >70% cell transfection results in vector productions with titers ranging 5×10^5 to 5×10^6/mL vector particles (*see* **Note 5**).

3.1.2. Titration of SFV-1 Vector by β-Galactosidase Activity Assay

Since the pCV7-9 vector carries the *lacZ* gene as a marker SFV-1 vector, stocks can be titered by infecting 293T cells and assaying for β-Gal activity.

1. In a 12-well tissue culture plate seed 2×10^5 293T cells and incubate cells at 37°C until the culture reach 80% confluence (24–48 h).
2. On the day of infection, prepare a 10-fold serial dilution of the vector stock using complete DMEM medium in a total volume of 300 μL for each dilution.
3. The cell culture at this point should be at least 80% confluent. Remove medium from the day-old cell culture and add each of the 300 μL vector diluents to each well. Incubate the cultures overnight at 37°C under 5% CO_2 conditions.
4. Replace the medium with fresh complete DMEM and incubate for 2 more days in a 37°C cell culture incubator.
5. Wash cells twice with 0.5 mL of 1× PBS.
6. Fix the cells by adding 0.5 ml/well of 0.05% gluteraldehyde. Incubate at room temperature for no longer than 15 min.

7. Remove the gluteraldehyde solution and gently rinse the cells three times with 1× PBS.
8. Add 0.5 mL fitered X-gal solution (1 mg/mL X-gal, 2 mM MgCl$_2$, 5mM K$_4$Fe[CN]$_6$.3H$_2$O, and 5 mM K$_3$Fe[CN]$_6$ in 1× PBS).
9. Incubate cells at 37°C for 2–15 h to visualize the blue-stained cells, and count positive cells under a light microscope.

3.2. Infection of Nonadherent Cells

For nonadherent cells such as primary and established lymphoid cells, we found higher transduction efficiency when infection is carried out by a spin inoculation method. The infection procedure provided for nonadherent cells is a modification of the method described previously *(68,69)*.

1. Seed established nonadherent cells at a density of 3.5×10^4 in 24-well culture plates. Primary cells such as peripheral blood lymphocytes are plated at a density of 5×10^5 are plated for infection.
2. The following day centrifuge the cells briefly at 1500g for 10 min.
3. Remove medium carefully and resuspend cells in 200–300 µL vector supernatant. Multiplicity of infection (MOI) should be at least 1 for efficient gene transduction.
4. Spin the resuspended cells for 1–2 h at 1000g at room temperature and maintain in a 37°C cell culture incubator.
5. After 8 h repeat steps 2–4 and incubate at 38°C overnight.
6. Spin cells at 1500g for 10 min the next day, remove supernatant carefully, and add fresh medium. Incubate cells at 37°C for 48 h.
7. Transfer 200 µL of resuspended cells to a microscope slide by cytospin for 10 min at room temperature.
8. Make a circle around the cytospun cells with a water-repellent pen.
9. Rinse cells with PBS twice and fix with 200 µL of 0.05% gluteraldehyde for 15 min at room temperature.
10. Rinse cells with PBS 3 times, and add X-gal solution. Incubate slides at 37°C in a humidified chamber overnight.
11. Place cover slip after rinsing with PBS and score blue cells for transduction under a light microscope.

4. Notes

1. Among several transfection protocols we tested, including the CaPO$_4$ procedure, the Lipofectamine method with the reagent from Gibco-BRL gives the least variability, and the transfection efficiency is reproducible.
2. Sodium butyrate has been shown to increase retroviral vector production by enhancing LTR or CMV promoter activity *(70–72)*.
3. Each cell line has a unique seeding concentration that is important to transfection efficiency. A preliminary test determining the right seeding density for each cell line may be required for optimum transfection efficiency.

4. Incubation time for sodium butyrate should not exceed 12 h. Cells transfected with the MuLV vector system survive 10 mM sodium butyrate concentration for 24 h. Incubation of SFV-1 vector-transfected cells in 10 mM sodium butyrate for 24 h, however, results in cell death.

5. Virus titer can be increased fivefold by combining cell-associated vector particles with that of supernatants. Foamy virus matures within the cell or by budding through the plasma membrane and is released to the culture medium. The particles within the cell can be obtained by three freeze-thaw cycles. Foamy virus virion formation occurs only in the presence of its envelope. The foamy virus, therefore, cannot be pseudotyped with the VSV-G envelope protein in order to concentrate the vector by centrifugation, as demonstrated for other retroviral vector systems. However, the foamy virus can be concentrated to more than a 100-fold without loss of infectivity. Therefore, if needed, SFV-1 vector can be concentrated by centrifugal ultrafiltration (Centricon, Amicon).

References

1. Coffin, J M., Hughes, S. H., and Varmus, H E. (1997) *Retroviruses.* Cold Spring Harbor Laboratory Press, Cold Spring Harbor, NY.
2. Miller, A. D., Miller, D. G., Garcia, J. V., and Lynch, C. M. (1993) Use of retroviral vectors for gene transfer and expression. Methods Enzymol. **217,** 581–599.
3. Vile, R. G. and Rusell, S. J. (1995) Retroviruses as vectors. *Br. Med. Bull.* **51,** 12–30.
4. Miller, D. G., Adam, M. A., and Miller, A. D. (1990) Gene transfer by retroviral vectors occurs only in cells that are actively replicating at the time of infection. *Mol. Cell Biol.* **10,** 4239–4242.
5. Roe, T., Reynolds, T. C., Yu, G., and Brown, P. O. (1993) Integration of murine leukemia virus DNA depends on mitosis. *EMBO J.* **12,** 2099–2108.
6. Springett, G. M., Moen, R. C., Anderson, S., Blaese, R. M., and Anderson, W. F. (1989) Infection efficiency of T lymphocytes with amphotropic retroviral vectors is cell cycle dependent. *J. Virol.* **63,** 3865–3869.
7. Lewis, P., Hensel, M., and Emerman, M. (1992) Human immunodeficiency virus infection of cells arrested in the cell cycle. *EMBO J.* **11,** 3053–3058.
8. Lewis, P. F. and Emerman, M. (1994) Passage through mitosis is required for oncoretroviruses bu– not for the human immunodeficiency virus. *J. Virol.* **68,** 510-516.
9. Poeschla, E. M., Wong-Staal, F., and Looney, D. J. (1998) Efficient transduction of nondividing human cells by feline immunodeficiency virus lentiviral vectors. *Nat. Med.* **4,** 354–357.
10. Mergia. A,, Leung, N. J., and Blackwell, J. (1996) Cell tropism of of the simian foamy virus type 1 (SFV-1). *J. Med. Primatol.* **25,** 2–7.
11. Hooks J. J. and Detrick-Hooks, B. (1981) Spumavirinae: foamy virus group infections. Comparative aspects and diagnosis, in *Comparative Diagnosis of Viral Disease,* vol. 4 (Kurstak, E. and Kurstak, C., eds.), Academic, New York, pp 599-618.

12. Weiss, R. A. (1988) Foamy retroviruses. A virus in search of a disease. *Nature (Lond.)* **333,** 497–498.
13. Flugel, R. M. (1991) Spumaviruses: a group of complex retroviruses. *J. AIDS* **4,** 739–750.
14. Mergia, A. and Luciw, P. A. (1991) Replication and regulation of primate foamy viruses. *Virology* **184,** 475–482.
15. Linial, M. (2000) Why aren't foamy viruses pathogenic? *Trends Microbiol.* **8,** 284–289.
16. Heneine, W., Switzer, W. M., Sandstrom, P., et al. (1998) Identification of a human population infected with simian foamy viruses. *Nat. Med.* **4,** 403–407.
17. Schweizer, M., Falcone, V., Gange, J., Turek R, and Neumann-Haefelin, D. (1997) Simian foamy virus isolated from an accidentally infected human individual. *J. Virol.* **71,** 4821–4824.
18. Mergia, A., Soumya., C. L., Kolson, D., Goodenow, M. M., and Ciccarone, T. The efficiency of simian foamy virus vector type-1 (SFV-1) in non-dividing cells and in human PBLs. *Virology* **28,** 243–252.
19. Kupiec, J., Kay, A., Hayat, M., et al. (1991) Sequence analysis of the simian foamy virus type 1 genome. *Gene* **101,** 185–194.
20. Flugel, R. M., Rethwilm, A., Maurer, B., and Darai, G. (1987) Nucleotide sequence analysis of the *env* gene and its flanking regions of the human spumaretrovirus reveals two novel genes. *EMBO J.* **6,** 2077–2084.
21. Maurer, B., Bannert, H., Darai, G., and Flugel, R. M. (1988) Analysis of the primary structure of the long terminal repeat and the *gag* and *pol* genes of the human spumaretrovirus. J. Virol. **62,** 1590–1597.
22. Renne, R., Mergia, A., Renshaw-Gegg, L. W., Neuman-Haefelin, D., and Luciw, P. A. (1993) Regulatory elements in the long terminal repeat (LTR) of simian foamy virus type 3. *Virology* **192,** 365–369.
23. Herchenroder, O., Renne, R., Loncar, D., et al. (1994) Isolation, cloning, and sequencing of simian foamy viruses from chimpanzees (SFVcpz): high homology to human foamy virus (HFV). *Virology* **201,** 187–199.
24. Renshaw, R. W. and Casey, J. W. (1994) Transcriptional mapping of the 3' end of the bovine syncytial virus genome. *J. Virol.* **68,** 1021–1028.
25. Winkler, I., Bodem, J., Haas, L., et al. (1997) Characterization of the genome of feline foamy virus and its proteins shows distinct features different from those of primate spumaviruses. *J. Virol.* **71,** 6727–6741.
26. Tobaly-Tapiero, J., Bittoun, P., Neves, M., et al. (2000) Isolation and characterization of an equine foamy virus. *J. Virol.* **74,** 4064–4073.
27. Mergia, A., Shaw, K. E. S., Pratt-Lowe, E., Barry, P. A., and Luciw, P. A. (1991) Identification of the simian foamy virus transcriptional transactivator gene (*taf*). *J. Virol.* **65,** 2903–2909.
28. Rethwilm, A., Otto, E., Baunach, G., Maurer, B., and Meulen, V. (1991) The transcriptional transactivator of human foamy virus maps to the bel 1 genomic region. *Proc. Natl. Acad. Sci. USA* **88,** 941–945.
29. Venkatesh, L. K., Theodorakis, P. A., and Chinnadurai, G. (1991) Distinct cis-

acting regions in the U3 regulate trans-activation of the human spumaretrovirus long terminal repeat by the viral bell gene product. *Nucleic Acids Res.* **19**, 3661–3666.

30. Keller, A., Partin, K. M., Lochelt, M., Bannert, H., et al. (1991) Characterization of the transcriptional *trans*-activator of human foamy virus. *J. Virol.* **65**, 2589–2594.

31. Herchenroder, O., Turek, R., Neumann-Haefelin, D., Rethwilm, A., and Schneider, J. (1995) Infectious proviral clones of chimpanzee foamy virus (SFVcpz) generated by long PCR reveal close functional relatedness to human foamy virus. *Virology* **214**, 685–689.

32. Renshaw, R. W., and Casey, J. W. (1994) Analysis of the 5' long terminal repeat of bovine syncytial virus. *Gene* **141**, 221–224.

33. Lochelt, M., Zentgraf, H., and Flugel, R. M. (1991) Construction of an infectious DNA clone of the full-length human spumaretrovirus genome and mutagenesis of the *bel* 1 gene. *Virology* **184**, 43–54.

34. Baunach, G., Maurer, B., Hahn, H., Kranz, M., and Rethwilm, A. (1993) Functional analysis of human foamy virus accessory reading frames. *J. Virol.* **67**, 5411–5418.

35. Giron, M. L., Rozain, F., Debons-Guillemin, M. C., et al. (1993) Human foamy virus polypeptides: identification of env and bel gene products. *J. Virol.* **67:** 3596–3600.

36. He, F., Sun, J. D., Garrett, E. D., and Cullen, B. R. (1993) Functional organization of the Bel-1 transactivator of human foamy virus. *J. Virol.* **67**, 1896–1904.

37. Yu, S. F. and Linial, M. L. (1993) Analysis of the role of the bel and bet open reading frames of human foamy virus by using a new quantitative assay. J. Virol. **67**, 6618-6624.

38. Muranyi W, Flugel RM. (1991) Analysis of splicing patterns of human spumaretrovirus by polymerase chain reaction reveals complex RNA structures. J. Virol. **65**, 727-735.

39. Mergia, A. (1994) Simian foamy virus type 1 contains a second promoter located at the 3' end of the env gene. *Virology* **199**, 219–222.

40. Hahn, H., Gerald, B., Brautigam, S., et al. (1994) Reactivity of primate sera to foamy virus Gag and Bet proteins. *J. Gen. Virol.* **75**, 2635–2644.

41. Saib, A., Koken, M. H., Spek, P. V. D., Peries, J.,and The, H. D. (1995) Involvement of a spliced and defective human foamy virus in the establishment of chronic infection. J. Virol. **69:** 5261-5268.

42. Bock, M., Heinkelein, M., Lindemann, D., and Rethwilm, A. (1998) Cells expressing the human foamy virus (HFV) accessory Bet protein are resistant to productive HFV superinfection. *Virology* **250**, 194–204.

43. Callahan, M. E., Switzer, W. M., Matthews, A. L., et al. (1999) Persistent zoonotic infection of a human with simian foamy virus in the absence of an intact orf-2 accessory gene. *J. Virol.* **73**, 9619–9624.

44. Lochelt, M., Muranyi, W., and Flugel, R. M. (1993) Human foamy virus genome possesses an internal, Bel-1-dependent and functional promoter. *Proc. Natl. Acad. Sci. USA* **90**, 7317–7321.

45. Lochelt, M., Flugel, R. M., and Aboud, M. (1994) The human foamy virus internal promoter directs the expression of the functional Bel 1 and Bet proteins early after infection. *J. Virol.* **68,** 638–645.

46. Campbell, M., Renshaw-Gegg, L., Renne, R., and Luciw, P. A. (1994) Characterization of the internal promoter of simian foamy viruses. *J. Virol.* **68,** 4811–4820.

47. Bodem, J., Lochelt, M., Winkler, I., et al. (1996) Characterization of the spliced pol transcript of feline foamy virus: the splice acceptor site of the pol transcript is located in gag of foamy viruses. *J. Virol.* **70,** 9024–9027.

48. Yu, S. F., Baldwin, D. N., Gwynn, S. R., Yendapalli, S., and Linial, M. L. (1996) Human foamy virus replication: a pathway distinct from that of retroviruses and hepadnaviruses. *Science* **271,** 1579–1582.

49. Enssle, J., Jordan, I., Mauer, B., and Rethwilm, A. (1996) Foamy virus reverse transcriptase is expressed independently from the Gag protein. *Proc. Natl. Acad. Sci. USA* **93,** 4137–4141.

50. Lochelt, M. and Flugel, R. M. (1996) The human foamy virus pol gene is expressed as a Pro-Pol polyprotein and not as a Gag-Pol fusion protein. *J. Virol.* **70,** 1033–1040.

51. Enssle, J., Fischer, N., Moebes, A., et al. (1997) Carboxy-terminal cleavage of the human foamy virus gag precursor molecule is an essential step in the viral life cycle. *J. Virol.* **71,** 7312–7317.

52. Yu, S. F., Sullivan, M. D., and Linial, M. L. (1999) Evidence that the human foamy virus genome is DNA. *J. Virol.* **73,** 1565–1572.

53. Mergia, A. and Wu, M. (1998) Characterization of provirus clones of simian foamy virus type 1 (SFV-1). *J. Virol.* **72,** 817–822.

54. Schmidt, M. and Rethwilm, A. (1995) Replicating foamy virus-based vectors directing high level expression of foreign genes. *Virology* **210,** 167–178.

55. Russell, D. W. and Miller, A. D. (1996) Foamy virus vectors. *J. Virol.* **70,** 217–222.

56. Hirata, R. K., Miller, A. D., Andrews, R. G., and Russell, D. W. (1996) Transduction of hematopoietic cells by foamy virus vectors. *Blood* **88,** 3654–3661.

57. Bieniasz, P. D., Erlwein, O., Aguzzi. A., Rethwilm, A., and McClure, M. O. (1997) Gene transfer using replication-defective human foamy virus vectors. *Virology* **235,** 65–72.

58. Wu, M,, Chari, S., Yanchis, T., and Mergia, A. (1998) Cis-acting sequences required for simian foamy virus type 1 (SFV-1) vectors. *J. Virol.* **72,** 3451–3454.

59. Erlwein, O., Bieniasz, P. D., and McClure, M. O. (1998) Sequences in pol are required for transfer of human foamy virus-based vectors. *J. Virol.* **72,** 5510–5516.

60. Heinkelein, M., Schmidt, M., Fischer, N., et al. (1998) Characterization of a cis-acting sequence in the Pol region required to transfer human foamy virus vectors. *J. Virol.* **72,** 6307–6314.

61. Fischer, N., Heinkelein, M., Lindemann, D., et al. (1998) Foamy virus particle formation. *J. Virol.* **72,** 1610–1615.

62. Trobridge, G. D. and Russel, D. W. (1998) Helper-free foamy virus vectors. *Hum. Gene Ther.* **9,** 2517–2525.

63. Wu, M. and Mergia, A. (1999) Packaging cell lines for simian foam virus type 1 (SFV-1) vectors. *J. Virol.* **73,** 4498–4501.

64. Schenk, T., Enssle, J., Fischer, N., and Rethwilm, A. (1999) Replication of a foamy virus mutant with a constitutively active U3 promoter and deleted accessory genes. *J. Gen. Virol.* **80,** 1591–1598.

65. Kim, V. N., Mitrophanous, K., Kingsman, S. M, and Kingsman, A. J. (1998) Minimal requirement for a lentivirus vector based on human immunodeficiency virus type 1. *J. Virol.* **72,** 811–816.

66. Naviaux, R. K., Costanzi, E., Haas, M., and Verma, I. M. (1996) The pCL vector system: rapid production of helper-free, high-titer, recombinant retroviruses. *J. Virol.* **70,** 5701–5705.

67. Soneoka, Y., Cannon, P. M., Ramsdale, E. E., et al. (1995) A transient three-plasmid expression system for the production of high titer retroviral vectors. *Nucleic Acids Res.* **23,** 628–633.

68. Bahner, I., Kearns, K., Hao, Q. L., Smogorzewska, E. M., and Kohn, D. B. (1996) Transduction of human CD34+ hematopoietic progenitor cells by a retroviral vector expressing an RRE decoy inhibits human immunodeficiency virus type 1 replication in myelomonocytic cells produced in long-term culture. *J. Virol.* **70,** 4352–4360.

69. Bunnell, B. A., Muul, L. M., Donahue, R. E., Blaese, R. M., and Morgan, R. A. (1995) High-efficiency retroviral-mediated gene transfer into human and nonhuman primate peripheral blood lymphocytes. *Proc. Natl. Acad. Sci. USA* **92,** 7739–7743.

70. Tanaka, J., Sadanari, H., Sato, H., and Fukuda, S. (1991) Sodium butyrate-inducible replication of human cytomegalovirus in a human epithelial cell line. *Virology* **185,** 271–280.

71. Olsen, J. C. and Sechelski, J. (1995) Use of sodium butyrate to enhance production of retroviral vectors expressing CFTR cDNA. *Hum. Gene Ther.* **6,** 1195–1202.

72. Sakoda, T., Kasahara, N., Hamamori, Y., and Kedes, L. (1999) A high-titer lentiviral production system mediates efficient transduction of differentiated cells including beating cardiac myocytes. *J. Mol. Cell Cardiol.* **31,** 2037–2047.

23

Recombinant Feline Immunodeficiency Virus Vectors

Preparation and Use

Michael A. Curran and Garry P. Nolan

1. Introduction

Traditional onco-retroviral vector packaging systems, typified by those based on Moloney murine leukemia virus (MMLV), have many appealing features for gene transfer. MMLV vectors stably integrate their genome into their target cells, can package and carry up to 6.5 kb of foreign insert, can be produced free of helper virus, and, because of their low immunogenicity, can be administered in vivo multiple times to the same organism. The major disadvantage of these vectors; however, is their inability to transduce nondividing cells *(1)*.

Lentiviral packaging systems, such as those based on the human and feline immunodeficiency viruses (HIV and FIV), possess all the strengths of the onco-retroviral systems coupled to the ability to transduce most nondividing and primary cells. This chapter focuses on the preparation and optional concentration of FIV vector particles, as well as on transduction of target cells with these viral preparations. Before moving on to the practical methods of FIV production, it is useful to begin with an understanding of FIV biology, particularly the elements of the virus that allow infection of nondividing cells. Although most cell types are susceptible to transduction with FIV vectors, there are a few known exceptions, which are also discussed. More thorough reviews of the biology of lentiviruses and lentiviral vectors can be found in the literature *(2,3)*.

From: *Methods in Molecular Medicine, Vol. 69, Gene Therapy Protocols, 2nd Ed.*
Edited by: J. R. Morgan © Humana Press Inc., Totowa, NJ

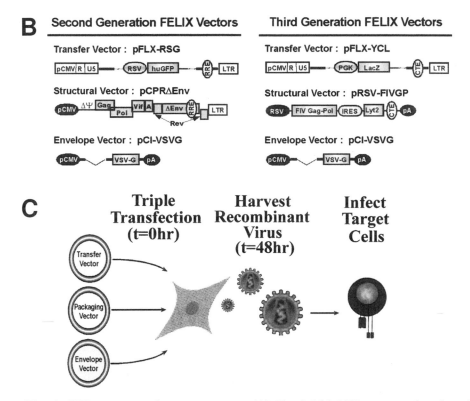

Fig. 1. FIV genome and vector systems. **(A)** The 9.4-kb FIV genome showing all ORFs and the protein constituents of the Gag-Pol polyprotein. **(B)** Second- and third-generation FIV-based FELIX vectors. Shown are a reporter transfer vector, the packaging vector, and a VSV-G envelope vector for both three-plasmid systems. (C) Schematic depicting the production of recombinant FIV for gene transfer using one of the triple transient transfection systems shown above.

1.1. FIV Genomic Organization and Life Cycle

The genomes of the simple (i.e., non-lenti, non-spuma) retroviruses contain three genes coding for the Gag, Pol, and Env proteins, which encode the structural, enzymatic, and surface proteins, respectively. In addition to these essential open reading frames (ORFs), lentiviral genomes typically contain several accessory genes that perform complex functions in both the production and infection stages of their life cycle. Within its 9.4-kb genome, FIV possesses three accessory genes, *vif* (increases infectivity of virus produced in lymphocytes), *ORFA/ORF2* (weak LTR transactivator), and *rev* (facilitates nucleocytoplasmic transport of intron-containing viral RNA), which encode proteins important in the viral life cycle *(4)* (**Fig. 1A**). Third-generation vector systems that eliminate these accessory proteins for the sake of added safety must, to some degree, compensate for their functions in virus production or suffer losses in infectious viral titer.

The FIV life cycle begins with transcription of the viral genome from the U3 promoter region of the 5' long terminal repeat (LTR), so named because it is present in tandem at both ends of the retroviral genome. Full-length transcripts arc terminated and polyadenylated at the U5 region of the 3' LTR. Completely spliced variants of this transcript give rise to the Rev and A proteins. Rev then returns to the nucleus, binds to the Rev responsive element (RRE) present at the 3' end of the *env* gene, and promotes nuclear export of partially spliced and unspliced viral RNA. Partially spliced transcripts give rise to the Vif and Env proteins, and unspliced transcripts yield the Gag and Gag-Pol (generated through ribosomal frameshifting) polyproteins and also serve as the genome for packaging into mature virions. The viral protease cleaves itself away from Gag-Pol and then processes Gag into matrix (MA), capsid (CA), and nucleocapsid (NC), and Pol into integrase (IN), reverse transcriptase (RT), and a dUTPase. Upon entering a target cell (notably via CXCR4 or CCR5), the FIV genome is reverse transcribed into double-stranded DNA by RT and transported to the nucleus in the context of the preintegration complex (MA, IN, RT), where it is integrated into the host cell genome.

The ability to transduce nondividing cells can probably be attributed to several features of FIV. Although specific NLS cites have not been mapped for FIV as precisely as they have for HIV, it is possible that FIV contains sequences in its MA and IN proteins that facilitate its entry into the nucleus via a karyopherin-mediated pathway analogous to HIV *(5,6)*. Also, by analogy to HIV and equine infectius anemia virus (EIAV), FIV probably also contains a central DNA flap region that facilitates its entry into the nucleus through an unknown mechanism *(7)*. Interestingly, FIV contains a dUTPase as part of its Pol polyprotein that is not present in HIV Pol. Mutation studies of the FIV and

EIAV dUTPases suggest that their function is to proofread the viral genome for uracil mis-incorporations, which can occur more frequently in metabolically inactive cells with low nucleotide pools *(8,9)*. Thus, the ability of FIV to infect nondividing cells efficiently in vivo probably results from the synergy of several viral components acting at multiple stages of the viral life cycle.

1.2. Development of FIV Vectors

Although the three labs responsible for developing FIV-based vector systems have taken slightly different approaches, most key steps in adapting FIV into a useful vector system are common to all three systems *(10–12)*. In all systems, a transfer vector (carries the gene of interest), a packaging vector (produces the FIV structural proteins), and an envelope vector (directs virion entry through binding to a cognate receptor on target cells) are transfected together into human 293T-cells to produce virus. Viral supernatants are collected 48 h posttransfection and added to target cells to allow infection by the vector particles present in the supernatants.

All FIV, and almost all retroviral, production systems use 293T cells as their packaging cells (i.e., the cells in which virus is produced) because of their high transfectability, parental species (i.e., human), and strong metabolic activity. The native FIV U3 promoter, however, exhibits very low activity in human cells, resulting in very little viral genome transcription following transfection into 293T cells *(13)*. For this reason, all FIV vectors contain the human cytomegalovirus (hCMV) enhancer/promoter, which is very strong in 293T cells, grafted in place of the native FIV U3 promoter. Since this hCMV promoter is present only in the 5' LTR of the vector, integrated FIV vector genomes will contain the native viral LTR at both the 5' and 3' ends of the genome (the 3' U3 region serves as the template for both LTRs during reverse transcription). This again creates a transcription problem, necessitating the use of an internal promoter to drive expression of the gene(s) of interest. The upstream hCMV promoter modification and insertion of an internal promoter are common transcriptional modifications necessary to make FIV viable as a vector.

Like the onco-retroviral vector systems, FIV packaging systems require expression of the viral Gag-Pol protein and an envelope protein (e.g., MMLV amphotropic, vesicular stomatitis virus G [VSV-G]) in order to produce infectious particles. Of the FIV accessory proteins, Vif is not needed by virtue of using 293T cells as producers, and replacement of the FIV U3 promoter with CMV makes A dispensable; however, all FIV vectors excluding the third-generation Nolan lab vectors require the viral Rev protein for efficient nuclear export of unspliced vector genome. In general, second-generation FIV packaging vectors produce some or all of the viral accessory proteins, whereas third-generation packaging constructs produce only Gag-Pol.

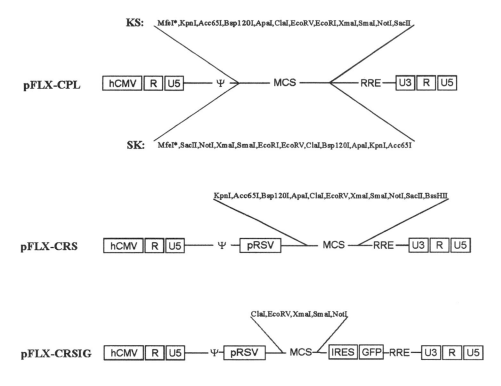

KS: MfeI*,KpnI,Acc65I,Bsp120I,ApaI,ClaI,EcoRV,EcoRI,XmaI,SmaI,NotI,SacII

pFLX-CPL | hCMV | R | U5 |——— Ψ ——— MCS ——— RRE—| U3 | R | U5 |

SK: MfeI*,SacII,NotI,XmaI,SmaI,EcoRI,EcoRV,ClaI,Bsp120I,ApaI,KpnI,Acc65I

KpnI,Acc65I,Bsp120I,ApaI,ClaI,EcoRV,XmaI,SmaI,NotI,SacII,BssHII

pFLX-CRS | hCMV | R | U5 |——— Ψ —| pRSV |——— MCS —RRE—| U3 | R | U5 |

ClaI,EcoRV,XmaI,SmaI,NotI

pFLX-CRSIG | hCMV | R | U5 |——— Ψ—| pRSV |——— MCS—| IRES | GFP |—RRE—| U3 | R | U5 |

* - cloning into this MfeI site reduces the packaging signal size but probably not enough to significantly effect titre.

Fig. 2. Second-generation FIV transfer vectors for cloning. Shown are maps of the pFLX-CPL series of vector, which contain only a multiple cloning site (MCS), the pFLX-CRS vector, which contains a promoter upstream of the MCS, and pFLX-CRSIG, which contains an IRES-eGFP cassette downstream of a promoter and MCS.

1.3. Second-Generation FIV Vectors

All second-generation FIV vector systems are based on transient triple trans-fection of a transfer vector, packaging vector, and envelope construct into 293T-cells. There are currently no stable FIV packaging cell lines. All vector names used in the remainder of the text refer to the FELIX system developed in the Nolan lab. From a practical point of view, the transfer vector is the most interesting, as it will carry the gene of interest to the target cell and control its expression following integration into the genome. As mentioned above, all FIV transfer vectors contain the hCMV promoter cloned upstream of the 5' regions necessary for replication and packaging (i.e., R, U5, and the packaging signal) (**Fig. 1B**). Optimal packaging of FIV vectors requires a region that extends into the Gag reading frame; consequently, most vectors include an FIV sequence up to the *Mfe*I (+875 bp into the provirus) site or maximally to the *Tth*III1 site (+921 bp). This large packaging region includes the major 5' splice donor as well as the Gag ATG, which have not been removed in most vectors currently available. The 3' end of all transfer vectors includes the 3' end of the *env* gene (which includes the RRE) and the complete 3' LTR.

The gene cassette of interest is inserted between the packaging region and the 3' region (RRE and LTR) described above. Positive control vectors such as pFLX-RSG contain a central Rous sarcoma virus (RSV) promoter driving expression of the green fluorescent protein (GFP). For most applications, the vectors pFLX-CPL or pFLX-CRS are used as backbones into which a particular gene or gene cassette of interest is inserted (**Fig. 2**). pFLX-CPL contains the multiple cloning site (MCS) from pBluescript II inserted in one of two orientations (KS or SK)—this vector is designed for cloning in promoter-gene combinations of interest, allowing them to be transferred over easily from standard plasmids or MMLV vectors using internal promoters. pFLX-CRS contains an RSV promoter upstream of the MCS from pBluescript in the KS orientation and is designed for cloning of cDNA inserts. A variant of pFLX-CRS that contains an internal ribosome entry site (IRES)-GFP marker cassette downstream of the MCS, pFLX-CRSIG, is also available.

The transfer vector is cotransfected into 293T-cells along with the packaging vector and envelope vector. The second-generation packaging vectors available—pCPRdEnv and pCFWdEnv—both contain a CMV promoter driving expression of all the FIV proteins except Env. They have been rendered packaging deficient by deletion of the entire LTR and packaging region up to 500 bp into the proviral genome. In pCPRdEnv, the Env protein has been neutralized by introduction of a point mutation early in the reading frame that truncates and frameshifts the protein, whereas in pCFWdEnv there is a deletion in *env* that removes part of the ORF and frameshifts the remainder. pCPRdEnv is derived from the PPR strain of FIV, whereas pCFWdEnv is derived from the 34TF10 strain. In general, pCPRdEnv and pCFWdEnv give approximately equivalent titers when used with the transfer vectors described above (all of which are based on FIV-34TF10).

FIV vectors can be pseudotyped with several amphotropic/polytropic envelopes, yielding high titres and broad host range. The best titers for all FIV vector system are achieved by using the glycoprotein VSV-G as the envelope protein *(14)*. In the FELIX system, this envelope is expressed off the CMV promoter on the pCI-VSVG plasmid. VSV-G pseudotyped virions can enter almost all eukaryotic cells (from *Drosophila* to human) and can be concentrated using high-speed centrifugation. FIV vectors also pseudotype effectively with the MMLV amphotropic envelope. In contrast, FIV vectors do not pseudotype well with the MMLV ecotropic envelope or with the Gibbon-Ape leukemia virus (GALV) envelope.

In summary, triple transfection of an FIV transfer vector (e.g., pFLX-RSG), a packaging vector (e.g., pCPRdEnv), and a VSV-G envelope vector (e.g., pCI-VSVG) into 293T-cells yields viral titers in the 10^6 IU/mL range. The particles can be used to transduce a wide variety of dividing and nondividing target cells permanently with the gene of interest carried in the transfer vector.

1.4. Third-Generation FIV Vectors

The second-generation FIV vectors yield good viral titres and are very useful for in vitro work and in vivo work in animal models. The potential for use of FIV vectors in human subjects, however, encouraged the creation of a system in which the packaging and transfer vectors contained less native FIV sequence and in which no viral accessory proteins were present during the packaging or transduction phases of the vector life cycle (**Fig. 1B**). The most significant changes in the third-generation system are in the packaging vector. pRSV-FIVGP, the third-generation packaging construct, expresses only the viral Gag-Pol protein with a linked Lyt2 surface marker. Since Gag-Pol contains splice donors, an RNA element known as the cytoplasmic transport element (CTE) has been included to facilitate nuclear export of intact Gag-Pol-IRES-Lyt2 mRNA. The CTE is a structured RNA element from Mason Pfizer monkey virus (MPMV, a type D retrovirus) that functions like the Rev-RRE systems of the lentiviruses, but without the need for a cognate Rev protein *(15)*.

The third-generation transfer vectors, typified by the LacZ reporter vector pFLX-YCL, contain a CTE in place of the FIV RRE to facilitate their nucleocytoplasmic transport. Interestingly, the CTE must be placed as far 3' as possible in the vector to provide efficienct nuclear export. As the third-generation vectors are rarely used for in vitro work or animal studies, no transfer vectors with convenient MCSes exist as of this writing. Titers of the third-generation system (e.g., pFLX-YCL, pRSV-FIVGP, pCI-VSVG) are in the 10^6 IU/mL range but are generally between 1.5-fold and 2-fold less than those of the second-generation system, reflecting the lack of efficiency of the CTE relative to the FIV Rev-RRE system.

1.5. FIV Vector Tropism

When they are pseudotyped with the VSV-G envelope, it is believed that FIV vector particles can enter almost any target cell type. In in vitro systems, FIV vectors infect G1/S- and G2/M- arrested cells with efficiencies similar to infection of dividing cells *(10,12)*. Primary cells transduced by FIV vectors include human aortic smooth muscle cells, human umbilical vein endothelial cells, human hepatocytes, human CD34+ progenitor cells, human T-cells, human dendritic cells, human pancreatic islet cells, stimulated murine T-cells, rat neurons, and rat pancreatic islet cells. In our hands, however, murine hematopoeitic progenitor cells and completely unstimulated (no cytokines or antibodies) human T-cells have yielded no significant transduction with FIV vectors. Also, FIV (and HIV) vectors seem to have a quantitative defect for infecting some murine (specifically murine, not rat) cells as titers measured for these vectors on some human cells fall off between 2- and 10-fold on similar murine cells. This defect may be even more pronounced on murine primary cells, but more work needs to be done to confirm this phenomenon. *See* www.stanford.edu/group/Nolan for updates.

2. Materials

2.1. Cell Culture

1. Human embryonic kidney 293T-cells. Cells should be grown in the DMEM described below and should be split 1/5 to 1/10 every 3–4 days.
2. Dulbecco's modified Eagle's medium (DMEM) with high glucose, L-glutamine, 110 mg/L sodium pyruvate, and pyridoxine hydrochloride (Gibco-BRL, cat. no. 11995-065). Optionally, supplement with penicillin-streptomycin (Gibco-BRL, cat. no. 15140-122) or with penicillin-streptomycin-amphotericin B (Gibco-BRL, cat. no. 15240-062).
3. Trypsin-EDTA solution 1× (Gibco-BRL, cat. no. 25300-054).
4. Fetal bovine/calf serum (FBS/FCS).
5. Tissue culture treated dishes (6 or 15 cm diameter).

2.2. Transfection Reagents

2.2.1. $Ca_3(PO_4)_2$ Transfection

1. Chloroquine: Stock is 50 mM in ddH$_2$O (Sigma, cat. no. C-6628).
2. CaCl$_2$: Stock is 2 M in ddH$_2$O, 2-μm filtered (Mallinkrodt, cat no. 4160).
3. HEPES sodium salt (Sigma, cat no. H-7006).
4. Na$_2$HPO$_4$ dibasic (Fisher, cat no. S374-500). Stock solution is 5.25 g dissolved in 500 mL ddH$_2$O (i.e., approx 74 mM).
5. 2× HBS: 8.0 g NaCl, 6.5 g HEPES, 10 mL Na$_2$HPO$_4$ stock solution. pH to 7.0 using NaOH or HCl. Bring volume up to 500 mL. Check pH again. The pH is very important; it must be exactly 7.0.

2.2.2. FugENE 6 Transfection

1. FuGENE 6 Transfection Reagent (Roche, cat no. 1 814 443).
2. Opti-MEM (Gibco-BRL, cat. no. 31985-070).

2.4. FIV Vector Plasmids

1. An FIV transfer vector containing a reporter gene cassette or promoter and gene of interest to be transferred to a target cell (e.g., pFLX-RSG). DNA should be prepared using a large-scale plasmid purification method that yields pure plasmid DNA and minimal endotoxin.

2. An FIV packaging plasmid vector (e.g., pCPRdEnv). DNA should be prepared as above.

3. A plasmid expressing an envelope with which FIV vectors readily pseudotype (e.g., pCI-VSVG). DNA should be prepared as above.

2.5. Virus Purification

1, 5-mL syringes.
2. 0.45-μm low-protein-binding syringe filters.
3. 15-mL sterile polypropylene tubes.

2.6. Virus Concentration

1. Tris-NaCl-EDTA buffer (TNE): 50 mM Tris-HCl, pH 7.8, 130 mM NaCl, 1 mM EDTA. Add 50 mL of 1 M Tris-HCl, pH 7.8, to 950 mL H$_2$O containing 7.6 g of NaCl, 0.4 g of EDTA and autoclave prior to storage at 4°C.

2. 50 mL Oak Ridge polyproylene copolymer tubes (Nalgene Nunc, cat. no. 3139-0050).

3. 0.45-μm low-protein-binding 115 mL Nalgene filters (cat. no. 245-0045).

2.7. Infection Reagents

1. Polybrene (hexadimethrine bromide, Sigma, cat. no. H9268). 1000× stock concentration is 5 mg/mL in PBS.

2. Retronectin (Panvera, cat. no. TAK T100).

3. Methods

3.1. Production of Recombinant Virus

Described below are the two methods for triple transfection of 293T-cells that have worked the best and most reproducibly in our experience. In each case, 293T-cells are transfected with 5 μg of packaging vector plasmid, 3 μg of transfer vector plasmid, and 2 μg of envelope vector plasmid; 48 h later, recombinant FIV viral supernatants are collected from these cells. Both the transfection methods described below and the ratio of packaging to transfer to envelope vector are open to further optimization. Conditions described below are for 6-cm dishes; however, larger or smaller dishes may be used providing all the DNA and reagent volumes are scaled appropriately based on the difference in surface area. The FuGENE 6 protocol is recommended for its consistently high yields and low toxicity; however, calcium phosphate transfection provides a much more economical alternative. Note that <50% transfection will generally not result is useful viral titers.

3.1.1. Triple Transfection Using CaPO$_4$

1. 293T-cells should be grown and maintained in DMEM with 10% FCS. The day prior to transfection split out 2–2.5 × 10^6 cells/6-cm dish. At the time of transfec-

tion, cells should be 65–85% confluent in 3 mL of DMEM with 10% FCS (*see* **Note 1**).

2. Add 2 µL of 50 m*M* chloroquine stock to plates prior to transfection (*see* **Note 2**).
3. To a 15-mL polypropylene tube, add 5 µg FIV packaging vector (e.g., pCPRdEnv), 3 µg transfer vector (e.g., pFLX-RSG), and 2 µg envelope vector (e.g., pCI-VSVG) (*see* **Note 3**).
4. Add 61 µL of 2 *M* CaCl2—try to wash the DNA down the tube.
5. Add 430 µL of sterile ddH$_2$O—must be room temperature.
6. Centrifuge tubes briefly at 1500 rpm (400–500*g*) to spin liquid down to bottom of tube.
7. Add 500 µL of 2× HBS and bubble for 2–10 s in the water/CaCl$_2$/DNA mixture. (Bubbling times vary greatly between batches of HBS and should be optimized for each one.)
8. Add the HBS/water/CaCl$_2$/DNA mix dropwise to the plated cells. Swirl the plate gently when done (*see* **Note 4**).
9. Incubate the cells at 37°C for 8 h.
10. Change media to 2–3 mL of fresh DMEM with 10% FCS.
11. Change media again in 16–24 h.
12. *Optional*: Move plates to 32°C approx 40 h posttransfection (*see* **Note 5**).

3.1.2. Triple Transfection Using FuGENE 6

1. 293T-cells should be grown and maintained in DMEM with 10%FCS. The day prior to transfection, split out 2–2.5 × 10^6 cells/6-cm dish. At the time of transfection, cells should be approx 50% confluent in 3 mL of DMEM with 10% FCS (*see* **Note 1**).
2. To a sterile 1.5-mL microfuge tube, add 266 µL of serum-free Opti-MEM (or DMEM).
3. Add 27 µL of FuGENE 6 directly to the Opti-MEM already in the tube without allowing it to touch the sides (*see* **Note 6**).
4. Incubate the FuGENE/media mixture for 5 min at room temperature.
5. To a second sterile 1.5-mL tube, add 5ug FIV packaging vector (e.g., pCPRdEnv), 3 µg of transfer vector (e.g., pFLX-RSG), and 2 µg of envelope vector (e.g., pCI-VSVG). If possible, add them directly to the bottom of the tube (*see* **Note 3**).
6. Add the FuGENE/media mixtures from **step 4** dropwise to the tube containing the DNA from **step 5**.
7. Tap the bottom of the tube to mix the contents. Do not vortex.
8. Incubate the FuGENE/DNA mixture for 15 min at room temperature.
9. Add the mixture dropwise to the plated cells. Swirl to distribute.
10. Incubate the cells at 37°C for 24–30 h.
11. Change the media to 2–3 mL of fresh DMEM with 10% FCS approx 24 h posttransfection.
12. *Optional*: Move the plates to 32°C approx 40 h posttransfection (*see* **Note 5**).

3.2. Collection of Viral Supernatants

1. Approximately 48 h posttransfection, remove viral supernatants from plates and add to sterile 15-mL polypropylene tubes.
2. If desired, add 2–3 mL of new media to each plate, and return the plates to the 32°C incubator for up to two additional collections.
3. Centrifuge the 15-mL tubes containing the viral supernatants for 5 min at 1000 rpm (300g) to remove cell debris.
4. *Optional*: Filter-clear supernatants through 5-mL syringes tipped with 0.45 μm low-protein-binding syringe filters. Never use 0.2 μm filters or filters that are not low-protein-binding or you may lose significant viral titer (*see* **Note 7**).
5. Cleared or cleared/filtered supernatants can now be used immediately or frozen at –80°C. (Freezing results in an approx twofold loss of titer) (*see* **Note 8**).

3.3. Concentration of Virus (Optional)

1. Prepare 15-cm dishes of 293T cells split with approximately 12.5×10^6 cells/dish in 18 mL of DMEM with 5% FCS or Opti-MEM with 1% FCS the day before transfection (*see* **Note 9**).
2. Transfection by the FuGENE 6 protocol is recommended using the following parameters: 1663 μL of Opti-MEM, 167 μL of FuGENE, 31 μg of packaging vector, 19 μg of transfer vector, and 12.5 μg of envelope vector.
3. Transfect *two* 15-cm dishes for each sample.
4. At 48 h posttransfection, collect supernatants from the two dishes and pool them into a sterile 50-mL polypropylene tube (final volume approx 36 mL).
5. Obtain a 115-mL 0.45 μm or 0.8 μm low-protein-binding filter (*see* **Note 10**).
6. Filter the 50-mL tube of pooled supernatant through the filter.
7. Pour the filtered supernatant into an autoclaved 50-mL polycarbonate Oak Ridge tube and screw on the top securely.
8. Spin Oak Ridge tubes containing supernatants at 50,000g for 1.5 h. Generally this is done using a Beckman JA20, JA21, or JA25.50 rotor spinning at approximately 20,000 rpm.
9. Pour off the supernatant from the tube. A small pellet should be visible on the wall near the bottom.
10. Resuspend pellet in 300 μL to 1 mL of TNE. Wash the lower wall thoroughly and pipette up and down to disperse the pellet (*see* **Note 11**).
11. *Optional*: Add 300 μL to 1 mL of a desired media to the TNE-virus solution (*see* **Notes 11** and **12**).
12. Use virus or freeze at –80°C in cryotubes.

3.4. Infection of Target Cells In Vitro

The most basic means of infecting target cells with a recombinant FIV viral supernatant is obvious—place the supernatant on the cells, place the cells at 32°C, add Polybrene, and change the media after 8–16 h. Several more refined methods exist, however, which have been shown to increase infectivity of retroviral supernatants on many different types of cells. Spinoculation, usually in the presence of a polycation such as Polybrene or protamine sulfate, refers to spinning plates of cells with virus on them. The exact mechanism by which this method works remains unclear, as the relatively low g forces involved should not significantly alter the movement of viral particles in solution. Transduction of hematopoeitic stem cells, dendritic cells, and possibly other types of cells increases significantly when infections are performed in Retronectin (recombinant fibronectin fragment CH296)-coated plates. Polybrene is the most frequently used infection catalyst; however, it can be toxic to several types of primary cells. When Polybrene toxicity becomes a problem, protamine sulfate (1000× stock is 8 mg/mL) or dioleoyl-trimethylammonium propane (DOTAP; Gibco) can be used as less toxic alternatives.

3.5. Basic Infection Using Recombinant FIV Supernatant

1. Plate out target cells to a desired density. (Infecting target cells at confluence is less efficient as virus-cell contacts are reduced.)
2. Aspirate media off cells and add the desired amount of viral supernatant (if adherent) or spin down cells and resuspend in viral supernatant (if nonadherent).
3. Add the desired volume of the target cells' normal media. Lower total volume results in higher infectivity, but be sure to add enough of your cells' native media that viability will not be compromised.
4. Add Polybrene to a final concentration of 5 µg/mL.
5. Incubate cells for 8–16 h at 32°C.

3.6. Spinoculation of Target Cells

1. Add viral supernatant to target cells in a 96-, 48-, 24-, 12-, or 6-well plate.
2. Add Polybrene to a final concentration of 5 µg/mL.
3. Make sure plates have a sufficient volume of viral supernatant and media that there won't be dry spots in the wells during centrifugation.
4. Secure plate lids by taping or parafilming.
5. Place plates on plate carriers for a centrifuge with a swinging bucket rotor. (We use a Sorvall RT6000B or Beckman Allegra 6R, but there are many suitable centrifuge/rotor combinations.)
6. Centrifuge at 2500 rpm (100g) for 90 min.
7. Remove sealant (tape or parafilm) from plate lids.
8. Place plates at 32°C 8–16 h.
9. Replace virus-containing media with target cells' normal media and move them to 37°C.

3.7. Infection Using Retronectin (Takara)

1. Plate target cells in a Retronectin-coated plate or dish. Plates can be purchased precoated from Panvera or coated with Retronectin prior to use. If coating plates,

follow the instructions included with the Retronectin and remember to use plates that are *not* tissue culture treated.

2. Aspirate media off cells and add the desired amount of viral supernatant (if adherent) or spin down cells and resuspend in viral supernatant (if nonadherent).
3. Add the desired volume of the target cells' normal media. Lower total volumes are better if not spinning plates; higher may be better to avoid drying if spinning.
4. *Optional*: Add Polybrene to a final concentration of 5 µg/mL. Little proof exists that using a polycation in addition to Retronectin significantly improves titers (*see* **Note 13**).
5. *Optional*: Spin cells 2500 rpm (100*g*) for 90 min. There is also little evidence that spinning with Retronectin gives superior transduction to use of Retronectin alone (*see* **Note 14**).
6. Incubate cells 8–16 h at 32°C.
7. Replace virus-containing media with target cells' normal media and move them to 37°C.

3.8. Measurement of Viral Titer

Viral titer can be measured on any cell line when the VSV-G envelope is used; therefore, we recommend using a cell line derived from the same species as the targets the viral supernatant is to be used to transduce. An additional possibility is to use 293 cells, which are readily infectable by FIV and can be considered to give a "best case" titer. Calculating titer with retroviral vectors is a matter of some debate, as different combinations of cell number, plate size, dilution of viral supernatant, and use of infection catalysts can give widely varied titer measurements for the same batch of virus. The method given here is meant to establish a useful working titer measurement that should provide a good reference titer regardless of the virus volume to media volume to cell number ratios used. Remember also that in a population of cells that is >33% infected, some cells will have multiple integration events.

1. Prepare viral supernatants using a transfer vector expressing a reporter gene that can be detected by fluorescence-activated cell sorting (FACS; i.e., GFP, β-Gal, or a surface marker that can be detected by an antibody).
2. Split 1×10^5 cells/well of a 12-well plate of the cell line to be used for measuring titer. For each sample to be titered, split out three wells. The ideal is for there to be approximately 1×10^5 cells/well at the time of infection; thus, if you are splitting the day before, you may want to seed 5 x 10^4 cells/well.
3. To the three wells allocated for each sample, add 100 µL of viral supernatant + 900 µL media to one, 500 µL virus + 500 µL media to the other, and 1 mL of virus to the last (*see* **Note 11**).
4. Add 1 µL of 1000× Polybrene to each well.
5. Spin plates for 1.5 h at 2500 rpm.
6. Incubate plates for 8–16 h at 32°C.

7. Remove virus and add fresh media.
8. Incubate plates for 36–48 h at 37°C.
9. Measure the percentage of cells infected by FACS analysis (e.g., on a BD FACScan).
10. Calculate titer for all three wells individually as follows:
    Well #1: %infected $\times$ 1 $\times$ 10^5 $\times$ 10 $\times$ 10 = titer IU/mL
    Well #2: %infected $\times$ 1 $\times$ 10^5 $\times$ 2 $\times$ 2 = titer IU/mL
    Well #3: %infected $\times$ 1 $\times$ 10^5 $\times$ 1 $\times$ 1 = titer IU/mL.
    The general method is %infected $\times$ no. of cells infected $\times$ fraction of 1 mL of virus used $\times$ final dilution in media.
11. Average all three titers to get a working titer. If virus volume is limiting, preparing and analyzing only the 100- and 500-µL wells still gives a good titer estimate.

4. Notes

1. 293T cells may be selected with 1 mg/mL G418 if they begin to lose their transfectability or if their growth rate slows.
2. The chloroquine is toxic to the 293T cells, limiting transfection time to 8 h. If longer incubation times are necessary, less chloroquine can be used.
3. Because of their repeated LTRs and large size, some FIV vectors may be more stable when prepared in stable bacterial strains such as SURE (Stratagene) or STBL2 (Gibco).
4. After applying the HBS/DNA mixture, a fine granular black precipitate should be visible above the cell monolayer. (This will become more pronounced throughout the 8 h.) Less fine precipitates with larger black particles can signal suboptimal transfection and greater toxicity.
5. Recombinant virus is more stable at 32°C; therefore, incubation of plates at 32°C prior to and/or during virus collection(s) may yield higher titers than incubation at 37°C.
6. For FuGENE transfections, in general, use 2.67 µL FuGENE 6/1 µg of DNA.
7. Filtration removes any remaining cell debris and helps reduce carry-over transfection when it is a problem.
8. Flash-freezing supernatants on dry-ice/ethanol or in liquid nitrogen provides the best viability. Alternatively, supernatants can be placed directly at –80°C.
9. Lower concentrations of serum are used when concentrating viral supernatants so that fewer precipitated serum proteins pellet with the virus particles.
10. Skipping filtration yields highest titers; however, unfiltered concentrated virus may be toxic to cells owing to residual VSV-G coated membrane fragments.
11. Concentrated virus may be toxic to some cells because of the fusogenic nature of VSV-G pseudotyped particles. It is recommended that the concentrated virus first be applied to the target cells in a variety of dilutions to establish an optimal range of transduction to toxicity.
12. Addition of media to the TNE/virus solution helps make the virus preparation less toxic for certain primary cells and provides additional liquid volume for resuspension.

13. Addition of polycationic molecules such as Polybrene is known to enhance retroviral transduction; however, little evidence exists that this effect is additive when using recombinant fibronectin.
14. Spin infection is known to enhance retroviral transduction; however, little evidence exists that this effect is additive when using recombinant fibronectin.

References

1. Miller, D. G., Adam, M. A., and Miller, A. D. (1990) Gene transfer by retrovirus vectors occurs only in cells that are actively replicating at the time of infection [published erratum appears in *Mol. Cell. Biol.* 1992: 433. *Mol. Cell. Biol.* **10,** 4239–4242.
2. Naldini, L. (1998) Lentiviruses as gene transfer agents for delivery to non-dividing cells. *Curr. Opin. Biotechnol.* **9,** 457–463.
3. Trono, D. (2000) Lentiviral vectors: turning a deadly foe into a therapeutic agent. *Gene Ther.* **7,** 20–23.
4. Tomonaga, K. and Mikami, T. (1996) Molecular biology of the feline immunodeficiency virus auxiliary genes. *J. Gen. Virol.* **77,** 1611–1121.
5. Bukrinsky, M. I., Sharova, N., Dempsey, M. P., et al. (1992) Active nuclear import of human immunodeficiency virus type 1 preintegration complexes. *Proc. Natl. Acad. Sci. USA* **89,** 6580–6584.
6. Gallay, P., Stitt, V., Mundy, C., Oettinger, M., and Trono, D. (1996) Role of the karyopherin pathway in human immunodeficiency virus type 1 nuclear import. *J. Virol.* **70,** 1027–1032.
7. Zennou, V., Petit, C., Guetard, D., et al. (2000) HIV-1 genome nuclear import is mediated by a central DNA flap. *Cell* **101,** 173–185.
8. Lerner, D. L., Wagaman, P. C., Phillips, T. R., et al. (1995) Increased mutation frequency of feline immunodeficiency virus lacking functional deoxyuridine-triphosphatase. *Proc. Natl. Acad. Sci. USA* **92,** 7480–7484.
9. Lichtenstein, D. L., Rushlow, K. E., Cook, R. F., et al. (1995) Replication in vitro and in vivo of an equine infectious anemia virus mutant deficient in dUTPase activity. *J. Virol.* **69,** 2881–2888.
10. Curran, M. A., Kaiser, S. M., Achacoso, P. L., and Nolan, G. P. (2000) Efficient transduction of non-dividing cells by optimized feline immunodeficiency virus vectors. *Mol. Ther.* **1,** 31–38.
11. Poeschla, E. M., Wong-Staal, F., and Looney, D. J. (1998) Efficient transduction of nondividing human cells by feline immunodeficiency virus lentiviral vectors. *Nat. Med.* **4,** 354–357.
12. Johnston, J. C., Gasmi, M., Lim, L. E., et al. (1999) Minimum requirements for efficient transduction of dividing and nondividing cells by feline immunodeficiency virus vectors. *J. Virol.* **73,** 4991–5000.
13. Miyazawa, T., Kawaguchi, Y., Kohmoto, M., Tomonaga, K., and Mikami, T. (1994) Comparative functional analysis of the various lentivirus long terminal repeats in human colon carcinoma cell line (SW480 cells) and feline renal cell line (CRFK cells). *J. Vete. Med. Sci.* **56,** 895–899.

14. Burns, J. C., Friedmann, T., Driever, W., Burrascano, M., and Yee, J. K. (1993) Vesicular stomatitis virus G glycoprotein pseudotyped retroviral vectors: concentration to very high titer and efficient gene transfer into mammalian and nonmammalian cells [see comments]. *Proc. Nat.l Acad. Sci. USA* **90,** 8033–8037.

15. Bray, M., Prasad, S., Dubay, J. W., et al. (1994) A small element from the Mason-Pfizer monkey virus genome makes human immunodeficiency virus type 1 expression and replication Rev-independent. *Proc. Nat.l Acad. Sci. USA* **91,** 1256–1260.

24

Lentivirus Vector Based on Simian Immunodeficiency Virus

Development and Use

Klaus Überla

1. Introduction

Vector systems based on human immunodeficiency viruses (HIVs) *(1)* allow an efficient and stable gene transfer into nondividing cells *(2,3)*. Because of the pathogenicity of the parental virus, the use of lentiviral vectors based on feline immunodeficiency virus *(4,5)*, equine infectious anemia virus *(6)*, or simian immunodeficiency virus (SIV) *(7,8)* has also been explored. SIV is closely related to HIV, and some strains of SIV can induce an AIDS-like disease in nonhuman primates. This animal model has been extensively used to study the pathogenesis of immunodeficiency virus infections and should also be valuable for a preclinical safety assessment of SIV-based vectors. Although SIV can replicate in human CD4+ cells, the course of laboratory-acquired infections suggests greatly reduced virulence in comparison with HIV *(9)*.

The aim of this chapter is to discuss various practical aspects of the development and use of SIV-based vectors. This is preceded by a short overview of the genome structure and replication cycle of the virus. More detailed information on the molecular biology of immunodeficiency viruses can be obtained from recent reviews *(10,11)*.

1.1. Genome Structure

The viral genome is flanked by two long terminal repeats (LTRs; **Fig. 1A**), which consist of U3, R, and U5 regions. The U3 region harbors the promoter

From: *Methods in Molecular Medicine, Vol. 69, Gene Therapy Protocols, 2nd Ed.*
Edited by: J. R. Morgan © Humana Press Inc., Totowa, NJ

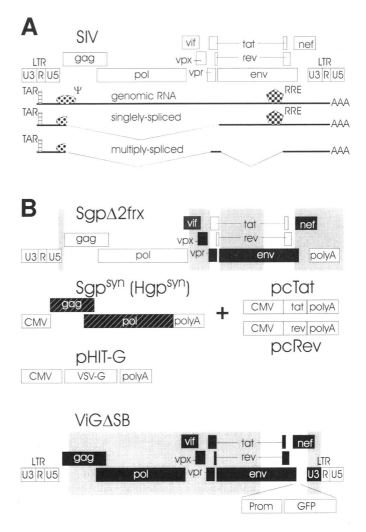

Fig. 1. Map of proviral DNA and mRNA transcripts of SIV **(A)** and maps of SIV vector and packaging constructs **(B)**. Deleted regions are shaded. Inactivated reading frames are marked in black. Hatched regions indicate codon optimization. ψ, assumed packaging signal; CMV, immediate early promoter and enhancer of human cytomegalovirus; polyA, heterologous polyadenylation signal; Prom, heterologous promoter; GFP, gene for the green fluorescence protein. Plasmids Sgpsyn *(13)*, Hgpsyn *(13)*, pcTat *(16)*, pcRev *(17)*, and pHIT-G *(20)* have been previously described.

and enhancer elements, which direct initiation of viral transcription at the 5'LTR. All viral transcripts start with the 5'R region and end with the 3'R region. Differential splicing and ribosomal frameshifting allow expression of various open reading frames. Five of them (*gag*, *pol*, *env*, *tat*, and *rev*) are particularly relevant for vector development. The *gag* reading frame codes for the matrix and capsid proteins, *pol* for viral enzymes, and *env* for the surface protein. TAT is a strong transcriptional activator; it binds to the *trans*-acting responsive region, TAR, at the 5'end of the viral RNA. Rev also binds to a secondary structure of the viral RNA, called Rev-responsive element (RRE), and mediates nuclear export of unspliced or singly spliced viral RNA.

1.2. Replication Cycle

The viral particle, which contains two copies of the unspliced viral RNA, binds via the Env protein to the cellular receptors, leading to fusion of the viral and cellular membrane. This delivers the viral core to the cytoplasm, where the core dissociates. The viral RNA is reverse transcribed by the particle-associated reverse transcriptase. During reverse transcription the 5'U5 and 3'U3 regions of the viral RNA serve as templates for the 3'U5 and 5'U3 regions, respectively. This ensures regeneration of two complete LTRs during DNA synthesis. After transport of the viral DNA into the nucleus, it integrates into the genome of the cell with the help of the viral integrase. The viral genome is then transcribed by cellular factors. During the early phase of the viral life cycle, multiply spliced transcripts coding for Tat and Rev are formed. Tat acts as a strong activator of viral transcription, leading to a positive feedback loop. If sufficient amounts of Rev have accumulated, Rev mediates export of unspliced and singly spliced transcripts to the cytoplasm, where they are translated into Gag, Gag-Pol, and Env. The viral particle then assembles at the plasma membrane. RNA secondary structures presumably located at the 5' untranslated region and in the beginning of *gag* act as packaging signals, leading to the incorporation of two copies of unspliced viral RNA into the particle. Finally, the viral particle buds off the membrane, leading to the release of infectious virus particles.

1.3. Development of SIV-Based Vectors

The key step in the development of lentiviral vectors is to split up the viral genome in three or more constructs, which will provide all functions necessary for a single round of infection. Usually, two packaging constructs are used: one codes for the surface protein (pHIT-G, **Fig. 1B**), and the second allows expression of *gag* and *gag-pol* (SgpΔ2frx or Sgp^{syn}, **Fig. 1B**). Both of them lack packaging signals and elements required for reverse transcription and integration. Therefore, cotransfection of both packaging constructs leads to the production of empty particles, which cannot transfer either *env, gag,* or *gag-pol.* If the packaging constructs are cotransfected with a vector construct (ViGΔSB, **Fig. 1B**), the encoded vector RNA can be packaged and transferred. The features of these three constructs are discussed separately in more detail below.

1.3.1. Selection of the Envelope Protein

The envelope protein (Env) of the vector particle is an important determinant of cell tropism. In the case of SIV-based vectors, the homologous Env protein would only provide entry into CD4+ cells, which also express the right coreceptor. Since heterologous cell surface proteins can also be incorporated into lentiviral vector particles, the SIV Env can be replaced by heterologous envelope proteins such as the amphotropic murine leukemia virus (MLV) Env or the G-protein of vesicular stomatitis virus (VSV-G) *(7)*. Both allow entry into a wide variety of different cells. The VSV-G has the additional advantage that its stable association with the vector particle allows purification and concentration *(12)*. The toxicity of VSV-G makes the establishment of stable packaging cell lines difficult. Toxic effects can also be observed at high doses on some target cells and might limit its application in vivo.

1.3.2. Minimizing the Vector Construct

Since the vector RNA encoded by the vector construct is transfered to target cells, it is desirable to eliminate as many viral sequences as possible. *cis*-acting sequences that cannot be deleted from the vector construct are promoter and enhancer elements in the 5'LTR, the packaging signal (ψ), and elements required for reverse transcription or integration such as the primer binding site, the 3' polypurine tract, the R regions, and the ends of the LTRs. Most vector constructs also require the RRE, since 5'*gag* sequences, which were included for better packaging, render the constructs Rev dependent (ViGΔSB, **Fig. 1B**). One exception is a minimal vector construct that only contains viral sequences present on the multiply spliced *nef* transcript *(7)*. The vector titers of this RRE-less vector are, however, approximately 10-fold lower than titers of vectors containing 5'*gag* sequences. Deletion of the promoter and enhancer elements of the U3 region of the 3'LTR leads to self-inactivating vectors, which have a reduced risk of vector mobilization by superinfection of transduced cells with HIV. The deletion in the 3'U3 region does not affect transcription of vector RNA in the vector producer cells since transcription is controlled by the 5'U3 region. During reverse transcription the 3'U3 region serves as a template for the 5'U3 region of the viral DNA in the target cells. If the promoter and enhancer elements of the 3'U3 region have been deleted, the 5'U3 region will also lack these elements in the target cells. Therefore, the packaging signals present in the 5'untranslated regions are no longer transcribed in the target cells, reducing the risk of vector mobilization by superinfection of transduced cells with HIV. Insertion of an internal promoter upstream of the therapeutic gene allows its expression in the target cells and offers the possibility of transcriptional targeting (ViGΔSB, **Fig. 1B**). Problems of promoter interference in target cells can also be circumvented by these self-inactivating vectors.

When different therapeutic genes are cloned into the same vector construct, the titers can vary considerably. Although the reasons for failures are not always understood, the following points should be considered when designing new vector constructs. The overall length of the vector should be kept as short as possible. It is recommended that the vector construct not exceed the length of the wild-type SIV genome, which is roughly 10 kb. The therapeutic gene is usually inserted in the *sense* orientation to avoid *antisense* inhibition phenomena during vector production. Polyadenylation signals must be absent, and elimination of cryptic splice sites can signficantly increase the titer of some vectors. Direct repeats should also be avoided, since strand transfer during reverse transcription can lead to the deletion of intervening sequences. If the therapeutic gene is to be inserted in the *antisense* orientation, it is advisable to use an internal promoter that has low activity in the vector producer cells.

1.3.3. Optimizing the gag-pol Expression Plasmid

The most efficient SIV *gag-pol* plasmids had initially preserved the functional organization of the SIV genome *(7)*. Sequences involved in packaging were deleted from the 5'untranslated region (SgpΔ2frx, **Fig. 1B**). The 3'LTR had been replaced by a heterologous polyadenylation signal. Deletions were also introduced into the *env* gene and the accessory genes *vif*, *vpr*, *vpx*, and *nef*. The Tat-TAR transcriptional activation pathway and the Rev-RRE nuclear export system had been left untouched for efficient expression of *gag-pol*.

One of the disadvantages of such *gag-pol* expression plasmids is that the *gag-pol* genes are flanked by two regions that are homologous to vector sequences. These regions of homology are *gag* sequences, which are maintained on the vector for better packaging, and the RRE, which is required on most SIV vector constructs for export of the unspliced vector RNA from the nucleus to the cytoplasm. Homologous recombination events between the *gag-pol* expression plasmid and the vector could therefore lead to transfer of *gag-pol* sequences into target cells. The risk for these homologous recombination events was limited by optimizing the codon usage of SIV *gag-pol* *(13)* without changing the amino acid sequence. The degree of nucleotide sequence identity to *gag* sequences present on the vector construct was thus reduced to 75%. The degree of homology between vector and *gag-pol* expression plasmid could be further minimized by packaging the SIV vector with a codon-optimized HIV-1 *gag-pol* expression plasmid *(13)*. As observed previously, optimization of lentiviral codon usage *(14,15)* also allowed Rev-independent expression of *gag-pol*. Therefore the RRE could also be omitted from the codon-optimized *gag-pol* expression plasmids, which finally only contain the *gag-pol* open reading frame without any other lentivirus-derived sequences. When the codon-optimized *gag-pol* expression plasmid is used to package vector constructs such as ViGΔSB, *tat* and *rev* expression plasmids (pcTat, pcRev *[16,17]*) are cotransfected for efficient transcription and nuclear export of the vector RNA, respectively.

2. Materials

2.1. Cell Culture

1. Cell culture medium: Dulbecco's modified Eagle's medium (DMEM)-Glutamax (Gibco-BRL cat. no. 61965-026, Life Technologies, Karlsruhe, Germany) + 100 U penicillin G/mL + 100 µg streptomycin/mL.
2. 293T cells (293ts/A1609 *[18]*) are cultured in DMEM + 10% fetal bovine serum (FBS). Cells are split 1/10 twice a week by trypsin-EDTA (Gibco-BRL cat. no. 25300-054, Life Technologies) treatment. Overgrowth should be avoided. Thaw fresh cells every 3–4 months.
3. 293A cells (Quantum Biotechnologies, Montreal, Canada) kept in DMEM + 5% FBS are split 1/5 twice a week.

2.2. Plasmids

The Qiagen (Hilden, Germany) Plasmid Mini Kit (cat. no. 12143) or the EndoFree Plasmid Maxi Kit (cat. no. 12362) are used to prepare plasmid DNA of the packaging and the vector constructs (**Fig. 1B** and *see* **Note 1**).

2.3. Generation of Vector Particles by Transient Transfection

1. 2× HBS: 10× solution: 8.18% NaCl, 5.94% Hepes, 0.2% Na_2HPO_4 (all w/v). Dilute to 2× HBS solution. Add 1 M NaOH to bring pH to 7.12. Filter-sterilize and store in aliquots at –20°C. 2 M $CaCl_2$.
2. Injectable water (B. Braun, Melsungen, Germany).
3. Carrier DNA: Sonicated calf thymus DNA (Gibco-BRL cat. no. 15633-019, Life Technologies).
4. 0.45 µm filter: Minisart (Sartorius cat. no. 17598K, Göttingen, Germany).

2.4. Concentration of Vector Supernatants

1. Serum-free medium: AIM-V medium (Gibco-BRL cat. no. 12055-083, Life Technologies).
2. Vivaspin 20 tubes (Sartorius cat. no. VS2041).

3. Methods

3.1. Generation of Vector Particles by Transient Transfection

Vector particles are prepared by transient cotransfection of the packaging constructs with the vector construct into 293T cells using the calcium phosphate coprecipitation method since stable packaging cell lines for SIV-based vectors are not yet available (*see* **Note 2**).

1. On the day prior to transfection plate 1×10^6 293T cells in 25-cm^2 tissue culture plates in 5 mL DMEM + 10% FBS, which should give about 80% confluency on the next day.
2. Replace the culture medium with fresh prewarmed (37°C) medium and incubate for 4–6 h at 37°C and 5% CO_2 (*see* **Notes 3** and **4**).
3. To set up the transfection cocktail, thaw all reagents at 37°C and place on ice.
4. Mix on ice 5 μg *gag-pol* expression plasmid, 2 μg VSV-G expression plasmid, and 5 μg vector plasmid with carrier DNA to a total amount of 15–20 μg DNA.
5. Add 31 μl 2 *M* $CaCl_2$ to the DNA and adjust the volume to 250 μl by adding injectable water.
6. For each tissue flask to be transfected, aliquot 250 μl 2 × HBS into a separate 1.5-mL reaction tube.
7. Place one tube at a time on a vortexer under the tissue culture hood and add the DNA-$CaCl_2$ solution drop by drop.
8. After a 10-min incubation period at room temperature, pipet the transfection cocktail into the medium of the 293T cell flasks, mix by gentle rocking, and place in the incubator.
9. Eighteen to 24 h after addition of the transfection cocktail, replace the medium gently with 5 mL fresh culture medium.
10. One and/or 2 days later, remove the supernatant of the transfected cells and clear by low-speed centrifugation ($370g$, 10 min).
11. Then slowly press the supernatant with a syringe through 0.45-μm filters and store in aliquots at –80°C. Filtered supernatants can be stored at 4°C, if infections are set up within a day after harvest (*see* **Note 5**).

3.2. Titration of Vectors

The titers of vectors expressing the green fluorescence protein (GFP) reporter gene are determined as follows.

1. One day prior to infection plate 5×10^4 293A cells in 1 mL DMEM + 5% FBS per well of a 24-well plate (*see* **Note 6**).
2. Thaw vector stocks briefly at 37°C and make 10-fold serial dilutions in DMEM + 5% FBS.
3. Remove the medium from the wells of the 24-well plate and add 200 μL of vector dilutions immediately.
4. After 4 h in the CO_2 incubator, add 1 mL prewarmed DMEM.

5. Two to 3 days later, count the number of GFP⁺ cells in an inverted fluorescence microscope.
6. To calculate the titer without having to view the entire well, first count the number of optical fields at a given magnification that can be fitted across the diameter of the well.
7. Then determine the average number of GFP⁺ cells per optical field at this magnification from a representative number of optical fields.
8. Multiply the number of GFP⁺ cells per optical field by the square of the number of optical fields fitting the diameter of the well, to give the number of GFP⁺ cells per well.
9. Multiplication by 5 and by the dilution factor gives the number of GFP-forming units per mL vector supernatant.

3.3. Concentration of Vectors

Vector supernatants can be easily concentrated approximately 100-fold by filtration through Vivaspin 20 tubes if the supernatants were harvested in serum-free medium. This is achieved as follows:

1. Replace the DMEM medium 24 h after addition of the transfection cocktail with the same volume of serum-free AIM-V medium.
2. Than harvest supernatants 24 and 48 h later and clear and filter as described in **Subheading 3.1**.
3. Sterilize the Vivaspin columns by washing with 70% ethanol and subsequent drying in the cell culture hood.
4. Add 20 mL of cleared and filtered vector supernatant to the upper chamber of the Vivaspin 20 tube and centrifuge at 2800g for 2 h.
5. Measure the volume of the remaining fluid with a 200-µL pipet.
6. If the volume is larger than 200 µL, continue centrifugation.
7. Once the desired concentration factor has been reached, the concentrated supernatant can be frozen in aliquots and titered as described in **Subheading 3.2**. The recovery rate is usually >90% (*see* **Note 7**).

4. Notes

1. We routinely use Qiagen columns to prepare plasmid DNA, but other isolation procedures giving pure DNA should work as well. If problems are encountered, the lipopoly saccharide (LPS)-free plasmid purification kit could be used. The plasmid DNA is eluted with injectable water, and we make sure that the DNA concentration of the plasmid preparation is around 1 µg/µL or higher. This reduces the effect of the buffer the DNA is dissolved in on the pH of the transfection cocktail.
2. Although a high percentage of transfected 293T cells can be obtained with a variety of transfection protocols, we obtained the highest vector titers with the calcium phosphate coprecipitation method. The number of plasmid molecules delivered to each transfected cell might be particularly high with this method,

allowing high expression and ensuring cotransfection of each cell with all packaging and vectors constructs.

3. The pH is critical for the formation of fine CaPO$_4$-DNA precipitates. Therefore, for each new batch of 2× HBS buffer, we prepare 3 aliquots with pH values of 7.10, 7.12, and 7.14, respectively. Transfection efficiencies are compared, and the best aliquot is subsequently used. To avoid shifts in pH, freshly thawed 2× HBS and 2 M CaCl$_2$ as well as injectable water are used to set up the transfection cocktail.

4. The medium is changed 4–6 h prior to transfection to standardize the pH of the medium at the time of transfection. Medium, that was stored with a lot of air above the liquid, was occasionally observed to give poor results. It is also advisable to keep the number of cells plated constant.

5. If transfection efficiency goes down, thawing fresh 293T cells often solves the problem.

6. For titration, 293A cells, which are a subclone of 293 cells, are used since they adhere very well to the plates and form good monolayers. Titers on 293A cells are higher than on those on 293T cells.

7. We previously concentrated vectors by ultracentrifugation as described *(19)*. However, in our hands recovery rates were rather variable. If vector supernatants have to be concentrated more than 100-fold, ultracentrifugation should be used, since higher concentration by filtration leads to viscous solutions.

References

1. Poznansky, M., Lever, A., Bergeron, L., Haseltine, W., and Sodroski, J. (1991) Gene transfer into human lymphocytes by a defective human immunodeficiency virus type 1 vector. *J. Virol.* **65,** 532–536.
2. Naldini, L., Blömer, U., Gallay, P., et al. (1996) In vivo gene delivery and stable transduction of nondividing cells by a lentiviral vector. *Science* **272,** 263–267.
3. Reiser, J., Harmison, G., Kluepfel Stahl, S., etal. (1996) Transduction of nondividing cells using pseudotyped defective high-titer HIV type 1 particles. *Proc. Natl. Acad. Sci. USA* **93,** 15266–15271.
4. Poeschla, E. M., Wong-Staal, F., and Looney, D. J. (1998) Efficient transduction of nondividing human cells by feline immunodeficiency virus lentiviral vectors. *Nat. Med.* **4,** 354–357.
5. Johnston, J. C., Gasmi, M., Lim, L. E., et al. (1999) Minimum requirements for efficient transduction of dividing and nondividing cells by feline immunodeficiency virus vectors. *J. Virol.* **73,** 4991–5000.
6. Mitrophanous, K., Yoon, S., Rohll, J., et al. (1999) Stable gene transfer to the nervous system using a non-primate lentiviral vector. *Gene Ther.* **6,** 1808–1818.
7. Schnell, T., Foley, P., Wirth, M., Münch, J., and Überla, K. (2000) Development of a self-inactivating, minimal lentivirus vector based on simian immunodeficiency virus. *Hum. Gene Ther.* **11,** 439–447.
8. Mangeot, P. E., Negre, D., Dubois, B., et al. (2000) Development of minimal lentivirus vectors derived from simian immunodeficiency virus (SIVmac251) and their use for gene transfer into human dendritic cells. *J. Virol.* **74,** 8307–8315.

9. Khabbaz, R. F., Heneine, W., George, J.R., et al. (1994) Brief report: infection of a laboratory worker with simian immunodeficiency virus. *N. Engl. J. Med.* **330,** 172–177.

10. Tang, H., Kuhen, K. L., and Wong-Staal, F. (1999) Lentivirus replication and regulation. *Annu. Rev. Genet.* **33,** 133–370.

11. Emerman, M. and Malim, M. H. (1998) HIV-1 regulatory/accessory genes: keys to unraveling viral and host cell biology. *Science* **280,** 1880–1884.

12. Yee, J. K., Miyanohara, A., LaPorte, P., et al. (1994) A general method for the generation of high-titer, pantropic retroviral vectors: highly efficient infection of primary hepatocytes. *Proc. Natl. Acad. Sci.USA* **91,** 9564–9568.

13. Wagner, R., Graf, M. , Bieler, K., et al. (2000) Rev-independent expression of synthetic gag-pol genes of HIV-1 and SIV: implications for the safety of lentiviral vectors. *Hum. Gene Ther.* **11,** 2403–2413.

14. Haas, J., Park, E. C., and Seed, B. (1996) Codon usage limitation in the expression of HIV-1 envelope glycoprotein. *Curr. Biol.* **6,** 315–324.

15. Schneider, R., Campbell, M., Nasioulas, G., Felber, B. K., and Pavlakis, G. N. (1997) Inactivation of the human immunodeficiency virus type 1 inhibitory elements allows Rev-independent expression of Gag and Gag/protease and particle formation. *J. Virol.* **71,** 4892–4903.

16. Cullen, B. R. (1986) Trans-activation of human immunodeficiency virus occurs via a bimodal mechanism. *Cell* **46,** 973–982.

17. Malim, M. H., Hauber, J., Fenrick, R., and Cullen, B. R. (1988) Immunodeficiency virus rev trans-activator modulates the expression of the viral regulatory genes. *Nature* **335,** 181–183.

18. DuBridge, R. B., Tang, P., Hsia, H. C., et al. (1987) Analysis of mutation in human cells by using an Epstein-Barr virus shuttle system. *Mol. Cell Biol.* **7,** 379–387.

19. Yee, J. K., Friedmann, T., and Burns, J. C. (1994). Generation of high-titer pseudotyped retroviral vectors with very broad host range, in *Methods in Cell Biology*, vol. 43, Academic, New York, pp. 99–112.

20. Fouchier, R. A. M., Meyer, B. E., Simon, J. H. M., Fischer, U., and Malim, M. H. (1997) HIV-1 infection of non-dividing cells: evidence that the amino-terminal basic region of the viral matrix protein is important for Gag processing but not for post-entry nuclear import. *EMBO J.* **16,** 4531–4539.

25

Cytoplasmic RNA Vector Derived from Nontransmissible Sendai Virus

Production and Use

Akihiro Iida and Mamoru Hasegawa

1. Introduction

Sendai virus (SeV) is an enveloped virus with a nonsegmented negative-strand RNA genome of 15,384 nucleotides; it is a member of the Paramyxoviridae family *(1–3)*. The virus is pneumotropic in rodent species such as mice and rats, but no pathogenicity has been reported in humans. SeV has been utilized for generation of hybrid cells between mammalian cells of different species in vitro by use of its cell fusion activity, and it has contributed to somatic cell genetics *(4)*. The inactivated SeV has been also used for generation of a fusogenic viral liposome (hemagglutinating virus of Japan, or Sendai virus [HVJ]-liposome), which aimed to improve the efficiency of liposome-mediated DNA transfer *(5,6)*.

One of the characteristics of SeV is that replication occurs exclusively in the cytoplasm of infected cells and does not go through a DNA phase; thus there is no concern for unwanted integration of foreign sequences into chromosomal DNA. No need to transfer genetic materials to nuclei of target cells might be another advantage. Therefore, we expect that SeV may be a safer and more efficient viral vector than existing vectors for application to human therapy in various fields including gene therapy.

Methods to rescue infectious SeV from cloned cDNA were established by two groups in 1995 and 1996, respectively *(7,8)*. Since then, such reverse genetics technology allowed construction of genetically engineered SeVs, which carry additional foreign genes, and opened the way for the development of gene transfer vectors (**Fig. 1**). The addendum type of the vectors prepared by

From: *Methods in Molecular Medicine, Vol. 69, Gene Therapy Protocols, 2nd Ed.*
Edited by: J. R. Morgan © Humana Press Inc., Totowa, NJ

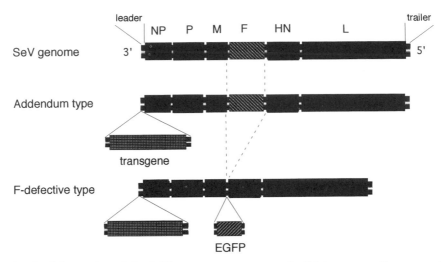

Fig. 1. Schematics of the SeV genome structures of wild type as well as recombinant addendum and F-defective types. The SeV genome is 15,384 nt long and its genes (NP, P, M, F, HN, and L) are in order from 3' to 5' in the nonsegmented negative-strand RNA. Addendum type has been constructed and reported previously *(9–14)*. In the F-defective type, the entire F gene has been removed by PCR strategy, and transgenes can be introduced in a unique *Not*I site between the leader sequence and the NP gene. The text describes the introduction of the enhanced green fluorescent protein (EGFP) reporter gene between the M and HN genes.

this method have shown high efficiency of gene transfer and expression of foreign proteins in mammalian cells in vitro and in vivo *(9–14)*. However, the recombinant SeVs constructed in these studies have contained all the viral structural genes and thus propagated virions that were fully infectious and capable of spreading in animal bodies.

We have recently succeeded in establishing a method of constructing nontransmissible SeV vectors capable of self-replication in infected cells but incapable of infecting neighboring cells. This novel type of vector has been shown to keep the same ability of gene introduction and expression and is thus expected to be useful in human therapy *(15)*.

1.1. Design of F-defective SeV Vectors

The SeV genome has a leader and a trailer region at each of its 3' and 5' ends, and contains six major genes, which are lined up in tandem on single negative-strand RNA (**Fig. 1**). Initiation of replication and transcription of the viral genes occurs in the leader region in infected cells. SeV genomic RNA and the three virus proteins, the nucleoprotein (NP), phosphoprotein (P), and large protein (L; the catalytic subunit of RNA polymerase), form a ribonucleoprotein complex (RNP), which acts as a template for transcription and replication. Matrix protein (M) engages in the assembly and budding of viral particles. Two envelope glycoproteins, hemagglutinin-neuraminidase (HN) and fusion protein (F), mediate attachment of virions and penetration of RNPs into infected cells. HN protein binds to the cell surface sialic acids and then liberates them to make asialo receptors. F protein is essential for the infectivity of SeV; it is synthesized as an inactive precursor protein, F_0, and then split into F_1 and F_2 by proteolytic cleavage of trypsin-like enzyme. In our intial design for nontransmissible SeV vectors, we deleted the F gene from the SeV genome of the attenuated Z strain *(16)* and replaced it with a reporter enhanced green fluorescence protein (EGFP) gene. EGFP expression was detectable in a single living cell, which allowed us to validate successful recovery of RNPs containing an F-defective RNA genome inside such cells.

For expressing additional foreign gene(s) of interest, the gene can be inserted in a unique *Not*I restriction enzyme cleavage site located between the start signal and the initiation ATG codon of the NP gene *(9)*. The gene can be amplified by polymerase chain reaction (PCR) with transcriptional end (E) and restart (S) signals connected by a trinucleotide intergenic sequence at the 3' end and *Not*I recognition sequences at both ends *(9)*. The length of the fragment should be a multiple of 6 bp, following rule of six for efficient SeV replication *(17,18)*.

1.2. Construction of the Packaging Cell Line

SeV F protein is required for the formation of infectious SeV particles. Therefore, recovery of SeV vectors from the RNA genome lacking the F gene has to be complemented with this gene in *trans*. We thus constructed an F-expressing packaging $LLC-MK_2$ cell line with a Cre/*loxP*-inducible expression system *(19)*. $LLC-MK_2$ cells were transfected with the plasmid pCALNdLw/F (with the F gene located under the stuffer *neo* sequence, flanked by a pair of *loxP* sequences), and stable Neo^r clones were isolated. The recombinant adenovirus vector AxCANCre expressing Cre recombinase *(20)* was infected into these neo^r clones at a multiplicity of infection (MOI) of 3. Seven of 15 clones expressed F protein inducibly; the clone that showed the highest F protein was designated $LLC-MK_2/F7$ and used as a packaging cell line for the F-defective SeV vectors.

1.3. Generation of F-defective SeV Vectors

For generating F-defective SeV vectors, we have devised the following procedure, which consists of two steps (**Fig. 2**). The first step is to recover RNPs of F-defective RNA genome in $LLC-MK_2$ cells using a plasmid bearing F-defective cDNA and three other plasmids expressing NP, P, and L proteins,

Step 1: Generation of functional RNP   Step 2: Recovery of F-defective SeV vector

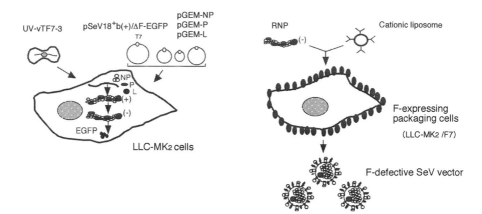

Fig. 2. System for generating the F-defective SeV vectors from SeV cDNAs carrying foreign genes. Step 1: LLC-MK$_2$ cells are transfected with SeV cDNA defective in the F gene and the three plasmids encoding viral proteins, NP, P, and L, under the transcriptional control of T7 RNA polymerase. The enzyme is provided by the recombinant vaccinia virus vTF7-3 previously inactivated by psoralen and long-wave UV light (UV-vTF7-3). Transfection resulted in the formation of RNP complexes containing an F-defective RNA genome. Step 2: The reconstituted functional RNPs in step 1 are transfected into F-expressing LLC-MK$_2$/F7 cells with cationic liposome (DOSPER). RNP complexes are rescued to generate infectious viral particles (SeV vector), which are released into the supernatant of the cells.

respectively. All these cDNAs were under transcriptional control of the T7 promoter, which can be driven by T7 RNA polymerase from the recombinant vaccinia virus vTF7-3 *(21)*. We confirmed the formation of functional RNPs by observing EGFP-expressing cells 3 days after transfection. Such cells were observed only when these four materials were cotransfected into LLC-MK$_2$ cells.

The second step is to transfect the RNPs into the F-expressing packaging cell line and to collect infectious particles (SeV vectors) from the supernatant. To raise the recovery efficiency of RNPs in the first step, we pretreated vTF7-3 with psoralen and long-wave UV irradiation. This treatment inactivated replication capability of the viruses without impairing infectivity and T7 RNA polymerase expression *(22)*. We estimated the recovery frequency by the method as described previously *(7)* using wild-type SeV cDNA and inoculating the diluted lysates of transfected cells into embryonic hen eggs. With untreated vTF7-3, 1 infectious unit was detected from 10^5 transfected cells *(7)*. However, with psoralen and long-wave UV-treated vTF7-3, 1 infectious unit was formed from 10^3 cells, indicating an improvement of nearly 100-fold. As to the F-defective SeV cDNA, the numbers of EGFP-expressing cells were scored to estimate the recovery efficiency of functional RNP. Under these conditions, such cells were detected approximately 1 in 10^5 transfected cells. The lysates containing functional RNPs were obtained by freeze and thaw cycles, mixed with cationic liposome, and transfected into F protein-expressing LLC-MK_2/F7 cells. The infectious virus particles were recovered in the culture supernatant, and the recovery efficiency was at least 1 infectious unit from 10^5 transfected cells. The particles were recovered only from LLC-MK_2/F7 cclls cultured with trypsin. When the cells were cultured in the absence of trypsin, scattered EGFP expression signals were detected, but they did not spread to neighboring cells. After passaging the supernatant to LLC-MK_2/F7 cells, the infectious titer of the particles reached $0.5-1.0 \times 10^8$ cell infectious units (CIU)/mL.

1.4. Introduction and Expression of the EGFP Gene by F-Defective SeV Vectors in Cultured Cells

SeV HN protein interacts with sialic acids, which exist in glycoproteins and glycolipids on the cell surface as the receptor, and thus permits the vectors to enter into a wide variety of cell types. Indeed, EGFP in the F-defective SeV vector can be expressed in various normal human cells including smooth muscle cells, hepatocytes, and lung microvascular endothelial cells, and in primary cultures of fetal rat hippocampal neurons. The infection are usually performed by incubating the cells for 1 h with an appropriate MOI of vector stock, after which the cells are incubated in complete medium.

2. Materials
2.1. Cell Culture

1. LLC-MK_2 cell line (rhesus monkey kidney; ATCC CCL-7) and LLC-MK_2/F7 cell line (*see* **Subheading 1.2.**): Cells were cultured at 37°C in a humidified 5% CO_2 atmosphere.
2. MEM + 10% FCS: Eagle's modified essential medium (MEM) supplemented with 10% heat-inactivated fetal calf serum (FCS), 100 U/mL penicillin, and 100 µg/mL streptomycin sulfate.
3. Cytosine arabinoside (Ara-C) stock solution: Dissolve Ara-C (Sigma) in H_2O at 40 mg/mL, filter-sterilize, and store at –20°C.
4. 0.25% (w/v) trypsin solution.
5. MEM + Ara-C + trypsin: MEM supplemented with Ara-C at 40 µg/mL (1:1000 volume of Ara-C stock solution) and trypsin at 7.5 µg/mL (3:1000 volume of 0.25% trypsin solution).
6. MEM + trypsin: MEM supplemented with trypsin at 7.5 µg/mL.

7. Phosphate-buffered saline (PBS): 170 m*M* NaCl, 3.4 m*M* KCl, 10 m*M* Na$_2$HPO$_4$, and 1.8 m*M* KH$_2$PO$_4$ (pH 7.2).

2.2. Psoralen and Long-Wave UV Light Inactivation of Recombinant Vaccinia Virus (UV-vTF7-3)

1. Recombinant vaccinia virus stock (vTF7-3) (*see* **Note 1**).
2. Bovine serum albumin fraction V solution (7.5% [w/v] BSA).
3. 0.1% BSA/PBS: PBS containing 0.1% (w/v) BSA.
4. STRATALINKER (Stratagene).
5. Psoralen solution: Dissolve psolaren (Aldrich) in dimethyl sulfoxide (DMSO) at 1 mg/mL and store 1-mL aliquots at –20°C in the dark.

2.3. Inducible Expression of F Protein in LLC-MK$_2$/F7 Cells

1. LLC-MK$_2$/F7 cells in 6-well tissue culture dishes.
2. Recombinant adenovirus AxCANCre (RIKEN Gene Bank, RDB1748).
3. Anti-SeV F protein monoclonal antibody (f-236) *(23)* (*see* **Note 2**).

2.4. Generation of F-Defective SeV Vectors

1. Monolayers of LLC-MK$_2$ cells in 100-mm tissue culture dishes.
2. Psoralen and long-wave UV-inactivated recombinant vaccinia virus vTF7-3 (UV-vTF7-3).
3. SuperFect transfection reagent (Qiagen).
4. DNA materials: The plasmids pSeV18$^+$b(+)/ΔF-EGFP (F-defective SeV cDNA with EGFP gene) *(15)* and pGEM-NP, pGEM-P, and pGEM-L, expressing NP, P, and L, respectively *(24)* (*see* **Note 3**).
5. Opti-MEM medium (Gibco-BRL).
6. Monolayers of F protein-induced LLC-MK$_2$/F7 cells in 24-well tissue culture dishes.
7. DOSPER liposomal transfection reagent (Roche Diagnostic).

3. Methods

3.1. Generation of F-Defective SeV Vectors

3.1.1. Preparation of Psoralen and Long-Wave UV-Inactivated Recombinant Vaccinia Virus

1. Adjust the titer of vTF7-3 to 1×10^8 pfu/mL with 0.1% BSA/PBS.
2. Add psoralen solution to the virus suspension at 0.3 µg/mL and keep on ice for 10 min.
3. Transfer 1-mL aliquots of the virus suspension to 35-mm tissue culture dishes.
4. Irradiate 350-nm long-wave UV light to the virus suspension for 20 min using STRATALINKER.
5. Transfer 0.5-mL aliquots of the virus solution to 2-mL cryogenic vials and keep at –80°C until use.

3.1.2. Inducible Expression of F Protein in LLC-MK₂/F7 Cells

1. Day 1: Plate 5×10^5 LLC-MK$_2$/F7 cells in 6-well tissue culture dishes and culture at 37°C overnight.
2. Day 2: Wash the cells once with MEM and infect the recombinant adenovirus AxCANCre at a MOI of 3. Tilt the dishes every 15 min to ensure uniform distribution of the virus adsorption and incubate at 37°C for 1 h. Wash the cells once with MEM, add 2 mL of MEM + 10% FCS, and incubate at 37°C for 2 days.
3. Day 4: Examine the expression of F protein by Western blotting using the anti-F protein monoclonal antibody f-236. The cells expressing F protein can be passaged within 20 generations in use for vector propagation.

3.1.3. Generation of F-Defective SeV Vectors from cDNA Stocks

1. Day 1: Plate 4×10^6 LLC-MK$_2$ cells in 100-mm-diameter tissue culture dishes and incubate at 37°C overnight to 100% confluency.
2. Day 2: Wash the cells once with MEM and infect them with vTF7-3 at an MOI of 2. Tilt the dishes every 15 min at room temperature for 1 h to ensure uniform distribution of the virus adsorption. Wash the cells twice with MEM and transfect with a mixture containing pSeV18$^+$b(+)/ΔF-EGFP (12 µg), pGEM-NP (4 µg), pGEM-P (2 µg), pGEM-L (4 µg), and 110 µL of SuperFect transfection reagent in 200 µL Opti-MEM, which was preincubated at room temperature for 10 min in a 15-mL polystyrene tube. Maintain the transfected cells in 3 mL of Opti-MEM with 3% FCS for 3 h, then wash them once with MEM and incubate in 10 mL MEM containing 40 µg/mL Ara-C for 60 h under 5% CO$_2$ conditions. (*See* **Notes 4** and **5**).
3. Day 5: Examine EGFP expression of the transfected cells under fluorescence microscopy using a 450–490- and 500–550-nm exitation bandpass filter to validate the recovery of RNPs inside of the cells. Scrape the transfected cells from the dish, collect by centrifugation at 1000*g* for 5 min, resuspend in 800 µL Opti-MEM, and lyse by three cycles of freezing and thawing. Mix the lysate with 600 µL of Opti-MEM and 200 µL of DOSPER in a 15-mL polystyrene tube for 15 min at room temperature. Transfect the F-expressing LLC-MK$_2$/F7 cells in 24-well tissue culture dishes with 200 µL each of the lysate mixture.
4. Day 6: 24 h after the transfection, wash the cells once with MEM and incubate in 0.5 mL MEM + Ara-C + trypsin for 3–6 days.
5. Days 9–12: Examine the spread of EGFP-expressing cells under fluorescence microscopy. Determine the vector titer by the method described in **Subheading 3.3.** The vector suspension can be passaged by addition of the supernatant to LLC-MK$_2$/F7 cells and incubation for 2–5 days in MEM + Ara-C + trypsin. After centrifugation of the supernatant at 1000*g* for 5 min, aliquot the vector suspension and store at –80°C as the vector stock. Determine the final vector titer as indicated in **Subheading 3.3.** after quick thawing of the stock in a tube at 37°C for 5 min (*see* **Notes 6** and **7**).

3.2. Propagation of F-Defective SeV Vectors

1. Day 1: Plate 5×10^6 cells of F protein-induced LLC-MK$_2$/F7 cells in 100-mm tissue culture dishes and incubate overnight to approximately 100% confluency.
2. Day 2: Wash the cells once with MEM and infect with the SeV vector stock at an MOI of 0.1–1. Tilt the dishes every 15 min to distribute the inoculum and incubate at 37°C for 1 h. Wash the cells once with MEM and add 10 mL MEM + trypsin. Incubate the dishes at 37°C under 5% CO$_2$ conditions.
3. Days 3–5: Harvest the supernatants every 24 h and replace with a fresh 10 mL of MEM + trypsin for 3 days. After centrifugation of the supernatant at 1000g for 5 min, aliquot the virus suspension, store at –80°C, and determine the vector titer.

3.3. Titration of F-Defective SeV Vectors by Cell Infectious Assay

1. Day 1: Plate 5×10^5 LLC-MK$_2$ cells in 6-well tissue culture dishes.
2. Day 2: Prepare a series of tenfold dilutions of a vector stock in 1 mL MEM. Add the 100 μL of each dilution of the stock plus 900 μL MEM to the cells in duplicate. Tilt the dishes every 20 min and incubate at 37°C for 1 h. Wash the cells once with MEM and add 2 mL MEM + 10% FCS.
3. Day 4: Count the number of EGFP-expressing cells under a fluorescence microscope in triplicate. The total average number in the well can be obtained by multiplying this number by total area/counted area. The titer can be expressed as CIU/mL.

3.4. Infection of F-Defective SeV Vectors In Vitro

1. Infect cells with a vector stock at an appropriate MOI. Tilt the dishes every 15 min and incubate at 37°C for 1 h. Wash the cells once with MEM and add 10 mL complete medium.
2. Incubate for several days and examine the expression of EGFP under a fluorescence microscope and/or the gene(s) of interest by methods such as RT-PCR, Western blotting, or enzyme-linked immunosorbent assay.

4. Notes

1. The recombinant vaccinia virus vTF7-3 may be obtained from Dr. B. Moss (NIAID, NIH, Bethesda, MD).
2. Anti-F protein monoclonal antibody f-236 can recognize F$_0$ protein but not F$_1$ or F$_2$ protein by Western blotting. The antibody can also be used for flow cytometric analysis of F-expressing cells.
3. All plasmid DNA samples should be prepared through cesium chloride gradients or commercial kits by ion-exchange chromatography.
4. The cytopathic effect by UV-vTF7-3 still exists and causes rounding and detaching of LLC-MK$_2$ cells. Care must be taken when washing and changing the medium.
5. The ratio of NP-, P-, and L-expressing plasmids is an important determinant for the efficient recovery of RNP from SeV cDNA, as described in the **ref.** *(7)*.

6. The recombinant vaccinia virus vTF7-3 exists in SeV vector stock in early stages of passage even after the virus has been inactivated by psoralen and long-wave UV light. To remove the residual vaccinia virus, SeV vectors should be propagated in the medium containing Ara-C for several passages. The vaccinia virus can be detected by plaque assay *(7)*.

7. Recovery of vectors is dependent on several factors such as cell density, transfection efficiency, and successful preparation of UV-vTF7-3. Improvements in the efficiency of the recovery procedure are in progress.

Acknowledgments

We thank Atsushi Kato and Yoshiyuki Nagai at the National Institute of Infectious Disease in Tokyo, Japan, and Hai-Ou Li, Ya-Feng Zhu, Makoto Asakawa, Hidekazu Kuma, Takahiro Hirata, Yasuji Ueda, Yun-Sik Lee, and Masayuki Fukumura at DNAVEC Research Inc. for contributing to the work.

References

1. Lamb, R. A. and Kolakofsky, D. (1996) *Paramyxoviridae*: the viruses and their replication,. In *Virology*, 3rd ed. (Fields, B. N., Knipe, D. M., and Howley, P. M., eds.), Lippincott-Raven, Philadelphia, pp. 1177-1204.
2. Conzelmann, K.-K. (1998) Nonsegmented nagative-strand RNA viruses: Genetics and manipulation of viral genomes. *Ann. Rev. Genet.* **32,** 123–162.
3. Nagai, Y. and Kato, A. (1999) Paramyxovirus reverse genetics is coming of age. *Microbiol. Immunol.* **43,** 613–624.
4. Okada, Y. and Tadokoro, J. (1962) Analysis of giant polynuclear cell formation caused by HVJ virus from Ehrlich's ascites tumor cells. *Exp. Cell. Res.* **26,** 98–128.
5. Kato, K., Nakanishi, M., Kaneda, Y., Uchida, T., and Okada, Y. (1991) Expression of hepatitis B virus surface antigen in adult rat liver. *J. Biol. Chem.* **26,** 3361–3364.
6. Dzau, V., Mann, M., Morishita, R., and Kaneda, Y. (1996) Fusigenic viral liposome for gene therapy in cardiovascular diseases. *Proc. Natl .Acad. Sci. USA* **93,** 11421–11425.
7. Kato, A., Sakai, Y., Shioda, T., et al. (1996) Initiation of Sendai virus multiplication from transfected cDNA or RNA with negative or positive sense. *Genes Cells* **1,** 569–579.
8. Garcin, D., Pelet, T., Calain, P., et al. (1995) A highly recombinogenic system for the recovery of infectious Sendai paramyxovirus from cDNA. *EMBO J.* **14,** 5773–5784.
9. Hasan, M. K., Kato, A., Shioda, T., et al. (1997) Creation of an infectious recombinant Sendai virus expressing the firefly luciferase from 3' proximal first locus. *J. Gen. Virol.* **78,** 2813–2820.
10. Moriya, C., Shioda, T., Tashiro, K., et al. (1998) Large quantity production with extreme convenience of human SDF-1α and SDF-1β by a Sendai virus vector. *FEBS Lett.* **425,** 105–111.

11. Yu, D., Shioda, T., Kato A., et al. (1997) Sendai virus-based expression of HIV-1 gp120: reinforcement by the V(-) version. *Genes Cells* **2**, 457–466.
12. Sakai, Y., Kiyotani, K., Fukumura, M., et al. (1999) Accommodation of foreign genes into the Sendai virus genome: sizes of inserted genes and viral replication. *FEBS Lett.* **456**, 221–226.
13. Toriyoshi, H., Shioda, T., Sato, H., et al. (1999) Sendai virus-based production of HIV type 1 subtype B and subtype E envelope glycoprotein 120 antigen and their use for highly sensitive detection of subtype-specific serum antibody. *AIDS Res. and Hum. Retroviruses* **12**, 1109–1120.
14. Yonemitsu Y., Kitson, C., Ferrari, S., et al. (2000) Efficient gene transfer to airway epithelium using recombinant Sendai virus. *Nat. Biotechnol.* **18**, 970–973.
15. Li, H-O., Zhu, Y-H., Asakawa, M., et al. (2000) A cytoplasmic RNA vector derived from nontransmissible Sendai virus with efficient gene transfer and expression. *J. Virol.* **74**, 6564–6569.
16. Shioda, T., Iwasaki, K., and Shibuta, H. (1986) Determination of the complete nucleotide sequence of Sendai virus genome RNA and the predicted amino acid sequence of the F, HN, and L proteins. *Nucleic Acids Res.* **14**, 1545–1563.
17. Calain, P. and Roux, L. (1993) The rule of six, a basic feature for efficient replication of Sendai virus defective interfering RNA. *J. Virol.* **67**, 4822–4830.
18. Korakofsky, D., Pelet, T., Garcin, D., et al. (1998) Paramyovirus RNA synthesis and requirement for hexamer genome length: the rule of six revisited. *J. Virol.* **72**, 891–899.
19. Arai, T., Matsumoto, K., Saitoh, K., et al. (1998) A new system for stringent, high-titer vesicular stomatitis virus G protein-pseudotyped retrovirus vector induction by introduction of Cre recombinase into stable prepackaging cell lines. *J. Virol.* **72**, 1115–1121.
20. Kanegae, Y., Takamori K., Sato, Y., et al. (1996) Efficient gene activation system on mammalian cell chromosome using recombinant adenovirus producing Cre recombinase. *Gene* **181**, 207–212.
21. Fuerst, T. R., Niles, E. G., Studier, F. W., and Moss, B. (1986) Eukaryotic transient-expression system based on recombinant vaccinia virus that synthesizes bacteriophage T7 RNA polymerase. *Proc. Natl. Acad. Sci. USA* **83**, 8122–8126.
22. Tsung, K., Yim, J. H., Marti, W., Buller, R. M. L., and Norton, J. A. (1996) Gene expression and cytopathic effect of vaccinia virus inactived by psoralen and long-wave UV light. *J Virol.* **70**, 165–171.
23. Segawa, H., Kato, M., Yamashita, T., and Taira, H. (1998) The role of individual cysteine residues of Sendai virus fusion protein in intracellular transport. *J. Biochem.* **123**, 1064–1072.
24. Curran, J., Boeck, R., and Kolakofsky, D. (1991) The Sendai virus P gene expresses both an essential protein and an inhibitor of RNA synthesis by shuffling modules via mRNA editing. *EMBO J.* **10**, 3079–3085.

26

Preparation of Helper-Dependent Adenoviral Vectors

Philip Ng, Robin J. Parks, and Frank L. Graham

1. Introduction
1.1. First-Generation Adenovirus Vector

Adenoviruses (Ads) are excellent mammalian gene transfer vectors because of their ability to infect efficiently a wide variety of quiescent and proliferating cell types from various species to direct high-level gene expression. Consequently, Ad vectors are extensively used as potential recombinant viral vaccines, for high-level protein production in cultured cells and for gene therapy *(1–4)*. First-generation Ad vectors typically have foreign DNA inserted in place of early region 1 (E1). E1-deleted vectors are replication deficient and are propagated in E1-complementing cells such as 293 *(5)*. Although these vectors remain very useful for many applications, it has become clear that transgene expression in vivo is only transient. Several factors contribute to this, including strong innate and inflammatory responses to the vector *(6,7)*, acute and chronic toxicity caused by low-level viral gene expression from the vector backbone *(8)*, and generation of anti-Ad cytotoxic T-lymphocytes caused by *de novo* viral gene expression *(9–12)* or processing of virion proteins *(13)*. Although high-level transient transgene expression afforded by first-generation Ad vectors may be adequate, or even desirable, for many gene transfer and gene therapy applications, the transient nature of expression kinetics renders these vectors unsuitable when prolonged, stable expression is required.

1.2. Helper-Dependent Adenovirus Vector

In principle, the simplest way to eliminate Ad-induced toxicity caused by expression of viral proteins is to remove all viral coding sequences from the

From: *Methods in Molecular Medicine, Vol. 69, Gene Therapy Protocols, 2nd Ed.*
Edited by: J. R. Morgan © Humana Press Inc., Totowa, NJ

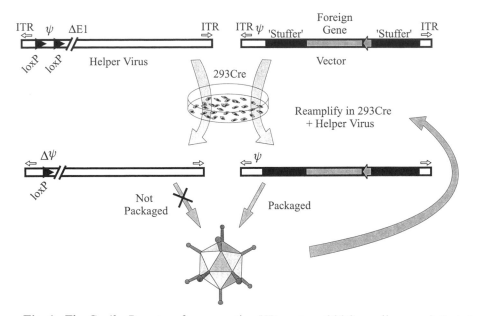

Fig. 1. The Cre/*lox*P system for generating HD vectors. 293Cre cells are coinfected with the HD vector and a helper virus bearing a packaging signal flanked by *lox*P sites. Cre-mediated excision of the packaging signal (ψ) renders the helper virus genome unpackageable but does not interfere with its ability to provide all the necessary *trans*-acting factors for propagation of the HD vector. The titer of the HD vector is increased by serial passage in helper virus-infected 293Cre cells. The HD vector need contain only those Ad *cis*-acting elements required for DNA replication (ITRs) and encapsidation (ψ); the remainder of the genome consists of the desired transgene and non-Ad stuffer sequences.

vector. In the absence of cell lines that contain and express the entire Ad genome, a "helper" Ad is required to provide all necessary functions in *trans* for propagation of these fully deleted, helper-dependent (HD) vectors. The only Ad sequences required in *cis* for viral replication are the inverted terminal repeats (ITRs) necessary for DNA replication and the packaging signal (ψ) necessary for encapsidation of the vector genome. These *cis*-acting elements amount to only 500 bp. HD vectors retain all the benefits of first-generation Ad vectors but have the added advantage of increased cloning capacity (37 kb), and it is now becoming clear that they offer the potential for reduced toxicity and prolonged, stable transgene expression *(14–21)*. The purpose of this chapter is to provide a detailed protocol for the preparation of HD vectors. The reader is referred elsewhere for excellent, comprehensive reviews on HD vectors *(22,23)*.

1.3. Overview of the Helper-Dependent Adenovirus Vector System

A critical feature of this system is a method of inhibiting helper virus propagation without interfering with its ability to provide helper functions for propagation of the HD vector. To date, the most efficient means of achieving this goal is the Cre/*lox*P method developed by Graham and co-workers (**Fig. 1**) *(24)*. In this system the packaging signal of the helper virus is flanked by *lox*P sites. Thus, following infection of 293 cells expressing Cre recombinase, such as 293Cre4 cells *(25)*, the packaging signal is excised from the helper virus by Cre-mediated site-specific recombination between the *lox*P sites, rendering the genome unpackageable. However, the ability of the viral DNA to replicate and provide helper functions in *trans* for propagation of a coinfected HD vector is not impaired. The titer of the HD vector is increased by serial coinfection of 293Cre cells with the HD vector and the helper virus. In the last step, the HD vector is purified by CsCl ultracentrifugation. Although a number of factors can influence the yield and purity of HD vectors (*see* **Subheading 3.**), typically about 10^{10}–10^{11} virus particles are produced per 150-mm dish of coinfected cells with a helper virus contamination level of ≤0.1% using the method described in **Subheading 3**.

2. Materials
2.1. Cell Culture
1. Low-passage 293 and 293Cre4 cells.
2. Complete medium: Minimum essential medium (MEM; Gibco-BRL cat. no. 61100) containing 10% fetal bovine serum (FBS) (heat inactivated), 100 U/mL penicillin/streptomycin, 2 m*M* L-glutamine, and 2.5 µg/mL fungizone.
3. G418.
4. Citric saline: 135 mM KCl, 15 m*M* sodium citrate, autoclave-sterilized

2.2. Helper-Dependent Adenovirus Vector

The helper-dependent vector contains the expression cassette of interest and 500 bp of *cis*-acting Ad sequences necessary for vector DNA replication (ITRs) and packaging (ψ). In addition to these minimal Ad sequences, Sandig et al. *(26)* have shown that inclusion of a small segment of noncoding Ad sequence from the E4 region adjacent to the right ITR increases vector yields, possibly by enhancing packaging of the vector DNA.

The size of the vector is an important consideration. For efficient packaging the Ad DNA should be between 75% (>27 kb) *(27)* and 105% (<37.8 kb) *(28)* of the wild type. Vector backbones outside this range may undergo rearrangements or may not be efficiently packaged, if at all. Thus, the vector must often include stuffer DNA if the genes or cassettes to be cloned are relatively small.

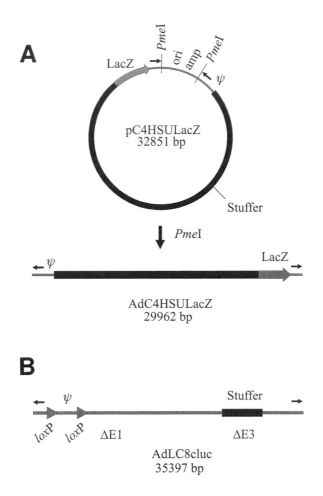

Fig. 2. (A) The plasmid pC4HSULacZ contains an HD vector genome bearing a LacZ reporter gene and was constructed by inserting the LacZ expression cassette into the HD vector plasmid backbone pC4HSU *(26)*. The HD vector genome is liberated from the bacterial sequences by restriction enzyme digestion, and HD vectors (AdC4HSULacZ) are generated following transfection of the linear vector DNA into 293Cre cells and infection with the helper virus. **(B)** The helper virus, AdLC8cluc *(24)* is a first-generation Ad virus (E1-deleted) bearing a packaging signal (ψ) flanked by *lox*P sites. Following infection of 293Cre cells, the helper virus genome is rendered unpackageable due to excision of the packaging signal by Cre-mediated site-specific recombination between the *lox*P sites; however, its ability to replicate and provide all necessary *trans*-acting factors for propagation of an HD vector is unimpaired. A firefly luciferase expression cassette, which serves as a stuffer sequence, is inserted in place of the E3 region to prevent the formation of replication-competent Ad (RCA; E1[+]). Small black arrows represent Ad ITR. ψ represents the packaging signal.

The choice of stuffer DNA is important with regard to vector stability and replication efficiency *(26,29)*. Although it remains unclear what constitutes a good stuffer, in general, noncoding eukaryotic DNA is preferable; repetitive elements and unnecessary homology with the helper virus should be avoided to ensure vector stability and prolonged transgene expression.

In addition to the absolute size of the HD vector, its relative size compared with the helper virus is an important consideration. This is because Cre-mediated selection against the helper virus, although efficient, is not absolute. If the genome size of the HD vector is sufficiently different from that of the helper virus, then the two species can be physically separated by CsCl ultracentrifugation based on their different buoyant densities. HD vectors between 28 and 31 kb are ideal because they are efficiently packaged and easily separated from residual helper viruses (35–37 kb) following CsCl ultracentrifugation.

The HD vector is initially constructed as a bacterial plasmid (**Fig. 2A**) such that the vector genome can be liberated from the bacterial sequences prior to transfection of 293Cre4 cells by restriction enzyme digestion (*see* **Subheading 3.2.**). The restriction enzyme sites are located immediately adjacent to the ITRs. Obviously, the restriction enzyme used should not cleave elsewhere in the vector genome. The HD vector plasmid DNA used for transfection should be of high quality. (DNA purification by CsCl gradient centrifugation is the method of choice in our laboratory.)

2.3. Helper Virus

The helper virus is a first-generation Ad (E1-deleted) with its packaging signal flanked by *lox*P sites (**Fig. 2B**). As with all first-generation Ad vectors propagated in 293 or 293Cre4 cells, the potential exists for the generation of replication-competent Ad (RCA; E1$^+$) as a consequence of homologous recombination between the helper virus and the Ad sequences present in 293 cells. To prevent the formation of RCA, a stuffer sequence is inserted into the E3 region to render any E1$^+$ recombinants too large to be packaged *(24)* . The size of the stuffer should be chosen such that the total size of the helper virus genome is <105% of the wild type but >105% following homologous recombination with Ad sequences in the 293 cells. The choice of sequence used as stuffer in the E3 region may be important, as it has been observed that some sequences result in poor virus propagation, probably because of interference with fiber expression (F. L. Graham; unpublished results). Although it remains unclear what sequences constitute a good E3 stuffer, in general, noncoding sequences with no homology to the HD vector are desirable. A number of sequences have been found to be stable and not to affect virus propagation adversely *(24,26,30)*.

It is important that helper virus stocks be derived from plaque-purified isolates since any variants present may be preferentially amplified during serial passage. High-titer stocks should be purifed by CsCl ultracentrifugation, dialyzed against 10 mM Tris-HCl, pH 8.0, supplemented with glycerol to 10%, and stored in aliquots at –70°C. Working stocks are diluted (to 2×10^8 plaque-forming units [pfu]/mL) in PBS^{2+} supplemented with glycerol to 10% and stored in small aliquots at –70°C (*see* **Note 1**).

2.4. Rescue and Amplification of HD Adenovirus Vector

1. Monolayers of 293Cre4 cells at 90% confluency in 60-mm dishes.
2. HEPES-buffered saline (HBS): 21 mM HEPES, 137 mM NaCl, 5 mM KCl, 0.7 mM Na$_2$HPO$_4$, 5.5 mM glucose, pH 7.1 (adjusted with NaOH), filter-sterilized. Store in small aliquots at 4°C in tightly sealed conical plastic tubes.
3. TE: 10 mM Tris-HCl, pH 8.0, 1 mM EDTA, pH 8.0, autoclave sterilized.
4. Salmon sperm DNA (2 µg/µL in TE).
5. 2.5 M CaCl$_2$ filter-sterilized. Store in small aliquots at 4°C in tightly sealed conical tubes.
6. Complete medium (*see* **Subheading 2.1.**).
7. Maintenance medium: MEM (Gibco-BRL, cat. no. 61100) containing 5% horse serum (HS) (heat inactivated), 100 U/mL penicillin/streptomycin, 2 mM L-glutamine, and 2.5 µg/mL fungizone.
8. HD vector plasmid DNA (*see* **Subheading 2.2.**).
9. Helper virus (diluted working stock) (*see* **Subheading 2.3.**).
10. Phosphate-buffered saline (PBS): 137 mM NaCl, 8.2 mM Na$_2$HPO$_4$, 1.5 mM KH$_2$PO$_4$, 2.7 mM KCl, autoclave-sterilized.
11. PBS^{2+}: PBS supplemented with 0.01 vol 68 mM sterile MgCl$_2$ and 0.01 vol 50 mM sterile CaCl$_2$.
12. Pronase solution: 20 mg/mL pronase in 10 mM Tris-HCl, pH 7.5. Preincubate at 56°C for 15 min followed by 37°C for 1 h. Aliquot and store at –20°C.
13. Pronase-sodium dodecyl sulfate (SDS) solution: 0.5 mg/mL pronase (above) in 0.5% SDS, 10 mM Tris-HCl, pH 7.4, 10 mM EDTA, pH 8.0.
14. 40% sucrose solution, filter-sterilize.
15. β-gal fix solution: 0.2% glutaraldehyde, 2% paraformaldehyde, 2 mM MgCl$_2$ in PBS (optional).
16. β-gal stain solution: 5 mM K$_4$Fe(CN)$_6$, 5 mM K$_3$Fe(CN)$_6$, 2 mM MgCl$_2$ in PBS (optional).
17. X-gal: 20 mg/mL in N,N'-dimethylformamide (optional).

2.5. Large Scale Isolation and Purification of Helper-Dependent Vector

1. 150-mm dishes of 90% confluent 293Cre4 cells. The number of dishes is dictated by the amount of HD vector desired.
2. Helper virus (high-titer stock) (*see* **Subheading 2.3.**).
3. 10 and 100 mM Tris-HCl, pH 8.0, autoclave sterilized.
4. 5% sodium deoxycholate, filter sterilize.

5. 2 M MgCl$_2$, autoclave sterilize.
6. DNAase I: 100 mg bovine pancreatic deoxyribonuclease I in 10 mL of 20 mM Tris-HCl, pH 7.4, 50 mM NaCl, 1 mM dithiothreitol, 0.1 mg/mL BSA, 50% glycerol. Aliquot and stored at –20°C.
7. CsCl solutions:

Density (g/mL)	CsCl (g)	10mM Tris HCl, pH 8.0
1.5	90.8	109.2
1.35	70.4	129.6
1.25	54.0	146.0

Dissolve CsCl into 10 mM Tris-HCl, pH 8.0, solution in the amounts indicated above to achieve the desired density solution, and filter sterilize. Weigh 1.00 mL to confirm density.
8. Glycerol, autoclave-sterilized.
9. Beckman SW41 rotor and ultraclear tubes.
10. Beckman SW 50.1 rotor and ultraclear tubes.
11. Slide-A-Lyzer dialysis cassettes (Pierce).

3. Methods

3.1. Cell Culture

Low-passage 293 and 293Cre4 cells are maintained in 150-mm dishes and split 1 to 2 or 1 to 3 when they reach 90% confluency (every 3–4 days). 293Cre4 cells are maintained under 0.4 mg/mL G418 selection. In general, an approx 90% confluent 150-mm dish of 293 or 293Cre4 cells is split into eight and six 60-mm dishes, respectively, for use the next day for transfections or infections.

1. Remove medium from 150-mm dish of cells.
2. Rinse monolayer with 5 mL citric saline (twice for 293 cells and once for 293Cre4 cells).
3. Remove citric saline from **step 2**, add 0.5 mL citric saline, and leave dish at room temperature until cells start to round up and detach from dish (no more than 15 min).
4. Tap the sides of the dishes to detach all cells.
5. Resuspend cells in complete medium (supplemented with 0.4 mg/mL G418 for 293Cre4 cells) and distribute into new dishes.

3.2. Helper-Dependent Vector Amplification

Amplification of the HD vector is initiated by transfection of 293Cre4 cells with the HD vector plasmid and infection with the helper virus (*see* **Note 2**). Subsequently, serial passages involving coinfection of 293Cre4 cells with the helper virus and the HD vector are performed to increase the titer of the HD vector. *Optional:* Ideally, the titer of the HD vector would be ascertained after

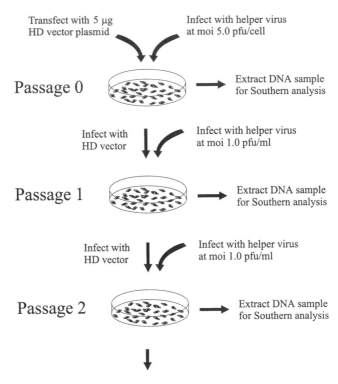

Fig. 3. Overview of the protocol for rescue and amplification of an HD vector by serial coinfection with a helper virus.

each serial passage to ensure that the amplification is progressing as expected. However, since most transgenes cannot be quickly and easily assayed, an HD vector containing a reporter gene can be amplified in parallel (*see* **Note 3**). Although the titer of the reporter vector does not necessarily reflect that of the vector of interest, this is nevertheless a useful control to ensure that the system is functioning properly. We use a LacZ reporter because vector titers can be quickly ascertained following each serial passage during vector amplification. The following protocol (**Fig. 3**) is for the rescue and amplification of one HD vector and should be modified accordingly for more vectors (e.g., to include control LacZ vector, or others).

1. Seed 293Cre4 cells into a 60-mm dish to reach 90% confluency in 1 or 2 days for transfection in complete medium (no G418).
2. Completely digest 5 µg of the HD vector plasmid with the appropriate restriction enzyme in a total volume of 25 µL and then heat to 65°C for 20 min. The digested

DNA can be stored at 4°C for transfection the next day. A small amount (0.5 μL) of the digested DNA is visualized by ethidium bromide staining following agarose gel electrophoresis to confirm complete digestion.

3. One hour prior to transfection, replace the medium from the 60-mm dish of 293Cre4 cells with 5 mL of freshly prepared complete medium without washing.

4. Set up the transfection reaction by adding 5 μg salmon sperm DNA to 0.5 mL HBS buffer in a polystyrene tube. Vortex the solution for 1 min at maximum setting. Then add the digestion reaction from **step 2** containing the HD vector and mix thoroughly but gently.

5. Add 25 μL of 2.5 M CaCl$_2$ dropwise with mixing. The solution should appear slightly cloudy. Incubate the solution at room temperature for 30 min and then apply 0.5 mL of the solution dropwise to the monolayer of 293Cre4 cells without removing the medium. Rock the dish to distribute the precipitate evenly and return the dish to the incubator for 6 h or overnight if more convenient.

6. Remove the medium from the transfected monolayer and wash twice with 1.0 mL freshly prepared complete medium. Immediately infect the transfected cells with the helper virus at a multiplicity of infection (MOI) of 5 pfu/cell in a total volume of 0.2 mL (supplemented with PBS^{2+}). Adsorb for 1 h in the incubator, rocking the dishes occasionally (every 10–15 min).

7. Following adsorption, add 5 mL of maintenance medium to the monolayer.

8. Complete cytopathic effect (CPE; >90% of the cells rounded up and detached from the dish) should be observed by about 48 h postinfection, at which time the cells are scraped into the medium. Then transfer 1 mL of the cell suspension to a microfuge tube for extraction of total DNA to monitor vector amplification (*see* **Subheading 3.3.1.**) and transfer the remainder into a 15-mL polypropylene tube and store at –70°C after adding 0.1 vol 40% sucrose.

9. Thaw the lysate at 37°C and use 0.4 mL to coinfect a 60-mm dish of 90% confluent 293Cre4 cells with helper virus at an MOI of 1 pfu/cell. Adsorb for 1 h in the incubator, occasionally rocking the dishes. Add 5 mL of maintenance medium to the monolayer 1 h postinfection. *Optional:* For the LacZ control vector, infect a 60-mm dish of 90% confluent 293 cells with an appropriate volume (start with 0.4 mL of the transfection/infection lysate and reduce volume accordingly as the titer increases with serial passage) of the lysate (no helper virus is added) as described above. Twenty-four hours later, remove the medium and fix the monolayer by adding 1 mL of β-gal fix solution and incubating the dish at 37°C for 5 min. Remove the β-gal fix solution and add 2 mL of β-gal stain solution supplemented with 0.5 mg/mL X-gal. Incubate the dishes at room temperature in the dark and determine the vector titer the next day by counting blue cells.

10. For serial passages of the vector, complete CPE should be observed by about 48 h postcoinfection (*see* **Note 4**), at which time the cells are scraped into the medium. Transfer 1 mL of the cell suspension to a microfuge tube for extraction of total DNA (*see* **Subheading 3.3.1.**), and transfer the remainder to a 15-mL polypropylene tube and store it at –70°C after adding 0.1 vol 40% sucrose. Repeat **steps 9** and **10** to increase the titer of the HD vector. Typically, the vector

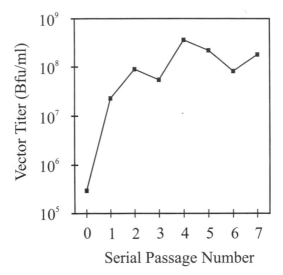

Fig. 4. Amplification of the HD vector AdC4HSULacZ by serial passage using AdLC8cluc-infected 293Cre4 cells. The titer (blue-forming units [bfu]/mL) of the HD vector was determined at each passage, as described in **Subheading 3.2.**

titer increases 10- to 100-fold per passage, and six to eight serial passages are performed (**Fig. 4**) (*see* **Note 5**).

3.3. Monitoring Helper-Dependent Vector Amplification

Once six to eight serial passages have been performed, a passage is chosen for large-scale preparation of the vector. Ideally, this passage should contain the maximum amount of HD vector and the minimum amount of helper virus. To determine which passage best meets these criteria, the amounts of vector DNA and helper DNA are determined by Southern blot analysis of the DNA extracted from each passage (*see* **Note 6** for an alternative procedure to expedite vector production).

3.3.1. Extraction of Total DNA

1. Spin lysate collected as in **Subheading 3.2.**, **steps 8** and **10**, in a microcentrifuge for 5 min at 750*g* to pellet infected cells.
2. Discard the supernatant and add 200 μL of pronase-SDS solution (*see* **Subheading 2.4.**) to pellet.
3. Resuspend the cell pellet and incubate overnight at 37°C.
4. Add 200 μL dH$_2$0 and precipitate DNA by adding 1 mL 95% ethanol. Mix by inverting the tube until a visible precipitate is formed.
5. Pellet the DNA by spinning in a microcentrifuge for 2 min at maximum speed.
6. Wash the DNA twice with 70% ethanol and dry.

7. Resuspend the DNA in an appropriate volume (usually 30–35 µL) of TE (*see* **Note 7**).

We use Southern blot hybridization analysis with a packaging signal probe to determine the relative amount of potentially packageable HD vector and helper virus genomes in each passage (**Fig. 5**) (*see* **Note 8**). PCR-based methods may also be used for this purpose (**26**). Large-scale preparation of the HD vector is performed on the passage with the highest HD vector to helper virus ratio (*see* **Subheading 3.4.**). If several passages meet this criterion, then the earliest passage should be chosen since the chance of rearrangement of either the vector or helper may increase with higher passage numbers.

3.4. Large Scale Preparation and Purification of Helper-Dependent Vector

Once the optimal passage has been determined as described in **Subheading 3.3.**, large-scale preparation of the HD vector can be performed. This is accomplished by coinfecting 150-mm dishes of 90% confluent 293Cre4 cells (seeded 1–2 days previously in nonselective complete medium) with 1 mL of lysate from the passage previous (relative to the optimum passage) and helper virus at an MOI of 1 pfu/cell. (High-titer stock is typically used to minimize the volume.) Cells are coinfected for 1 h in the incubator and the dishes are rocked every 10–15 min, following which 15 mL of maintenance medium is added to each dish. The amount of HD vector desired dictates the number of dishes coinfected. As more dishes are required, more lysate from the previous passage will be required. This necessitates repeating and scaling up the appropriate earlier passages to generate enough inoculum (*see* **Note 9**). Complete CPE should be observed 48 h postinfection, at which time the cells are scraped into the medium and the cell suspension is harvested for extraction and purification of the HD vector. The protocol given below is for processing up to 20 150-mm dishes and should be scaled up accordingly for more dishes.

1. Harvest cells and medium into a suitable centrifuge bottle and centrifuge at 750*g* for 10 min.
2. Discard supernatant and resuspend cell pellet with 10 mL of 100 m*M* Tris-HCl, pH 8.0, and transfer into a 50-mL conical tube. Samples can be stored at –70°C.
3. Thaw samples if necessary and add 1 mL 5% sodium deoxycholate to lyse cells. Incubate at room temperature for 30 min with occasional mixing. The lysate should become highly viscous.
4. Add 100 µL 2 *M* MgCl$_2$ and 50 µL DNAase I to digest the cellular and unpackaged viral DNA. Incubate at 37°C for 1 h mixing every 10 min. The viscosity should be greatly reduced.

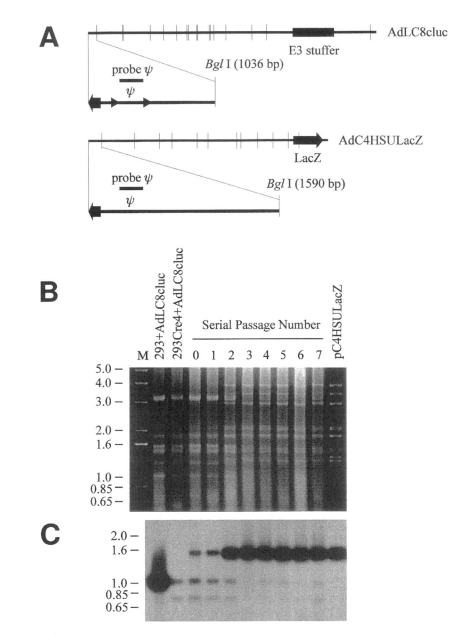

Fig. 5. Analysis of amplification of AdC4HSULacZ by serial passage using AdLC8cluc-infected 293Cre4 cells. **(A)** *BglI* restriction map of AdC4HSULacZ and AdLC8cluc. Probe ψ is composed of two synthetic oligonucleotides:

5' CCG GTG TAC ACA GGA AGT GAC AAT TTT CGC GCG GTT TTA GGC GGA TGT TGT AGT AAA TTT GGG CGT AAC CGA GTA AGA TTT GGC CAT TTT CGC GGG AAA ACT GAA TAA GAG GAA GTG AAA TCT GAA TAA TTT TGT GTT ACT CAT AGC GCG TAA T 3'
and

5. Spin the lysate at maximum speed for 15 min at 5°C in a Beckman tabletop centrifuge.
6. Meanwhile, prepare CsCl step gradient (1 Beckman SW41 Ti ultraclear tube for each 5 mL of lysate): Add 0.5 mL of 1.5 g/mL solution to tube. Gently overlay with 3 mL of 1.35 g/mL solution. Gently overlay 3 mL of 1.25 g/mL solution.
7. Apply 5 mL of supernatant from **step 5** to the top of the gradient. If necessary, top up the tubes with 100 mM Tris-HCl, pH 8.0.
8. Spin at 35,000 rpm in a Beckman SW41 Ti rotor (151,000g) for 1 h at 10°C.
9. Collect virus band (should be at 1.25/1.35 interface) with a needle and syringe by piercing the side of the tube. The volume collected is unimportant at this stage, so try to recover as much of the virus band as possible. Pool virus bands, if more than one tube was used, into a Beckman SW 50.1 ultraclear tube.
10. Fill the tube with 1.35 g/mL solution and centrifuge in a Beckman SW 50.1 rotor at 35,000 rpm (151,000g) for 16–24 h at 4°C.
11. To collect the virus band, puncture the side of the tube just below the virus band with a needle and syringe. Collect the virus band in the smallest volume possible and transfer to a Slide-A-Lyzer dialysis cassette. Dialyze at 4°C against three changes of 500 mL 10 mM Tris-HCl, pH 8.0, for at least 24 h total.
12. Transfer the virus into a suitable vial and add sterile glycerol to a final concentration of 10%. Store at –70°C.

Typically, only one virus band is visible after **step 11**. However, on occasion, more than one band may be observed. In this case, the HD vector is expected to be the higher band (lower buoyant density), and the helper virus is expected to be the lower band (higher buoyant density) because of differences

---

5' A TTA CGC GCT ATG AGT AAC ACA AAA TTA TTC AGA TTT CAC TTC CTC TTA TTC AGT TTT CCC GCG AAA ATG GCC AAA TCT TAC TCG GTT ACG CCC AAA TTT ACT ACA ACA TCC GCC TAA AAC CGC GCG AAA ATT GTC ACT TCC TGT GTA CAC CGG 3'.

These are 100% homologous to the packaging signals of AdC4HSULacZ and AdLC8cluc. (**B**) Approximately equivalent amounts of total DNA (as determined by OD$_{260}$) extracted from each passage were digested with *Bgl*I, and the DNA fragments were separated by electrophoresis through a 1% agarose gel. 293 + AdLC8cluc represents total DNA extracted from AdLC8cluc-infected 293 cells, and 293Cre4 + AdLC8cluc represents total DNA extracted from AdLC8cluc-infected 293Cre4 cells. Serial passage number 0 refers to the initial transfection/infection. pC4HSULacZ was digested with *Pme*I and *Bgl*I. M, 1 kb PLUS DNA Marker (Gibco-BRL). (**C**) DNA from the agarose gel depicted in (B) was transferred to a nylon membrane and subjected to Southern blot hybridization analysis using the [α-^{32}P]dCTP random prime-labeled probe ψ. The band visible between 0.85 and 0.65 kb is not packaged and may represent the excised circular DNA bearing the packaging signal from AdLC8cluc (P. Ng and F. Graham, unpublished results).

in their genome size (*see* **Subheading 2.2.**). All bands should be collected for analysis (*see* **Note 10**). If necessary, **step 10** may be repeated to achieve higher purity.

3.5. Helper-Dependent Vector Characterization

Following purification of the HD vector, several basic characterizations should be performed: 1) the concentration of virus particles should be determined; 2) the DNA structure should be analyzed to confirm that the vector has not rearranged; and 3) the level of helper virus contamination should be ascertained.

The concentration of vector particles in the purified preparation is determined spectrophotometrically as follows:

1. Dilute (usually 20-fold) purified virus with TE supplemented with SDS to 0.1%. Set up blank the same except add virus storage buffer (10 mM Tris-HCl, pH 8.0, supplemented with glycerol to 10%) instead of virus.
2. Incubate for 10 min at 56°C.
3. Vortex sample briefly.
4. Determine OD_{260}.
5. Calculate the number of particles/mL based on the extinction coefficient of wild-type Ad as determined by Maizel et al. *(31)* and corrected for the size difference between wild-type Ad and the HD vector as follows:

$$(OD_{260})(\text{dilution factor})(1.1 \times 10^{12})(36)/(\text{size of vector in kb})$$

Virion DNA is extracted and analyzed to confirm the proper structure and the level of helper virus contamination as follows:

1. Add an appropriate volume (25–100 µL depending on the vector concentration) of the purified virus to 0.4 mL of pronase-SDS solution and incubate overnight at 37°C to lyse the virions and digest virion proteins.
2. Precipitate virion DNA by adding 1/10 vol 3 M NaAcetate, pH 5.2, and 2.5 vol 95% ethanol and incubate at –20°C for 15–30 min.
3. Spin in a microcentrifuge for 10–15 min at maximum speed.
4. Discard the supernatant and wash the DNA pellet twice with 70% ethanol.
5. Dry the DNA pellet and resuspend in an appropriate volume of TE.

A sample of the vector DNA is digested with the appropriate diagnostic restriction enzyme(s) and the structure of the DNA is confirmed by ethidium bromide staining following agarose gel electrophoresis (*see* **Note 11**). To determine the level of helper virus contamination, the DNA can be transferred to a suitable membrane for Southern blot analysis with the packaging signal probe, and the relative intensities of the HD vector and helper virus-derived bands can be compared by phosphoimager analysis. Alternatively, PCR-based methods can be used *(26)*. The amount of helper virus can also be ascertained by determining the number of pfu/mL following titration on 293 cells. However, a comparable unit of measurement for the HD vector cannot be obtained by this method.

In addition to these standard characterizations, the biologic activity of the HD vector, with respect to expression of the transgene of interest, should also be determined. This will be unique to the vector in question, and appropriate assays should be employed in each case.

4. Notes

1. It is essential that the helper virus titer be accurately known. Thus, duplicate, independent titrations should be performed on both the high-titer and diluted working stocks.
2. High transfection efficiency is important for achieving a high initial concentration of vector for subsequent amplification. We use the calcium-phosphate coprecipitation method *(32)*, but other methods may also be suitable.
3. The reporter vector and the vector of interest should be as similar as possible (i.e., same backbone, site, and orientation of transgene insertion, similar size, and so on).
4. If complete CPE is not observed by 48 h post infection, then the titer of the helper virus is probably lower than expected. If complete CPE was observed for previous passages, then discard the current working stock of diluted helper virus and repeat the passage using a different aliquot. Otherwise retitrate the helper virus and start the amplification over.
5. The titer of a LacZ vector typically reaches a peak of 10^8–10^9 blue-forming units (bfu)/mL. With subsequent passages, the titer may fluctuate (cf. **Fig. 4**) but should not drop and remain below 10^7 bfu/mL.
6. We have observed that the amplification kinetics of a LacZ control vector closely parallel that of many HD vectors when the conditions in **Note 3** have been met (unpublished results). Therefore, to expedite vector production, large-scale vector preparation (*see* **Subheading 3.4.**) may be performed on the passage number of the HD vector that is equivalent to the passage number of the LacZ control vector when peak titer is first reached.
7. DNA prepared by this method does not dissolve easily. Heat the DNA at 65°C with occasional vortexing at maximum setting until dissolved.
8. Do not misinterpret results from eithium bromide-stained gels (**Fig. 5B**); the presence of helper virus-specific DNA fragments does not necessarily imply the presence of infectious helper virions since viral DNA from which the packaging signal has been excised can still replicate. For example, in **Figure 5B**, comparable amounts of helper virus DNA are observed in the control lanes (293 + AdLC8cluc and 293Cre4 + AdLC8cluc). However, Southern analysis with a packaging signal probe, which hybridizes equally to the packaging signals of both the HD vector

and the helper virus, reveals that the amount of potentially packageable helper virus genomes is drastically lower (**Fig. 5C**). Furthermore, if total DNA is analyzed, as in **Figure 5**, this method cannot accurately determine the level of helper contamination since all DNA molecules containing the packaging signal will hybridize to the probe but may or may not be packaged. Also, any differences in packaging efficiencies will render such comparisons invalid. In contrast, if DNA extracted from virions is used for this analysis (e.g., from CsCl-purified virus; *see* **Subheading 3.4.**), then direct comparison between the intensity of the HD vector and helper virus-derived bands should be an accurate measure of the level of helper virus contamination. Although an effort should be made to load the same amount of DNA in each lane, there nevertheless may be variability, thus rendering comparisons between lanes not strictly valid.

9. The amplification kinetics for a given HD vector are fairly predictable. Therefore, once this has been established, scale-up can begin at the appropriate passage for future amplifications of that HD vector to avoid repeating and scaling up previous passages.

10. The higher virus band is collected first and, without removing the syringe from the tube, the lower band collected next with a second syringe. Care must be taken to collect the vector from the gradient with minimal contamination from the helper, even if only one band is visible.

11. If the amount of extracted virion DNA is low, then the DNA fragments may be undetectable by eithium bromide staining. In this case, the DNA can be transferred onto a suitable membrane for Southern blot analysis using the HD vector plasmid DNA as the probe.

Acknowledgments

This work was supported by grants from the National Institutes of Health, the Medical Research Council of Canada (MRC), the National Cancer Institute of Canada (NCIC), and by Merck Research Laboratories. P. N. is supported by an MRC Postdoctoral Fellowship.

References

1. Berkner, K. L. (1988) Development of adenovirus vectors for expression of heterologous genes. *Biotechniques* **6**, 616–629.
2. Graham, F. L. and Prevec, L. (1992) Adenovirus-based expression vectors and recombinant vaccines, in *Vaccines: New Approaches to Immunological Problems* (Ellis, R. W., ed.), Butterworth-Heinemann, Boston, pp. 363–389.
3. Hitt, M., Addison, C. L., and Graham, F. L. (1997) Human adenovirus vectors for gene transfer into mammalian cells. *Adv. Pharmacol.* **40**, 137–206.
4. Hitt, M. M., Parks, R. J., and Graham, F. L. (1999) Structure and genetic organization of adenovirus vectors, in *The Development of Human Gene Therapy* (Friedman, T., ed.), Cold Spring Harbor Laboratory Press, Cold Spring Harbor, NY, pp. 61–86.

5. Graham, F. L., Smiley, J., Russell, W. C., and Nairn, R. (1977) Characteristics of a human cell line transformed by DNA from human adenovirus 5. *J. Gen. Virol.* **36,** 59–72.

6. Wolf, G., Worgall, S., Van, R. N., Song, W. R., Harvey, B. G., and Crystal, R. G. (1997) Enhancement of in vivo adenovirus-mediated gene transfer and expression by prior depletion of tissue macrophages in the target organ. *J. Virol.* **71,** 624–629.

7. Worgall, S., Wolff, G., Falck-Pedersen, E., and Crystal, R. G. (1997) Innate immune mechanisms dominate elimination of adenoviral vectors following in vivo administration. *Hum. Gene Ther.* **8,** 37–44.

8. Morral, N., O'Neal, W., Zhou, H., Langston, C., and Beaudet, A. (1997) Immune responses to reporter proteins and high viral dose limit duration of expression with adenoviral vectors: comparison of E2a wildtype and E2a deleted vectors. *Hum. Gene Ther.* **8,** 1275–1286.

9. Dai, Y., Schwartz, E. M., Gu, D., et al. (1995) Cellular and humoral immune responses to adenoviral vectors containing factor IX gene: tolerization of factor IX and vector antigens allows for long-term expression. *Proc. Natl. Acad. Sci. USA* **92,** 1401–1405.

10. Yang, Y., Nunes, F. A., Berencsi, K., et al. (1994) Cellular immunity to viral antigens limits E1-deleted adenoviruses for gene therapy. *Proc. Natl. Acad. Sci. USA* **91,** 4407–4411.

11. Yang, Y., Li, Q., Ertl, H. C., and Wilson, J. M. (1995) Cellular and humoral immune responses to viral antigens create barriers to lung-directed gene therapy with recombinant adenoviruses. *J. Virol.* **69,** 2004–2015.

12. Yang, Y., Xiang, Z., Ertl, H. C., and Wilson, J. M. (1995) Upregulation of class I major histocompatibility complex antigens by interferon gamma is necessary for T-cell-mediated elimination of recombinant adenovirus-infected hepatocytes in vivo. *Proc. Natl. Acad. Sci. USA* **92,** 7257–7261.

13. Kafri, T., Morgan, D., Krahl, T., et al. (1998) Cellular immune response to adenoviral vector infected cells does not require de novo viral gene expression: implications for gene therapy. *Proc. Natl. Acad. Sci. USA* **95,** 11377–11382.

14. Chen, H-H., Mack, L. M., Kelly, R., et al. (1997) Persistence in muscle of an adenoviral vector that lacks all viral genes. *Proc. Natl. Acad. Sci. USA* **94,** 1645–1650.

15. Morral., N., Parks, R. J., Zhou, H., et al. (1998) High doses of a helper-dependent adenoviral vector yield supraphysiological levels of α1-antitrypsin with negligible toxicity. *Hum. Gene Ther.* **9,** 2709–2716.

16. Morsy, M. A., Gu, M., Motzel, S., et al. (1998) An adenoviral vector deleted for all viral coding sequences results in enhanced safety and extended expression of a leptin transgene. *Proc. Natl. Acad. Sci. USA* **95,** 7866–7871.

17. Schiedner, G., Morral, N., Parks, R. J., et al. (1998) Genomic DNA transfer with a high-capacity adenovirus vector results in improved in vivo gene expression and decreased toxicity. *Nat. Genet.* **18,** 180–183.

18. Morral, N., O'Neal, W., Rice, K., et al. (1999) Administration of helper-dependent adenoviral vectors and sequential delivery of different vector serotype for

long-term liver directed gene transfer in baboons. *Proc. Natl. Acad. Sci. USA* **96,** 12816–12821.

19. Balague, C., Zhou, J., Dai, Y. et al. (2000) Sustained high-level expression of full-length human factor VIII and restoration of clotting activity in hemophilic mice using a minimal adenovirus vector. *Blood* **95,** 820–828.

20. Cregan, S. P., MacLaurin, J., Gendron, T. F., et al. (2000) Helper-dependent adenovirus vectors: their use as a gene delivery system to neurons. *Gene Ther.* **14,** 1200–1209.

21. Maione, D., Wiznerowicz, M., Delmastro, P., et al. (2000) Prolonged expression and effective readministration of erythropoietin delivered with a fully deleted adenoviral vector. *Hum. Gene Ther.* **11,** 859–868.

22. Kochanek, S. (1999) High-capacity adenoviral vectors for gene transfer and somatic gene therapy. *Hum. Gene Ther.* **10,** 2451–2459.

23. Parks R. J. (2000) Improvements in adenoviral vector technology: overcoming barriers for gene therapy. *Clin. Genet.* **58,** 1–11.

24. Parks, R. J., Chen, L., et al. (1996) A helper-dependent adenovirus vector system: removal of helper virus by Cre-mediated excision of the viral packaging signal. *Proc. Natl. Acad. Sci. USA* **93,** 13565–13570.

25. Chen, L., Anton, M., and Graham, F. L. (1996) Production and characterization of human 293 cell lines expressing the site-specific recombinase Cre. *Somat. Cell Mol. Genet.* **22,** 477–488.

26. Sandig, V., Youil, R., Bett, A. J., et al. (1999) Optimization of the helper-dependent adenovirus system for production and potency in vivo. *Proc. Natl. Acad. Sci. USA* **97,** 1002–1007.

27. Parks, R. J. and Graham, F. L. (1997) A helper-dependent system for adenovirus vector production helps define a lower limit for efficient DNA packaging. *J. Virol.* **71,** 3293–3298.

28. Bett, A. J., Prevec, L., and Graham, F. L. (1993) Packaging capacity and stability of human adenovirus type 5 vectors. *J. Virol.* **67,** 5911–5921.

29. Parks, R. J., Bramson, J. L., Wan, Y., Addison, C. L., and Graham, F. L. (1999) Effects of stuffer DNA on transgene expression from helper-dependent adenovirus vectors. *J. Virol.* **73,** 8027–8034.

30. Parks, R. J., Evelegh, C. M., and Graham, F. L. (1999) Use of helper-dependent adenoviral vectors of alternative serotypes permits repeat vector administration. *Gene Ther.* **6,** 1565–1573.

31. Maizel, J. V., White, D., and Scharff. M. D. (1968) The polypeptides of adenovirus. I. Evidence of multiple protein components in the virion and a comparison of types 2, 7a, and 12. *Virology* **36,** 115–125.

32. Graham, F. L. and van der Eb, A. J. (1973) A new technique for the assay of infectivity of human adenovirus 5 DNA. *Virology* **52,** 456–467.

27

Construction of First-Generation Adenoviral Vectors

Philip Ng and Frank L. Graham

1. Introduction

Adenoviruses (Ads) possess several features that make them attractive as mammalian gene transfer vectors. They can efficiently infect a wide variety of quiescent and proliferating cell types from various species to direct high-level viral gene expression, their 36-kb double-stranded DNA genome can be manipulated with relative ease by conventional molecular biology techniques, and they can be readily propagated and purified to yield high-titer preparations of very stable virus. Consequently, Ads have been extensively used as vectors for recombinant vaccines, for high-level protein production in cultured cells and for gene therapy *(1–4)*.

1.1. Adenovirus Biology

The adenovirus is an icosohedral, nonenveloped capsid containing a linear double-stranded DNA genome of approx 36 kb. Of the approximately 50 serotypes of human Ad, serotype 2 (Ad2) and serotype 5 (Ad5) of subgroup C have been the most extensively characterized (reviewed in **ref. 5**). The 36-kb genome is flanked by inverted terminal repeats (ITRs), which are the only sequences required in *cis* for DNA replication. A *cis*-acting packaging signal, required for encapsidation of the genome, is located near the ITR at the left end (relative to the conventional map of Ad). The Ad genome can be divided into two regions with respect to the onset of viral DNA replication (**Fig. 1**).The early region genes, E1A, E1B, E2, E3, and E4, are expressed before DNA replication. The

From: *Methods in Molecular Medicine, Vol. 69, Gene Therapy Protocols, 2nd Ed.*
Edited by: J. R. Morgan © Humana Press Inc., Totowa, NJ

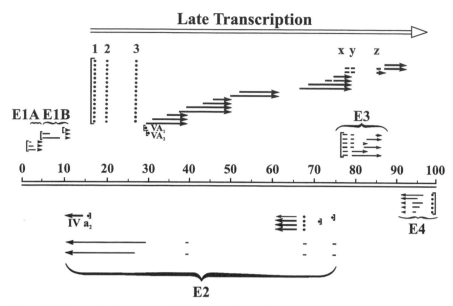

Fig. 1. Transcription map of human adenovirus serotype 5. The 100 map unit (approx 36 kb) genome is divided into four early region transcription units, E1–E4, and five families of late mRNA, L1–L5, which are alternative splice products of a common late transcript expressed from the major late promoter (MPL) located at 16 mu. Four smaller transcripts are also produced, pIX, IVa, and VA RNAs I and II. Not shown are the 103-bp inverted terminal repeats located at the termini of the genome involved in viral DNA replication and the packaging signal located from nucleotides 190 to 380 at the left end of the genome involved in encapsidation.

E1A transcription unit is the first Ad sequence expressed during viral infection and encodes two major E1A proteins that are involved in transcriptional regulation of the virus and stimulation of the host cell to enter S phase. The two major E1B proteins are necessary for blocking host mRNA transport, stimulating viral mRNA transport, and blocking E1A-induced apoptosis. The E2 region encodes proteins required for viral DNA replication and can be divided into to subregions, E2a and E2b. E2a encodes the 72-kDa DNA-binding protein, and E2b encodes the viral DNA polymerase and terminal protein precursor. The E3 regions encodes at least seven proteins, most of which are involved in evasion of the host immune system. This region is dispensable for virus growth in cell culture. The E4 region encodes at least six proteins, some of which function in facilitating DNA replication, enhancing late gene expression, and decreasing host protein systhesis. The late region genes, L1–L5, are expressed after DNA replication from a common major late promoter and are generated from alternative splicing of a single transcript. Most of the late mRNAs encode virion structural proteins.

Virus infection begins with the Ad fiber protein binding to specific primary receptors on the cell surface *(6,7)* followed by a secondary interaction between the virion penton base and $\alpha_v\beta_3$ and $\alpha_v\beta_5$ integrins *(8)*. The efficiency with which Ad binds and enters the cell is directly related to the level of primary and secondary receptors present on the cell surface *(9,10)*. Penton-integrin interaction triggers Ad internalization by endocytosis; it then escapes from the early endosome into the cytosol prior to lysosome formation *(11,12)*. The virion is sequentially disassembled during translocation along the microtubule network toward the nucleus, where the viral DNA is released into the nucleus *(13)*. Viral DNA replication, beginning 6–8 h post infection, and assembly of progeny virions occur in the nucleus. The entire life cycle takes about 24–36 h, generating about 10^4 virions per infected cell. In humans, Ads generally cause only relatively mild, self-limiting illness in immunocompetent individuals. Ads have never been implicated as a cause of malignant disease in their natural host. The reader is referred to an excellent review by Shenk *(5)* for a more comprehensive discussion of the adenovirus.

1.2. Adenovirus Vectors

Typically, Ads are modified into vectors by replacing the E1 region with the foreign DNA of interest. This modification serves two important purposes. First, deletion of E1 increases the cloning capacity to approx 5 kb. Second, it renders the vectors replication deficient, which is an important consideration with respect to safety for human gene therapy. These replication-deficient vectors can be propagated in E1-complementing cell lines such as 293 *(14)*. The E3 region can also be deleted from the vector as it is not required for virus propagation in culture. The combination of E1 and E3 deletions results in a cloning capacity of approx 8 kb, a size that is more than adequate for most expression cassettes.

1.3. Overview of the Two-Plasmid Rescue System

This chapter provides detailed protocols for construction of E1-substituted Ad vectors by site-specific recombination between two plasmids cotransfected into E1-complementing cells such as 293 (**Fig. 2**) *(15–17)*. One of the plasmids used in this method contains the entire Ad5 genome except for certain key deletions. Essential features of this Ad genomic plasmid include an ITR junction, a deletion spanning the E1 region and packaging signal, a target sequence for site-specific recombination, and a cassette for expression of the site-specific recombinase. Inclusion of the recombinase expression cassette allows site-specific recombination-mediated vector rescue to be performed in any

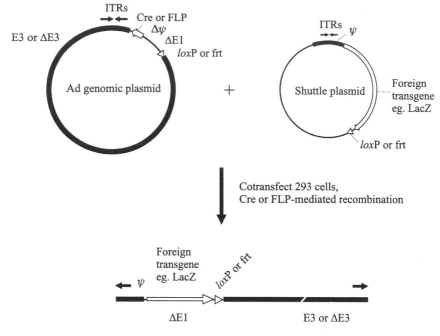

Fig. 2. Construction of Ad vectors by Cre- or FLP-mediated site-specific recombination following cotransfection of 293 cells with an Ad genomic plasmid (*see* **Fig. 3**), and a shuttle plasmid (*see* **Fig. 4**). Ad and bacterial plasmid sequences are represented by thick and thin lines, respectively, and *lox*P or *frt* sites are represented by "▷."

E1-complementing cell. The recombinase expression cassette is not incorporated into the recombinant vector due to its location in the plasmid with respect to the recombinase target site (**Fig. 2**). The second plasmid, called a shuttle, contains an ITR junction, a packaging signal, a polylinker region for insertion of foreign sequences, and a target sequence for site-specific recombination. The ITR junctions serve as efficient origins of Ad DNA replication (*18*) so that following cotransfection into 293 cells, the plasmids can undergo Ad-mediated DNA replication. However, the deletion of the packaging signal renders the Ad genomic plasmid noninfectious. The presence of the ITR junction in the shuttle plasmid significantly increases the efficiency of vector rescue (*16*). Following cotransfection in 293 cells, site-specific recombination between the two plasmids results in the generation of a recombinant Ad vector, which can replicate and be packaged into infectious virions, resulting in plaques on the monolayer of 293 cells. Since neither plasmid alone is capable of producing infectious virus, only the desired recombinant vector is generated following cotransfection. Vector rescue can be achieved by either of two site-specific recombination systems with comparable high efficiencies: the bacteriophage P1 Cre/*lox*P system *(15,16)* or the yeast FLP/*frt* system *(17)*.

2. Materials
2.1. Preparation of Plasmid DNA

1. Plasmid DNA: All plasmids described in this chapter and their sequences can be obtained from Microbix Biosystems (www.microbix.com).
2. Sterile Luria-Bertani (LB) broth (Lennox; Difco) and LB-agar plates supplemented with 50 μg/mL ampicillin.
3. Solution I: 10 mM EDTA, pH 8.0, 50 mM glucose, 25 mM Tris-HCl, pH 8.0, prepared from sterile stock solutions.
4. Solution II: 1% sodium dodecyl sulfate (SDS), 0.2 N NaOH, freshly prepared.
5. Solution III: 3 M potassium acetate, 11.5% glacial acetic acid, autoclave-sterilized.
6. Isopropanol.
7. TE: 10 mM Tris-HCl, pH 8.0, 1 mM EDTA pH 8.0, autoclave-sterilized.
8. Pronase stock solution: 20 mg/mL pronase in 10 mM Tris-HCl, pH 7.5. Preincubate at 56°C for 15 min followed by 37°C for 1 h. Aliquot and store at −20°C.
9. Pronase-SDS solution: 0.5 mg/mL pronase (above) in 0.5% SDS, 10 mM Tris-HCl, pH 7.4, 10 mM EDTA, pH 8.0.
10. CsCl (biotechnology grade).
11. 10 mg/mL ethidium bromide.

2.1.1. Ad Genomic Plasmids

A variety of Ad genomic plasmids are available for construction of vectors (**Fig. 3**). The plasmids pBHGloxE3Cre and pBHGfrtE3FLP are used to generate vectors bearing a wild-type E3 region by Cre-mediated and FLP-mediated recombination, respectively. The maximum foreign DNA insert that can be rescued into an E1-deleted vector with a wild-type E3 region is approx 5 kb owing to the size constraints of Ad *(19)*. The plasmids pBHGloxΔE3(X1)Cre and pBHGfrtΔE3(X1)FLP have a 1864-bp deletion in the E3 region and thus permit rescue into vectors of foreign sequences up to approx 7.2 kb by Cre-mediated and FLP-mediated recombination, respectively. The plasmids pBHGloxΔE1,3Cre and pBHGfrtΔE1,3FLP have a 2653-bp deletion in the E3 region and permit rescue into vectors of foreign sequences up to approx 8 kb by Cre-mediated and FLP-mediated recombination, respectively. These latter two plasmids offer maximum cloning capacity, but vectors bearing this larger E3 deletion may grow slightly more slowly and result in lower yields (approx twofold) than vectors bearing the wild-type or smaller E3 deletion (F. L. Gra-

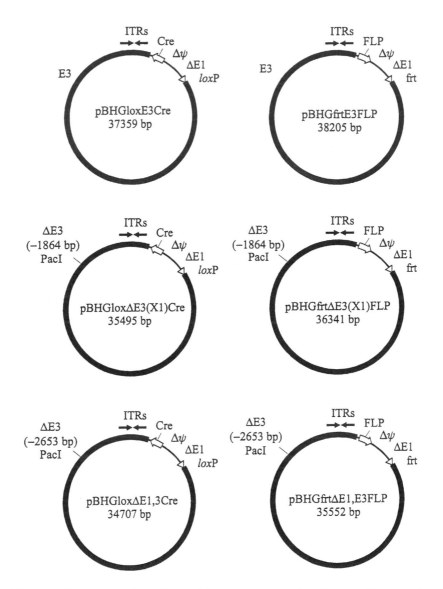

Fig. 3. Ad genomic plasmids used for vector rescue by site-specific recombination in vivo. The plasmids pBHGloxE3Cre, pBHGloxΔE3(X1)Cre, and pBHGloxΔE1,3Cre are used to rescue vectors bearing a wild-type E3 region, a 1864-bp deletion or a 2653-bp deletion of E3, respectively, by Cre-mediated recombination. The plasmids pBHGfrtE3FLP, pBHGfrtΔE3(X1)FLP, and pBHGfrtΔE1,3FLP are used to rescue vectors bearing a wild-type E3 region, a 1864-bp deletion, or a 2653-bp deletion of E3, respectively, by FLP-mediated recombination. The unique *Pac*I restriction enzyme site in pBHGloxΔE1,3Cre, pBHGloxΔE3(X1)Cre, pBHGfrtΔE3(X1)FLP, and pBHGfrtΔE1,3FLP permit insertion of foreign sequences into the E3 deletion. Ad and bacterial plasmid sequences are represented by thick and thin lines, respectively, and *lox*P or *frt* sites are represented by "▷."

ham, unpublished results). The presence of a unique *Pac*I site in pBHGfrtΔE1,3FLP, pBHGloxΔE1,3Cre, pBHGloxΔE3(X1)Cre, and pBHGfrtΔE3(X1)FLP permits insertion of foreign sequences for rescue into the E3 deletion if desired (**Fig. 3**). The choice of which Ad genomic plasmid to use is dictated by the size of the foreign sequence to be rescued, whether a wild-type or a deleted E3 region is desired in the vector, and which site-specific recombination system is preferred/necessitated.

2.1.2. Shuttle Plasmids

A variety of small shuttle plasmids are available for insertion of foreign sequences and rescue into Ad vectors by Cre- or FLP-mediated recombination (**Fig. 4**). The shuttle plasmids pDC311, pDC312, pDC511, and pDC512 are designed for the insertion of expression cassettes for rescue into vectors, the first two by Cre-mediated recombination and the last two by FLP-mediated recombination. The shuttle plasmids pDC315, pDC316, pDC515, and pDC516 are designed for the insertion of coding sequences for rescue into vectors, the first two by Cre-mediated recombination and the last two by FLP-mediated recombination. In these plasmids, the site of insertion is flanked by a murine cytomegalovirus (MCMV) immediate early promoter and the SV40 polyadenylation signal, thus permitting high-level transgene expression. The choice of which shuttle plasmid to use is dictated by whether expression from a strong viral promoter is desired, by the orientation of the polylinker, and by the site-specific recombination system desired for vector rescue. We have observed that higher expression levels are obtained when the transcription orientation of the transgene is in the same direction as E1 and that the MCMV immediate early promoter is stronger in most cell lines than its more commonly used human counterpart *(20)*.

2.2. Cell Culture

1. Low-passage 293 cells (Microbix Biosystems).
2. 293N3S cells (Microbix Biosystems).
3. Complete medium: minimum essential medium (MEM; Gibco-BRL, cat. no. 61100) containing 10% fetal bovine serum (FBS) (heat inactivated), 100 U/mL penicillin/streptomycin, 2 mM L-glutamine, and 2.5 µg/mL fungizone.
4. Joklik's modified-MEM (Gibco-BRL, cat. no. 22300) supplemented with 10% horse serum (heat inactivated).
5. Citric saline: 135 mM KCl, 15 mM sodium citrate, autoclave-sterilized.

2.3. Cotransfection

1. Monolayers of low-passage 293 cells at 80% confluency in 60-mm dishes.
2. HEPES-buffered saline (HBS): 21 mM HEPES, 137 mM NaCl, 5 mM KCl, 0.7 mM Na$_2$HPO$_4$, 5.5 mM glucose, pH 7.1 (adjusted with NaOH), filter-sterilized. Store at 4°C in small aliquots in tightly sealed conical tubes.

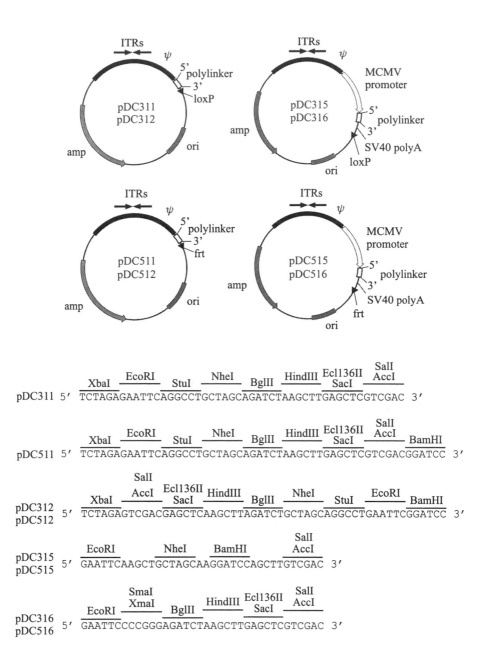

Fig. 4. Shuttle plasmids and their polylinker sequences used for vector rescue by in vivo site-specific recombination. The shuttle plasmids pDC311 (3276 bp), pDC312 (3288 bp), pD315 (3913 bp), and pDC316 (3913 bp) are used to rescue vectors by Cre-mediated recombination. The shuttle plasmids pDC511 (3277 bp), pDC512 (3277 bp), pDC515 (3957 bp), and pDC516 (3957 bp) are used to rescue vectors by FLP-mediated recombination. The plasmids pDC311, pDC312, pDC511, and pDC512 are used for insertion of expression cassettes (inserts with a promoter/enhancer and

3. Salmon sperm DNA (2 μg/μL in TE).
4. 2.5 M CaCl$_2$, filter-sterilized.
5. Complete medium (*see* **Subheading 2.1.**).
6. 2× maintenance medium: 2× MEM (Gibco-BRL, cat. no. 61100) supplemented with 10% horse serum (heat inactivated), 200 U/mL penicillin/streptomycin, 4 mM L-glutamine, 5 μg/mL fungizone, and 0.2% yeast extract.
7. 1% agarose solution, autoclave-sterilized. Store at room temperature and melt in a microwave oven prior to use.
8. Ad genomic plasmid DNA (*see* **Subheading 2.2.1.**).
9. Shuttle plasmid DNA with desired foreign sequence inserted (*see* **Subheading 2.2.2.**).
10. Phosphate-buffered saline (PBS): 137 mM NaCl, 8.2 mM Na$_2$HPO$_4$, 1.5 mM KH$_2$PO$_4$, 2.7 mM KCl, autoclave-sterilized.
11. PBS^{2+}: PBS supplemented with 0.68 mM sterile MgCl$_2$ and 0.5 mM sterile CaCl$_2$.
12. Glycerol, autoclave-sterilized.

2.4. Analysis of Recombinant Vectors

1. 90% confluent 60-mm dishes of 293 cells.
2. TE (*see* **Subheading 2.1.**).
3. Complete medium (*see* **Subheading 2.1.**).
4. Maintenance medium: MEM (Gibco-BRL, cat. no. 61100) containing 5% horse serum (heat inactivated), 100 U/mL penicillin/streptomycin, 2 mM L-glutamine, and 2.5 μg/mL fungizone.
5. PBS^{2+} (*see* **Subheading 2.3.**).
6. Pronase-SDS solution (*see* **Subheading 2.1.**).

2.5. Titration of Adenovirus

1. 80–90% confluent 60-mm dishes of 293 cells.
2. PBS^{2+} (*see* **Subheading 2.3.**).
3. 1% agarose solution (*see* **Subheading 2.3.**).
4. 2× maintenance medium (*see* **Subheading 2.3.**).
5. Glycerol, autoclave-sterilized.

2.6. Preparation of High-Titer Viral Stocks (Crude Lysate)

1. PBS^{2+} (*see* **Subheading 2.3.**).
2. Glycerol, autoclave-sterilized.

Fig. 4. *Continued*
polyadenylation signal as well as coding sequence). The plasmids pDC315, pDC316, pDC515, and pDC516 bear a polylinker flanked by the murine cytomegalovirus immediate early promoter/enhancer and SV40 polyadenylation signal and are used for insertion of coding sequences (e.g., cDNAs).

Specific materials, in addition to those above, are required for virus preparation from either 293 cells or 293N3S cells as listed below.

2.6.1. Preparation of High-Titer Viral Stocks (Crude Lysate) from Cells in Monolayer

1. 150-mm dishes of 80–90% confluent 293 cells.
2. Maintenance medium (*see* **Subheading 2.4.**).

2.6.2. Preparation of High-Titer Viral Stocks (Crude Lysate) from Cells in Suspension

1. 3–4 L 293N3S cells (Microbix Biosystems).
2. Joklik's modified MEM supplemented with 10% horse serum (heat inactivated) (*see* **Subheading 2.2.**).
3. 1% sodium citrate.
4. Carnoy's fixative: Add 25 mL glacial acetic acid to 75 mL methanol.
5. Orcein solution: Add 1 g orcein dye to 25 mL glacial acetic acid plus 25 mL dH_2O and filter through Whatman no. 1 paper.

2.7. Purification of Adenovirus by CsCl Banding

1. 10 and 100 mM Tris-HCl, pH 8.0, autoclave-sterilized.
2. 5% sodium deoxycholate, filter-sterilize.
3. 2 M $MgCl_2$, autoclave-sterilized.
4. DNAase I: 100 mg bovine pancreatic deoxyribosenuclease I in 10 mL of 20 mM Tris-HCl, pH7.4, 50 mM NaCl, 1 mM dithiothreitol, 0.1 mg/mL bovine serum albumin, 50% gyclerol. Aliquot and stored at –20°C).
5. CsCl solutions:

Density (g/ml)	CsCl (g)	10 mM Tris-HCl, pH 8.0 (g)
1.5	90.8	109.2
1.35	70.4	129.6
1.25	54.0	146.0

Dissolve CsCl into 10 mM Tris-HCl, pH 8.0, solution in the amounts indicated above to achieve the desired density solution, and filter sterilize. Weigh 1.00 mL to confirm density.

6. Glycerol, autoclave-sterilized.
7. Beckman SW 41 Ti and SW50.1 rotor and ultraclear tubes.
8. Slide-A-Lyzer dialysis cassettes (Pierce).

2.8. Characterization of Adenoviral Vector Preparations

1. All materials listed in **Subheading 2.5.**
2. TE (*see* **Subheading 2.1.**).
3. 10 mM Tris-HCl, pH 8.0.

4. 10% SDS
5. Pronase-SDS solution (*see* **Subheading 2.1.**)
6. 3 *M* sodium acetate, pH 5.2, autoclave-sterilized.
7. 95% and 70% ethanol.
8. A549 cells.
9. α-MEM (Gibco-BRL, cat. no. 12000) supplemented with 10% FBS (heat inactivated).
10. α-MEM supplemented with 5% horse serum (heat inactivated).

3. Methods

Figure 5 presents an overview of the steps involved in the production of recombinant Ad vectors. Briefly, 293 cells are cotransfected with the Ad genomic plasmid and the shuttle plasmid. The recombinant vector, generated by site-specific recombination in vivo between the two plasmids, forms a plaque in the cell monolayer. The plaques are isolated, the virus is expanded, and the vector DNA is extracted for confirmation by restriction enzyme digestion. The vector is plaque-purified by titration, and a high-titer stock is generated. The vector is then purified by CsCl banding and characterized with respect to concentration, DNA structure, level of RCA contamination, and transgene expression. The remainder of this chapter provides detailed protocols for each of these steps. Although it is recommended that the steps outlined in **Figure 5** be followed, acceptable alternatives, where applicable, are provided in **Subheading 3.9.** to expedite vector production.

3.1. Preparation of Plasmid DNA

The foreign DNA of interest is inserted into an appropriate shuttle plasmid and transformed into *E. coli* by conventional molecular biology techniques. The following is a protocol for the preparation of high-quality plasmid DNA for cotransfection.

1. Inoculate 5 mL of LB supplemented with 50 µg/mL ampillicin with bacteria bearing the desired plasmid in the morning. Incubate culture at 37°C with shaking (*see* **Note 1**).
2. Inoculate 500 mL of LB supplemented with 50 µg/mL ampillicin with the above culture in the late afternoon. Incubate overnight at 37°C with shaking (*see* **Note 2**).
3. Transfer culture to a centrifuge bottle and pellet bacteria by spinning at 6000*g* for 10 min at 4°C.
4. Resuspend bacterial pellet in 40 mL of cold solution I so that no cell clumps are visible.
5. Add 80 mL of freshly prepared solution II, and mix thoroughly but gently by swirling. This should produce a relatively clear, viscous lysate.

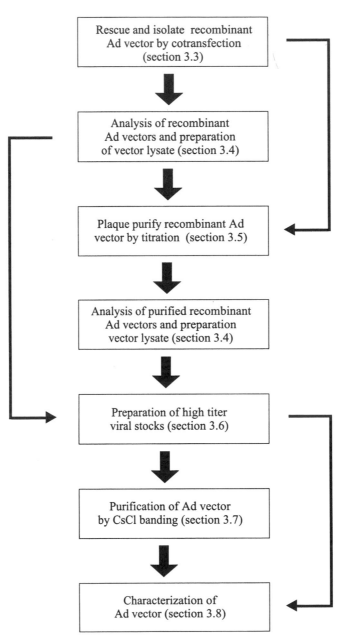

Fig. 5. Overview of the steps involved in rescue, propagation, purification, and characterization of Ad vectors. The recommended steps are indicated by thick arrows. Acceptable alternatives to expedite vector production are indicated by thin arrows (*see* **Subheading 3.9.**).

6. Add 40 mL of cold solution III, mix thoroughly but gently by swirling, and incubate for 20 min on ice. A white precipitate should form, and the viscosity should be greatly reduced.
7. Add 10 mL of dH_2O, and centrifuge at 4°C for 10 min at 6000g.
8. Collect the supernatant by filtering it through two to three layers of cheesecloth into a centrifuge bottle.
9. Precipitate plasmid DNA by adding 100 mL (0.6 vol) of isopropanol, mix well, and incubate for 30 min at room temperature.
10. Pellet plasmid DNA by centrifuging at 4°C for 10 min at 6000g.
11. Pour off the supernatant and drain at room temperature for 15 min. Keeping the bottle upside down, wipe inside the rim with a clean KimWipe™ to remove all residual isopropanol.
12. Dissolve plasmid DNA pellet in 5 mL TE and transfer to 50-mL conical tube.
13. Add 2 mL pronase-SDS solution, mix well, and incubate for 30 min at 37°C.
14. Add 8.6 g CsCl, mix to dissolve completely, and incubate on ice for 30 min.
15. Centrifuge at maximum speed in a Beckman tabletop centrifuge for 30 min at 5°C. Slowly collect the supernatant using a 10-mL syringe and 16-gage needle, avoiding as much of the pellicle as possible.
16. Transfer to a Beckman VTi 65.1 ultracentrifuge tube. Add 25 µL of 10 mg/mL ethidium bromide and fill the tube with light parafin oil. Seal the tube and mix by inversion.
17. Centrifuge in a Beckman VTi 65.1 rotor at 55,000 rpm (288,000g) for 10–14 h at 14°C.
18. Remove tube and support it with a stand. The supercoiled plasmid DNA band should be the thicker, lower red band in the gradient, but it should not be at the very bottom of the tube. The upper red band (if present) is nicked plasmid or chromosomal DNA. Puncture the top of the tube to allow entry of air. Collect plasmid DNA through the side of the tube with a 3-mL syringe and 18-gage needle by puncturing the side of tube just below the band. (*see* **Note 3**).
19. Transfer plasmid DNA to a 15-mL polypropylene tube containing 5 mL isopropanol, which has been saturated with CsCl in TE, and mix immediately to extract the ethidium bromide into the solvent layer. Allow the phases to separate and remove the ethidium bromide-solvent (pink) layer. Repeat extraction until the solvent layer is completely colorless.
20. Add TE to bring the volume up to 4 mL, add 8 mL cold 95% ethanol, and mix by inversion to precipitate the DNA.
21. Centrifuge at top speed for 15 min to pellet DNA. Discard supernatant and wash pellet twice with 5 mL 70% ethanol.
22. Remove as much of the 70% ethanol as possible, allow the pellet to dry, and resuspend with an appropriate volume of TE. Ideally, the concentration should be 1–2 µg/µL.
23. Determine the plasmid DNA concentration by OD_{260}, digest a sample with appropriate diagnostic restriction enzymes, and confirm the structure by agarose gel electrophoresis.

3.2. Cell Culture

Low-passage 293 cells maintained in 150-mm dishes, are split 1 to 2 or 1 to 3 when they reach approx 90% confluency (every 2–3 days). In general, for cotransfections, a confluent 150-mm dish of 293 cells is split into eight 60-mm dishes for use the next day (*see* **Note 4**).

1. Remove medium from 150-mm dish of 293 cells.
2. Rinse monolayer twice with 5 mL citric saline.
3. Remove citric saline from **step 2**, add 0.5 mL citric saline, and leave the dish at room temperature until cells start to round up and detach from the dish (no more than 15 min).
4. Tap the side of the dishes to detach all cells.
5. Resuspend cells with complete medium and distribute into new dishes.

293N3S are suspension-adapted 293 cells and can be used for large-scale vector production instead of 293 cells because of their greater ease of handling. 293N3S cells are grown in Joklik's modified MEM supplemented with 10% horse serum (heat inactivated) and should be diluted 1 to 2 or 1 to 3 when the density reaches 5×10^5 cells/mL. All cell culture reagents should be prewarmed to 37°C prior to use.

3.3. Cotransfection

Typically, a large number of plaques are generated by cotransfecting a single 60-mm dish of 293 cells with 2 μg of the shuttle plasmid and 2 μg of the Ad genomic plasmid by site-specific recombination (**Table 1**). However, a number of factors can influence the efficiency of vector rescue including the quality of the DNA, the efficiency of transfection, and especially the state of the 293 cells. Another important consideration is that the plaques be well isolated. Therefore, it is recommended that a range of DNA amounts be cotransfected to ensure that the vector is rescued and also that the plaques are well isolated. The infectious Ad genomic plasmid pFG140 *(18)* is used as a control for transfection efficiency and under optimal conditions should yield up to approx 100 plaques per 0.5 μg. Below is a typical protocol in which four 60-mm dish of 293 cells are cotransfected with 0.5, 2, and 5 μg of each plasmid (Fig. 6).

1. Label four 60-mm dishes "A," four dishes "B," four dishes "C," and two dishes "D" and seed them with 293 cells to reach approx 80% confluency in 1–2 days for cotransfection.
2. In the late afternoon, 1 h prior to cotransfection, the medium from the 60-mm dishes of 293 cells is replaced with 5 mL of freshly prepared complete medium without washing.
3. Meanwhile, combine in a 50-mL conical tube 8 mL of HBS and 40 μl of salmon sperm DNA and vortex at maximum setting for 1 min.

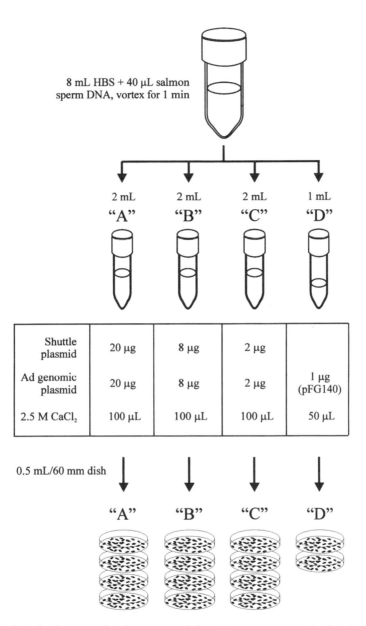

Fig. 6. Standard cotransfection protocol for Ad vector rescue by in vivo site-specific recombination.

Table 1
Vector Rescue Efficiency by In Vivo Site-Specific Recombination[a]

Shuttle plasmid	Ad genomic plasmid	Average plaques/60-mmdish
pCA35loxΔITR	pBHGloxΔE1,3Cre	43
	pBHGloxΔE3(X1)Cre	63
	pBHGloxE3Cre	48
pCA35frtΔITR	pBHGfrtΔE1,3FLP	41
	pBHGfrtΔE3(X1)FLP	27
	pBHGfrtE3FLP	25
	pFG140 [b]	103

[a] 60-mm dishes of 293 cells were cotransfected with 2 μg of each plasmid, and plaques were counted at 10 days post cotransfection.

[b] 60-mm dishes of 293 cells were transfected with 0.5 μg of pFG140, and plaques were counted at 10 days post cotransfection.

4. Aliquot 2 mL each of the above solution to three polystyrene tubes labeled "A," "B," and "C"and 1 mL into a fourth polystyrene tube labeled "D."

5. To tube "A" add 2 μg of shuttle plasmid DNA and 2 μg of Ad genomic plasmid DNA. (This will result in 0.5 μg of each plasmid per dish.) To tube "B" add 8 μg of shuttle plasmid DNA and 8 μg of Ad genomic plasmid DNA (2 μg of each plasmid per dish). To tube "C" add 20 μg of shuttle plasmid DNA and 20 μg of Ad genomic plasmid DNA (5 μg of plasmid per dish). To tube "D" add 1 μg of pFG140 DNA. Mix each tube thoroughly but gently.

6. Add 100 μL of 2.5 M CaCl$_2$ dropwise to tubes "A," "B," and "C" with gentle mixing. Add 50 μL of 2.5 M CaCl$_2$ dropwise to tube "C" with gentle mixing. The solutions should become slightly cloudy. Incubate at room temperature for 30 min.

7. Apply 0.5 mL of the contents in tube "A" dropwise to the monolayer in each of the dishes labeled "A" without removing the medium. Do the same for tubes "B," "C," and "D". Rock the dishes to distribute the precipitate evenly, and return to the incubator.

8. Next day, melt 1% agarose solution in a microwave oven and allow it to equilibrate to 45°C in a waterbath. Equilibrate 2× maintenance medium to 37°C. Once the two solutions have equilibrated to the desired temperature, prepare overlay solution by combining 70 ml of melted 1% agarose and 70 mL of 2× maintenance medium.

9. Remove the medium from the cotransfected dishes and add 10 mL of the overlay prepared in **step 8** to each dish (*see* **Note 5**).

10. Allow the overlay to solidify at room temperature (10–15 min), and then return the dishes to the incubator. Plaques should begin to appear as early as 5 days post cotransfection and will continue to appear until about 12–14 days post-cotransfection.

11. Ten days post cotransfection, pick well-isolated plaques from the monolayer by punching out agar plugs with a sterile cotton-plugged Pasteur pipet attached to a rubber bulb (*see* **Note 6**).

12. Transfer the agar plugs into 0.5 mL PBS^{2+} supplemented with glycerol to 10% in a suitable vial. Vortex briefly and store at $-70°C$.

It is recommended that mock-transfected controls be included for comparison until the researcher gains familiarity with the appearance and morphology of Ad plaques.

3.4. Analysis of Recombinant Vectors

Once plaques have been isolated, they are expanded for extraction of vector DNA for analysis and to yield a working vector stock as follows:

1. Seed 60-mm dishes of 293 cells (one per plaque) to reach approx 90% confluency on the day of use.

2. Thaw virus plaque picks and vortex briefly. Remove medium from 60-mm dishes of 293 cells and add 0.2 mL of the plaque pick. Adsorb for 1 h in the incubator, rocking the dishes every 10–15 min.

3. Following adsorption, add 5 mL of maintenance medium and return dishes to incubator until complete cytopathic effect (CPE) is observed ($\geq$90% cells rounded up and detached from dish, usually 3–4 days post infection) (*see* **Note 7**).

4. Once complete CPE is reached, the dishes can be processed as follows: leave dishes undisturbed in a laminar flow hood for 20–30 min to allow cells in suspension to settle. Gently remove medium with a pipet, leaving the majority of the cell in the dish. Transfer the medium into a suitable vial, supplement with glycerol to 10%, and store at $-70°C$. The medium should contain a significant amount of virus (10^7–10^8 plaque-forming units [pfu]/mL) and can be used for plaque purification of the vector (*see* **Subheading 3.5.**) (or can be used in preliminary experiments or for further vector expansion, as described in **Subheading 3.6.**).

5. Add 0.5 mL pronase-SDS solution and incubate dishes at 37°C overnight.

6. Transfer lysate to a microfuge tube and add 1 mL 95% ethanol. Precipitate DNA by inverting the tube several times.

7. Pellet DNA by spinning in a microcentrifuge (maximum speed for 2 min) and wash pellet twice with 70% ethanol. Let the pellet dry and resuspend in an appropriate volume of TE (approx 50–100 µL). Dissolve DNA by heating at 65°C with occasional vortexing.

8. Digest 5–10 µL of the DNA with an appropriate restriction enzyme and analyze the DNA structure by ethidium bromide staining following agarose gel electrophoresis to verify the DNA structure of the recombinant virus is correct.

Once the DNA structure of the vector has been verified, it is purified by titration (*see* **Subheading 3.5.**).

3.5. Titration of Adenovirus

The procedure outlined below is used to plaque purify recombinant vectors and also to determine the concentration of vector stocks (*see* **Note 8**).

1. Seed 60-mm dishes of 293 cells to reach approx 80–90% confluency in 1–2 days for titration.
2. Prepare serial dilutions of the recombinant virus in PBS^{2+} (10^{-2}–10^{-6} for samples prepared in **Subheading 3.4.** and 10^{-4}–10^{-10} for samples prepared in **Subheading 3.6.** and **3.7.**).
3. Remove medium from 60-mm dishes of 293 cells and infect with 0.2 mL of the diluted samples. Return dishes to the incubator and adsorb for 1 h, rocking the dishes every 10–15 min.
4. During the adsorption period, melt 1% agarose solution in a microwave oven and equilibrate to 45°C in a waterbath. Equilibrate 2× maintenance medium to 37°C.
5. Following 1-h adsorption, combine equal volumes of molten agarose solution with 2× maintenance medium, mix well, and gently overlay dishes with 10 mL (*see* **Note 5**).
6. Allow overlay to solidify for 10–15 min and return dishes to the incubator.
7. Plaques should start to appear about 4 days post infection and should be counted 12 days post infection. For isolation of recombinant virus by plaque purification, well-isolated plaques should be picked according to **steps 11** and **12** of **Subheading 3.3.** around 10 days post infection. The plaque-purified vectors are expanded according to **Subheading 3.4.** and used as inoculum for the preparation of high-titer viral stocks (*see* **Subheading 3.6.**).
8. Determine the vector concentration in pfu/mL as follows:

$$\text{titer} = (\text{number of plaques})(\text{dilution factor})/(\text{infection volume})$$

Calculate the titer from dishes bearing 20–70 plaques.

3.6. Preparation of High-Titer Viral Stocks (Crude Lysate)

Because most of the virus remains associated with the infected cells until very late in infection (i.e., until the cells lyse), high-titer stocks can be prepared easily by concentrating infected 293 cells. The following protocol describes the production of high-titer virus preparations using either monolayers of 293 cells or suspension cultures of 293N3S cells. However, because of the greater ease of handling suspension cultures, 293N3S cells are preferable for the production of very large amounts of high-titer viral stocks. The following describes protocols for the preparation of crude lysates of high-titer vector stocks that are suitable for most experiments. Prior to the preparation of high-titer stocks, confirm that enough inoculum is available and if not, prepare an intermediate-scale virus stock by infecting two to three 150-mm dishes of 293 cells.

3.6.1. Preparation of High-Titer Viral Stocks (Crude Lysate) from Cells in Monolayer

1. Set up 150-mm dishes of 293 cells to be 80–90% confluent at the time of infection. The number of dishes is dictated by the amount of vector desired (20–30 dishes can be easily handled).
2. Dilute virus sample prepared as in **Subheading 3.5., step 7**, 1:8 with PBS^{2+}.
3. Remove medium from the 293 cells and add 1 mL of the diluted virus sample prepared in **step 2** to each 150-mm dish of cells (MOI of 1–10 pfu/cell).
4. Adsorb for 1 h in the incubator, rocking the dishes every 10–15 min. Following adsorption, add 25 mL maintaince medium and return dishes to the incubator. Examine daily for signs of cytopathic effect.
5. When cytopathic effect is nearly complete, i.e., most cells rounded but not yet detached, harvest by scraping the cells into the medium and centrifuging the cell suspension at 800g for 15 min.
6. Discard the supernatant and resuspend the cell pellet in 2 mL PBS^{2+} + 10% glycerol per 150-mm dish infected. Freeze (–70°C) and thaw the crude virus stock prior to characterization of the vector (*see* **Subheading 3.8.**). Store aliquots at –70°C.

3.6.2. Preparation of High-Titer Viral Stocks (Crude Lysate) from Cells in Suspension

1. Grow 293N3S cells to a density of 2–4 × 10^5 cells/mL in 4 L complete Joklik's modified MEM supplemented with 10% horse serum. Centrifuge cell suspension at 750g for 20 min, saving half of the conditioned medium. Resuspend the cell pellet in 0.1 vol fresh medium, and transfer to a sterile 500-mL bottle containing a sterile stir bar.
2. Add virus at an MOI of 10–20 pfu/cell and stir gently at 37°C. After 1 h, return the cells to the 4-L spinner flask and bring to the original volume using 50% conditioned medium and 50% fresh medium. Continue stirring at 37°C.
3. Monitor infection daily by inclusion body staining as follows:
   a. Remove a 5-mL aliquot from the infected spinner culture. Spin for 10 min at 750g and resuspend the cell pellet in 0.5 mL of 1% sodium citrate.
   b. Incubate at room temperature for 10 min; then add 0.5 mL Carnoy's fixative and fix for 10 min at room temperature.
   c. Add 2 mL Carnoy's fixative, spin for 10 min at 750g, aspirate, and resuspend the pellet in a few drops of Carnoy's fixative. Add 1 drop of fixed cells to a slide and air-dry for about 10 min; then add 1 drop orcein solution and a cover slip and examine using a microscope. Inclusion bodies appear as densely staining nuclear structures resulting from accumulation of large amounts of virus and viral products at late times post infection. A negative control should be included in initial tests.
4. When inclusion bodies are visible in 80–90% of the cells (1 1/2–2 1/2 days depending on the input MOI), harvest by centrifugation at 750g for 20 min in sterile 1-L bottles. Combine pellets in a small volume of medium, and spin again.

5. Discard supernatant and resuspend pellet in 20 mL PBS^{2+} supplemented with 10% glycerol. Freeze (–70°C)- thaw, then aliquot, store at –70°C and characterize vector as described in **Subheading 3.8.**

3.7. Purification of Adenovirus by CsCl banding

Many experimental studies can be carried out using virus in the form of crude infected cell lysates prepared as described in **Subheading 3.6.** However, for some experiments, particularly for animal work, it is usually desirable to use purified virus. The following protocol describes a method for purifying vectors by CsCl banding obtained from 4 L of infected 293N3S cells, or 30 × 150-mm dishes of 293 cells.

1. Prepare crude cell lysate from infected cells as follows:
   a. For 30 × 150-mm dishes prepared as in **Subheading 3.6.1.**: When complete CPE is evident, scrape the cells into the medium, transfer the cell suspension to a centrifuge bottle, and spin for 10 min at 750g. Resuspend cell pellet in 15 mL 0.1 M Tris-HCl, pH 8.0. Sample can be stored at –70°C.
   b. For 3-L spinner cultures prepared as in **Subheading 3.6.2.**: When inclusion bodies are visible in 80–90% of the cells, harvest cells by centrifugation at 750g for 20 min in sterile bottles. Resuspend pellet in 15 mL 0.1 M Tris-HCl, pH 8.0. Samples can be stored at –70°C.
2. Thaw sample and add 1.5 mL 5% Na deoxycholate for each 15 mL of cell lysate. Mix well and incubate at room temperature for 30 min. This results in a highly viscous suspension.
3. Add 150 µL 2 M MgCl$_2$ and 75 µL DNase I solution for each 15 mL of cell lysate, mix well, and incubate at 37°C for 60 min, mixing every 10 min. The viscosity should be reduced significantly.
4. Spin at 3000g for 15 min at 5°C in the Beckman tabletop centrifuge.
5. Meanwhile, prepare CsCl step gradients (one SW 41 Ti ultraclear tube for each 5 mL of sample): Add 0.5 mL of 1.5 g/mL solution to each tube. Gently overlay with 3.0 mL of 1.35 g/mL solution. Gently overlay this with 3.0 mL of 1.25 g/mL solution.
6. Apply 5 mL of supernatant from **step 4** to the top of each gradient. If necessary, top up tubes with 0.1 M Tris-HCl, pH 8.
7. Spin at 35,000 rpm in a Beckman SW 41 Ti rotor (151,000g) at 10°C, for 1 h.
8. Collect virus band (should be at 1.25 d/1.35 d interface) with a needle and syringe by piercing the side of the tube. The volume collected is unimportant at this stage, so try to recover as much of the virus band as possible. If more than one tube was used, pool virus bands into a Beckman SW 50.1 ultraclear tube.
9. Top up tubes with 1.35 g/mL CsCl solution if necessary and centrifuge in a SW50.1 rotor at 35,000 rpm (115,000g), 4°C, for 16–20 h.
10. To collect the virus band, puncture the side of the tube just below the virus band

with a needle and syringe. Collect the virus band in the smallest volume possible and transfer to a Slide-A-Lyzer dialysis cassette. Dialyze at 4°C against three changes of 500 mL 10 m*M* Tris-HCl, pH 8.0, for at least 24 h total.

11. After dialysis, transfer the virus to a suitable vial and add sterile glycerol to a final concentration of 10%. Store the purified virus at –70°C in small aliquots.

3.8. Characterization of Adenoviral Vector Preparations

Before the recombinant vector is used for experimentation, several basic characterizations should be performed: 1) the concentration should be determined; 2) the DNA structure should be confirmed; and 3) the presence and level of RCA should be ascertained.

The concentration in pfu/mL is determined by titration on 293 cells as described in **Subheading 3.5.** The concentration of virus particles, based on DNA content at OD_{260}, can also be determined spectrophotometrically as follows:

1. Dilute (usually 20-fold) purifed virus with TE supplemented with SDS to 0.1%. Set up blank the same except add virus storage buffer (10 m*M* Tris-HCl, pH 8.0, supplemented with glycerol to 10%) instead of virus.
2. Incubate for 10 min at 56°C.
3. Vortex sample briefly.
4. Determine OD_{260}.
5. Calculate the number of particles/mL, based on the extinction coefficient of wild-type Ad as determined by Maizel et al. *(21)* as follows:

$$(OD_{260})(\text{dilution factor})(1.1 \times 10^{12})$$

The DNA structure of the recombinant vector should be confirmed following large scale preparation. Virion DNA can be extracted from CsCl banded virus for analysis as follows:

1. Add an appropriate volume (approx 25 µL depending on the concentration of the virus) of the purified virus to pronase-SDS solution to a final volume of 0.4 mL and incubate overnight at 37°C to lyse the virions and digest virion proteins.
2. Precipitate virion DNA by adding 1/10 vol 3 *M* sodium acetate, pH 5.2, and 2.5 vol 95% ethanol and incubate at –20°C for 15–30 min.
3. Spin in microcentrifuge for 10–15 min at maximum speed.
4. Discard supernatant and wash DNA pellet twice with 70% ethanol.
5. Dry DNA pellet and resuspend in an appropriate volume of TE.

For crude preparations, viral DNA can be extracted following infection of 293 cells as described in **Subheading 3.4.** A sample of the vector DNA is digested with the appropriate diagnostic restriction enzyme(s), and the structure of the DNA is analyzed by agarose gel electrophoresis.

293 cells contain nts 1–4344 bp of Ad5 DNA *(22)* with consequent homology flanking the expression cassette of most first-generation vectors. Therefore, the possibility exists that homologous recombination between the Ad vector and the Ad sequences present in 293 cells may result in the formation of E1$^+$ replication-competent Ad (RCA) (*see* **Note 9**). Thus, it is important to determine whether the vector preparations are contaminated with RCA and if so, at what level. A non-E1-complementing cell line, such as A549, can be used for this purpose since it supports replication of RCA but not the E1-deleted Ad vector. The protocol is as follows:

1. Seed two 60-mm dishes and one 150-mm dish of A549 cells in α-MEM supplemented with 10% FBS for each vector preparation to be tested
2. When the monolayers are 80–90% confluent, remove the medium and infect one 60-mm dish with 10^6 pfu and the other 60-mm dish with 10^7 pfu. (Make up the volume of the inoculum to 0.2 mL with PBS^{++}). Infect the 150 mm dish with 10^8 pfu (make up the volume of the inoculum with 1 mL PBS^{2+}.) Adsorb for 1 h in the incubator, rocking the dishes every 10–15 min.
3. Following adsorption, add 5 and 25 mL of α-MEM supplemented with 10% horse serum to each 60-mm dish and 150-mm dish, respectively, and return dishes to incubator.
4. Infection of A549 cells with high-titer Ad, even in the absence of RCA, will frequently result in complete CPE of the initial A549 monolayer because of toxicity of viral proteins in the inoculum and multiplicity-dependent "leaky" viral replication. Harvest the monolayers 7 days post infection or when complete CPE is observed, whichever occurs first. Harvest the monolayer by scraping the cells into the medium. Transfer 3 mL of the lysate into a suitable vial, supplement with glycerol to 10%, and store at –70°C.
5. Set up four 150-mm dishes of A549 cells, one for each of the lysates collected in **step 4** and one control.
6. When the monolayers reach 80–90% confluency, infect one dish each with 1 mL of the 10^6, 10^7, and 10^8 lysates collected in **step 4**. "Infect" the control dish with 1 mL of PBS^{2+}. Adsorb for 1 h in the incubator, rocking the dishes every 10–15 min.
7. Following adsorption, add 25 mL of α-MEM supplemented with 10% horse serum to each dish, and return dishes to incubator.
8. Replace the medium every 5 days if necessary.
9. Observe monolayers daily for signs of CPE by comparing the infected dishes with the control dish. If the infected dishes are indistinguishable from the control dish (no CPE) 21 days post infection, then the lysate collected at **step 4** was free of RCA and thus, the original aliquot of the vector preparation used in **step 2** was also RCA free (or below the limits of detection). This provides an estimate of the level of RCA in the vector stock (e.g., <1 RCA/10^7 pfu if the monolayers infected with the 10^6 and 10^7 lysates in **step 4** show no CPE 21 days post infection). If the

dishes exhibit CPE (usually evident 14 days post infection or sooner) then the original aliquot of the vector preparation used in **step 2** was contaminated with RCA.

10. To verify the presence of RCA, extract DNA from the dishes showing signs of CPE in **step 9** and analyze it by restriction enzyme digestion and agarose gel electrophoresis.
    a. When complete CPE is evident, scrape the monolayer into the medium, remove 5 mL of the cell suspension, and pellet cells at top speed in a tabletop centrifuge.
    b. Discard the supernatant, add 0.5 mL pronase-SDS solution, and incubate at 37°C overnight.
    c. Add 1 mL 95% ethanol and mix to precipitate DNA. Pellet DNA by centrifugation and wash twice with 70% ethanol. Dry pellet and resuspend in an appropriate volume of TE (approx 100 μL).
    d. Digest an aliquot with an appropriate diagnostic restriction enzyme and analyze by agarose gel electorphoresis. RCA will have a left end structure identical to wild-type Ad owing to the presence of E1 sequences.

Alternative approaches based on Southern blot hybridization or quantitative PCR can also be employed for the detection of RCA *(23)*.

In addition to these standard characterizations, the biologic activity of the vector, with respect to the transgene of interest, should also be determined. This will be unique to the vector in question, and the appropriate assays should be employed in each case.

3.9. Alternative Procedures to Expedite Vector Production

It is recommended that the steps outlined in the preceding sections be followed, as they have been well proved. However, since only the correct recombinant vector should be generated following cotransfection *(15–17)*, several alternative procedures are acceptable to expedite vector production (**Fig. 3**).

1. Once the vector has been rescued following cotransfection (*see* **Subheading 3.3.**), it can be immediately titrated for plaque purification, and the DNA structure can be checked afterwards.
2. Although plaque purification is strongly recommended (*see* **Subheading 3.5.**), especially if large quantities of the vector are to be generated for extensive experimentation, this step is not absolutely essential since all infectious virus generated should be the desired recombinant. In this case, high-titer stocks can be generated directly from the plaques isolated following cotransfection.
3. As mentioned in **Subheading 3.7.**, vector purification by CsCl banding, although recommended, may not be necessary for many experiments.
4. It is strongly recommended that the recombinant vector be isolated from individual plaques following cotransfection using the method described in **Subhead-**

ing **3.3.** However, vector production can be expedited by omitting **steps 8–12** in
Subheading 3.3. In this case, following overnight cotransfection, remove the
medium from the monolayers and add 5 mL of maintenance medium. If complete
CPE is observed within 7 days post cotransfection, then proceed from **step 4** in
Subheading 3.4. If complete CPE is not observed by 7 days post cotransfection,
then proceed as described in **Note 7**.

4. Notes

1. For the large Ad genomic plasmid, bacterial cultures should be started from well-
   isolated colonies picked from a bacterial plate <1 week old.
2. For higher yields of plasmid DNA, richer medium such as Super Broth (SB) can
   be used (LB broth supplemented with 22 g/mL peptone, 15 g/mL yeast extract, 1
   g/mL D-glucose, and 0.005 N NaOH).
3. Keep tubes covered with foil or otherwise in the dark except when recovering
   plasmid DNA bands. Do not expose to fluorescent or UV light more than necessary.
4. 293 cells should never be allowed to become overconfluent, should not be seeded
   too thinly, and should have regular medium changes between splits (twice weekly
   if they are not growing rapidly enough to permit splitting every 2–4 days). It is
   recommended that a sufficient number of ampoules of the cells be frozen and
   stored in liquid N_2 to permit initiation of new cultures when the passage number
   of the lab stocks reaches 40–45 passages or when the cells are no longer behaving
   well under agar overlays (*see* **Subheading 3.3.** and **3.5.**). Higher passage or poorly
   adherent cells may be suitable for growth of virus, but the properties of the cells
   are more critical for plaque assays and cotransfections.
5. It is important to perform this step quickly (to prevent the overlay solution from
   solidifying prematurely) but gently (to prevent disturbing the monolayer).
6. Although it is tempting to isolate the plaques as soon as they are visible, it is
   recommended that they be isolated at about 10 days post cotransfection. This
   ensures that the plaques chosen are well isolated and that no other overlapping
   plaques will form in close proximity.
7. It is important that the DNA be extracted following complete CPE so that vector
   DNA bands are clearly visible above the background of cellular DNA. If complete CPE is not reached by 5–6 days post infection (most likely because of low
   multiplicity) then scrape the monolayer into the medium, transfer the cell suspension into a suitable vial, and supplement with glycerol to 10%. Freeze (–70°C)-
   thaw the cell suspension and infect 60-mm dishes of 90% confluent 293 cells with
   0.2–0.4 mL as described in **Subheading 3.4.** Complete CPE should be reached
   within 6 days, and the vector DNA can be extracted.
8. To determine vector concentration accurately, titrations should be performed in
   duplicate or triplicate. The number of plaques should vary in direct proportion to
   the dilution factor; otherwise, repeat the titration, making sure that the samples
   are thoroughly mixed when setting up the serial dilutions.
9. The frequency with which Ad vectors recombine with Ad sequences in 293 cells

is unknown, but in general $E1^+$ RCA replicate faster than $E1^-$ vectors. Consequently, the proportion of RCA increases with prolonged propagation of the vectors in 293 cells. To minimize RCA contamination, vectors should not be serially propagated indefinitely. It is recommended that large-scale vector preparations be initiated from a stock prepared following plaque purification (*see* **Subheading 3.5.**). If the original plaque-purified stock is exhausted, plaque purification can be repeated.

Acknowledgments

This work was supported by grants from the National Institutes of Health, the Medical Research Council of Canada (MRC), the National Cancer Institute of Canada (NCIC), and Merck Research Laboratories. P. N. is supported by an MRC Postdoctoral Fellowship.

References

1. Berkner, K. L. (1988) Development of adenovirus vectors for expression of heterologous genes. *Biotechniques* **6,** 616–629.
2. Graham, F. L. and Prevec, L. (1992) Adenovirus-based expression vectors and recombinant vaccines, in *Vaccines: New Approaches to Immunological Problems*(Ellis, R. W., ed.), Butterworth-Heinemann, Boston, pp. 363–389.
3. Hitt, M., Addison, C. L., and Graham, F. L. (1997) Human adenovirus vectors for gene transfer into mammalian cells. *Adv. Pharmacol.* **40,** 137–206.
4. Hitt, M. M., Parks,R. J., and Graham, F. L. (1999) Structure and genetic organization of adenovirus vectors, in *The Development of Human Gene Therapy* (Friedman, T., ed.), Cold Spring Harbor Laboratory Press, Cold Spring Harbor, NY, pp. 61–86.
5. Shenk, T. (1996) Adenoviridae: the viruses and their replication, in *Fields Viology* (Knipe, B. N. and Howely, D.M., eds.), Lipponcott-Raven, Philadelphia, PA, pp. 2111–2148.
6. Bergelson, J. M., Cunningham, J. A., Droguett, G., et al. (1997) Isolation of a common receptor for coxsackie B viruses and adenoviruses 2 and 5. *Science* **275,** 1320–1323.
7. Hong, S. S., Karayan, L., Tournier, J., Curiel, D. T., and Boulanger, P. A. (1997) Adenovirus type 5 fiber knob binds to MHC class I alpha2 domain at the surface of human epithelial and B lymphoblastoid cells. *EMBO J.* **16,** 2294–2306.
8. Wickham, T. J., Mathias, P., Cheresh, D. A., and Nemerow, G. R. (1993) Integrins alpha v beta 3 and alpha v beta 5 promote adenovirus internalization but not virus attachment. *Cell* **73,** 309–319.
9. Wickham, T. J., Segal, D. M., Roelvink, P. W., et al. (1996) Targeted adenovirus gene transfer to endothelial and smooth muscle cells by using bispecific antibodies. *J. Virol.* **70,** 6831–6838.
10. Goldman, M., Su, Q., and Wilson, J. M. (1996) Gradient of RGD-dependent entry of adenoviral vector in nasal and intrapulmonary epithelia: implications for gene therapy of cystic fibrosis. *Gene Ther.* **3,** 811–818.

11. Mellman, I. (1992) The importance of being acidic: the role of acidification in intracellular membrane traffic. *J. Exp. Biol.* **172,** 39–45.

12. Leopold, P. L., Ferris, B., Grinberg, I., et al.(1998) Fluorescent virions: dynamic tracking of the pathway of adenoviral gene transfer vectors in living cells. *Hum. Gene Ther.* **9,** 367–378.

13. Greber, U. F., Willetts, M., Webster, P., and Helenius, A. (1993) Stepwise dismantling of adenovirus 2 during entry into cells. *Cell* **75,** 477–486.

14. Graham, F. L., Smiley, J., Russell, W. C., and Nairn, R. (1977) Characteristics of a human cell line transformed by DNA from human adenovirus 5. *J. Gen. Virol.* **36,** 59–72.

15. Ng, P., Parks, R. J., Cummings, D. T., et al. (1999) A high efficiency Cre/*lox*P-based system for construction of adenoviral vectors. *Hum. Gene Ther.* **10,** 2667–2672.

16. Ng, P., Parks, R. J., Cummings, D. T., Evelegh, C. M., and Graham, F. L.(2000) An enhanced system for construction of adenoviral vectors by the two plasmid rescue method. *Hum. Gene Ther.* **11,** 693–699.

17. Ng, P., Cummings, D. T, Evelegh, C. M., and Graham, F. L. (2000) The yeast recombinase flp functions effectively in human cells for construction of adenovirus vectors. *Biotechniques* **29,** 524–528.

18. Graham, F. L. (1984) Covalently closed circles of human adenovirus DNA are infectious. *EMBO J.* **3,** 2917–2922.

19. Bett, A. J., Prevec, L., and Graham, F. L. (1993) Packaging capacity and stability of human adenovirus type 5 vectors. *J. Virol.* **67,** 5911–5921.

20. Addison, C. L., Hitt, M., Kunsken, D., and Graham, F. L. (1997). Comparison of the human versus murine cytomegalovirus immediate early gene promoters for transgene expression by adenoviral vectors. *J. Gen. Virol.* **78,** 1653–1661.

21. Maizel, J. V., White, D., and Scharff, M. D. (1968) The polypeptides of adenovirus. I. Evidence of multiple protein components in the virion and a comparison of types 2, 7a, and 12. *Virology* **36,** 115–125.

22. Louis, N., Evelegh, C., and Graham, F. L. (1997) Cloning and sequencing of the cellular/viral junction from the human adenovirus type 5 transformed 293 cell line. *Virology* **233,** 423–429.

23. Lochmuller, H., Jani, A., Haurd, J., et al. (1994) Emergence of early region 1-containing replication-competent adenovirus in stocks of replication-defective adenovirus recombinants (ΔE1+ΔE3) during multiple passages in 293 cells. *Hum. Gene Ther.* **5,** 1485–1491.

28

Preparation of Ovine Adenovirus Vectors

Peter Löser, Daniel Kümin, Moritz Hillgenberg, Gerald W. Both, and Christian Hofmann

1. Introduction

Human adenovirus vectors have been used widely as gene delivery vehicles. However, one major problem related to the use of these vectors is the presence of a preexisting immunity to human adenoviruses in a majority of the population. We *(1–3)* and others *(4–6)* are therefore developing nonhuman adenoviruses as gene therapy vectors. Among these, vectors derived from the ovine adenovirus isolate 287 (OAV) have been found to deliver genes with high efficiency to several mammals. This chapter introduces this novel vector system and describes the methodology of generating OAV vectors.

1.1. Structure of the OAV Genome

Genome arrangement and base composition strongly distinguish OAV from the mastadenoviruses (which include all known human adenoviruses), and analyses of hexon and protease genes have shown that OAV is phylogenetically distant from human adenoviruses. Therefore, OAV has been grouped together with egg drop syndrome (EDS) virus and several subtypes of bovine adenovirus in the proposed genus of *Atadenovirus (7,8)*. **Figure 1** shows the OAV genome organization. As in all adenoviruses, the genome contains overlapping early and late transcription units, and homologies to the E2 genes as well as to most of the late genes of mastadenoviruses have been found *(9,10)*. The most striking differences from the mastadenoviruses are, however, the lack of typical E1A/B and E3 regions. The putative E4 region differs in its genomic location, and limited homologies to mastadenoviruses have been found for only two open reading frames (ORFs). Genes coding for homologs of core protein

From: *Methods in Molecular Medicine, Vol. 69, Gene Therapy Protocols, 2nd Ed.*
Edited by: J. R. Morgan © Humana Press Inc., Totowa, NJ

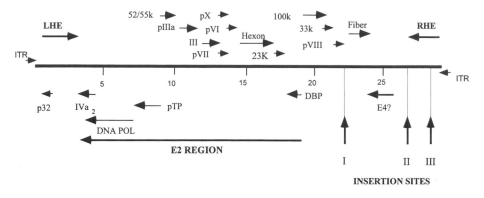

Fig. 1. Scheme of OAV genome organization. The location of early regions LHE, RHE, E2, and E4 are shown in bold arrows. Open reading frames with homologs in other adenoviruses (except p32) are shown. Sites I–III for insertion of transgene cassettes are indicated.

V, structural protein IX, and virus-associated RNA are absent from the OAV genome, but an additional polypeptide is encoded in the strand complementary to the E1 genes of mastadenoviruses *(11–13)*. In agreement with the lack of an E1A/B region is the finding that, in contrast to Ad5 E1A/B sequences, OAV sequences are unable to transform mammalian cells in vitro *(14)*.

1.2. Recombinant OAV as a Gene Transfer Vector

The first rescue of OAV recombinants was described in 1996 *(15)*. Two sites for integration of foreign DNA were identified. These are located between the ORFs for protein VIII and fiber (site I in **Fig. 1**) and at the distal unique *Sal*I site of the OAV genome (site II in **Fig. 1**). Mutations and/or insertions in either site did not affect rescue and growth of the respective recombinant viruses. In addition, a third integration site was created by insertion of a short polylinker at bp 26,676 of the OAV genome and subsequent deletion of related, apparently redundant regions between this polylinker and the *Sal*I site. Recombinants containing this deletion show only a marginal reduction in virus titer *(16)* (site III in **Fig. 1**). Using these sites, several transgenes were introduced into the OAV genome, and recombinant viruses were rescued. Genes include those coding for the rotavirus isolate SA11-derived VP7sc (OAV204), 45W antigen from *Taenia ovis* (OAV205), human pancreatic alkaline phosphatase (OAV216), human α_1-antitrypsin (OAV-haat), and green fluorescent protein (OAV217A) *(1,2,16,17)*. The size restrictions for recombinant OAV genomes seem to differ from those for human adenovirus vectors. Insertion of 4,3 kbp additional DNA without a compensating deletion of OAV sequences still allowed rescue and growth of the recombinant virus, indicating that OAV can package at least 114% of its genome size *(17)*. Deletion of the redundant region and the use of site III for transgene insertion further increased the theoretical packaging capacity by about 2 kbp.

Efficient infection of cell lines of different origin was shown with OAV recombinants. However, the spectrum of cell lines susceptible to infection with OAV vectors differs markedly from that infected by human adenovirus vectors. For example, C2C12, CHO, and HeLa cells were infected to a higher degree by OAV-haat than by a human adenovirus vector (Ad5-haat) containing an identical expression cassette, whereas little infection by OAV-haat was detected on 293 and A549 cells *(18)*. This indicates that OAV uses a receptor different from that utilized by human adenovirus type 5-based vectors and is in accordance with the fact that exchange of the OAV fiber cell binding domain with that of Ad5 results in an alteration of the cellular tropism of OAV *(19)*.

In vivo data obtained with OAV-derived vectors so far are promising: Systemic administration of OAV-haat to mice as well as local delivery to liver and skeletal muscle resulted in high levels of reporter gene expression *(1–3)*. Interestingly, in all animal models used so far, no preference of the vector for the liver was observed, as has been reported for human adenovirus vectors in rodents *(20)*. Most importantly, generation of anti-Ad5 neutralizing antibody titers in mice, which were comparable to those present in the human population, prevented successful gene transfer with hAd vectors but had no effect on OAV-mediated gene delivery *(1)*. High reporter gene expression was also observed in rats and rabbits after systemic delivery of OAV vectors *(18)*. Taken together, these data make OAV a promising tool for gene delivery in vaccination and gene therapy applications. Several studies are in progress to characterize further the molecular biology and the in vivo characteristics as well as safety aspects of this novel vector system.

2. Materials

2.1. OAV Genomic Plasmids

pOAV100 was the first plasmid containing a full-length OAV genome for convenient manipulation *(15)*. As with all such plasmids, the infectious viral genome can be released by digestion with *Kpn*I. Insertion of a short polylinker between the ORFs for pVIII and fiber (site I) resulted in pOAV200, and insertion of the polylinker at bp 26,676 gave rise to pOAV600. Deletion of the region between this polylinker and the unique *Sal*I site of pOAV600 produced pOAV603 *(16)*. Schemes of the plasmids are outlined in **Figure 2**.

2.2. OAV Shuttle Plasmids

Recombination of the gene of interest into the OAV genome requires shuttle plasmids. These plasmids typically contain sequences that flank the insertion

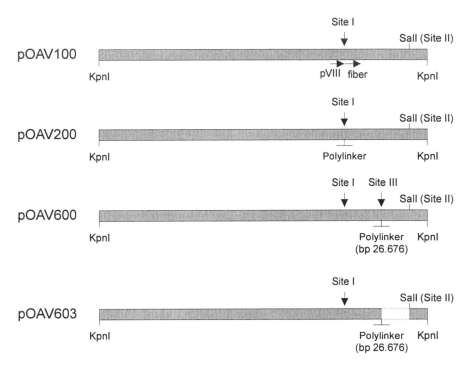

Fig. 2. Schemes of selected genomic plasmids. pOAV100 *(15)* contains the full-length OAV genome. The position of insertion site I between the open reading frames for the pVIII and fiber genes is indicated. Site II is identical to the unique *Sal*I site at bp 28,673 of the OAV genome. Insertion of a short polylinker into site I resulted in pOAV200. Insertion of the same polylinker at bp 26,676 (site III) of the OAV genome gave rise to pOAV600. Deletion of the apparently redundant region between the polylinker in pOAV600 and the *Sal*I site resulted in pOAV603 *(16)*. Infectious virus DNA can be obtained by *Kpn*I digestion of either plasmid.

site and a polylinker for convenient cloning of the gene of interest. Examples of shuttle plasmids for recombination into site I and site II of OAV are shown in **Figure 3**.

2.3. Manipulation of the OAV Genome

1. Luria-Bertani (LB) medium.
2. Agar plates.
3. Selection agents: Ampicillin, chloramphenicol.
4. Competent *E. coli* cells from strain BJ5183 and HB101, DH5, JM109, or XL-1.
5. Enzymes for manipulation of DNA: restriction endonucleases, Klenow enzyme, T4 DNA ligase, and others.
6. Kits and/or solutions for preparation of plasmid DNA from *E. coli*.

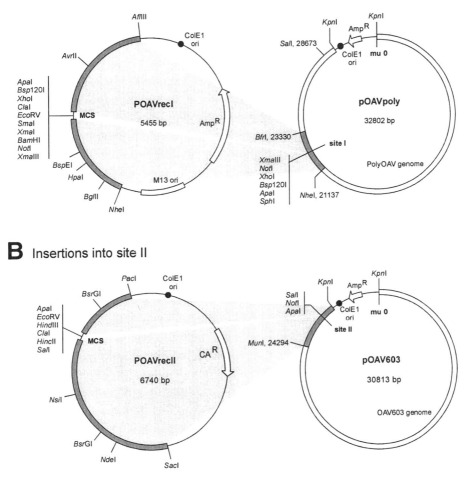

Fig. 3. Shuttle plasmids for recombination into site I (**A**, pOAVrecI) *(1)* and site II (**B**, pOAVrecII) *(18)* of the OAV genome. pOAVrecI contains the OAV sequences flanking site I in pOAV200 (bp 21,137–22,128 and bp 22,129–23,330 of the OAV genome, respectively), and pOAVrecII contains the regions flanking site II in pOAV603 (bp 24,294–26,675 and bp 28,673–29,574 of the OAV genome, respectively). Both plasmids contain a multiple cloning site (MCS) for convenient insertion of the transgene cassette and unique restriction enzyme sites for releasing fragments for recombination. The genomic OAV plasmids pOAVpoly (a derivate of pOAV200) and pOAV603 *(16)* are shown to the right. The scheme of recombination is indicated.

7. Kits and/or solutions for gel purification of DNA fragments.
8. Phenol and chloroform solutions.

2.4. Cell Culture

2.4.1. General

1. CSL503 and/or HVO156 cells.
2. Dulbecco's modified Eagle's medium (DMEM), containing 200 mM glutamine, 100 mg/L streptomycin, and 60–100,000 U/L penicillin.
3. Tissue culture dishes (30-, 6-, 100-, and 150-mm diameter).
4. Fetal calf serum (FCS).

2.4.2. Generation of Recombinant Virus

1. Monolayers of CSL503 or HVO156 cells (about 60% confluent).
2. Tissue culture dishes.
3. Falcon polystyrene tubes (Becton-Dickinson, Heidelberg, Germany).
4. Infectious recombinant viral DNA (*see* **Subheading 3.1.2.**).
5. Cell culture grade water.
6. Transfection buffer: 273.8 mM NaCl, 10 mM KCl, 1,4 mM Na$_2$HPO$_4$, 38.4 mM HEPES, pH adjusted to 6.75.
7. 2.5 M CaCl$_2$.
8. 15- and 50-ml polypropylene Falcon tubes (Becton-Dickinson).

2.4.3. Large-Scale Production and Purification of Recombinant OAV

1. Monolayers of CSL503 or HVO156 cells (about 75% confluent).
2. Tissue culture dishes (150-mm diameter).
3. Centrifuge bottles (500-mL, Nalgene, Rochester, NY).
4. 14 × 95 mm ultra clear^{TM} tubes (Beckman, München, Germany).
5. CsCl solutions at densities of 1.25, 1.35, and 1.5 g/cm^3.
6. NAP^{TM}-25 columns (Amersham-Pharmacia, Freiburg, Germany).
7. Purification buffer: 10 mm Tris-HCl, pH 7.8, 135 mM NaCl, 3 mM KCl, 1 mM MgCl$_2$, 10% v/v glycerol.

3. Methods

3.1. Generation of Recombinant Viral Genomes

3.1.1. Generation of Recombinant Viral Genomes by Direct Cloning

Foreign DNA can be cloned directly into the site of interest. Cleavage of pOAV200, pOAdV600, and pOAdV603 at sites I, III, or II, respectively, can be performed using *Apa*I and/or *Not*I. pOAV600 and pOAV603 can also be cut by *Sal*I. The expression cassette containing the gene of interest can be ligated into these sites for transformation into *E. coli* and selection of ampicillin-resistant clones. Whereas *Sal*I and *Not*I can be blunt-ended by the large Klenow fragment of DNA polymerase I, *Apa*I overhangs must be removed by T4 DNA polymerase. For greater convenience, *Bsp*120I (MBI fermentas, Vilnius, Lithuania) can be used to create blunt ends.

3.1.2. Generation of Recombinant Viral Genomes by Recombination in E. coli BJ 5183

Since cloning of large DNA constructs can be difficult, recombination of genes of interest into the OAV genome might be more effective. In the following, recombination of an expression cassette into site III of OAV603 is described. This follows a method described for human adenovirus vectors *(21)* and was applied to OAV with several reporter genes including human α_1-antit-rypsin (haat), human factor IX, green fluorescent protein (gfp), and the *E. coli* lacZ gene.

1. Ligate the gene of interest (including promoter region and polyA signal) into the polylinker of pOAVrecII (**Fig. 3b** and *see* **Note 1**).
2. Transform ligation products into XL1-blue (Stratagene, Amsterdam, Netherlands) and select for chloramphenicol-resistant clones.
3. Produce sufficient DNA to verify the plasmid structure and for the recombination procedure (10–20 µg is sufficient).
4. Release the recombination fragment containing the transgene cassette and flanking OAV sequences from the plasmid by restriction with appropriate endonucleases. The overlapping OAV sequences should extend at least 400 bp on either side of the expression cassette. For example, digestion with *Bsr*G1 will produce sufficient overlap. Alternatively, *Pac*I and *Nsi*I or *Nde*I might be used. Ensure that the expression cassette does not contain a *Kpn*I site or one of the sites used for releasing the fragment.
5. Gel-purify the recombination fragment and determine the DNA concentration. About 100 ng of gel-purified fragment is needed for recombination. Store at – 20°C until use.
6. Digest pOAV603 with *Sal*I. Gel-purify the linearized DNA fragment, aliquot, and store at –80°C (*see* **Note 2**).
7. Cotransform about 10 ng of the pOAV603 fragment together with 100 ng of the pOAVrecII fragment into competent *E. coli* BJ5183 cells and select ampicillin-resistant transformants.
8. Prepare minipreps from 1.5 mL LB using standard procedures. Remove bacterial protein by extraction with phenol/chloroform to avoid degradation of plasmid DNA by bacterial endonucleases. Dissolve DNA in 30 µL TE/RNase and check the plasmid size on an agarose gel.
9. Transform 5 µL of one miniprep into a recA minus *E. coli* strain (DH5, HB-101, and XL1 blue have been used successfully) and select ampicillin-resistant transformants.
10. Prepare maxiprep DNA and confirm the presence of the transgene cassette within

the recombinant OAV plasmid. For further procedures, about 50 µg of DNA will be needed.

11. Digest 20–30 µg of recombinant OAV plasmid with *Kpn*I to release infectious viral DNA. Perform a phenol/chloroform extraction followed by two chloroform extractions.
12. Precipitate DNA with ethanol, wash with 70% ethanol, and dissolve the pellet in about 50 µL TE for at least 12 h at 4°C. Check the integrity on a 0.6% agarose gel and estimate the DNA concentration. This DNA will be used for transfection of CSL503 or HVO156 cells.

3.2. Production of Recombinant Virus

3.2.1. Cell Lines for Generation of Recombinant OAV

The sheep fetal lung fibroblast line CSL503 *(22)* reportedly grows for 50–70 doublings (about 20–25 passages) and has been used to produce recombinant OAV. This cell line grows well under standard conditions (DMEM containing 10% FCS, 200 m*M* glutamine, 100 mg/L streptomycin, and 100,000 U/L ampicillin) at 37°C and 5% CO_2, but in our hands shows signs of senescence at higher passages (>12). Transfection of CSL503 cells was performed using cationic lipids *(15)*. As an alternative, the cell line HVO156, derived from embryonic sheep skin *(18)* has been used for production of recombinant OAV. Virus titers with this cell line are slightly higher than those obtained with CSL503, and cells can be grown for 40 passages (about 100–120 doublings) without loss in virus yield. HVO156 can be grown under the same conditions as CSL503 cells, but the best transfection results were obtained using the $CaCl_2$ coprecipitation method described below.

3.2.2. Generation of Recombinant OAV Using HVO156 Cells

1. One day prior to transfection, seed HVO156 at a density of 2×10^4 cells/cm² onto 30-mm dishes. At the time of transfection, a confluency of about 60% is desirable. Six hours before transfection, replace the medium with 2.5 mL of fresh DMEM/10% FCS.
2. Mix 5 µg of infectious virus DNA (from **Subheading 3.1.2, item 12**) with water to a volume of 112.5 µL in a 12-mL polystyrene tube. Add 12.5 µL of 2,5 *M* $CaCl_2$ and mix thoroughly.
3. Add 125 µL of transfection buffer drop by drop under permanent mixing (*see* **Note 3**).
4. Add the transfection mixture (250 µL) to the medium and return cells to the incubator.
5. After 12–18 h, wash the cells once with medium, and then add 2.5 mL of fresh medium containing 10% FCS.
6. A cytopathic effect (CPE) will become visible after 8–21 days. It is not necessary to change medium during this period.

7. Harvest the cells and supernatant, submit the cell suspension to three cycles of freezing and thawing, and spin down the cell debris. The supernatant will contain the A_0 (amplification round 0) generation of the recombinant virus.
8. Produce A_1, A_2, and A_3 generations of recombinant OAV by infecting permissive cells on 60-, 100-, and 150-mm plates, respectively, with about 10% of the previous virus generation. Infection is done in a small volume of medium overnight. Replace infection medium with 5, 10, or 30 mL, fresh DMEM/10% FCS, respectively. CPE becomes visible within 2 days, and harvest of virus should be performed at 72–96 h post infection. Harvest and process virus as in **step 7**. Passage A_3 of the recombinant virus should give enough material for large-scale production.

3.2.3. Large-Scale Production and Purification of OAV Stocks

Large-scale production of OAV is performed following the protocols for hAd vector production. The cell lysate containing the viral particles undergoes two subsequent CsCl gradients and salt is removed by gel chromatography. Alternatively, virus can be dialyzed against the desired storage buffer.

1. Seed 20 150-mm plates with CSL503 or HVO156 cells at a density of 3×10^4 cells/cm^2.
2. The next day, remove medium and replace with 10 mL of fresh medium. Add 400 µL of virus lysate from the A_3 generation and allow the cells to stand for at least 6 h in the incubator. Add medium to a volume of 30 mL and incubate cells for another 3 days.
3. Collect cells and supernatant in a Nalgene 500-mL centrifuge bottle. Spin down the cells ($400g$) and carefully remove supernatant. Recentrifuge supernatant at $700g$ and collect cells. Resuspend cells in a total volume of 12 mL serum-free DMEM and submit suspension to three steps of freezing and thawing. Spin down cell debris in a 15-mL Falcon tube at maximum speed.
4. Prepare CsCl gradients in Beckman 14×95 mm ultra clear™ tubes by underlaying 2.5 mL of 1.25 g/cm^3 CsCl with 2.5 mL of 1.35 g/cm^3 and 2 mL of 1.5 g/cm^3 CsCl. Carefully overlay the gradient with 6 mL of the cleared virus suspension from **step 3** and ultracentrifuge for at least 2 h at 30,000 rpm ($114,000g$) and 8°C in a Beckman SW40 rotor. Two bands will become visible after centrifugation, the lower one at the junction of the 1.25 and 1.35 g/cm^3 layers contains the recombinant OAV.
5. Carefully transfer the band containing the virus to a fresh 50-mL Falcon tube, and add CsCl (1.35 g/cm^3) to a total volume of 13 ml. Transfer virus suspension to a Beckman 14×95-mm ultra clear™ tube and centrifuge for at least 10 h under the same conditions as before. A single virus band should appear.
6. Remove the virus-containing band and apply it to a NAP™-25 column equilibrated according to the manufacturer's instructions with purification buffer. Apply the sample and elute virus using the same buffer. Aliquot virus, freeze, and store at –80°C (*see* **Note 4**).

3.3.4. Determination of Virus Titer

The number of viral particles (opu) can be determined according to the method of Maizel et al. *(23)* or Mittereder et al. *(24)* by measuring the absorption at 260 nm. The number of infectious particles can be determined by a classical $TCID_{50}$ assay on HVO156 or CSL503 cells. Typical titers range between 10^9 and 10^{11} infectious particles per mL (*see* **Note 5**). The yields of virus from CSL503 and HVO156 cells are typically in the range of 1,000–10,000 particles/cell, and the opu/IU ratio is usually between 10 and 40.

4. Notes

1. A problem related to the rescue of recombinant OAV may be the presence of cryptic splice donor or acceptor sites within the transgene cassette. Particularly for site I, these may interfere with correct processing of genes such as fiber that flank the integration site. Therefore, evaluation of more than one site for transgene insertion might alleviate this problem. Site III in particular appears to be located between transcription units *(12)*.

2. Avoid using glass milk-based kits for isolation of the large DNA fragment from agarose gel since it might be sheared because of its size. A good alternative is the use of a phenol extraction method: Cut out the gel fragment and slice it into small pieces. Put the gel pieces into an Eppendorf tube and add approximately the same volume of TE-saturated phenol. Vortex and allow to stand on the bench for 10 min. Transfer the Eppendorf tube to liquid nitrogen and leave it for another 10 min. Centrifuge for 20 min in a tabletop centrifuge at maximum speed, extract the supernatant with phenol/chloroform and chloroform, and precipitate the DNA with ethanol. After washing with 70% ethanol, dissolve DNA pellet in Tris-EDTA for at least 12 h at 4°C.

3. The transfection efficiency will be critical for successful rescue of OAV recombinants. Therefore, the pH of the transfection buffer has to be adjusted very carefully (use two pH meters if available). Moreover, rapid addition of the transfection buffer to the $CaCl_2$-DNA solution will result in large precipitates, which might be harmful to cells and will reduce transfection efficiency.

4. Purification buffer is used for equilibration of the NAP™-25 column as well as for elution of virus and storage. Alternatively, virus can be dialyzed against a 500-fold volume of this buffer for at least 6 h at 4°C. As with human adenoviruses, repeated thawing and freezing of virus can result in loss of activity and should therefore be avoided. Storage on dry ice will also result in loss of virus infectivity because of pH change in the storage buffer.

5. Plaque assays for titration and isolation of OAV recombinants using CSL503 and HVO156 cells are not always successful. This might be caused by inhibition of virus spread by agar compounds. However, the isolation of single virus plaques is not necessary because OAV recombinants are derived from cloned plasmid DNA, and homologous recombination between the viral genome and sequences of the producer cells, as was described for human adenovirus vectors on 293 cells, do

not occur. All recombinant OAV produced so far are replication competent in CSL503 and HVO156 cells but replicate abortively in other cell types.

References

1. Hofmann, C., Löser, P., Cichon, G., et al. (1999) Ovine adenovirus vectors overcome preexisting humoral immunity against human adenoviruses in vivo. *J. Virol.* **73,** 6930–6936.
2. Löser, P., Cichon, G., Jennings, G. S., Both, G., and Hofmann, C. (1999) Ovine adenovirus vectors promote efficient gene delivery in vivo. *Gene Ther. Mol. Biol.* **4,** 33–43.
3. Löser P., Hillgenberg, M., Arnold, W., et al. (2000) Ovine adenovirus vectors mediate efficient gene transfer to skeletal muscle. *Gene Ther.* **7,** 1491–1498.
4. Michou, A.-I., Lehrmann, H., Saltyk, M., and Cotten, M. (1999) Mutational analysis of the avian adenovirus CELO, which provides a basis for gene delivery vectors. *J. Virol.* **73,** 1399–1410.
5. Rasmussen, U. B., Benchabi, M., Meyer, V., Schlesinger, Y., and Schughart, K., (1999) Novel human gene transfer vectors: evaluation of wild-type and recombinant animal adenoviruses in human-derived cells. *Hum. Gene Ther.* **10,** 2587–2599.
6. Kremer E. J., Boutin, S., Chillon, M., and Danos, O. (2000) Canine adenovirus vectors: an alternative for adenovirus-mediated gene transfer. *J. Virol.* **74,** 505–512.
7. Harrach, B., Meehan, B. M., Benkö, M., Adir, B. M., and Todd, D. (1997) Close phylogenetic relationship between egg drop syndrome virus, bovine adenovirus serotype 7, and ovine adenovirus strain 287. *Virology* **229,** 302–306.
8. Harrach, B. and Benkö, M. (1998) Phylogenetic analysis of adenovirus sequences. Proof of the necessity of establishing a third genus in the adenoviridae family, in *Adenovirus Methods and Protocols* (Wold, W. S. M., ed.), Humana, Totowa, NJ, pp. 309–339.
9. Vrati, S., Boyle, D. B., Kockerhans, R., and Both, G. W. (1995) Sequence of ovine adenovirus 100k hexon assembly, 33k, pVIII and fiber genes: early region E3 is not in the expected location. *Virology* **209,** 400–408.
10. Vrati, S., Brookes, D. E., Boyle, D. B., and Both, G. W. (1996) Nucleotide sequence of ovine adenovirus tripartite leader sequence and homologous of IVa2, DNA polymerase and terminal proteins. *Gene* **177,** 35–41.
11. Vrati, S. Brookes, D. E., Strike, P., et al. (1996) Unique genome arrangement of an ovine adenovirus: identification of new proteins and proteinase cleavage sites. *Virology* **220,** 186–199.
12. Khatri, A. and Both, G. W. (1998) Identification of transcripts and promoter regions of ovine adenovirus OAV287. *Virology* **245,** 128–141.
13. Venktesh, A., Watt, F., Xu, Z. Z., and Both, G. W. (1998) Ovine adenovirus (OAV287) lacks a virus-associated RNA gene. *J. Gen. Virol.* **79,** 509–516.
14. Xu, Z. Z., Nevels, M., MacAvoy, E. S., et al. (2000) An ovine adenovirus vector lacks transforming ability in cells that are transformed by Ad5 E1A/B sequences. *Virology* **270,** 162–172.

15. Vrati, S., Macavoy, E. S., Xu, Z. Z., et al. (1996). Construction and transfection of ovine adenovirus genomic clones to rescue modified viruses. *Virology* **220,** 200–203.
16. Xu, Z. Z., Hyatt, A., Boyle, D. B., and Both, G. W. (1997) Construction of ovine adenovirus recombinants by gene insertion or deletion of related terminal region sequences. *Virology* **230,** 62–71.
17. Khatri, A., Xu, Z. Z., and Both, G. W. (1997) Gene expression by atypical recombinant ovine adenovirus vectors during abortive infection of human and animal cells in vitro. *Virology* **239,** 226–237.
18. Löser, P. and Hofmann, C., unpublished results.
19. Xu, Z. Z. and Both, G. W. (1998). Altered tropism of an ovine adenovirus carrying the fiber protein cell binding domain of human adenovirus type 5. *Virology* **248,** 156–163.
20. Herz, J. and Gerard, R. D. (1993). Adenovirus-mediated transfer of low density lipoprotein receptor gene acutely accelerates cholesterol clearance in normal mice. *Proc. Natl. Acad. Sci. USA* **90,** 2812–2816.
21. Chartier, C., Degryse, E., Gantzer, M., et al. (1996) Efficient generation of recombinant adenovirus vectors by homologous recombination in *Escherichia coli. J. Virol.* **70,** 4805–4810.
22. Pye, D. (1989) Cell lines for growth of sheep viruses. *Austr. Vet. J.* **66,** 231–232.
23. Maizel, J. V., White, D. O., and Scharff, M. D. (1968). The polypeptides of adenovirus. I: evidence for multiple protein components in the virion and a comparison of types 2, 7a and 12. *J. Virol.* **35,** 115–125.
24. Mittereder, N., March, K. L., and Trapnell, B. C. (1996) Evaluation of the concentration and bioactivity of adenovirus vectors for gene therapy. J. Virol. **70,** 7498–7509.

29

Highly Purified Recombinant Adeno-Associated Virus Vectors

Preparation and Quantitation

Bruce Schnepp and K. Reed Clark

1. Introduction

Adeno-associated virus (AAV) is a nonpathogenic, replication-defective parvovirus that is being developed as a vector for human gene transfer. The recent interest in recombinant (r)AAV has been driven by the unexpected finding that these simple vectors can efficiently transduce a variety of postmitotic cells in vivo, resulting in robust long-term gene expression. Efficient in vivo gene transfer via rAAV transduction requires reasonably high multiplicities of infection, estimated to be between 10^3 and 10^5 DNA-containing particles per cell depending on the cell type targeted. Moreover, based on large animal studies, a clinical dose in humans will require 10^{12}–10^{14} rAAV vector particles, depending on the level of therapeutic protein expression needed for treatment efficacy *(1)*. Therefore, the ability to produce pure, high-titer rAAV is critical for clinical applications. This chapter discusses the isolation of stable rAAV producer cell lines and the associated downstream methods of vector purification and quantitation.

1.1. Requirements for rAAV Production

There are three essential components for rAAV production:

1. The first element is the rAAV vector, which is a transgene expression cassette flanked by AAV inverted terminal repeats (ITRs). The AAV ITRs are located at the ends of the AAV genome and contain all the *cis* sequence information neces-

From: *Methods in Molecular Medicine, Vol. 69, Gene Therapy Protocols, 2nd Ed.*
Edited by: J. R. Morgan © Humana Press Inc., Totowa, NJ

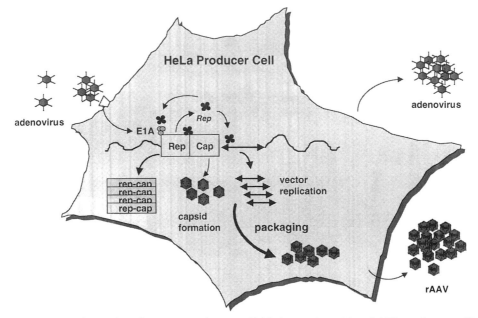

Fig. 1. Schematic of rAAV producer cell biology. A stable rAAV producer cell contains 5–20 integrated copies of the pTP-D6deltaNot plasmid (or its derivative). rAAV is produced by infecting cells with adenovirus, which induces rep-cap gene amplification (≥100-fold) and rep protein production. Rep proteins then transactivate the p40 cap gene promoter and are also required for rAAV vector replication. The single-stranded rAAV genome (both polarities) is encapsidated into preformed virions within the nucleus. After development of adenovirus-induced cytopathic effect, rAAV is purified from contaminating adenovirus using heat inactivation and affinity chromatography.

sary for vector DNA replication and encapsidation *(2,3)*. As a result, this vector contains only 6% of the wild-type genome and lacks all viral coding sequences. During rAAV synthesis from stable cell lines, the rAAV vector genome is excised from the host cell chromosome and replicated, and the single-stranded genome is packaged into preformed AAV virions.

2. The two AAV genes (*rep* and *cap*), which encode replication and AAV structural proteins, are expressed in *trans* to provide required AAV helper functions. The physical separation of these genes and the ITR packaging signal is critical for avoiding the illegitimate formation of wild-type AAV. The *rep* gene possesses two promoters (p5 and p19), from which four Rep proteins are expressed (Rep78, Rep68, Rep52, and Rep40). The p5-derived Rep proteins (Rep78 or Rep68) are essential for rAAV replication. Three capsid proteins (VP1, VP2, and VP3) are synthesized following transcription from the *rep*-inducible p40 cap gene promoter.

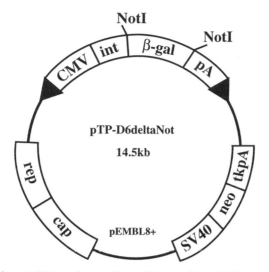

Fig. 2. Tripartite rAAV producer plasmid for stable cell line production. The base vector (pTP-D6deltaNot) contains three components necessary for stable cell line production. 1. A gene expression cassette flanked by AAV ITRs (arrowheads) containing the IE CMV promoter/enhancer, SV40 large T-antigen intron, and polyadenylation signal, all derived from plasmid pCMVß (Clontech). Alternative cDNAs can be readily replaced using *Not*I restriction sites. 2. Rep and cap helper genes are supplied in *trans* under control of the endogenous p5, p19, and p40 promoters. 3. For stable cell line selection, the plasmid also possesses the neomycin resistance gene under the control of SV40 promoter and thymidine kinase polyadenylation signal.

3. Lastly, because AAV and recombinant derivatives are replication defective, these viruses require helper virus gene products from adenovirus or herpes virus for efficient viral production. The helper virus (or specific helper virus genes) is necessary to provide a favorable intracellular milieu for robust rAAV replication and capsid viral synthesis.

1.2. Stable Cell Line Production

Several methods are currently in use for the large-scale production of rAAV using packaging cell lines *(4–9)*. We have developed a simplified method for generating rAAV based on stable HeLa cell lines containing integrated copies of the AAV rep and cap genes and an rAAV vector *(4)*. To generate rAAV, the cells are simply infected with wild-type adenovirus, which results in induction of the integrated p5 rep promoter, which in turn initiates an expression cascade resulting in rAAV synthesis. This process is shown schematically in **Figure 1**. rAAV producer cell lines are isolated by cloning the desired cDNA sequence into an expression cassette that is flanked by AAV ITRs. Also present on this producer plasmid (pTP) is a selectable marker gene (neomycin resistance) and the native *rep-cap* helper genes (**Fig. 2**). Following plasmid transfection of HeLa cells, stable cell lines are isolated and screened for viral production by adenovirus infection, and yields are quantitated by real-time PCR *(10)*. Depending on the producer cell line, viral yields between 5×10^3 and 2×10^4 rAAV DNA-containing particles per cell are typical. Importantly, recombinant virus generated in this manner appears to be free of replication-competent wild-type AAV (<1 IU of wild type $AAV/10^{11}$ DRP rAAV). As with all stable cell line production methods, the process is amenable to high-density cell growth, and we routinely culture $>10^{10}$ producer cells per vector production run using a small-scale adherent cell bioreactor (Corning Cell Cube). Assuming a particle per cell yield of 10^4, this stable cell line approach can yield $>10^{14}$ rAAV particles per vector preparation.

1.3. Vector Purification

A significant advance in rAAV vector technology has been the recent development of refined chromatography purification methods that render traditional isopycnic CsCl gradient purification obsolete *(10–12)*. Importantly, vector purity, biological potency, and process throughput are all increased using these chromatographic methods. We have achieved large-scale (≥1 L) vector purification from clarified cell lysates in <1 day. This approach uses a commercial heparin sulfate resin as an affinity ligand for viral purification. Using a single-pass purification procedure, rAAV recoveries are on average ≥70%, and purity is estimated to be ≥95%. Column capacity is $\geq 10^{14}$ DNA-containing particles. This method also appears to produce lower DNA-containing particle/infectious particle ratios (≤100:1) than that observed using CsCl purification. Thus, this simplified method allows rapid vector purification that is essentially free of contaminating cellular and helper virus proteins.

1.4. Vector Titration

No standard method exists for quantitating rAAV vector preparations, which has resulted in some confusion in the field. Essentially, rAAV can be measured in four ways:

1. A physical particle titer that enumerates the total number of viral particles (whether containing DNA or not) present in a rAAV preparation. A particle titer can be determined by direct counting of particles by electron microscopy, or by using a commercial enzyme-linked immunosorbent assay (ELISA) *(13)*.
2. The second method determines the number of DNase-resistant vector genomes (encapsidated viral genomes) present in the rAAV vector stock. As we will detail, the advantage to this measurement is that titers can now be rapidly generated in approximately 4 h by using quantitative polymerase chain reaction (PCR) *(10)*.

Moreover, comparable DNase-resistant genome titers can be derived for all vector preparations, regardless of the transgene or promoter used in the construct.

3. The third method determines the proportion of viral particles that can infect and *replicate* in a target cell line. This method, commonly referred to as a replication center assay, determines the number of infectious units in a rAAV preparation by intracellular amplification of the rAAV genome in the presence of wild-type AAV and adenovirus *(14)*. Detection of rAAV replication foci was accomplished by filter hybridization of infected cell monolayers. Herein, we have extended this approach using a stable cell line (C12) expressing the rep gene to facilitate replication-based titration, combined with real-time PCR detection of the amplified vector genomes.

4. Lastly, a transduction (expression) titer measures the number of virus particles competent to infect and express the transgene in the target cell. However, widely varying titers (over 3 orders of magnitude) can be generated from the same vector stock by altering the target cell line or the intracellular environment (e.g., concurrent helper adenovirus infection or cell confluency). Ideally, for each rAAV production method, the viral yields should be expressed in the form of infectious units/cell and DNase-resistant vector genome particles/cell. If these values are known, each rAAV vector preparation can be defined in terms of the DNase resistant particle/infectivity ratio. We have found that for most column purified rAAV preparations this ratio is 10–100:1.

2. Materials

2.1. Generation of Producer Cell Lines

2.1.1. Generation of rAAV Producer Plasmid (pTP) Containing the Desired Transgene (cDNA)

1. Qiaquick-Spin columns (Qiagen).
2. Tripartite rAAV producer plasmid pTP-D6deltaNot or similar construct (may be obtained from authors).
3. Dulbecco's modified Eagle's medium (DMEM) containing 100 U/mL penicillin, 100 µg/mL streptomycin, and 10% fetal bovine serum (FBS).
4. Ultra-Competent DH5-alpha *E. coli* cells (Gibco-BRL).
5. SuperFect Transfection Reagent (Qiagen).
6. *Msc*I and *Sma*I restriction enzymes (New England Biolabs).

2.1.2. Producer Cell Line Generation Using Superfect Transfection Reagent

1. Low-passage HeLa cells obtained from American Type Culture Collection (cat. no. CCL-2) split 1:20 every 4–5 days in DMEM media.
2. DMEM media (*see* **Subheading 2.1.1.**).
3. Hanks' balanced salt solution (HBSS).
4. 100 mg/mL geneticin stock solution dissolved in 100 m*M* HEPES, pH 7.4 (Gibco-BRL).

5. 2% trypsin solution.
6. Phosphate-buffered saline (PBS): 1.0% NaCl, 0.025% KCl, 0.14% Na_2HPO_4, 0.025% KH_2PO_4 (all w/v), pH 7.4.
7. SuperFect Transfection Reagent (*see* **Subheading 2.1.1.**).
8. Tissue culture dishes: 6-well, 96-well, 100-mm plates (Falcon).
9. CsCl-purified adenovirus type 5 stock (10^{10} plaque-forming units [pfu]/mL).
10. 10% deoxycholic acid stock solution (Sigma).
11. DNase I (Gibco-BRL, cat. no. 18047-019).
12. DNase digestion buffer: 150 mM NaCl, 1 mM $MgCl_2$, 20 mM Tris-HCl, pH 8.0, 0.5% deoxycholic acid, 750 U/mL DNase I.
13. 96-well dot-blot apparatus (convertible filtration manifold system; Gibco-BRL).
14. Positively charged nylon membrane (Boehringer-Mannheim).
15. Denaturation buffer: 4 M NaOH, 50 mM EDTA, 10 µg/mL herring sperm.
16. 2× SSC solution: 0.3 M NaCl, 0.03 M sodium citrate, pH 7.0.
17. UV crosslinker (Hybilinker HL-2000, UVP Lab Products).
18. DIG High Prime DNA Labeling and Detection Starter Kit II (Boehringer Mannheim).
19. X-ray film (Kodak X-OMAT).

2.1.3. Mid-Scale Production Screen

1. Tabletop centrifuge (Beckman GS-6R).
2. T-175 tissue culture flasks.
3. 250-mL conical tubes (Corning).
4. Adenovirus type 5 stock (*see* **Subheading 2.1.2.**).
5. TMN: 20 mM Tris-HCl, pH 8.0, 1 mM $MgCl_2$, 150 mM NaCl.
6. 10% deoxycholic acid stock solution (*see* **Subheading 2.1.2.**).
7. Benzonase nuclease (EM Industries, cat. no. 1.016979M).
8. UNIFLO-25 1.2-µm syringe filter (Schleicher & Schuell, cat. no. 02360).

2.2. Vector Purification

2.2.1. Cell Lysate Preparation and Column Chromatography

1. Benzonase nuclease (*see* **Subheading 2.1.3.**).
2. 10% deoxycholic acid stock solution (*see* **Subheading 2.1.2.**).
3. Polycap 75 SPF Serum Capsule (Whatman Scientific, cat. no. 6705-7500).
4. Masterflex (Cole-Parmer) or similar peristaltic pump (7016 pump head).
5. Silicon pump tubing (L/S 16 platinum tubing, Cole-Parmer, cat. no. 96410-16).
6. Biocad Sprint high-performance liquid chromatography (HPLC) system (Applied Biosystems) or similar fast protein liquid chromatography (FPLC)/HPLC system.
7. POROS HE20/M heparin sulfate column (Applied Biosystems).
8. Column equilibration buffer: 20 mM Tris-HCl, pH 8.0, 100 mM NaCl.
9. SYPRO-Orange stain (Bio-Rad, cat. no. 170-3120).
10. 10% HCl-Tris Ready gel (Bio-Rad, cat. no. 161-1155).
11. SYPRO-Orange Broad Range Molecular Marker (Bio-Rad, cat. no. 161-0332).

12. Slide-A-Lyzer dialysis cassettes (Pierce Chemical, 10K MWCO).
13. Sterile glycerol (50% v/v).

2.3. Vector Quantitation

2.3.1. Quantitation of DNase-Resistant Vector Genomes

1. Taqman 7700 Sequence Detector (PE Applied Biosystems).
2. TaqMan PCR Core Reagent Kit (contains all reagents for TaqMan PCR reaction except gene-specific primers/probe set and DNA template) (PE Applied Biosystems, cat. no. N808-0228).
3. 1× PCR buffer: 50 mM KCl, 10 mM Tris-HCl, pH 8.3, 5 mM MgCl$_2$, and 0.01% (w/v) gelatin.
4. MicroAmp Optical 96-well/tube plate (PE Applied Biosystems, cat. no. N801-0560).
5. MicroAmp Optical caps, 8 caps/strip (PE Applied Biosystems, cat. no. N801-0935).
6. Gene-specific PCR primers and dual fluorescent oligonucleotide (PE Applied Biosystems).
7. PicoGreen DNA Quantitation Kit (Molecular Probes).
8. Tissue culture plate: 24-well (Falcon).
9. DMEM media (*see* **Subheading 2.1.1.**).
10. C12 cells (may be obtained from authors).
11. Adenovirus type 5 stock (*see* **Subheading 2.1.2.**).
12. DNase I (*see* **Subheading 2.1.2.**).
13. Proteinase K, 10 mg/mL (Gibco-BRL, cat. no. 25530-015).
14. Qia-Amp blood isolation kit (Qiagen).
15. Qiagen Vacuum Manifold for Qiagen spin columns.

3. Methods

3.1. Generation of Recombinant AAV Producer Cell Lines

This section details the generation and isolation of optimal rAAV producer cell lines. The method is predicated on the generation of a producer plasmid that contains the desired cDNA cloned into an expression cassette that is flanked by AAV ITRs. Once the function of the rAAV producer plasmid is confirmed, stable cell line isolation using HeLa cells is begun. Based on our experience, HeLa cells yield the highest titer of rAAV per cell. A two-step, high-throughput screening process is described, whereby 192 cell clones are selected and analyzed to isolate superior producer cell lines (>5000 rAAV particles/cell). A schematic of the cell isolation process is shown in **Figure 3**.

3.1.1. Generation of rAAV Producer Plasmid (pTP) Containing the Desired Transgene (cDNA)

1. Clone the desired cDNA into rAAV producer plasmid pTP-D6deltaNot using the unique *Not*I restriction sites (**Fig. 2**). This places the cDNA downstream of the

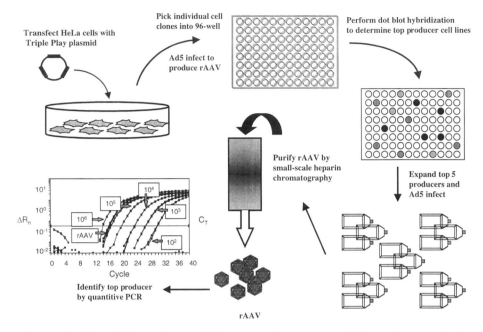

Fig. 3. Stable rAAV producer cell line screening strategy. HeLa cells are initially transfected with a functional pTP producer plasmid, and the resulting neomycin-resistant cell colonies are picked into 96-well plates. Subsequently, a test plate is adenovirus infected to induce rAAV production. Cells producing high levels of rAAV are identified using a high-throughput dot-blot hybridization protocol with a probe specific to the rAAV genome. The top 5 producer lines are expanded for midscale rAAV screening, whereby rAAV from each line is purified by heparin affinity chromatography. The yield of purified rAAV particles from each line is quantified using real-time PCR to determine the rAAV particles per cell.

immediate early cytomegalovirus (IE CMV) promoter and between the AAV ITRs.

2. Transform DH5-alpha *E. coli* and screen 40 ampicillin-resistant colonies by size using a supercoiled quick screen (*see* **Note 1**). Usually 10% of colonies contain the desired cDNA insert.

3. Check AAV ITR integrity by restriction enzyme digestion with *Sma*I and *Msc*I, which cut within the ITRs, and a second restriction enzyme digestion to orient the cDNA sequence 5'–3' relative to the promoter (*see* **Note 2**).

4. Select two putative clones, and transfect plasmid pTP DNA into HeLa cells using Superfect Transfection Reagent (*see* **Subheading 3.1.2.**).

5. Two days post transfection, confirm transgene expression by either immunofluorescence or Western blot analysis on the transfected cell lysate.

3.1.2. Producer Cell Line Generation Using Superfect Transfection Reagent

1. Once a functional producer plasmid clone is identified, the isolation of stable cell lines is begun by seeding 2×10^5 low-passage HeLa cells into a 6-well plate in DMEM media.
2. The next day, dilute 2 µg of supercoiled producer plasmid DNA (dissolved in TE, pH 7.4) with HBSS in a total volume of 100 µL in a 1.5-mL microfuge tube. Mix and spin down.
3. Add 10 µL of SuperFect Reagent to the DNA solution. Mix by pipeting up and down several times, and incubate the mixture at room temperature for 10 min (*see* **Note 3**).
4. While the transfection mixture is incubating, remove growth media from HeLa cells and wash cells once with 2 mL of PBS.
5. Add 1 mL of DMEM media (including serum and antibiotics) to the transfection mixture, pipet up and down 2 times, and immediately add the mixture over the entire HeLa cell well in a dropwise manner. Incubate for 2 h at 37°C in a 5% CO_2 incubator.
6. Aspirate media from the cells, wash 2× with PBS, and then add 2 mL of DMEM containing serum and antibiotics. Incubate overnight.
7. The next day, trypsinize transfected cells and plate in 100-mm tissue culture dishes in 10 mL of media at the following dilutions: 5 dishes at 5×10^3 cells, and 5 dishes at 7.5×10^3 cells, 5 dishes at 10^4 cells. Move dishes gently from side to side to distribute cells evenly.
8. After 2 days, aspirate cell media and replace with DMEM media supplemented with 700 µg/mL Geneticin (G418).
9. Continue incubating cells for 6–8 days in selective media until cell colonies appear (*see* **Note 4**).
10. When colonies are 2–3 mm in diameter, aspirate selective media from cells and wash with 10 mL of HBSS to remove cell debris. Add 8 mL of fresh selective media to dishes, and with a P-1000 pipetman (containing a sterile filter tip) depress the plunger and place the pipet tip over a single colony. Rotate the pipet to loosen cells and slowly release pressure on plunger to aspirate cells into tip along with 200 µL of media. Transfer cells into a single well of a 96-well plate. Pick 2 × 96-well plates in this manner.
11. Incubate cells in 96-well plates for approximately 4–6 days until 50% of wells are confluent (*see* **Note 5**).
12. Using a multichannel pipet, remove media and trypsinize cells in 20 µL volume. Add 150 µL of selective media, and resuspend cells well by pipeting up and down several times. Place 30 µL of the cell suspension into a 96-well "master plate" containing 150 µL of fresh selective media. Place 100 µL of the cell suspension into a separate 96-well "test plate" containing 100 µL of fresh selective media. Incubate both sets of plates at 37°C.
13. After overnight incubation of the test plates, aspirate media using a multichannel pipet, and add 100 µL of 2% DMEM containing 10^6 pfu of adenovirus 5 per well.

14. Incubate test plates at 37°C for 36 h. Then carefully remove media (not detaching rounded cells) and add 100 μL of DNase digestion buffer to each well (*see* **Note 6**). Incubate at 37°C for 30 min.

15. Add 10 μL of denaturation buffer to each well and incubate at 65°C for 30 min.

16. Transfer cell lysates to a dot-blot apparatus on positively charged nylon membrane prewetted in 2× SSC. Apply lysate to apparatus and rinse wells with 100 μL of 0.4 *M* NaOH.

17. Remove nylon membrane, rinse twice with 2× SSC for 5 min, and allow membrane to air-dry for 15 min. Crosslink DNA to membrane using a UV crosslinker apparatus.

18. Hybridize membrane using a digoxigenin (or radioactive) labeled DNA probe specific for the transgene. Detect hybridization by chemiluminescence and expose to X-ray film.

19. Expand from master plate the 10 individual cell clones exhibiting the strongest hybridization and proceed to the midscale screen (*see* **Note 7**).

3.1.2. Midscale Production Screen

1. Expand putative rAAV producer cells until sufficient cells are present to seed three T-175 tissue culture flasks at a density of 6×10^6 cells/flask in DMEM media containing 2% FBS.

2. The next day, dilute an adenovirus stock in HBSS to a concentration of 2×10^8 pfu/mL and add 1 mL/flask.

3. Harvest cells at maximum complete cytopathic effect (CPE; all cells rounded with 10% floating), which is usually 48 h post infection. The cells should detach easily from the flask with a gentle tap. Pellet cells in a tabletop centrifuge at 2000 rpm ($1000g$) for 10 min at 4°C.

4. Resuspend the cell pellet in 6 mL of TMN.

5. Add deoxycholic acid to a final concentration of 0.5% and mix the lysate well (do not vortex).

6. Add Benzonase nuclease to a final concentration of 50 U/mL, mix, and incubate the lysate at 37°C for 30 min with intermittent mixing every 10 min to ensure complete cell lysis.

7. Clarify the lysate at 3500 rpm ($2000g$) for 30 min at 4°C to remove debris. Transfer the clarified lysate to a new 50-mL conical tube, and heat-inactivate adenovirus at 56°C for 30 min with intermittent vortexing every 10 min.

8. Quick freeze in a dry ice/ethanol bath and store at −20°C overnight (*see* **Note 8**).

9. After thawing, remove the floculent precipitate by centrifugation at 3500 rpm ($2000g$) for 30 min at 4°C. Filter the supernatant through a 1.2-μm syringe filter.

10. Purify small-scale vector preparation by heparin sulfate column chromatography (*see* **Subheading 3.2.**).

11. Analyze small-scale preparations by quantitative PCR to derive the DNase-resistant vector genomes per cell (*see* **Subheading 3.3.1.**). Calculate the particle per cell yield for each cell line. The optimal cell line should produce >5000 rAAV vector particles per cell.

12. Expand the optimal producer cell line (in the presence of Geneticin) for large-scale rAAV vector production (*see* **Note 9**).

3.2. Vector Purification

3.2.1. Cell Lysate Preparation and Column Chromatography

rAAV-containing cell lysates for heparin column chromatography are easily prepared by detergent lysis, followed by the addition of an endonuclease to reduce lysate viscosity. Finally, the lysate is filtered through a 1-μm serum filter prior to column application.

1. Resuspend rAAV producer cells in TMN at a cell density of 5×10^6 cells/mL. Add deoxycholic acid and Benzonase to a final concentration of 0.5% and 50 U/mL, respectively, and incubate for 30 min at 37°C with intermittent mixing.
2. Remove cellular debris by tabletop centrifugation at 3500 rpm (2000*g*) for 15 min at 4°C. Heat the clarified lysate to 56°C for 45 min to inactivate adenovirus and then freeze at –20°C overnight.
3. Thaw the lysate at 37°C and remove the floculent precipitate by centrifugation (2000*g*).
4. Filter the supernatant through a 1-μm serum capsule filter unit using a peristaltic pump at a flow rate of 10 mL/min. The cell lysate is ready for column application.
5. Connect a POROS HE20/M heparin column (1.7-mL bed volume) to a Biocad Sprint HPLC system (Applied Biosystems) or similar FPLC/HPLC system.
6. Equilibrate the column with 20 mL of equilibration buffer at a flow rate of 5 mL/min.
7. Apply the clarified cell lysate (≤1.2 L) to the column at a flow rate of 3–5 mL/min at ambient temperature using the internal Biocad system pumps. The system back pressure should not be allowed to exceed 800 psi, and lysate conductivity should be between 12 and 16 mS.
8. After sample loading, the column is washed with 60 mL of equilibration buffer at 3 mL/min. The A_{280} should return to baseline (≥0.005).
9. Elute rAAV by application of a linear NaCl gradient from 0.1 to 1 *M* (pH 8.0) at a flow rate of 3 mL/min over 20 column volumes (approx 34 mL). Collect 1-mL gradient fractions, with the single dominant protein peak eluting at approx 0.35 *M* NaCl.
10. To determine peak virus containing fractions, run a small aliquot (20 μL) of individual fractions on a 10% sodium dodecyl sulfate-polyacrylamide gel electrophoresis (SDS-PAGE) gel followed by fluorescent staining (SYPRO-Orange). Identify those fractions that contain the highest levels of the three AAV capsid proteins (87-kDa VP1, 73-kDa VP2, and 61-kDa VP3).
11. Pool the peak virus-containing fractions (typically three) and dialyze against three changes of 20 m*M* Tris-HCl, pH 8.0, 2 m*M* MgCl$_2$, 150 m*M* NaCl, and store in aliquots at –80°C in 5% glycerol (*see* **Note 10**).

12. Determine rAAV vector yield using the titration methods detailed in **Subheading 3.3**.
13. Clean heparin column with 10 mM Tris-HCl (pH 8.0), 2.4 M NaCl at a flow rate of 5 mL/min followed by column equilibration with 20 mL of equilibration buffer prior to storage at 4°C (*see* **Note 11**).

3.3. Vector Quantitation

3.3.1. Quantitation of DNase-Resistant Vector Genomes

In this section we describe a rapid, reproducible, and sensitive PCR-based method to measure <u>D</u>Nase-<u>r</u>esistant rAAV particles (DRPs). The DRP Taqman assay is essentially a modified dot-blot protocol whereby purified rAAV is serially diluted and sequentially digested with DNase I and proteinase K. A portion of the treated sample is then subjected to quantitative PCR analysis using the PE Applied Biosystems Prism 7700 sequence detector system *(15,16)*. Comparison with a plasmid standard curve (using system software) allows accurate quantitation of the initial vector genome copy number.

1. Prepare 10^{-3}, 10^{-4}, and 10^{-5} serial dilutions of a rAAV viral stock in 1× PCR buffer.
2. Digest 100 µL of each rAAV serial dilution with 1 µL DNase I (350 U) for 30 min at 37°C (*see* **Note 12**).
3. Place a second 100-µL aliquot of the rAAV serial dilutions into separate microfuge tubes and maintain at 4°C. These serve as minus DNase I control samples.
4. After DNase digestion, add 10 µg of proteinase K (1 µL) to all samples and incubate at 50°C for 1 h.
5. Inactivate the proteinase K by a 20-min incubation at 95°C.
6. Add 2.5 µL of the treated rAAV vector dilutions to 22.5 µL of a standard 1X Taqman PCR master mix (PE Applied Biosystems) that contains the appropriate Taqman primers and probe sequences at their working concentrations to a 96-well/tube Taqman PCR plate (*see* **Note 13**).
7. Concurrently, set up a DNA plasmid standard curve in triplicate using 10^{2}–10^{6} AAV genome equivalents. To prepare a plasmid DNA standard for Taqman PCR, first accurately determine the plasmid DNA concentration by using a DNA-specific fluorescent stain (PicoGreen DNA Quantitation Kit; Molecular Probes).
8. Next, prepare a plasmid stock of 10^{10} copies/mL in TE and aliquot at 50 µL/tube. Store these aliquots at –20°C and use this stock to prepare 10-fold serial dilutions that will allow quantitation of between 5×10^{1} and 5×10^{6} plasmid molecules per Taqman reaction (*see* **Note 14**).
9. Perform a standard 40-cycle Taqman PCR amplification (50°C, 2 min, 1 cycle; 95°C, 10 min, 1 cycle; 95°C, 15 s, 60°C, 1 min, 40 cycles) using the Taqman 7700 PCR Sequence Detector (*see* **Note 15**).

10. Using the system software, calculate the initial vector particle titer by comparing the amplification profile of the sample unknowns with the plasmid standard curve. The final titer is derived by multiplying the vector particle value by the dilution factor and then by 400 to derive a titer per mL. An example of a typical amplification profile is shown in the bottom left of **Figure 3** (*see* **Note 16**).

3.3.2. Infectious rAAV Titration Using Quantitative PCR

A second assay, developed for infectious rAAV titration, also uses a quantitative PCR approach. We have previously demonstrated the utility of a stable *rep-cap*-expressing HeLa cell line (C12) for replication-based rAAV titration *(12,17)*. The assay is based on the ability of rAAV to replicate in C12 cells following coinfection with adenovirus. Using quantitative PCR, an end-point titer is generated based on the highest stock dilution that results in detectable rAAV vector genome amplification when compared with input vector (no adenovirus). The detection of rAAV replication at an end-point dilution could therefore be used to derive an infectious end-point titer ($TCID_{50}$). At 36 h post infection, total cellular DNA is isolated from each infected well and subjected to quantitative PCR. Because of the extreme sensitivity of PCR-based methods, low levels of rAAV input DNA are amplified in the absence of adenovirus infection. However, at some dilutional end point, the input rAAV vector genomes will become titratable and thus undetectable by PCR, whereas cells infected by rAAV (in the presence of adenovirus) will replicate and result in an increase of $\geq$10-fold in the rAAV vector genome levels over input.

1. Seed two 24-well plates with 7×10^4 C12 cells/well in DMEM media. One plate will be adenovirus infected, and the other plate will serve as a control for input virus.
2. The next day, prepare 10-fold serial dilutions of the rAAV stock from 10^{-7} to 10^{-12} using DMEM media containing 2% FBS.
3. Working with the minus Ad5 plate first, aspirate the media from the plate and infect cells with 0.5 mL of each viral dilution (6 wells total).
4. Aspirate the media from the adenovirus (+) plate. Starting with the highest dilution, infect 4 wells with 0.5 mL from each viral dilution (24 wells total).
5. Add 100 µL of DMEM media (2% FBS) containing 3×10^6 pfu of adenovirus type 5 to each well of the adenovirus (+) plate.
6. As a negative control, place 0.5 mL of 2% FBS DMEM on C12 cells alone.
7. After 36 h of incubation, the cells are ready for DNA isolation using a QIAamp Blood Isolation Kit-250. Starting with the adenovirus (-) plate, aspirate the media from the wells, and add 200 µL of 1× PBS to each well.
8. Add 25 µL of Qiagen Proteinase K and 200 µL of buffer AL. Gently rock the plate to ensure complete lysis.
9. Transfer the lysate from each well into a 1.5-mL microfuge tube, and incubate at 70°C for 10 min.

10. Add 210 μL of 100% ethanol to each tube and vortex. Load DNA/ethanol mixture onto DNA spin columns that are hooked up to a Qiagen vacuum manifold. Allow the DNA/ethanol to enter the column completely, then wash with 500 μL of buffer Qiagen wash buffer (AW), and repeat.

11. Transfer each spin column to a new 1.5-mL microfuge tube, and elute DNA by the addition of 100 μL of buffer Qiagen elution buffer (AE) preheated to 70°C. Wait 1 min and then spin at 8000 rpm (6000*g*) for 1 min. Store DNA at 4°C prior to use.

12. Repeat **steps 7–11** for adenovirus-infected samples.

13. Prepare the Taqman 1× master mix by adding the appropriate concentrations of primers and probe specific for your transgene or promoter to the supplied 2× Taqman master mix.

14. Add 22.5 μL of the Taqman master mix to each tube of the 96-tube/plate. Add 2.5 μL of the DNA sample (approx 500 ng) to each well and cap wells (*see* **Note 17**).

15. Perform a standard 40-cycle Taqman amplification on the Taqman 7700 Sequence Detector.

16. A standard curve is not necessary for this assay, although 10^5 plasmid targets should be run to serve as a positive control for the Taqman PCR reaction.

17. To determine an end-point infectious titer for the rAAV preparation, compare the amplification curve for each adenovirus-infected well with the input amplification profile at that particular dilution of the minus adenovirus sample. A cycle threshold difference of 3 or greater is indicative of input rAAV viral replication, and the well is scored positive. After scoring the wells, calculate a $TCID_{50}$ end-point titer using the Reed-Muench formula (*see* **Note 18**).

4. Notes

1. Because of possible plasmid instability, grow all bacteria harboring AAV ITR-containing plasmids at 32°C. Rapid analysis of potential recombinant clones can be accomplished by screening the bacterial colonies by plasmid size using a supercoiled DNA plasmid ladder (Gibco-BRL).

2. ITR sequences can spontaneously rearrange or delete upon passage into bacteria. Therefore, all ITR-containing clones should be digested with *Sma*I and *Msc*I, both of which cut within the ITRs.

3. We have used other transfection compounds (dioleoyl-trimethyl ammonium propane [DOTAP], Effectene, and $CaPO_4$) with similar results.

4. Selective media should be changed every 4–5 days depending on accumulation of cellular debris.

5. The cells will grow at different rates; in general, 25% of cells are fast growing, 40% are medium growing, 25% are slow growing, and 10% will die when transferred.

6. The DNase treatment is optional and can be omitted. Inclusion of this step helps to reduce the number of putative rAAV vector-producing cell lines.

7. Typically, 10 cell lines are expanded and $\geq 10^7$ cells from each are frozen (90% FBS + 10% dimethyl sulfoxide [DMSO]) and stored in liquid N_2. The top five lines are expanded for midscale screening.

8. Freezing of the lysate results in cellular protein precipitation that greatly increases the purity of the rAAV after column chromatography; therefore this step should not be omitted.

9. We have achieved large-scale producer cell growth ($\geq 10^{10}$) using the Corning Cell Cube; alternatively, cells can be grown in T175 tissue culture flasks (100–400). One to 3 days after seeding, cells are adenovirus infected at a multiplicity of infection (MOI) = 100. Harvest cells at maximum adenovirus CPE. Cell viability should be $\geq 80\%$.

10. We have found the Pierce Slide-a-Lyzer dialysis cassettes ideal for buffer exchange.

11. Column life is approximately 10 L of clarified lysate. We pack our own heparin columns using an Applied Biosystems packing device as instructed by the manufacturer. To extend the life of the column, more vigorous cleaning regimens can be used. We have found that a 0–100% isopropanol gradient or 0.5 N NaOH treatment (20–40 mL total volume) is useful to remove bound lipids and lipoproteins from the column.

12. DNase digestion, proteinase K incubation, and inactivation can all be conveniently performed in a preprogrammed PCR thermocycler.

13. Taqman PCR primers and probe sequences are identified using proprietary PE Applied Biosystems software. Working concentrations of the primers and probe are determined empirically using a plasmid containing the transgene sequence.

14. When calculating virus DRPs, remember that plasmid DNA is double-stranded, so 5×10^1 plasmid copies correspond to 10^2 genome copies of rAAV.

15. The Taqman 7700 is a 96-well thermocycler that utilizes a laser optic system, which provides periodic excitation to each well of the thermocycler. The fluorescence signal is generated by the PCR amplification process and occurs as a result of the 5'-3' nuclease activity of *Taq* polymerase. Fluorescence is achieved by the use of a dual-labeled (5' and 3' ends) fluorogenic oligonucleotide (referred to as the probe) that anneals specifically to the amplified PCR product. When the two fluorochromes are in physical proximity (connected by the oligonucleotide), the laser-excited fluorescent signal from the 5' reporter fluorescein is quenched (absorbed) by the 3' fluor. During the elongation phase of the PCR cycle, *Taq* polymerase degrades the annealed oligonucleotide, resulting in the release of the reporter fluorochrome into the medium and a consequent increase in reporter dye fluorescence. The increase in the reporter dye fluorescence is proportional to the number of PCR cycles. It is important to always perform replicates of 3 for each sample or standard. Also run three PCR reactions with no template (NTC) to control for reagent contamination. NTC reactions should fail to amplify, which results in a threshold cycle (C_T) ≥ 39.

16. A normalized fluorescence value (ΔR_n) is calculated by the system software and reflects the change in reporter fluorescence over time. ΔR_n is plotted as a function of cycle progression; the first cycle that is above background fluorescence is defined as the threshold cycle (C_T). A plot of the input plasmid copy number versus the C_T value yields a standard curve with a dynamic range of ≥ 5 orders of

magnitude. Routine detection of plasmid target input DNA is typically between 10 and 10^6 molecules per reaction

17. All samples should be run in duplicate; cells alone and adenovirus-infected C12 cells should also be run concurrently with NTC samples. All negative control samples should yield $C_T \geq 39$.

18. To standardize this titration method, positive sample amplification (plus adenovirus) was defined as possessing an average C_T value (two replicates) that exceeded the corresponding minus adenovirus control sample by three or more PCR cycles ($\Delta C_T \geq 3$). Based on numerous plasmid DNA amplification profiles, a 3–4 cycle differential represents an approximately 10-fold difference in starting template concentration.

References

1. Grimm, D. and Kleinschmidt, J. A. (1999) Progress in adeno-associated virus type 2 vector production: promises and prospects for clinical use. *Hum. Gene Ther.* **10**, 2445–2450.

2. McLaughlin, S. K., Collis, P., Hermonat, P. L., and Muzyczka, N. (1989) Adeno-associated virus general transduction vectors: analysis of proviral structures. *J. Virol.* **62**, 1963–1973.

3. Samulski, R. J., Chang, L. S., and Shenk, T. (1989) Helper-free stocks of recombinant adeno-associated viruses: normal integration does not require viral gene expression. *J. Virol.* **63**, 3822–3828.

4. Clark, K. R., Voulgaropoulou, F., Fraley, D. M., and Johnson, P. R. (1995) Cell lines for the production of recombinant adeno-associated virus. *Hum. Gene Ther.* **6**, 1329–1341.

5. Tamayose, K., Hirai, Y., and Shimada, T. (1996) A new strategy for large-scale preparation of high-titer recombinant adeno-associated virus vectors by using packaging cell lines and sulfonated cellulose column chromatography. *Hum. Gene Ther.* **7**, 507–513.

6. Liu, X. L., Clark, K. R., and Johnson, P.R. (1999) Production of recombinant adeno-associated virus vectors using a packaging cell line and a hybrid recombinant adenovirus. *Gene Ther.* **6**, 293–299.

7. Inoue, N. and Russell, D. W. (1998) Packaging cells based on inducible gene amplification for the production of adeno-associated virus vectors. *J. Virol.* **72**, 7024–7031.

8. Gao, G. P., Qu, G., Faust, L. Z., et al. (1998) High-titer adeno-associated viral vectors from a Rep/Cap cell line and hybrid shuttle virus. *Hum. Gene Ther.* **9**, 2353–2362.

9. Conway, J. E., Rhys, C. M., Zolotukhin, I., et al. (1999) High-titer recombinant adeno-associated virus production utilizing a recombinant herpes simplex virus type I vector expressing AAV-2 Rep and Cap. *Gene Ther.* **6**, 986–993.

10. Clark, K. R., Liu, X., McGrath, J. P., and Johnson, P. R. (1999) Highly purified recombinant adeno-associated virus vectors are biologically active and free of detectable helper and wild-type viruses. *Hum. Gene Ther.* **10**, 1031–1039.

11. Zolotukhin, S., Byrne, B. J., Mason, E., et al. (1999) Recombinant adeno-associated virus purification using novel methods improves infectious titer and yield. *Gene Ther.* **6,** 973–985.

12. Grimm, D., Kern, A., Rittner, K., and Kleinschmidt, J. A. (1998) Novel tools for production and purification of recombinant adeno-associated virus vectors. *Hum. Gene Ther.* **9,** 2745–2760.

13. Grimm, D., Kern, A., Pawlita, M., et al. (1999) Titration of AAV-2 particles via a novel capsid ELISA: packaging of genomes can limit production of recombinant AAV-2. *Gene Ther.* **6,** 1322–1330.

14. Yakobson, B., Koch, T., and Winocour, E. (1987) Replication of adeno-associated virus in synchronized cells without the addition of a helper virus. *J. Virol.* **61,** 972–981.

15. Gibson, U. E., Heid, C. A., and Williams, P. M. (1996) A novel method for real time quantitative RT-PCR. *Genome Res.* **6,** 995–1001.

16. Holland, P. M., Abramson, R. D., Watson, R., and Gelfand, D. H. (1991) Detection of specific polymerase chain reaction product by utilizing the 5'-3' exonuclease activity of *Thermus aquaticus* DNA polymerase. *Proc. Natl. Acad. Sci. USA* **88,** 7276–7280.

17. Clark, K. R., Voulgaropoulou, F., and Johnson, P. R. (1996) A stable cell line carrying adenovirus inducible rep and cap genes allows for infectivity titration of adeno-associated virus vectors. *Gene Ther.* **3,** 1124–1132.

30

High-Titer Stocks of Adeno-Associated Virus from Replicating Amplicons and Herpes Vectors

Ajay R. Mistry, Mahesh De Alwis, Elisabeth Feudner, Robin R. Ali, and Adrian J. Thrasher

1. Introduction

The adeno-associated virus (AAV) is a nonpathogenic member of the *Parvoviridae* family (for review, *see* **ref.** *1*) Recently this virus has gained considerable interest and has been developed as a gene delivery vector *(2)*. Six primate AAV serotypes (designated AAV types 1–6) have so far been identified and characterized in the literature *(3,4)*. The most extensively studied of these isolates is AAV type 2. The vast majority of the transduction studies have been carried out using recombinant vectors (rAAV) based on serotype 2. These studies have shown that rAAV2 has the ability to transduce a wide range of both dividing and nondividing cells, achieving efficient long-term gene expression in vivo in a variety of tissues including retina *(5)*, muscle *(6)*, central nervous system *(7)*, and liver *(8)*. The range of tissues transduced by recombinant AAV vectors based on other serotypes is currently being investigated by several laboratories *(9)*. It is hoped that rAAV vectors produced from these serotypes may prove to be useful for the transduction of tissues that are poorly infected by AAV2.

A major problem associated with the use of rAAV has been the difficulty in producing large quantities of high-titer stock *(10,11)*. This has become an important issue as vectors based on rAAV2 have now reached the stage at which they are starting to be used in human clinical trials *(12)*. This chapter describes the use of the herpes simplex virus type 1 (HSV-1) amplicons that have been developed in our laboratory to attain high-titer stocks of rAAV2 *(13)*. A brief description of the background and basis of the system is given in the first section.

From: *Methods in Molecular Medicine, Vol. 69, Gene Therapy Protocols, 2nd Ed.*
Edited by: J. R. Morgan © Humana Press Inc., Totowa, NJ

1.1. AAV Genome Structure and Organization

AAV2 has a single-stranded DNA genome that is 4675 nt long *(14)*. The genome is flanked by inverted terminal repeat sequences (ITRs) that are each 145 nt. The first 125 nucleotides of this sequence are palindromic, which allows it to fold back into a hairpin-like structure. The ITRs are required for AAV2 DNA replication *(1)* and also contain the sequences necessary to package viral DNA into particles *(15)*. The genome contains two large open reading frames with a single polyadenylation signal. The open reading frame in the 5' half of the AAV genome encodes four nonstructural regulatory proteins (Rep 78, Rep 68, Rep 57, and Rep 40), which have overlapping amino acid sequences; the open reading frame in the 3' half encodes three structural proteins (Vp1, Vp2, and Vp3) in which the amino acid sequences also overlap. The open reading frames are transcribed by three promoters (p5, p19, and p40). The p5 promoter intiates production of a transcript that unspliced encodes Rep 78 and spliced, encodes Rep 68. The p19 is also involved in the transcription of the Rep proteins; it initiates the production of a transcript that is either translated intact to produce Rep 52 or spliced to yield Rep 40. The p40 promoter is involved in the transcription of the genes encoding the AAV structural proteins (Vp1, Vp2, and Vp3). It produces two transcripts by alternative splicing that are at different levels. Vp1 is translated from the minor species, and Vp2 and Vp3 are translated from the major species from different start codons within the same transcript.

1.2. AAV Life Cycle and Conventional Production of rAAV

AAV2 belongs to the *Dependovirus* genus of the *Paroviridae* family, so-called because it usually requires a helper virus coinfection to allow productive AAV infection to occur. The helper viruses required for this productive lytic process include the adenovirus and HSV. In the absence of helper virus, AAV infects human cell lines and establishes latency as a provirus by integrating into the genome with a high preference for a specific site on human chromosome 19q13.1 *(16–18)*. If the cells containing the AAV provirus are subsequently infected with helper virus, the integrated AAV can be rescued from the chromosome and can enter the productive lytic process *(19)*. Vectors that are deleted for Rep lose the potential for site-preferential integration.

Elucidation of the molecular mechanisms involved in the life cycle of AAV is important for its development as a gene therapy vector, especially in terms of recombinant AAV production. The requirement of a helper virus for productive infection of AAV adds to the complexity of the process and has made it more difficult to understand the mechanisms involved. Nevertheless, several systems have been developed to package and produce rAAV. The conventional rAAV production system relies on transient transfection of permissive cells and utilizes three components:

1. A plasmid containing the rAAV vector genome. It contains two ITR sequences that flank the cDNA encoding the exogenous gene. The ITR sequences are the minimal *cis* elements necessary for replication and packaging of the recombinant gene *(20)*. The gene is present within an expression cassette. This is essentially a promoter located before the beginning of the gene that will drive its transcription and a polyadenylation signal at the 3' end of the gene that will stabilize the mRNA transcripts produced. The size of this expression cassette should be between 3.5 and 4.7 kb.
2. A helper plasmid containing the AAV2 genome encoding the *rep* and *cap* genes. The four Rep proteins and three Cap proteins are expressed from this plasmid and provide the components in *trans* required for the assembly of rAAV.
3. Wild-type adenovirus type 5, which provides other helper functions required for the lytic cycle.

The two plasmids are transfected into an appropriate cell line such as 293 or HeLa cells *(21,22)*, and these cells are subsequently infected with the helper adenovirus. This results in the production of infectious rAAV containing the recombinant exogenous gene as the single-stranded vector genome. The rAAV particles can be separated from contaminating adenoviral particles by density gradient centrifugation. However, the inefficiency of this process results in low levels of rAAV virus production, as well as residual contamination with adenoviral components (protein and infectious virus) that may be the cause of inflammatory processes observed in vivo *(23)*. More recently, improvements in transfection efficiency, development of optimized packaging plasmids, and the production of cell lines that stably maintain *rep-cap* genes have simplified and improved recovery of recombinant virus *(10,11,24)*. In addition, the minimal adenovirus components necessary for helper function are now known, and genes encoding these factors have been cloned into a single plasmid *(25,26)*. This plasmid can be cotransfected along with the helper and ITR plasmids and eliminates the need to carry out the helper virus infection step in the process of making rAAV, thereby eliminating the chance of adenovirus contamination.

1.3. Improving the rAAV Production Systems

Although the technical improvements outlined above have achieved some success in raising rAAV titers, none of these systems can achieve the levels of virus that can be produced during a wild-type AAV infectious lytic cycle.

Investigations have been carried out to identify the mechanisms that may limit the levels of rAAV production *(27)*. The main findings from these studies showed that during a wild-type infection the numbers of transcriptionally functional templates available for expression of the Rep78/68, Rep52/40, and capsid gene products are amplified as the whole genome is replicated. During the production of rAAV, the helper plasmid containing the *rep-cap* genes was found not to replicate after transfection. Thus the lower titer of rAAV particles was thought to be caused by insufficient expression of the Rep and Cap viral proteins.

Two types of strategies have been applied in an attempt to overcome this problem. The first strategy is based on the substitution of natural AAV promoters with alternative heterologous elements that increase the transcription levels of the *rep* and *cap* genes, resulting in higher levels of Rep and Cap proteins *(24,28,29)*. Increased synthesis of AAV capsid proteins has been shown to improve yields of rAAV *(29)*. However, this approach has not proved satisfactory, as overexpression of Rep gene products has also been shown to decrease rAAV DNA replication and to downregulate expression of the *cap* genes *(24,29)*. These results indicate that the control and regulation of the *rep* and *cap* gene transcription processes by the endogenous promoters (p5, p19, and p40), which allow coordinated production of Rep and Cap proteins, is an important aspect for rAAV production.

The second strategy is the use of replicating plasmids that contain sequences encoding the *rep* and *cap* genes. The basis of the system is that the plasmid, once transfected into the cell line, will replicate, resulting in an increase in the numbers of *rep* and *cap* genomic cassettes and thus overcoming the problem of low *rep-cap* gene levels. Transcription of the *rep* and *cap* genes is still under the control of the endogenous promoters, therefore allowing coordinated production of the Rep and Cap proteins. The system we use to achieve the replication processes is the HSV-1 amplicon plasmids in conjunction with HSV-1 as the helper virus. In contrast to the use of helper adenovirus, replication of these amplicon plasmids containing the *rep* and *cap* genomic cassettes, along with HSV-1 as the helper virus, did not markedly inhibit the expression of the Rep proteins *(13)*. These amplicons are considered in more detail below.

1.4. HSV Amplicons

HSV-1 has generated substantial interest as a vector to deliver genes to the nervous system because of its natural tropism for neuronal cells. There are two types of HSV-1-based vectors: 1) those produced by inserting the exogenous gene into a backbone virus genome, which are known as recombinant HSV vectors; and 2) HSV amplicon virions, which are produced by inserting the exogenous gene into an amplicon plasmid that is subsequently replicated and then packaged into virion particles (using HSV-1 as a helper virus). These HSV-1 amplicon virions were first described as contaminating defective genomes composed of multiple repetitions of a partial HSV-1 sequence orga-

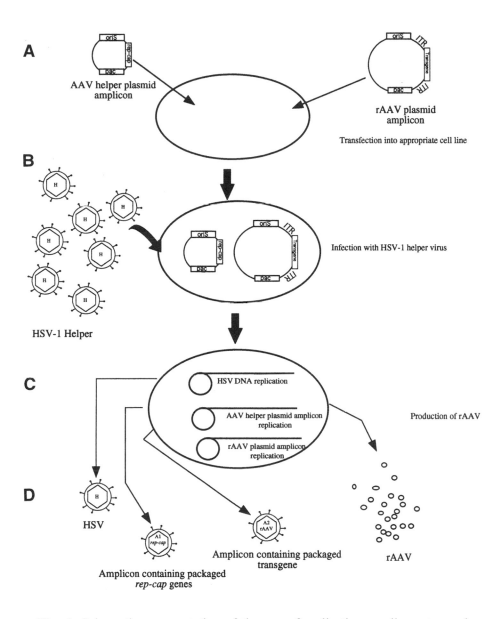

Fig. 1. Schematic representation of the use of replicating amplicons to produce rAAV. **(A)** A helper plasmid containing the AAV2 *rep* and *cap* genes and an rAAV plasmid containing transgene flanked by AAV2 ITRs have been cloned into separate amplicon vectors. These vectors each contain an HSV-1 packaging (pac) and an origin or replication (oriS) signals. The two plasmids are transfected into an appropriate cell line. They are then infected with HSV-1 helper virus. **(B)** This allows replication of the amplicon plasmid that results in the expression of Rep and Cap proteins and the packaging of the transgene to produce rAAV2. **(C)** In addition to the rAAV2, this process also produces HSV-1 helper virus and HSV-1 amplicon virions, which contain packaged *rep-cap* sequence and rAAV genome **(D)**.

nized in a head-to-tail fashion (concatemers). The monomers were shown to consistently include at least one origin of viral DNA replication (oriS) and a DNA cleavage/packaging sequence (pac). Cloning of these two elements in bacterial plasmids was shown to be sufficient to direct concatemeric packaging of the plasmid sequence into HSV-1 amplicon virions in the presence of complementing packaging functions (helper) provided by HSV-1.

1.5. Use of HSV Amplicons to Produce rAAV

To exploit the HSV amplicon system for the production of rAAV, the AAV genome containing the *rep* and *cap* genes is cloned into an amplicon plasmid. The DNA encoding the rAAV vector genome, i.e., exogenous gene within an expression cassette flanked by two ITR sequences, is also cloned into a separate amplicon plasmid. These two plasmids are transfected into an appropriate cell line and then subsequently infected with an HSV-1 helper virus. This allows the amplicons to replicate, resulting in amplification of the DNA encoding the *rep* and *cap* genome and the rAAV genome. The Rep and Cap proteins produced are then utilized to package the rAAV genome to produce the rAAV particles. In addition to the rAAV particles, this process also produces HSV-1 helper virus and HSV-1 amplicon virions, which contain the packaged *rep-cap* sequence and rAAV genome. The rAAV itself has to be purified away from the cellular material, the HSV-1 helper virus, and the HSV amplicon virions. **Figure 1** shows a schematic representation of this system, which we use.

Researchers have also shown that the HSV-1 amplicons that have packaged *rep* and *cap* sequences along with HSV-1 helper functions can themselves be used to produce rAAV. These amplicons were used to infect a cell line that had been transfected with a rAAV plasmid vector to produce rAAV particles *(30)*. In addition, HSV-1 itself has been engineered to express the AAV2 *rep* and *cap* genes. This recombinant HSV-1 was shown to be sufficient in providing the HSV-1 helper functions and the *rep-cap* sequence in *trans* for the production of rAAV2 from a cell line that had an integrated AAV2 provirus *(31)*. Attempts have also been made to produce HSV-1 amplicon/AAV hybrid constructs *(32)*. These constructs were essentially amplicon vectors that containing a transgene flanked by AAV2 ITRs along with the AAV2 *rep* genes under the control of their own endogenous promoters. The transgene in this amplicon was shown to be packaged into rAAV by using the amplicon to infect an appropriate cell line in the presence of an AAV helper plasmid and helper virus.

1.6. Use of DISC-HSV-1 as the Helper Virus

Production of the HSV-1 and HSV-1 amplicons in addition to the rAAV may be considered a safety hazard. Although the scheme devised to purify the rAAV away from the contaminating HSV-1 and HSV-1 amplicon should be sufficient in terms of safety for most applications, concern may still be raised about the presence of these contaminants, especially if the rAAV is to be used for clinical applications or in large animal models. For these reasons the HSV-1 helper virus used in our production system is DISC-HSV-1 (disabled infectious single-cycle virus). This is a genetically deleted herpesvirus that lacks the gene for essential glycoprotein H (gH) but can be propagated to very high titer as an infectious virus in cells expressing gH in *trans*. In permissive noncomplementing cell lines (gH⁻), the virus replicates effectively but is noninfective when released.

2. Materials

2.1. Cell Culture

1. Baby hamster kidney cells (BHK). These cells are split 1/6 every 2–3 days. They should not be allowed to overgrow and should be of low passage. The BHK cells can be obtained from European Collection of Animal Cell Cultures (ECACC, cat. no. 85011423). Production of rAAV2 using the BHK cells and the DISC-HSV-1 helper should ideally be carried out in BHK cell lines that have a deletion in glycoprotein H (gH⁻) for the reasons explained in **Subheading 1.6.** However, BHK cell lines that are not gH⁻ can also be used to produce rAAV2 using the system described here.
2. BHK-21 media (Gibco-BRL). To a bottle containing 500 mL of BHK-21 media, add the following under sterile conditions to produce B10 media:
    a. 5 mL antibiotic-antimycotic (Gibco-BRL).
    b. 50 mL of fetal calf serum (FCS), heat inactivated at 56°C for 1 h.
    c. 25 mL Tryptose Phosphate Broth (Gibco-BRL).
    d. 5 mL L-glutamine (Gibco-BRL).
3. Phosphate-buffered saline (PBS): 1.0% NaCl, 0.025% KCl, 0.14% Na_2HPO_4, 0.025% KH_2PO_4 (all w/v), pH 7.2, sterile.
4. Tissue culture dishes (150-mm diameter).
5. 50-mL polystyrene tubes (Falcon): 17×100 mm.
6. Trypsin-EDTA solution (Gibco-BRL).

2.2. Recombinant AAV Plasmid Vectors

The plasmid vectors used in this system are shown in **Figure 2**. The rationale behind the use of the helper plasmid pW7-Helper (**Fig. 2b**), which consists of the AAV2 *rep* and *cap* genes cloned into an HSV amplicon plasmid (pW7-Basic), has been described in **Subheadings 1.3.** and **1.4.** The rAAV amplicon vector that contains the foreign gene cassette is produced in two steps. The first step involves placing the foreign gene cassette between two ITRs by subcloning it into the pTRD-Basic vector (**Fig. 2c**). The expression cassette flanked by the ITRs is then cut out by digestion with the *Pac*I restriction

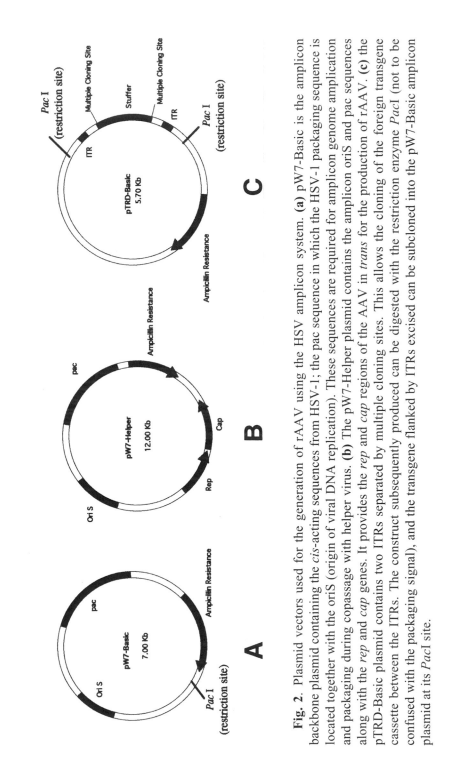

Fig. 2. Plasmid vectors used for the generation of rAAV using the HSV amplicon system. (**a**) pW7-Basic is the amplicon backbone plasmid containing the *cis*-acting sequences from HSV-1; the pac sequence in which the HSV-1 packaging sequence is located together with the oriS (origin of viral DNA replication). These sequences are required for amplicon genome amplication and packaging during copassage with helper virus. (**b**) The pW7-Helper plasmid contains the amplicon oriS and pac sequences along with the *rep* and *cap* genes. It provides the *rep* and *cap* regions of the AAV in *trans* for the production of rAAV. (**c**) the pTRD-Basic plasmid contains two ITRs separated by multiple cloning sites. This allows the cloning of the foreign transgene cassette between the ITRs. The construct subsequently produced can be digested with the restriction enzyme *Pac*I (not to be confused with the packaging signal), and the transgene flanked by ITRs excised can be subcloned into the pW7-Basic amplicon plasmid at its *Pac*I site.

enzyme and then subcloned into the amplicon plasmid vector pW7-Basic, which has a *Pac*I restriction site (**Fig. 2a**).

The plasmids used for transfections are prepared by purification on an ion-exhange resin column (e.g., Quiagen Maxi purification kit).

2.3. Generation of Recombinant Virus

1. Monolayer of BHK cells at approximately 80% confluency.
2. Opti-MEM medium (Gibco-BRL).
3. Lipofectin reagent (Gibco-BRL).
4. Cyclized integrin targeting peptide 6 (KKKKKKKKKKKKKKKKKGACRRET AWACG) at 0.1 mg/mL made up in Opti-MEM media. This peptide can be custom ordered from most peptide manufacturers. We obtain ours from Zinsser Analytic (Maidenhead, UK).
5. DNA samples (recombinant AAV plasmid vector and helper plasmid), column-purified (*see* **Notes 1–3**).
6. HSV-1 helper virus: DISC-HSV-1 (*see* **Subheading 1.6.**).

2.5. Isolation and Purification of Recombinant Virus

1. TMN Buffer: 50 mM Tris-HCl, pH 8.0, 5 mM MgCl$_2$, and 0.15 M NaCl.
2. Benzonase (Merck).
3. 60% iodixanol solution OptiPrep (Sigma).
4. 2 M NaCl solution.
5. PBS MK buffer: 1× PBS, 1 mM MgCl$_2$, and 2.5 mM KCl.
6. Phenol red (0.5% stock solution).
7. Ultracentrifuge tube, 25 × 89 mm (Beckman).
8. Peristaltic pump (e.g., Pharmacia model P1).
9. 10-mL syringe and needle (0.6 × 30 mm).
10. 2.5 mL prepacked heparin-agarose column (Sigma).
11. Elution buffer (PBS MK + 1 M NaCl).
12. Protein concentrators with a molecular weight cutoff of 10,000 Daltons (Amicon Centricon filter devices).
13. Storage buffer: 20 mM Tris-HCl, pH 8.0, 0.15 M NaCl, and 2 mM MgCl$_2$.

3. Methods
3.1. Generation of Recombinant Virus

This section describes the methods for transfecting BHK cells with the two-amplicon plasmid followed by infection with helper HSV to produce rAAV particles. The transfection efficiency is critical to the yield of rAAV obtained. We found that the most efficient system to use to transfect BHK cells was the three-component Lipofectin/integrin targeting peptide/DNA (LID) system *(33)*.

We routinely produce rAAV from ten 150-mm plates containing BHK cells during a single prep. The transfection procedure described below is for a single

150-mm dish containing BHK cells. This procedure has to be repeated a further 9 times so that 10 plates in total are transfected. This is important since the procedures described in subsequent sections on the purification of rAAV are based on the use of cells harvested from 10 transfected 150-mm dishes.

3.1.1 . Transfection of DNA Using LID Reagent

The day before transfection

1. Plate out 4.5×10^6 of BHK cells into a 150-mm round tissue culture dish in a total volume of 20 mL.
2. Repeat this procedure for a further nine plates.

The following day:

The procedures described below are for a single 150-mm plate containing BHK cells; they must be repeated so that 10 plates in total are transfected (*see* **Note 4**).

1. We routinely transfect a total of 60 μg of DNA into a 150-mm dish (*see* **Note 5**). For vectors cloned into amplicons, the ratio of the two vectors is 1 (recombinant AAV plasmid vector):3 (*rep/cap* helper plasmid). Therefore, when using 60 μg of DNA, the amounts of plasmids used are as follows: 15 μg (recombinant AAV plasmid vector):45 μg (helper plasmid).
2. Place the appropriate amounts of each vector into a polystyrene tube to give a final total of 60 μg of DNA. Add 6 mL of Opti-MEM to this tube.
3. Into a separate polystyrene tube place 45 μL (1 mg/mL) of Lipofectin and then add 6 mL of Opti-MEM to it.
4. Add 2.4 mL (at a concentration of 0.1 mg/mL) of the integrin targeting peptide to the tube containing the Lipofectin and mix together by inverting the tube several times (*see* **Note 6**).
5. Add the DNA to the polystyrene tube containing the Lipofectin and peptide. Mix by inverting the tube several times. Incubate the tube at room temperature for 1 h to allow time for the complexes to form.
6. Aspirate the medium from the cells and wash the monolayer twice with Opti-MEM medium.
7. Add the LID mixture to the cells.
8. Incubate at 37°C under normal 5% CO_2 conditions for 5 h.
9. Aspirate the media and replace with B10 media containing HSV-1 at 10–20 IU/cell.
10. Continue incubation for 32–35 h (*see* **Note 7**).

3.2. Isolation and Purification of Recombinant Virus

The isolation and purification of the recombinant AAV from the infected BHK cells can be divided into four stages:

Stage 1. Release of rAAV particles from within the infected BHK cells.
Stage 2. Purification of rAAV away from the vast majority of contaminating cellular and viral material by use of a discontinuous iodixanol gradient.
Stage 3. Further purification of the rAAV by heparin affinity chromatography.
Stage 4. Concentration and dialysis of the rAAV by using protein centrifugal concentrators.

3.2.1. Stage 1: Release of rAAV Particles from within the Infected BHK Cells

This stage of the process involves harvesting the infected cells by low-speed centrifugation, resuspension in the hypotonic TMN buffer, lysing the cells by three cycles of freeze thawing, and finally treating the cell lysate obtained with Benzonase to degrade the genomic DNA.

1. Scrape the infected cells and media from the 10 plates into five 50-mL polystyrene tubes.
2. Collect the cells by low-speed centrifugation (250g for 10 min).
3. Resuspend the cell pellets from the 10 plates in a total volume of 15 mL of TMN buffer. At this stage the cells can be stored at −70°C until needed.
4. Carry out three cycles of freeze-thawing in dry ice/ethanol and a 37°C water bath. The cells should be mixed by vortexing in between the cycles.
5. Add 50 U of Benzonase per 1 mL lysate from the 10 plates and incubate at 37°C for 30 min.
6. Clarify the lysate by centrifugation at 2700g for 20 min and retain the virus-containing supernatant.

3.2.2. Stage 2: Iodixanol Density Gradient

Use of the discontinuous iodixanol gradient as an alternative to the CsCl gradient centrifugation has considerably shortened the centrifugation period and in our hands has led to increases in the final titer of purified rAAV. This section describes in detail the procedures we use to make the discontinuous iodixanol gradient and load the clarified lysates onto it. It has mainly been based on the work of Zolotukhin and colleagues *(34)*.

1. In polystyrene tubes make up the following iodixanol solutions by adding the appropriate volumes of solutions, indicated in **Table 1**:
   20 mL of 15% iodixanol
   24 mL of 25% iodixanol
   15 mL of 40% iodxianol
   10 mL of 60% iodixanol
2. Pipet all the cleared lysate (approximately 15 mL) into the bottom of a 25 × 89-mm ultracentrifuge tube. Avoid creating air bubbles.
3. Connect the inlet tube from the peristaltic pump to the 15% iodixanol solution

Table 1
Volumes Required to Make Up Solutions for Discontinuous Iodixanol Gradient

	Iodixanol (mL)			
	15%	*25%*	*40%*	*60%*
60% Iodixanol stock	5	10	10	10
10× PBS MK	2	2.4	1.5	—
2 *M* NaCl	10	—	—	—
Phenol red (0.5 % stock)	—	525 µL	—	30 µL
ddH₂O	3	11.1	3.5	—
Final volume	20	24	15	10

and pump a small volume of the 15% iodixanol into the tubes to ensure that the dead volume contains this solution.

4. Place the outlet tube from pump into the lysate and allow 9 mL of this solution to enter the centrifuge tube slowly (*see* **Note 8**).

5. Remove the inlet tube from the 15% iodixanol solution and place it into the 25% solution. Place the outlet tube into the bottom of the centrifuge tube and allow 6 mL of this solution to enter the centrifuge tube.

6. Repeat this procedure first with the 40% iodixanol solution, allowing 5 mL to enter the centrifuge tube, and then with the 60% solution, allowing enough solution to enter so that the tube is filled to the top (5–6 mL). A discontinuous gradient is formed (*see* **Fig. 3a** and **Note 9**).

8. Seal the tube and place in a ultracentrifuge for 1 h, spinning at 250,000*g* (without brake) at 18°C.

9. Carefully remove the tube from the centrifuge and place in a clamp stand. The rAAV is located in the 40% iodixanol solution. This 40% band is located between the 60% solution and the 25% solution boundaries, which are marked by the phenol red (**Fig. 3b**). Insert a needle (0.6 × 30 mm) at the top of the tube to allow air to enter the tube. Using a 10-mL syringe with a needle, carefully insert the needle at the boundary between the 60% iodixanol solution and the 40% solution. Carefully draw out 5 mL of the 40% solution.

10. Add 6 mL of PBS-MK to this eluted 40% iodixanol solution. Put this solution through a 0.1-µm filter (25-mm diameter). This will remove any contaminating particles larger than 0.1 µm such as HSV-1 helper.

3.2.3. Stage 3: Further Purification of rAAV Particles Using a Heparin Column

1. Remove the existing column buffer from a 2.5-mL prepacked heparin-agarose column by spinning the column in a 50-mL polystyrene tube at 15*g* for 5 min.

2. Equilibrate the column by adding 5 mL of 1× PBS-MK and then spinning at 15*g* for 5 min. Repeat this procedure 3 more times so that a total volume of 20 mL of PBS-MK has passed through the column.

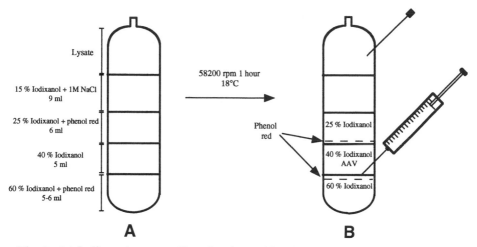

Fig. 3. (a) Iodixanol step gradient for the purification of rAAv. The volumes given above are for the 25 ¥ 89-mm ultracentrifuge tubes. The lysate is first pipeted into the bottom of the tube. Nine milliliters of the 15% iodixanol is then pumped into the bottom of the tube. This is followed by pumping 6 mL of the 25% iodixanol, then the 40% iodixanol, and finally 5–6 mL of 60% iodixanol solution to top up the tube. The tube is then sealed and subjected to centrifugation at 58,200 rpm (250,000 g) for 1 h at 18°C. (b) After the 1 h spin, the rAAV is ready to be retrieved from the 40% iodixanol solution. The red or yellow tinge in the 60% and the 25% steps is caused by the phenol red and will help locate the 40% step. The virus is collected by first puncturing a top region of the tube with a needle and then puncturing just below the 40% step with a needle and drawing out approximately 5 mL from this region using a syringe.

3. Add 10 mL of the filtered virus solution to the column, and allowing it to enter by gravity flow.
4. Wash the column by allowing 10 mL of PBS-MK to flow through the column.
5. Elute the virus by adding 6 mL of PBS-MK/1 *M* NaCl solution. The first 2 mL of the elution can be discarded (dead volume).

3.2.4. Stage 4: Concentration and Dialysis of the rAAV by Using Protein Centrifugal Concentrators

The final stage of the process is to concentrate the virus and exchange the solution the virus is in for a more physiologic one (storage buffer). To do this, we simply spin the virus in a centrifugal filter device with a molecular weight cutoff of 10,000 Daltons and then resuspend the virus in 2 mL of storage buffer. The virus is then spun again in the filter device, which is adequate to remove the vast majority of the elution buffer.

1. Place the eluted virus in 2 Centricon filter devices (2 mL each). Spin at 5000*g* for 50 min at 4°C.

2. Discard the elution and add 2 mL of storage buffer to each centricon device. Spin again at 5000*g* for 50 min.
3. Pool the concentrated virus (approximately 150 µL) and store at –70°C.

4. Notes

1. The AAV terminal repeats may be unstable in plasmids propagated in several commonly used laboratory strains of *E. coli.* Host strains carrying *recA,* as well as *recJ* and *recB* mutations, such as SURE strain cells (Stratagene), are recommended.
2. To ensure that the ITRs in the plasmid that are used for transfections are functional and not deleted, we routinely check the DNA by digesting the plasmid with the restriction enzyme *Sma*I. The *Sma*I sites are located within the ITRs; therefore digestion with the *Sma*I enzyme will cut out the foreign gene expression cassette, indicating that the ITRs are likely to be functional.
3. Use only high-quality DNA for transfections. Either CsCl-banded DNA or DNA prepared using high-grade commercial kits is adequate.
4. The amounts of peptide and Lipofectin used here have been optimized for BHK cells; they may need changing if different cell lines are used.
5. The 1:3 ratios of the two plasmid vectors have been determined systematically in our laboratory. This AAV production system can also be utilized to produce virus from recombinant AAV DNA that is not cloned into an amplicon backbone (albeit with lower titers). We have found that when using such a plasmid the best titers are obtained by using a ratio of 1 (recombinant AAV plasmid vector):1 (helper plasmid).
6. The order in which the LID components are made up are critical for the efficiency of transfection achieved. It is therefore important to mix the peptide and lipid first and then add this mixture to the DNA.
7. During this incubation period the cells are monitored as they undergo the cytopathic process (24–36 h). The cells are harvested just as they start lifting from the plate.
8. The volume that has been pumped into the centrifuge tube can be monitored simply by determining the flow rate of the pump and then allowing the appropriate volume to enter the tube within a timed period.
9. When making up the discontinuous gradient, ensure that the outlet tube from the pump is always at the bottom of the ultracentrifuge tube while the solution is entering the tube.

References

1. Berns, K. I. (1995) Parvoviridae: the viruses and their replication, in *Fundamental Virology*, 3rd ed. (Fields, B. N., Knipe, D. M., and Howley, P. M., eds.), Lippincott-Raven, Philadelphia.
2. Kotin, R. M. (1994) Prospects for the use of adeno-associated virus as a vector for human gene therapy. *Hum. Gene Ther.* **5**, 793–801.
3. Bantel-Schaal, U. and zur Hausen, H. (1984) Characterization of the DNA of a defective human parvovirus isolated from a genital site. *Virology* **134**, 52–63.

4. Blacklow, N. R., Hoggan, M. D., Kapikian, A. Z., Austin, J. B., and Rowe, W. P. (1968) Epidemiology of adenovirus-associated virus infection in a nursery population. *Am. J. Epidemiol.* **88,** 368–378.

5. Ali, R. R., Reichel, M. B., Thrasher, A. J., et al. (1996) Gene transfer into the mouse retina mediated by an adeno-associated viral vector. *Hum. Mol. Genet.* **5,** 591–594.

6. Kessler, P. D., Podsakoff, G. M., Chen, X., et al. (1996) Gene delivery to skeletal muscle results in sustained expression and systemic delivery of a therapeutic protein. *Proc. Natl. Acad. Sci. USA* **93,** 14082–14087.

7. McCown, T. J., Xiao, X., Li, J., Breese, G. R., and Samulski, R. J. (1996) Differential and persistent expression patterns of CNS gene transfer by an adeno-associated virus (AAV) vector. *Brain Res.* **713,** 99–107.

8. Koeberl, D. D., Alexander, I. E., Halbert, C. L., Russell, D. W., and Miller, A. D. (1997) Persistent expression of human clotting factor IX from mouse liver after intravenous injection of adeno-associated virus vectors. *Proc. Natl. Acad. Sci. USA* **94,** 1426–1431.

9. Rutledge, E. A., Halbert, C. L., and Russell, D. W. (1998) Infectious clones and vectors derived from adeno-associated virus (AAV) serotypes other than AAV type 2. *J. Virol.* **72,** 309–319.

10. Clark, K. R., Voulgaropoulou, F., Fraley, D. M., and Johnson, P. R. (1995) Cell lines for the production of recombinant adeno-associated virus. *Hum. Gene Ther.* **6,** 1329–41.

11. Clark, K. R., Voulgaropoulou, F., and Johnson, P. R. (1996) A stable cell line carrying adenovirus-inducible rep and cap genes allows for infectivity titration of adeno-associated virus vectors. *Gene Ther.* **3,** 1124–1132.

12. Kay, M. A., Manno, C. S., Ragni, M. V., et al. (2000) Evidence for gene transfer and expression of factor IX in haemophilia B patients treated with an AAV vector. *Nat. Genet.* **24,** 257–261.

13. Zhang, X., De Alwis, M., Hart, S. L., et al. (1999) High-titer recombinant adeno-associated virus production from replicating amplicons and herpes vectors deleted for glycoprotein H. *Hum. Gene Ther.* **10,** 2527–2537.

14. Srivastava, A., Lusby, E. W., and Berns, K. I. (1983) Nucleotide sequence and organization of the adeno-associated virus 2 genome. *J. Virol.* **45,** 555–564.

15. McLaughlin, S. K., Collis, P., Hermonat, P. L., and Muzyczka, N. (1988) Adeno-associated virus general transduction vectors: analysis of proviral structures. *J. Virol.* **62,** 1963–1973.

16. Kotin, R. M., Linden, R. M., and Berns, K. I. (1992) Characterization of a preferred site on human chromosome 19q for integration of adeno-associated virus DNA by non-homologous recombination. *EMBO J.* **11,** 5071–5078.

17. Kotin, R. M., Siniscalco, M., Samulski, R. J., et al. (1990) Site-specific integration by adeno-associated virus. *Proc. Natl. Acad Sci. USA* **87,** 2211–2215.

18. Samulski, R. J., Zhu, X., Xiao, X., et al. (1992) Targeted integration of adeno-associated virus (AAV) into human chromosome 19 [published erratum appears in *EMBO J.* 1992; **11,** 1228]. *EMBO J.* 1991; **10,** 3941–3950.

19. Berns, K. I. and Linden, R. M. (1995) The cryptic life style of adeno-associated virus. *Bioessays* **17,** 237–245.
20. Xiao, X., Xiao, W., Li, J., and Samulski, R. J. (1997) A novel 165-base-pair terminal repeat sequence is the sole cis requirement for the adeno-associated virus life cycle. *J. Virol.* **71,** 941–948.
21. Hermonat, P. L. and Muzyczka, N. (1984) Use of adeno-associated virus as a mammalian DNA cloning vector: transduction of neomycin resistance into mammalian tissue culture cells. Proc. *Natl. Acad. Sci. USA* **81,** 6466–6470.
22. Samulski, R. J., Chang, L. S., and Shenk, T. (1987) A recombinant plasmid from which an infectious adeno-associated virus genome can be excised in vitro and its use to study viral replication. *J. Virol.* **61,** 3096–3101.
23. Monahan, P. E., Samulski, R. J., Tazelaar, J., et al. (1998) Direct intramuscular injection with recombinant AAV vectors results in sustained expression in a dog model of hemophilia. *Gene Ther.* **5,** 40–49.
24. Li, J., Samulski, R. J., and Xiao, X. (1997) Role for highly regulated rep gene expression in adeno-associated virus vector production. *J. Virol.* **71,** 5236–5243.
25. Xiao, X., Li, J., and Samulski, R. J. (1998) Production of high-titer recombinant adeno-associated virus vectors in the absence of helper adenovirus. *J. Virol.* **72,** 2224–2232.
26. Matsushita, T., Elliger, S., Elliger, C., et al. (1998) Adeno-associated virus vectors can be efficiently produced without helper virus. *Gene Ther.* **5,** 938–945.
27. Fan, P. D. and Dong, J. Y. (1997) Replication of rep-cap genes is essential for the high-efficiency production of recombinant AAV. *Hum. Gene Ther.* **8,** 87–98.
28. Flotte, T. R., Barraza-Ortiz, X., Solow, R., et al. (1995) An improved system for packaging recombinant adeno-associated virus vectors capable of in vivo transduction. *Gene Ther.* **2,** 29–37.
29. Vincent, K. A., Piraino, S. T., and Wadsworth, S. C. (1997) Analysis of recombinant adeno-associated virus packaging and requirements for rep and cap gene products. *J. Virol.* **71,** 1897–1905.
30. Conway, J. E., Zolotukhin, S., Muzyczka, N., Hayward, G. S., and Byrne, B. J. (1997) Recombinant adeno-associated virus type 2 replication and packaging is entirely supported by a herpes simplex virus type 1 amplicon expressing Rep and Cap. *J. Virol.* **71,** 8780–8789.
31. Conway, J. E., Rhys, C. M., Zolotukhin, I., et al. (1999) High-titer recombinant adeno-associated virus production utilizing a recombinant herpes simplex virus type I vector expressing AAV-2 Rep and Cap. *Gene Ther.* **6,** 986–993.
32. Johnston, K. M., Jacoby, D., Pechan, P. A., et al. (1997) HSV/AAV hybrid amplicon vectors extend transgene expression in human glioma cells. *Hum. Gene Ther.* **8,** 359–370.
33. Hart, S. L., Arancibia-Carcamo, C. V., Wolfert, M. A., et al. (1998) Lipid-mediated enhancement of transfection by a nonviral integrin-targeting vector. *Hum. Gene Ther.* **9,** 575–85.
34. Zolotukhin, S., Byrne, B. J., Mason, E., et al. (1999) Recombinant adeno-associated virus purification using novel methods improves infectious titer and yield. *Gene Ther.* **6,** 973–985.

31

Herpes Simplex Virus/Adeno-Associated Virus Hybrid Vectors for Gene Transfer to Neurons

Preparation and Use

Lauren C. Costantini, Cornel Fraefel, Xandra O. Breakefield, and Ole Isacson

1. Introduction

Gene transfer to the central nervous system (CNS) has shown major advances in recent years, with the development of novel vector systems and progress in basic virology *(1–12)*. To improve gene transfer to CNS neurons, we have combined the critical elements of herpes simplex virus-1 (HSV-I) amplicons and recombinant adeno-associated virus (AAV) vectors to construct a hybrid amplicon vector, and then packaged the vector into HSV-1 virions via a helper virus-free system. These HSV/AAV hybrid amplicon vectors have shown efficient transduction and stability of transgene expression in neurons (and other nondividing cell types *[13]*), both in culture and after intracerebral injection with no apparent toxicity or immune response, as well as extended transgene expression in dividing cells *(14)*. Before detailing the hybrid amplicon vectors, a short description of the two vectors upon which the hybrid amplicon vectors are based is given.

1.1. Characteristics of HSV-1 and AAV

One of the most comprehensively studied viruses for gene transfer is HSV-1. The approximately 300-nm enveloped virus contains an icosahedral capsid encasing 152 kb of double-stranded DNA encoding over 80 genes, 38 of which are dispensable for virus replication in culture *(15)*. The virion shows high infectivity for neurons and glia, as well as many other cell types *(16)*, and

From: *Methods in Molecular Medicine, Vol. 69, Gene Therapy Protocols, 2nd Ed.*
Edited by: J. R. Morgan © Humana Press Inc., Totowa, NJ

enters the cell by fusion of the envelope with the plasma membrane. The capsid is transported along microtubules to the nucleus, where the viral DNA circularizes and then replicates, enters latency, or is degraded.

Two types of vectors are derived from HSV-1: recombinant virus vectors and amplicon vectors (for review, *see* **ref. *17***), the latter of which is utilized in the hybrid amplicon vectors. The HSV-1 amplicon plasmid bears an HSV-1 origin of DNA replication, *ori$_s$*, and a DNA cleavage/packaging signal, *pac*. *ori* allows the amplicon to be replicated by a rolling-circle mechanism, producing long concatenates of DNA that are subsequently cut into unit-length HSV-1 genomes at *pac* during packaging into capsids in the presence of HSV-1 helper functions *(18,19)*. Amplicon DNA is thereby packaged into HSV-1 virion, and after infection of cells, the amplicon concatenate assumes an extrachromosomal state in the infected cell nucleus. When packaged free of helper virus, these amplicon vectors show essentially no toxicity or antigenicity, as they express no virus genes. Since the combined size of *ori$_s$* and *pac* is <1 kb, amplicon plasmids can accommodate at least 22 kb of foreign DNA (and theoretically up to 150 kb using high-capacity cloning plasmids) *(20,21)*; with current packaging methods, they can be grown to high titers (10^8 transduction units (TU)/mL).

The nonpathogenic AAV has a smaller capsid (diameter 20–24 nm) and contains a single-stranded approx 4.7-kb DNA genome. Vectors based on AAV typically contain the AAV inverted terminal repeats (ITRs) needed for signaling extrachromosomal replication and integration of transgenes at multiple sites in the host cell genome *(22–25)*. AAV-based vectors have a 4.5-kb transgene capacity *(26)*. In wild-type AAV infections, the *rep* gene encodes a set of proteins that mediate the amplification of the ITR-flanked genome and facilitates site-specific integration into the human genome on chromosome 19q13.3 *(27–29)*. Integration of transgenes delivered by AAV vectors appears to be random, unless Rep proteins 78 and/or 68 are present, in which case it is site-specific into human chromosome 19q13.3 *(25,27–31)*. Long-term expression of transgenes from AAV-based vectors can be facilitated both by integration and by maintenance as an episomal element within the host cell nucleus *(25,32,33)*.

1.2. HSV/AAV Hybrid Amplicon Vectors

By combining key elements from HSV-1 and AAV vectors, we have produced HSV/AAV hybrid amplicon vectors (**Fig. 1**) *(14)* with the high infection efficiency of both dividing and nondividing cells, retrograde transport along neurites to the cell nucleus *(34,35)*, and the large transgene capacity of HSV-1, as well as the stable gene transduction and potential integration capacity of AAV. Using the HSV-1 amplicon backbone, which contains signals for propagation in bacteria, and HSV-1 *ori$_s$* and *pac*, the transgene is inserted, flanked

HSV/AAV hybrid amplicon

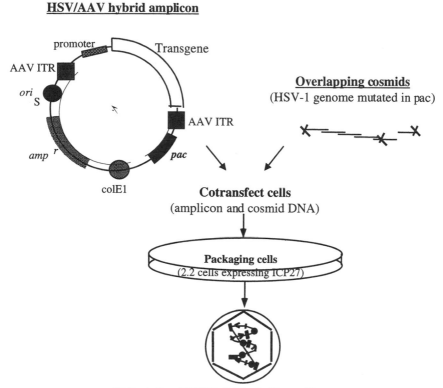

Helper-free HSV/AAV hybrid amplicon vector

Fig. 1. Helper virus-free packaging procedure. Amplicon plasmid DNA is cotransfected into permissive cells in culture with an overlapping set of cosmids that span the HSV-1 genome and provide HSV structural signals but are mutated in *pac* sequences *(19)*. Amplicon DNA is packaged into viral particles, and vectors with no contaminating helper virus are released.

by AAV ITR sequences. Hybrid amplicon vectors have been produced both with and without the AAV *rep* gene (inserted outside the ITR-flanked transgene cassette) to determine its role in mediating sustained transgene expression *(14,36)*. These vectors have been constructed with over 20 kb of transgene sequences *(37,38)*, and grown to high titers (10^8 TU/mL).

Packaging of these hybrid vectors is carried out via a helper virus-free system *(19)*. In one modality, overlapping sets of cosmids that contain the entire HSV-1 genome deleted for *pac* and transfected into permissive cells undergo homologous recombination to form circular replication-competent virus genomes, which then produce proteins needed for viral DNA replication and virion packaging. The reconstituted virus genomes cannot themselves be packaged because of deletion of their own *pac* signals, yet they still provide helper functions required for the packaging of cotransfected amplicon DNA. The resulting amplicon vector is structurally a herpes virion, in a preparation with essentially no contaminating helper virus. Alternatively to the *pac*-minus cosmid set, a bacterial artificial chromosome (BAC or F-plasmid) containing the entire HSV-1 genome with *pac* signals and essential genes deleted, can be used as the packaging-defective HSV-1 helper DNA *(39–41)*.

1.3. Neuronal Gene Transfer via HSV/AAV Hybrid Amplicon Vectors

The HSV/AAV hybrid amplicon vectors have unique capabilities for transfection of neural cells. First, they are highly efficient at gene delivery to neurons, astrocytes, and endothelial cells, thus reducing the number of virions that must be injected. Second, their large transgene capacity allows incorporation of substantial *cis*-acting regulatory sequences, which are believed to be necessary for targeted, regulated, and sustained transgene expression. Third, they can be delivered to the brain parenchyma via the cerebrospinal fluid (CSF), across the blood-brain barrier, or by direct injection *(36,42)*. Their ability to move within neurons by rapid retrograde transport allows access to cell bodies far removed from the point of virion entry. Limitations to delivery in the CNS by these (and many other) vectors include low titers, which can increase the volume injected (that in itself can cause damage); poor diffusibility through the extracellular space; and some toxicity of the virion itself at high ratios of virions per cell.

The efficacy, stability, and cytopathic effects of HSV/AAV hybrid amplicon vectors for gene delivery to CNS neurons have been analyzed, both in culture and in the rat brain. Transduction efficiency in primary neuronal cultures from fetal rat ventral mesencephalon by HSV-derived amplicon vectors was significantly higher than for AAV and adenovirus (Ad) vectors at the same multiplicity of infection (MOI) (**Fig. 2**) *(36)*. Furthermore, hybrid vectors mediated longer expression of the transgene in neurons, compared with AAV and Ad vectors. One month after injection of hybrid amplicon vectors into rat striatum, transduction efficiency was similar to and in many cases higher than that of standard HSV-1 amplicon vectors. Transgene-expressing neurons were observed in the striatum (over 8100 striatal cells at 30 days post infection; **Fig. 3**), with 70% of transduced cells being neurons using the hybrid amplicon vector, compared with 50% using the standard HSV-1 amplicon vector, suggesting extended transgene expression with the former. Transgene expression was also observed within the substantia nigra (over 500 nigrostriatal neurons) through retrograde transport ipsilateral to the injection (**Fig. 3**). It is noteworthy that there was no evidence of an immune response (analyzed via specific

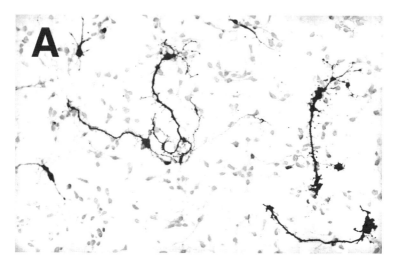

Fig. 2. Gene transfer to primary neurons with HSV/AAV hybrid amplicon vectors. Primary VM neurons expressing transgene (GFP immunostaining) 2 days after infection with HyRG.

immune-response markers for T-cells and microglia) or any inflammation caused by these vectors *(36)*.

Thus these hybrid vectors retain the retrograde transport, ability to infect both dividing and nondividing cells, and large transgene capacity of HSV-1, as well as the safety and long-term transgene expression of AAV.

2. Materials

2.1. HSV/AAV Hybrid Amplicon Components

1. Standard HSV-1 amplicon plasmid such as pHSVPrPUC (Howard_Federoff@ URMC.rochester.edu) or pHSVlac *(43)* (Geller_a@al.tch.harvard.edu, Breakefield@ helix.mgh.harvard.edu)
2. Cloned AAV genome, e.g., pAV2 *(44)* as source of *rep* genes and ITRs. The AAV sequence is in the database (GenBank Acc. no. NC 001401) (*see* **Note 1**).
3. Plasmid containing a transgene cassette, e.g., cytomegalovirus-green fluorescent protein (CMV-GFP) from pEGFPN3 (Clontech).
4. Material and equipment for cloning in *E. coli* (restriction enzymes and ligase, polymerase chain reaction [PCR], competent *E. coli* [DH10B, Gibco or SURE, Stratagene; Maniatis protocols]).

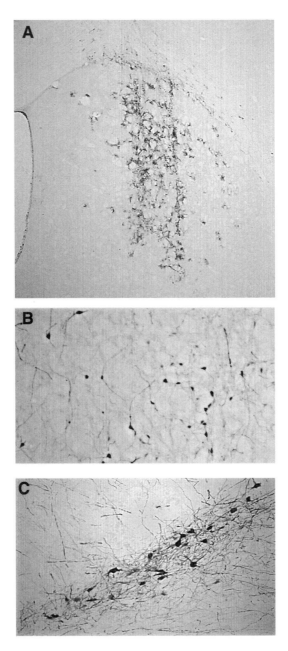

Fig. 3. Gene transfer to neurons in vivo with HSV/AAV hybrid amplicon vectors. (**A**) Striatal cells expressing transgene (GFP immunostaining) 1 month after intrastriatal injection of HyRG. (**B**) Fluorescence immunostaining for GFP transgene illustrating neuronal morphology of transduced striatal cells 1 month after intrastriatal injection of HyRG. (**C**) Neurons within the substantia nigra expressing transgene (GFP immunostaining) 1 month after intrastriatal injection of HyRG. Note the large soma and morphology typical of nigrostriatal dopamine neurons.

2.2. Packaging of HSV/AAV Hybrid Amplicon Vector

2.2.1. HSV-1 Cosmid DNA for Transfection

1. *E. coli* clones of HSV-1 cosmid set C6Δa48Δa, which includes cos6Δa, cos14, cos28, cos48Δa, and cos56. Cosmid clones are stored at –80°C in SOB medium supplemented with 7% dimethyl sulfoxide (DMSO).
2. SOB medium containing 50 µg/mL ampicillin; SOB^{amp}/liter—20 g tryptone, 5 g yeast extract, 0.5 g NaCl, H_2O to 950 mL, and 10 mL 250 mM KCl solution. Adjust the pH to 7.0 with 5 N NaOH and autoclave. Just before use, add 5 mL sterile 2 M $MgCl_2$ and 1 mL of 50 mg/mL ampicillin.
3. Qiagen-tip 500 Plasmid Purification Kit, which includes columns and buffers P1, P2, P3, QBT, QC, and QF (Qiagen).
4. Isopropanol.
5. 70% (v/v) ethanol.
6. TE buffer, pH 7.5: 10 mM Tris-HCl, pII 7.5, 1 mM EDTA. Store at room temperature (RT).
7. Restriction endonucleases *Dra*I, *Kpn*I, and *Pac*I.
8. High molecular weight DNA standard (Gibco, cat. no. 15618-010).
9. 1-kb DNA ladder (Gibco, cat. no. 15615-016).
10. Electrophoresis-grade agarose.
11. TAE electrophoresis buffer: 40 mM Tris-acetate, 2 mM Na_2EDTA, pH 7.5. Store at RT.
12. Ethidium bromide solution (1 mg/mL in H_2O).
13. 25:24:1 (v/v) phenol/chloroform/isoamyl alcohol.
14. 24:1 (v/v) chloroform/isoamyl alcohol.
15. 100% ethanol.
16. 3 M sodium acetate, pH 5.5.
17. Falcon 2059 tubes (14 mL, polypropylene).
18. 1.8-mL cryogenic vials.
19. 250-mL polypropylene tubes (Herolab, Wiesloch, Germany), 30-mL centrifuge tubes (polypropylene, Herolab, Wiesloch, Germany), Eppendorf tubes.

2.2.2. Helper Virus-Free Amplicon Stocks

Alhough any cell line that supports HSV-1 replication (including VERO, BHK21, or 293 cells) can be utilized for the packaging procedure, this protocol has been optimized for 2-2 cells, a derivative of VERO cells that expresses the HSV-1 immediate-early (IE) 2 gene ICP27 *(45)* (*see* **Notes 2** and **3**).

1. 2-2 cells (rmsandri@uci.edu).
2. Dulbecco's modified Eagle's medium (DMEM; Gibco).
3. Fetal bovine serum (FBS).
4. G418 (Geneticin; Gibco).
5. 0.25% trypsin/0.02% EDTA (Gibco).

6. Opti-MEM I (Gibco).
7. *Pac*I-digested cosmid DNA of set C6Δa48Δa (*see* **Subheading 2.2.1.**).
8. HSV/AAV hybrid amplicon DNA (maxiprep [Qiagen Column] isolated from *E. coli*).
9. LipofectAMINE (Gibco).
10. 10, 30, and 60% sucrose solutions (w/v in phosphate-buffered saline [PBS]).
11. PBS: 137 mM NaCl, 2.7 mM KCl, 4.3 mM Na$_2$HPO$_4$, 1.4 mM KH$_2$PO$_4$. Filter sterilize and store at RT or 4°C.
12. 75-cm^2 tissue culture flasks.
13. 60-mm-diameter tissue culture dishes.
14. 15-mL conical tubes (polypropylene; Greiner).
15. 0.45-μm syringe filters (Sarstedt polyethersulfone membrane filters, cat. no. 83.1826).
16. 20-mL disposable syringes.
17. 30-mL centrifuge tubes (Beckman Ultra-Clear 25 × 89 mm and 14 × 95 mm).

2.2.3. Titration of Amplicon Stocks

Quantification of infectious vector particles in an amplicon stock is determined before and after purification/concentration. Twenty-four to 48 h after infection, the cells expressing the transgene are counted microscopically to calculate the titer (TU/mL vector stock).

1. VERO (clone 76; ECACC, cat. no. 85020205), BHK (clone 21; ECACC, cat. no. 85011433), or 293 (ATCC, cat. no. 1573).
2. DMEM supplemented with 10% or 2% FBS (no antibiotics).
3. 4% (w/v) paraformaldehyde solution, pH 7.0. Under a hood, add 20 g paraformaldehyde to 300 mL H$_2$O and heat to 60°C with constant stirring. Dropwise, add 2 N NaOH until the solution becomes clear. Allow it to cool down and check the pH. (If necessary, add more NaOH to bring the pH to 7.0–7.5.) Add 100 mL of 0.5 M sodium phosphate buffer (pH 7.0) and then H$_2$O to a final volume of 500 mL. Filter the solution through a 0.45-μm bottle top filter (ZAPCAP S, cat. no. 10443401, Schleicher & Schuell, Dassel, Germany) and store at 4°C.
   Caution: Paraformaldehyde is a carcinogen and toxic.
4. X-gal staining solution: 20 mM K$_3$Fe(CN)$_6$, 20 mM K$_4$Fe(CN)$_6$, 2 mM MgCl$_2$, in PBS, pH 7.5. Filter-sterilize and store at 4°C. Before use, equilibrate to 37°C and per mL add 20 μL of a 50 mg/mL solution of 5-bromo-4-chloro-3-indolyl-D-galactopyranoside. Store X-gal solution in DMSO in 1-mL aliquots at –20°C.
5. Goat serum Triton (GST) solution: 2% (v/v) goat serum, 0.2% (v/v) Triton X-100, in PBS. Store at 4°C.
6. Primary antibodies to marker proteins and fluorescent secondary antibodies to primary antibody.
7. 24-well tissue culture plates.

2.2.4. Gene Transfer in Culture

1. Primary neuronal culture suspension diluted in DMEM (Gibco) containing heat-inactivated (in 56°C water bath for 30 min) horse serum (10%), glucose (6.0 mg/mL), penicillin (10,000 U/mL), streptomycin (10 mg/mL; Sigma), and glutamine (2 mM; Gibco).
2. Glass cover slips (Bellco, cat. no. 1943-10012).
3. Poly-L-lysine (PLL, Sigma, cat. no. P6282): Suspend 5 mg lyophilized PLL in 50 mL sterile water to a give concentration of 100 µg/mL. This 2× solution can be frozen at –20°C for storage. The appropriate amount of PLL can be diluted 1:1 with 0.2 M sodium borate buffer. For 250 mL: 3.09 g boric acid, 25 mL 1 M NaOH, 125 mL double-distilled water; pH 8.2. Then make to 250 mL with double-distilled water.
4. 24-well tissue culture trays (Falcon, cat. no. 353047).
5. Defined medium: DMEM as described in **step 1** above, with N2 cocktail (Gibco, cat. no. 17502-030) replacing the horse serum (1 mL N2/100 mL medium).
6. 4% paraformaldehyde (*see* **Subheading 2.2.3.**, **item 3**)/4% sucrose in PBS (*see* **Subheading 2.2.2.**, **item 11**).

2.2.5. Gene Transfer In Vivo

1. Stereotaxic frame (Kopf 900).
2. Anesthesia/preanesthesia 1:1 mixture of acepromazine and atropine; 10 min later, 2:1 mixture of ketamine and xylazine.
3. Hand drill and sterile drill bits.
4. 2 µL Hamilton syringe and beveled needle (0160826, 1-inch PT2).
5. Sterile 70% ethanol and 0.9% saline.
6. Sterile surgical supplies and instruments including scalpel with blade, sutures, scissors, and gauze.
7. Purified vector stocks of hybrid amplicon vector can be kept on ice for approximately 3 h.
8. 4% paraformaldehyde.
9. 20% sucrose.

3. Methods
3.1. HSV/AAV Hybrid Amplicon Vector Construction

1. Construct hybrid vector plasmids using standard cloning procedures in *E. coli*.
2. Use a standard amplicon plasmid such as pHSVPrPUC or pHSVlac as the backbone to insert 1) the ITR-transgene-ITR cassette and, if desired, 2) the AAV *rep* genes.
3. Use appropriate restriction enzymes to insert the transgene cassette between the AAV ITRs in pAV2 (e.g., *Dra*III and *Sna*BI digest removes *rep* and cap genes but leaves ITRs behind). From the resulting clone, isolate the ITR-transgene-ITR cassette by restriction digest (in pAV2, the AAV genome is flanked by *Bgl*II sites) and insert into a unique restriction site in the standard amplicon plasmid;

use linkers if necessary or perform blunt-end ligations. Insert AAV *rep* genes into a unique restriction site outside the ITR transgene cassette for the *rep*-hybrid amplicon.

3.2. Packaging of HSV/AAV Hybrid Amplicon Vector

3.2.1. Preparation of HSV Cosmid DNA for Transfection

The quality of DNA is the most important factor to achieve higher transfection efficiency and titers of packaged amplicon stocks. Because the cosmids of set C6Δa48Δa contain high-copy colE1 plasmid origins of DNA replication, special care must be taken to prevent random mutations during amplification of these clones. Cosmid DNA is extracted and purified from large bacterial cultures using a modified Qiagen-tip 500 protocol. To characterize the purified cosmid DNA preparations, they are digested with restriction enzymes (*Dra*I, *Kpn*I) and separated by electrophoresis. HSV-1 inserts are excised via digestion with *Pac*I, and the fragments are then purified by phenol extraction.

1. For each of the five clones of HSV-1 cosmid set C6Δa48Δa, prepare a Falcon 2059 tube containing 5 mL SOB medium supplemented with 50 μg/mL ampicillin (SOB^amp), and inoculate each with a loop used to scrape the surface of frozen cultures of the cosmid clones. After incubating the culture for 8 h in a shaker at 37°C, transfer 1 mL into a 2-L flask containing 300 mL SOB^amp and incubate further with shaking for another 12–16 h (*see* **Note 4**).
2. Remove several milliliters of the bacterial culture for long-term storage. Add 70 μL DMSO to 1 mL bacterial culture in a 1.8-mL cryogenic vial, mix well, and freeze at –80°C.
3. Extract cosmid DNA using a modified Qiagen-tip 500 protocol. Pellet the bacterial culture remaining from **step 2** by centrifugation for 10 min at 4°C and 5000*g* using a Sorvall GSA rotor. Pour the media off and leave the tube (250-mL polypropylene) inverted on a paper towel for 1–2 min to drain all liquid.
4. Resuspend the pellet in 15 mL buffer P1 and add 15 mL buffer P2. Mix by inverting the tube 4–6 times, and incubate for 5 min at RT.
5. Add 15 mL buffer P3 and mix immediately by inverting the tube 6 times. After incubation on ice for 20 min, invert the tube once more and centrifuge for 30 min at 4°C and 15,000*g* using a GSA rotor.
6. Equilibrate a Qiagen-tip 500 column with 10 mL buffer QBT, and allow the column to empty by gravity flow. Label the column with the name of the clone and drape a small piece of Kimwipe tissue on the column to remove cell debris.
7. Carefully filter the supernatant from **step 5** through a Kimwipe into the Qiagen-tip 500 column, and allow the liquid to enter the resin by gravity flow.
8. Wash the column 2 times with 30 mL buffer QC, and then elute DNA with 15 mL prewarmed (65°C) buffer QF into a 30-mL centrifuge tube.
9. Precipitate the DNA with 10.5 mL (0.7 vol) isopropanol, and immediately centrifuge for 30 min at 4°C and 16,000*g* using a Sorvall SS34 rotor.
10. Very carefully remove the supernatant from **step 9** and mark the location of the

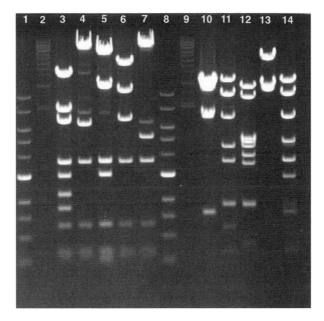

Fig. 4. Restriction fragment pattern of cosmids. *Lane 1*, 1 Kb ladder; *lane 2*, high-molecular-weight marker; *lanes 3–7*, *Dra*I digest of cos6Δa, cos14, cos28, cos48Δa, and cos56; *lanes 8 and 9*, markers; *lanes 10–14, Kpn*I digest of the cosmids.

pellet. Rinse surface of pellet with chilled 70% ethanol and, if necessary, repellet as in **step 9**.

11. Aspirate supernatant completely (but avoid drying), resuspend the pellet in 200 µL TE buffer (pH 7.5), transfer the DNA solution into a microcentrifuge tube, and store at 4°C (*see* **Note 5**).

12. Determine the absorbance of the DNA solution from **step 11** at 260 nm (A_{260}) and 280 nm (A_{280}) using a UV spectrophotometer. A value of 1.0 for A_{260} is equivalent to 50 µg/mL of double-stranded DNA. Also, pure DNA preparations have an A_{260}/A_{280} value of at least 1.8.

13. In two parallel reactions, digest 2 µg of the cosmid DNA from **step 11** with 10 U of *Dra*I and *Kpn*I for 2 h at 37°C. Separate the fragments overnight by electrophoresis on a 0.4% agarose gel at 40 V in TAE electrophoresis buffer. Compare the restriction fragment patterns with those shown in **Figure 4** (*see* **Note 6**).

14. After characterization of the DNA (**steps 12** and **13**), store aliquots of 10–50 µg at –20°C.

15. Pool 10 µg of each of the five cosmids into a microcentrifuge tube and digest for 3 h at 37°C with 50 U of *Pac*I in a total volume of 100 µL. Electrophorese an aliquot (1–2 µL) of the reaction mixture on a 0.4% agarose gel to confirm completion of the digest (*see* **Figure 4** and **Note 7**).

16. Extract DNA in the reaction mixture of **step 15** first with 100 µL (1 vol) phenol/

chloroform/isoamyl alcohol (25:24:1; v/v), and then with 100 μL (1 vol) chloroform/isoamyl alcohol (24:1; v/v). Finally, precipitate DNA overnight at –20°C with 250 μL (2.5 vol) 100% ethanol and 10 μL (0.1 vol) 3 M sodium acetate (pH 5.5). The DNA precipitate should be clearly visible at this point. Do not vortex— gently tap the tubes to avoid damaging large DNA fragments (see **Note 8**).

17. Spin tubes for 10 min at 16,000g. Carefully remove the supernatant, and rinse the pellet surface once with 70% ethanol. Allow the pellet to dry and resuspend (with minimal pipeting) in 100 μL TE buffer (pH 7.5).

18. Measure DNA concentration as described in **step 12**, and store aliquots of 10 μg at –20°C until transfection.

3.2.2. Preparation of Helper Virus-Free Amplicon Stocks

1. Maintain 2-2 cells at 37°C in humidified 5% CO_2 in DMEM supplemented with 10% FBS and 500 μg/mL G418. Split the cells twice a week approx 1:5 in fresh medium (20 mL) into a new 75-cm^2 tissue culture flask.

2. On the day before transfection, trypsinize the culture for 10 min, count cells using a hemocytometer, and plate at 1.2×10^6 cells per 60-mm-diameter tissue culture dish in 3 mL DMEM containing 10% FBS.

3. Per 60-mm-diameter tissue culture dish (maximum of 6) to be transfected, add 100 μL Opti-MEM I into two 15-mL conical tubes. To one tube, add 0.6 μg amplicon DNA and either a) 2 μg of the PacI-digested cosmid DNA mixture (0.4 μg of each of the five clones) or b) 2 μg fHSVΔpac DNA; to the other tube, add 12 μL LipofectAMINE.

4. Combine the two tubes from **step 3**. Mix well (without vortexing), and incubate for 45 min at RT (**Fig. 1**).

5. Wash the cultures prepared the day before (**step 2**; cells should be confluent at the time of transfection) once with 2 mL Optimem I. For each dish to be transfected, add 1.1 mL Opti-MEM I into the tube from **step 4** containing the DNA-LipofectAMINE mixture (1.3 mL total vol/dish). Remove all medium from the culture, add the DNA-LipofectAMINE mixture, and incubate the cells for 5.5 h at 37°C in a humidified, 5% CO_2 incubator.

6. Aspirate the transfection mixture, and wash the cells 3 times with 2 mL Opti-MEM I. After aspirating the last wash, add 3.5 mL DMEM supplemented with 6% FBS, and incubate cells for 2–3 days at 34°C in a humidified, 5% CO_2 incubator (see **Note 9**).

7. Scrape transfected cells from **step 6** into the medium using a rubber policeman. Transfer the suspension into a 15-mL conical tube and freeze-thaw 3 times using a dry ice-ethanol bath and a 37°C waterbath.

8. Place the tube containing the cells into a beaker with ice water. Submerge the tip of the sonicator probe approx 0.5 cm into the cell suspension, and sonicate for 20 s with 20% output energy (see **Note 10**).

9. Centrifuge for 10 min at 4°C and 1400g to remove cell debris, and filter the supernatant through a 0.45-μm syringe filter attached to a 20-mL disposable syringe into a new 15-mL conical tube. After taking a sample for titration (see

below), freeze the remaining stock in aliquots in a dry ice-ethanol bath and store at −80°C, or concentrate (*see* **steps 10** and **11**), or purify and concentrate (*see* **steps 12-14**).

10. To concentrate the stock further, transfer the vector solution from **step 9** into a 30-mL centrifuge tube and spin for 2 h at 4°C and 20,000g using a Sorvall SS34 rotor.

11. Resuspend the pellet in a small volume (e.g., 300 µL) of 10% sucrose (w/v; in PBS), divide into aliquots (e.g., 30 µL), freeze aliquots in a dry ice-ethanol bath, and store at −80°C. Before freezing, take a sample of the stock for titration (*see* **Subheading 3.2.3.**).

12. To purify the stock further, prepare a sucrose gradient in a Beckman Ultra-Clear 25 × 89-mm centrifuge tube by layering the following solutions, starting from the bottom of the tube: a) 3 mL 10% sucrose (w/v; in PBS); b) 7 mL 30% sucrose; and c) 7 mL 60% sucrose. Carefully add vector stock from **step 9** (up to 20 mL) on top of the gradient and centrifuge for 2 h at 4°C and 100,000g using a Beckman SW28 rotor.

13. Collect the virus band at the interphase between the 30% and 60% sucrose layers (after aspirating the 10% and 30% sucrose layers on top, the interphase between the 30% and 60% sucrose layers appears as a cloudy band when viewed with a fiberoptic illuminator) and transfer into a Beckman Ultra-Clear 14 × 95-mm centrifuge tube. After mixing in 15 mL PBS, centrifuge for 1 h at 4°C and 100,000g using a Beckman SW40 rotor to pellet vector particles.

14. Resuspend the pellet in a small volume (e.g., 300 µL) of 10% sucrose (w/v; in PBS), divide into aliquots (e.g., 30 µL), freeze the aliquots in a dry ice-ethanol bath, and store at −80°C. Before freezing, take a sample of the stock for titration (*see* **Subheading 3.2.3.**).

3.2.3. Titration of Amplicon Stocks

The titers expressed as TU/mL are relative, and do not accurately reflect numbers of infectious vector particles per mL. Transduction efficiencies and titration of vectors are influenced by a variety of factors, including the cells used for titration, the specific promoter regulating reporter gene expression, the transgene itself, and the sensitivity of a particular detection method.

1. Plate cells (e.g., VERO 76, BHK 21, or 293 cells) at a density of 1.0×10^5/well in a 24-well tissue culture plate in 0.5 mL DMEM containing 10% FBS.

2. One day later, aspirate the medium, wash each well once with PBS, and expose cells to the samples collected from vector stocks from **steps 9, 11**, or **14** above (e.g., 0.1, 1, or 5 µL diluted in 250 µL DMEM containing 2% FBS). Incubate cells at 37°C for 1–2 days.

3. Remove the inoculum, fix cells for 20 min at RT with 250 µL 4% paraformaldehyde (pH 7.0), and wash the fixed cells 3 times with PBS.

4. If the enhanced (E)GFP (Clontech) gene has been utilized as the reporter gene,

expression can be assessed in unfixed cells by counting fluorescent cells using an inverted fluorescence microscope.

5. If the *E. coli lacZ* gene has been utilized as a reporter gene, add 250 μL X-gal staining solution per well of the 24-well tissue culture plate from **step 3**, and incubate at 37°C for 4–12 h (depending on the cell type and the promoter-regulating expression of the transgene); stop the staining reaction by washing the cells 3 times with PBS, and count blue cells using an inverted light microscope.

6. Both reporter genes can also be detected via fixing the cultures and staining with antibody against the gene product using standard immunocytochemical techniques (polyclonal GFP antibody, Clontech).

7. Determine the vector titer as TU/mL by multiplying the number of the transgene-positive cells by the dilution factor.

3.3. Gene Transfer in Culture

HSV/AAV hybrid amplicon vectors have been shown to transduce neurons in primary cultures obtained from a variety of brain regions. Theoretically, any primary cell culture protocol should support successful gene transfer with these vectors; however, this protocol has been optimized for gene transfer to primary ventral mesencephalic cells from embryonic day 15 rat.

1. Place one sterile glass cover slip into each well of a 24-well tray, and add 500 μL PLL to the well; place at 37°C overnight; aspirate PLL and gently rinse each well 3 times with sterile filtered water; allow wells to dry, then add 500 μL serum-containing medium to each well, and place in humidified 37°C/5% CO_2 atmosphere for at least 1 h prior to adding primary cell suspension.

2. Plate 500 μL of primary cell suspension (1×10^6 cells/mL) onto glass cover slips precoated with PLL in each well of 24-well trays.

3. At 1 h in culture, aspirate unattached cells and gently add 1 mL of fresh serum-containing medium.

4. One day later, aspirate medium and gently replace with defined medium.

5. One day later, aspirate the medium, wash each well once with PBS, and expose cells to 1 mL of vectors (diluted stocks to desired MOI).

6. Allow transfection for 90 min at 37°C, then aspirate vector, and rinse wells 3× with defined medium, leaving 1 mL medium in wells and returning to incubator.

7. At desired time points, cultures can be imaged for GFP fluorescence, or fixed for 1 h with 4% paraformaldehyde/4% sucrose in PBS and analyzed via standard immunocytochemical procedures (*see* **steps 4-7** in **Subheading 3.2.3.**).

3.4. Gene Transfer In Vivo

1. Anesthesia: Inject (i.m.) preanesthesia of 1:1 mixture of PromAce and Atropine (0.1 mL of each, mixed, per 300-g rat); 10 min later inject (i.m.) 2:1 mixture of Ketaset and Rompun (0.2 mL Ketaset plus 0.1 mL Rompun, mixed, per 300-g rat) (*see* **Note 11**).

2. Place animals into stereotactic frame, setting incisor bar at 0.0.
3. Clean skull with 70% ethanol, and slowly cut an opening down midline of scalp, starting at the level of the eyes and extending back to the level just behind the ears.
4. Injections into adult rodent striatum have been made into four or seven sites (measured from midline and Bregma), with seven sites showing a larger area of transduction within the striatum and a larger number of nigral neurons transduced (via retrograde transport). For four site injections (1.5 μL/site, total of 6 μL injected), use the following coordinates: AP: +1.5 to –0.3; L: –2.5 to –3.0; V: –4.5 to –5.0. For seven site injections (1 μL/site; total of 7 μL injected) use: AP: +1.6 to –0.5; L: –2.5 to –4.0; V: –4.5 to –5.5. *(46)*.
5. Measure the sites, and drill all holes into the skull (being careful not to drill into brain tissue) before filling needle with vector.
6. Fill Hamilton needle from the bottom with vector just before inserting needle into brain.
7. Slowly lower needle into tissue, measuring ventral site from the dura; injections should be made at a rate of 0.5 μL/min (*see* **Note 13**).
8. Needle should be left in place after each injection for 1 min and then slowly pulled up and immediately cleaned with a sterile gauze pad, ejecting any vector preparation left in the needle.
9. Skull should be cleaned with sterile saline; the animal should then be sutured and placed on a warming pad or under a warming light until recovery from anesthesia.
10. At desired time points, animals are perfused with 4% paraformaldehyde; brains are postfixed for 8 h, cryoprotected in 20% sucrose overnight, and cut coronally at 40 μm on a sliding microtome; sections are analyzed via standard immunohistochemical procedures.

4. Notes

1. HSV/AAV hybrid amplicon plasmids with and without *rep* are available through Cornelf@vetvir.unizh.ch)
2. For packaging of amplicons, cells can be transfected by using the calcium phosphate technique, but lipofection, as described in this protocol, consistently results in higher transfection and packaging efficiencies.
3. The gene for enhanced green fluorescent protein (EGFP) is an ideal reporter gene for HSV/AAV hybrid amplicon vector because of its ability to image packaging and transfection efficiency in living cells during the entire process. Alternatively, the *E. coli lacZ* gene can also be utilized.
4. It is often convenient to grow the preculture (5 mL) during the day and the larger culture overnight. To avoid modifying the cosmid, avoid exceeding the given incubation times or colony-purifying the clones.
5. There are two types of replication-competent, packaging-defective HSV-1 genomes available: a) cosmid sets, such as C6Δa48Δa; and b) bacterial artificial

chromosomes (BAC), such as fHSVΔpac *(39)*. Prepare fHSVΔpac DNA from cultures grown overnight at 37°C in LB medium containing 12.5 μg/mL chloramphenicol. Extract DNA by alkaline lysis and purify over Qiagen-tip 500 columns and cesium chloride equilibrium centrifugation.

6. Treat gel with care; 0.4% gels are very delicate.

7. This step is not necessary for fHSVΔpac; digestion with *Pac*I releases the HSV-1 insert (35–40 kb) from the cosmid backbone (approx 7 kb) because of *Pac*1 restriction sites flanking the HSV-1 inserts in the cosmid. This facilitates homologous recombination following transfection.

8. The subsequent transfection into mammalian cells requires that all further procedures be carried out under sterile conditions.

9. Two days after transfection, at least 50% of the cells should show cytopathic effects.

10. Sonicating the suspension using a probe sonicator further disrupts cell membranes and liberates cell-associated vector particles. Cellular debris can be toxic to cells, however.

11. The HSV/AAV hybrid vector reported in Costantini et al. *(36)* was sequenced after publication, which revealed an error in the pHyRGN amplicon, since the transgene cassette was inadvertently inserted into the AAV *rep* gene at position 724 (bp). Furthermore, there is a partial loss of the AAV ITR, in both amplicons. Sequencing was carried out by Drs. Andreas Jacobs, Sam Wang, Cornel Fraefel, and Paul Allen.

12. Various ages of rodents (from neonates through adults) and several brain regions have shown positive gene transfer with hybrid amplicons; the procedure listed below is optimal for intrastriatal injections into adult male Sprague-Dawley rats.

13. To decrease the likelihood of clogging the needle, fill the needle just prior to injection, and then eject a small drop before inserting needle into brain.

Acknowledgments

These studies were supported by PHS grant NINDS P50NS39793, a NINDS Morris K. Udall Parkinson's Disease Center of Excellence (to OI), McLean Hospital, and PHS grant NINDS NS2429 (to X.O.B.), Massachusetts General Hospital and the Swiss National Science Foundation (to C. F.).

References

1. Lam, P. and Breakefield, X. O. (2000) Hybrid vector designs to conrol the delivery, fate and expression of transgenes. *J. Gene Med.*, **2**, 395–408.

2. Costantini, L. C., Bakowska, J. C., Breakefield, X. O., and Isacson, O. (2000) Gene therapy in the CNS. *Gene Ther.* **7**, 93–109.

3. Lawrence, M. S., Ho, D. Y., Sun, G. H., Steinberg, G. K., and Sapolsky, R. M. (1996) Overexpression of Bcl-2 with herpes simplex virus vectors protects CNS neurons against neurological insults *in vitro* and *in vivo*. *J. Neurosci.* **16**, 486–496.

4. Bowers, W., Howard, D., and Federoff, H. (1997) Gene therapeutic strategies for neuroprotection: implications for Parkinson's disease. *Exp. Neurol.* **144,** 58–68.
5. Davidson, B. L. and Bohn, M. C. (1997) Recombinant adenovirus: a gene transfer vector for study and treatment of CNS diseases. *Exp. Neurol.* **144,** 125–130.
6. Verma, I. and Somia, N. (1997) Gene therapy: promises, problems, and prospects. *Nature* **389,** 239–242.
7. Mandel, R., Rendahl, K., Spratt, S., et al. (1998) Characterization of intrastriatal recombinant adeno-associated virus-mediated gene transfer of human tyrosine hydroxylase and human GTP-cyclohydrolase I in a rat model of Parkinson's disease. *J. Neurosci.* **18,** 4271–4284.
8. Miyoshi, H., Blömer, U., Takahashi, M., Gage, F. H., and Verma, I. M. (1998) Development of a self-inactivating lentivirus vector. *J. Virol.* **72,** 8150–8157.
9. Oligino, T., Poliani, P. L., Wang, Y., et al. (1998) Drug inducible transgene expression in brain using a herpes simplex virus vector. *Gene Ther.* **5,** 491–496.
10. Rendahl, K., Leff, S., Otten, G., et al. (1998) Regulation of gene expression in vivo following transduction by two separate rAAV vectors. *Nat. Biotechnol.* **16,** 757–761.
11. Xiao, X., Li, J., and Samulski, R. J. (1998) Production of high-titer recombinant adeno-associated virus vectors in the absence of helper adenovirus. *J. Virol.* **72,** 2224–2232.
12. Fraefel, C., Jacoby, D. R., and Breakfield, X. O. (2000) Recent developments on herpes simplex virus type 1-based amplicon vector systems. *Adv. Virus Res.* **55,** 425–452.
13. Fraefel, C., Jacoby, D. R., Lage, C., Hilderbrand, H., et al. (1997) Gene transfer into hepatocytes mediated by helper virus-free HSV/AAV hybrid vectors. *Mol. Med.* **3,** 813–825.
14. Johnston, K., Jacoby, D., Pechan, P., et al. (1997) HSV/AAV hybrid amplicon vectors extend transgene expression in human glioma cells. *Hum. Gene Ther.* **8,** 359–370.
15. Ward, P. L. and Roizman, B. (1994) Herpes simplex genes: the blueprint of a successful human pathogen. *Trends Genet.* **10,** 267–274.
16. Roizman, B. and Sears, A. E. (1996) Herpes simplex viruses and their replication, in *Fields Virology* (Fields, B. N., Knipe, D. M., and Howley, P. M., eds.), Lippincott-Raven, Philadelphia, pp. 2231–2296.
17. Glorioso, J. C., Bender, M. A., Goins, W. F., DeLuca, N., and Fink, D. J. (1995) Herpes simplex virus as a gene-delivery vector for the central nervous system, in *Viral Vectors*, Kaplitt, M. G. and Loewy, A. D., eds., Academic, San Diego, pp. 1–23.
18. Spaete, R. and Frenkel, N. (1982) The herpes virus amplicon: a new eucaryotic defective-virus cloning-amplifying vector. *Cell* **30,** 295–304.
19. Fraefel, C., Song, S., Lim, F., et al. (1996) Helper virus-free transfer of herpes simplex virus type 1 plasmid vectors into neural cells. *J. Virol.* **70,** 7190–7197.
20. Kwong, A. D. and Frenkel, N. (1985) Efficient expression of chimeric chicken ovalbumin gene amplified with defective virus genomes. *Virology* **142,** 421–425.

21. Sena-Esteves, M., Saeki, Y., Fraefel, C., and Breakefield, X. O. (2000) HSV-1 amplicon vectors—simplicity and versatility. *Mol. Ther.* **2**, 9–15.

22. Xiao, X., McCown, T. J., Li, J., et al. (1997) Adeno-associated virus (AAV) vector antisense gene transfer *in vivo* decreases GABA(A) alpha1 containing receptors and increases inferior collicular seizure sensitivity. *Brain Res.* **756**, 76–83.

23. Yang, G. Y., Zhao, Y. J., Davidson, B. L., and Betz, A. L. (1997) Overexpression of interleukin-1 receptor antagonist in the mouse brain reduces ischemic brain injury. *Brain Res.* **751**, 181–188.

24. Duan, D., Sharma, P., Yang, J., et al. (1998) Circular intermediates of recombinant adeno-associated virus have defined structural characteristics responsible for long-term episomal persistence in muscle tissue. *J. Virol.* **72**, 8568–8577.

25. Wu, P., Phillips, M. I., Bui, J., and Terwilliger, E. F. (1998) Adeno-associated virus vector-mediated transgene integration into neurons and other nondividing cell targets. *J. Virol.* **72**, 5919–5926.

26. Muzyczka, N. (1992) Use of adeno-associated virus as a general transduction vector for mammalian cells. *Curr. Top. Microbiol./Immunol.* **158**, 97–129.

27. Weitzman, M. D., Kyostio, S. R., Kotin, R. M., and Owens, R. A. (1994) Adeno-associated virus (AAV) Rep proteins mediate complex formation between AAV DNA and its integration site in human DNA. *Proc. Nat.l Acad. Sci. USA* **91**, 5808–5812.

28. Balague, C., Kalla, M., and Zhang, W. W. (1997) Adeno-associated virus Rep78 protein and terminal repeats enhance integration of DNA sequences into the cellular genome. *J. Virol.* **71**, 3299–3306.

29. Walker, S. L., Wonderling, R. S., and Owens, R. A. (1997) Mutational analysis of the adeno-associated virus Rep68 protein: identification of critical residues necessary for site-specific endonuclease activity. *J. Virol.* **71**, 2722–2730.

30. Kotin, R. M., Linden, R. M., and Berns, K. I. (1992) Characterization of a preferred site on human chromosome 19q for integration of adeno-associated virus DNA by non-homologous recombination. *EMBO J.* **11**, 5071–5078.

31. Yang, C. C., Xiao, X., Zhu, X., et al. (1997) Cellular recombination pathways and viral terminal repeat hairpin structures are sufficient for adeno-associated virus integration *in vivo* and *in vitro*. *J. Virol.* **71**, 9231–9247.

32. Shelling, A. N. and Smith, M. G. (1994) Targeted integration of transfected and infected adeno-associated virus vectors containing the neomycin resistance gene. *Gene Ther.* **1**, 165–169.

33. Kearns, W. G., Afione, S. A., Fulmer, S. B., et al. (1996) Recombinant adeno-associated virus (AAV-CFTR) vectors do not integrate in a site-specific fashion in an immortalized epithelial cell line. *Gene Ther.* **3**, 748–755.

34. Sodeik, B., Ebersold, M., and Helenius, A. (1997) Microtubule-mediated transport of incoming herpes simplex virus 1 capsids to the nucleus. *J. Cell Biol.* **136**, 1007–1021.

35. Bearer, E. L., Breakefield, X. O., Schuback, D., Reese, T. S., and LaVail, J. H. (2000) Retrograde axonal transport of herpes simplex virus: evidence for a single mechanism and a role for tegument. *Proc. Natl. Acad. Sci. USA* **97**, 8146–8150.

36. Costantini, L. C., Jacoby, D. R., Wang, S., Fraefel, C., Breakefield, X. O. and Isacson, O. (1999) Gene transfer to the nigrostriatal system by hybrid herpes simplex virus/adeno-associated virus amplicon vectors [published erratum appears in *Hum. Gene Ther.* 2000; **10,** 11: 981]. *Hum. Gene The.r* **11,** 2481–94.

37. Wang, S., Fraefel, C., and Breakefield, X. O. HSV-1 amplicon vectors. *Methods in Enzymol.*, Phillips, I., ed. in press.

38. Sena-Esteves, M., Hamp, J. A., Camp, S., and Breakefield, X. O. Generation of stable retrovirus packaging cell lines after transduction with HSV/hybrid amplicons, in preparation.

39. Saeki, Y., Ichikawa, T., Saeki, A., et al. (1998) Herpes simplex virus type 1 DNA amplified as bacterial artificial chromosome in *Escherichia coli*: rescue of replication-competent virus progeny and packaging of amplicon vectors. *Hum. Gene Ther.* **9,** 2787–2794.

40. Stavropoulos, T. A. and Strathdee, C. A. (1998) An enhanced packaging system for helper-dependent herpes simplex virus vectors. *J. Virol.* **72,** 7137–7143.

41. Saeki, Y., Fraefel, C., Ichikawa, T., Breakefield, X. O., and Chiocca, E. A. (2001) Elimination of helper virus regeneration in packaged HSV-1 amplicon vector preparations by utilizing an ICP27-deleted and oversized HSV-1 helper DNA in a bacterial artificial chromosomal. *Mol. Ther.* **3,** 591–601.

42. Hampl, J., Brown., A., Rainov, N. and Breakefield, X. O. (2000) Methods for gene delivery to neural tissue, in *Methods in Genomic Neuroscience* (China, H. R. and Moldin, S. O., eds.), CRC Press, Boca Raton, in press.

43. Geller, A. I. and Breakefield, X. O. (1988) A defective HSV-1 vector expresses *Escherichia coli* beta-galactosidase in cultured peripheral neurons. *Science* **241,** 1667–1669.

44. Laughlin, C. A., Tratschin, J. D., Coon, H., and Carter, B. J. (1983) Cloning of infectious adeno-associated virus genomes in bacterial plasmids. *Gene* **23,** 65–73.

45. Smith, I. L., Hardwicke, M. A., and Sandri-Goldin, R. M. (1992) Evidence that the herpes simplex virus immediate early protein ICP27 acts post-transcriptionally during infection to regulate gene expression. *Virology* **186,** 74–86.

46. Paxinos, G., Watson, C., eds. (1986) *The Rat Brain in Stereotaxic Coordinates.* Academic, San Diego.

32

Development of Replication-Defective Herpes Simplex Virus Vectors

William F. Goins, David M. Krisky, Darren P. Wolfe, David J. Fink, and Joseph C. Glorioso

1. Introduction

A greater understanding of the molecular, biochemical, and genetic factors involved in the progression of a specific disease state has led to the development of genetic therapies using direct gene transfer to ameliorate the disease condition or correct a genetic defect *in situ*. Effective gene therapy approaches require delivery strategies and vehicles that 1) efficiently deliver the therapeutic gene(s) to a sufficient number of dividing or nondividing cells to achieve the desired therapeutic effect; 2) persist long term within the cell without disturbing host cell functions; and 3) can regulate the level and duration of therapeutic gene expression for diseases that may either require high-level transient transgene expression or continuous low-level synthesis of the therapeutic product. Numerous viral and nonviral vectors have been employed to treat a variety of genetic and acquired diseases. Each vector system has its own particular advantages and disadvantages that will suit it to a specific therapeutic application.

Herpes simplex virus type 1 (HSV-1) possesses a number of practical advantages for in vivo gene therapy to the nervous system and other tissues. HSV-1 can infect a wide variety of both dividing and postmitotic cell types and can be propagated to high titers on complementing cell lines. It also has a large genome size (152 kb), which allows the virus to accommodate large *(1)* or numerous therapeutic gene sequences (>35 kb) *(2)*. In addition, the natural biology of HSV-1 infection involves long-term persistence of the viral genome in a latent, nonintegrated state in neuronal cell nuclei *(3–5)* and other postmitotic cell types in the absence of viral protein synthesis, genome integration, or interference

From: *Methods in Molecular Medicine, Vol. 69, Gene Therapy Protocols, 2nd Ed.*
Edited by: J. R. Morgan © Humana Press Inc., Totowa, NJ

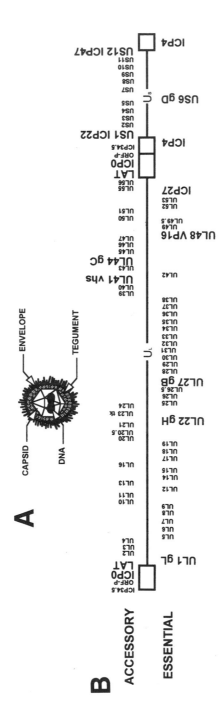

Fig. 1. HSV-1 virion structure and genome organization. (**A**) Schematic illustration of the HSV virion, showing the capsid, tegument, and glycoprotein-containing lipid envelope. (**B**) Schematic representation of the HSV genome, showing the unique long (U_L) and unique short (U_S) segments, each bounded by inverted repeat (IR) elements. The locations of the essential genes, which are required for viral replication in vitro, and the nonessential or accessory genes, which may be deleted without affecting replication in vitro, are indicated. The IE gene and other important gene loci are in capitals.

with host cell biology. Since the virus does not disrupt normal host cell biology or express viral antigens during latency, cells harboring latent virus will not be attacked by the host's immune system and should allow persistence of the viral genome for the lifetime of the host, obviating the need for repeat dosing of the virus vector. During latency the viral genome is transcriptionally silent except for expression of a set of viral latency-associated transcripts (LATs) *(6–11)*. Since the LATs are not required for establishment or maintenance of this latent state *(12–18)*, it should be possible to delete these genes and replace them with the desired therapeutic gene to drive expression of this gene product from an otherwise quiescent genome using the latency and neuronal cell-specific promoter complex resident within the vector genome.

1.1. The HSV-1 Life Cycle

The HSV-1 particle (**Fig. 1A**) is composed of 1) an envelope containing 10 glycoproteins that allow the virus to attach to and infect cells by fusion of the virus envelope with the cell surface membrane (for review, *see* **ref.** *19–21*); 2) a tegument matrix inside the envelope composed of viral structural components involved in shutoff of host protein synthesis *(22–25)*, activation of immediate early viral gene expression, and assembly functions *(26–31)*; 3) a regular icosahedral-shaped nucleocapsid *(32)*; and 4) a linear double-stranded DNA genome containing approximately 85 open reading frames *(33–36)*. The 152-kb viral genome is segmented (**Fig. 1B**), with each of its long (U_L) and short (U_S) unique segments flanked by inverted repeats (IRs). The viral functions have been categorized as to whether they are essential for virus replication in cell culture or are accessory (nonessential) functions that contribute to virus replication and spread in vivo. The viral genes are arranged in such a manner that many of the essential functions or accessory genes tend to be clustered within the genome. Additionally, since very few HSV-1 genes are spliced and the virus possesses a highly evolved recombination system, it is relatively easy to manipulate and engineer HSV-1 for purposes of gene transfer by deleting individual genes or blocks of genes that may play a role in vector toxicity.

During normal infection, HSV-1 infects the skin following direct contact and undergoes a productive (lytic) infection in skin fibroblasts and epithelial cells (**Fig. 2A**). The initial binding of virus to the host cell involves the interaction of glycoproteins B and C (gB and gC) with cell surface heparan sulfate moieties *(37–39)*. This is followed by the binding of gD to specific cell surface receptors HveA *(40–45)* and HveC *(45–49)*, which are members of the tumor necrosis factor-α/nerve growth factor (TNF-α/NGF) receptor superfamily and the immunoglobulin superfamily, respectively. Binding of gD to HveA/HveC initiates the fusion event between the host cell surface membrane and the viral envelope *(50–53)*, a process that also requires the functions of gB *(54)* and the

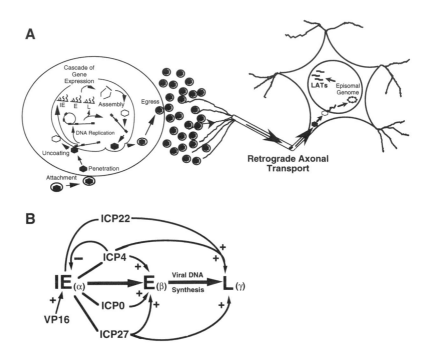

Fig. 2. HSV-1 life cycle. **(A)** Schematic diagram of the HSV-1 lytic replication cycle in vivo. HSV-1 binds to and enters mucosal or epithelial cells and proceeds through the lytic replication pathway. Following uncoating, the virion is transported to the nucleus, where the linear viral genome is injected through a nuclear pore. The lytic cascade of viral gene expression ensues, with subsequent replication of the viral genome. The resulting progeny virus particles can encounter and fuse with the cell surface membrane of peripheral nerve termini, which innervate the site of primary infection. The viral nucleocapsid then travels via retrograde axonal transport to the neuronal cell body, where virus can either proceed through the highly regulated cascade of lytic gene expression, or enter latency, during which the viral lytic gene program is interrupted and the latency-associated transcripts (LATs) are the sole viral RNAs expressed. In response to a wide variety of stimuli, the virus is capable of reactivating from the latent state, entering the lytic portion of the HSV life cycle, at which time progeny virions can either be transported back to the site of the primary infection or the virus may enter the central nervous system. **(B)** Temporal cascade of HSV-1 gene expression detailing the roles of HSV-1 immediate early (IE) gene transactivators ICP4, ICP27, ICP22, and ICP0 and the IE promoter stimulatory molecule VP16. The IE (or α) genes are expressed immediately upon infection in the absence of *de novo* protein synthesis. The VP16 (αTIF, Vmw65) virus tegument protein interacts with the cellular factor Oct-1 to regulate expression of the IE genes positively by binding to their promoters. The IE gene products ICP4, ICP27, and ICP0 are responsible for activating early (E or β) genes. Following viral DNA replication, the ICP4, ICP22, and ICP27 IE polypeptides regulate the expression of the late (L or γ) genes.

gH/gL complex *(55–57)*. Following uncoating of the particle, cellular molecular motors transport the viral nucleocapsid to the nuclear membrane, where the viral linear dsDNA genome is injected through a pore into the host cell nucleus.

Once within the nucleus, the lytic replication cycle of the virus takes place *(58,59)*, with the viral genes expressed in a highly regulated cascade *(60,61)* of coordinated gene expression consisting of three stages: immediate early (IE or α), early (E or β), and late (L or γ) (**Fig. 2B**). Three of the viral IE genes are transcriptional activators that induce expression of E and L genes *(62–69)*. Early gene functions participate in viral DNA replication, which must proceed in order for late gene expression to occur *(70,71)*. The late genes encode largely structural products comprising the nucleocapsid, tegument, and viral envelope glycoproteins. Viral particles are assembled within the nucleus and bud from the nuclear membrane; particle maturation proceeds during migration through the Golgi apparatus, followed by egress from the cell. Following focal replication of the virus in these permissive cell types, the virus invades the nervous system by directly infecting axon terminals of local sensory neurons of the peripheral nervous system (PNS) *(72,73)*. The viral nucleocapsids are transported in a retrograde manner back to the neuronal cell body, where the virus can either replicate or enter latency. During latency the linear viral genome circularizes, becomes methylated, and forms a higher order nucleosomal structure *(3–5,74)*. At this point the lytic genes become inactive and the LATs are readily detected by *in situ* hybridization using LAT-specific riboprobes *(8–11,17)*. The virus can remain within the latent state for the lifetime of the individual, or it can be induced to reactivate from latency by a variety of stimuli, resulting in the resumption of the lytic cycle and the subsequent synthesis of progeny virions, which may traverse the axon by anterograde transport, establishing an active infection at or near the site of primary infection.

1.2. Engineering HSV-1 Vectors

The optimal HSV vector should be 1) safe and completely devoid of replication-competent virus; 2) noncytotoxic; 3) incapable of affecting normal host cell biology; 4) able to persist in the neuronal cell body in a nonintegrated state; and 5) capable of expressing the therapeutic gene(s) to appropriate levels at the proper time(s). The three major considerations in the design of HSV-1 vectors concern the elimination of the cytotoxic properties of the virus, the ability to target binding and entry of the virus to the specific cells or tissues of interest, and the development of promoter systems for proper expression of the therapeutic gene. Considerable effort has been expended in addressing these issues of vector design. In this work we will concentrate on the engineering of HSV-1 vectors deleted for essential gene functions that display reduced toxicity following infection.

1.2.1. Solving the Problem of HSV-1 Cytotoxicity

Since UV-irradiated virus displays substantially reduced cytotoxicity in vitro *(75,76)* and disruption of viral IE gene expression by interferon also reduces toxicity, it is presumed that the cytotoxicity of HSV-based vectors results from the expression of HSV-1 gene products. The fundamental approach to designing HSV-1 vectors with reduced toxicity is to remove the essential IE genes (**Fig. 1B**) of the virus, as well as several nonessential genes whose products interfere with host cell metabolism and are part of the virion (tegument) structure. Deletion of the two essential IE genes that encode the infected cell proteins 4 (ICP4) and 27 (ICP27) blocks early and late gene expression *(62,77,78)* (**Fig. 2B**). These deletions require that the missing functions be supplied *in trans*, using a complementing cell line *(62,77,79)* to propagate virus. To ensure that recombination does not occur between the defective virus and the viral sequences present within the complementing cell line during propagation, it is essential that these sequences not share homology with sequences present within the viral genome and that the deletion of the IE genes from the virus exceed the limits of the complementing sequences. Since many of these viral IE genes are toxic to cells, expression of the complementing genes must be inducible upon infection with the defective virus. This is achieved through the use of HSV-1 IE promoters to drive expression of these toxic transactivating genes from the cell, since these promoters have been shown to respond to the HSV-1 transactivator VP16 (Vmw65 or α-TIF) *(27–31)*, a virion tegument component that accompanies the viral DNA molecule into the nucleus of the infected cell. VP16 recognizes a consensus sequence (octomer-TAATGARAT) located at various sites in all HSV-1 IE promoters; together with the cellular transcription factors octomer binding protein one (Oct-1) and HCF (also termed C1, VCAF-1, and CFF) *(80–84)*, VP16 transactivates the IE gene promoters *(85–89)*. In the absence of virus infection, the complementing IE genes in the cell chromosome are silent. However, upon infection with the replication-defective mutant, the VP16/Oct-1/HCF complex transactivates the IE promoters upstream of the complementing viral sequences in the cell line, thereby inducing expression of the necessary products for propagation of the deletion virus.

A third IE gene of interest is ICP0, which is both cytotoxic and capable of promiscuous transactivation of a variety of cellular genes *(64,90–92)*; it also enhances the level of expression of other viral genes *(90)*. ICP0 appears to collaborate cooperatively with ICP4, for example, to increase the activity of this key viral function *(93,94)*, although ICP0 is not a promoter binding protein and thus appears to stimulate an event prior to direct promoter activation *(95)*. Although ICP0 is a nonessential viral function, deletion of ICP0 results in decreased viral titers *(67,96)*, and propagation of high-titer stocks of virus deleted for ICP0 in conjunction with the essential IE genes will require the production of a cell line capable of complementing ICP0 as well as ICP4 and ICP27. Generation of such a line has been extremely difficult, since even low-level synthesis of ICP0 is toxic to the host cell *(97)*. The two remaining IE genes, ICP22 (which affects the phosphorylation of RNA polymerase II) *(98)* and ICP47 (which affects the processing of MHC class I antigens) *(99–101)*, may also need to be deleted depending on the specific therapeutic application.

In addition to the viral IE gene functions, an additional gene remains a target of interest for removal to reduce toxicity. The infecting virus carries in with the particle a tegument component that has a virion-associated host shutoff (vhs) activity *(22,23)*. vhs appears to interfere nondiscriminately with mRNA stability *(24)*. Removal of the UL41 (vhs) gene does not affect viral replication *(25)* but enhances the health of the cell upon infection with nonreplicating viral mutants *(102)*.

Deletion of a combination of the four IE regulatory genes (ICP4, ICP22, ICP27, and ICP0) along with UL41 (vhs) has yielded vectors that display diminished toxicity for a variety of cell types in culture *(103–106)*. Recombinants deleted for both ICP4 and ICP27 *(103–106)* were less toxic than the ICP4/ICP27 single IE gene deleted viruses d120 *(77)* or 5dl1.2 *(107)*. The removal of three of the IE gene (ICP4, ICP22, and ICP27) dramatically reduced cytotoxicity *(103,106)* and led to increased duration of vector-mediated transgene expression in neurons in culture *(103)*. Recombinants that failed to express any of the five IE genes including ICP0 essentially shut down viral gene expression upon infection of noncomplementing cells, rendering the virus safe and noncytotoxic *(105)*. However, transgene expression from these recombinants was transient or undetectable in many cell types *(105)*, owing to deletion of the ICP0 gene *(97,105)*. Thus, it may prove necessary to retain ICP0 to achieve expression of the therapeutic gene product. For some specific therapeutic applications, the retention of expression of ICP47, which has recently been shown to downregulate MHC class I antigen expression *(99–101)*, should provide an additional level of protection from immune surveillance at the initiation of infection in vivo.

1.2.2. Construction of Mutant Viruses Deleted for IE and Other Toxic Functions

The first generation of replication-defective mutant viruses consisted of mutants deleted for the essential ICP4 gene. One of these recombinants, designated d120, can be grown on a complementing cell line (E5) that expresses ICP4 *in trans* upon infection with the ICP4 deletion mutant virus *(77)*. We have subsequently introduced the β-galactosidase (*lac*Z) reporter gene, under control of the strong human cytomegalovirus (HCMV) immediate early gene promoter into the thymidine kinase gene locus of the genome of this replication-defective mutant, in order to identify rapidly cells and tissues infected with this vector to evaluate the potential uses of such vectors. We have shown that the transgene was transiently expressed in a variety of cell types in culture and in various tissues in vivo *(108,109)*. Long-term expression of the transgene both in culture and in some cells in vivo may be affected by the virus-induced toxicity, leading to death of the cells in culture and/or immune recognition and clearance in vivo. Therefore, it was clear that additional IE genes must be removed to reduce viral toxicity further.

To generate second-generation replication-defective viruses deleted for multiple IE essential gene functions, it was necessary to construct a cell line to complement the essential IE gene functions. A cell line was constructed to complement both ICP4 and ICP27 by transfecting Vero cells with a plasmid containing the HSV-1 sequences coding for ICP4 and ICP27 along with a neomycin expression (SV2-neo) cassette for rapid selection of individual clones. To eliminate the chance of homologous recombination and rescue of the mutant viruses during propagation in the complementing line, the plasmid containing the coding regions for ICP4 and ICP27 was engineered to avoid overlap of these sequences with the deletions present within the virus. One clone, designated 7b, was isolated after multiple rounds of drug selection; this clone was able to complement the growth of the IE deletion mutants d120 (ICP4⁻) *(77)* and 5dl1.2 (ICP27⁻) *(107)*.

A second-generation double-mutant virus deleted for both ICP4 and ICP22 was engineered using the 7b cell line (**Fig. 3**). This recombinant, DHZ.1, was engineered by recombining the linearized plasmid pB5, containing the HCMV IE promoter-*lacZ* BGHpA cassette surrounded by ICP22 flanking sequences, into the ICP22 gene locus of the ICP4 replication-defective mutant d120. Positive recombinants were propagated and isolated on the 7b complementing cell line owing to the production of blue plaques following X-gal staining. The DHZ.1 recombinant fails to replicate or produce the blue plaque phenotype on normal noncomplementing (Vero) cells. To engineer a third-generation triple IE gene-deleted vector, we simply crossed DHZ.1 (ICP4⁻/ICP22⁻) with the ICP27 mutant 5dl1.2 *(107)*. Recombinants deleted for the three IE genes (**Fig. 3**) were selected for the production of blue plaques on the 7b double IE complementing cell line compared with viruses that would only plaque on either the ICP4-expressing line (E5) or the ICP27-expressing line (N23). The triple IE gene-deleted vector THZ.1 was employed in a genetic cross with the UL41 gene mutant ΔSma *(110)* to engineer a vector deleted for ICP4, ICP27, and ICP22 and the vhs gene product (**Fig. 3**). The structure of each recombinant vector was confirmed by Southern blot analysis.

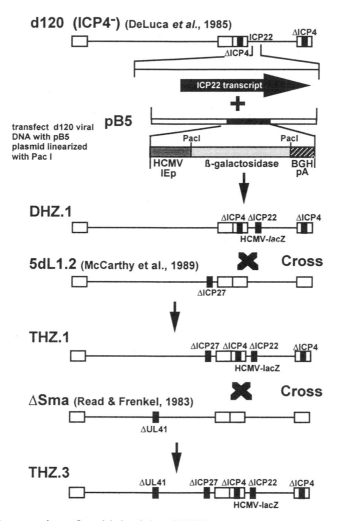

Fig. 3. Construction of multiply deleted HSV vectors. The construction of multiple gene deletion mutants can be accomplished by either standard homologous recombination or by genetic cross. The ICP4⁻/ICP22⁻ double IE gene deletion mutant (DHZ.1) was constructed by cotransfection of d120 *(77)* viral DNA with the plasmid pB5 (in which ICP22 sequences were replaced with an HCMV IEp-*lacZ* BGHpA cassette) in the 7b (ICP4/ICP27) complementing cell line. The desired ICP4⁻/ICP22⁻ recombinants were identified by the production of a blue plaque phenotype on 7b cells following X-gal staining, and their genome structure was confirmed by Southern blot analysis. The triple IE gene deletion mutant THZ.1 was produced by crossing DHZ.1 with the single ICP27 deletion mutant 5dl1.2 *(107)*. Progeny viruses that were *lacZ* positive and replicated on 7b (ICP4/ICP27) cells but not E5 (ICP4) or N23 (ICP27) cells were selected and confirmed by Southern blot analysis. The mutant THZ.3 was engineered by crossing THZ.1 with the UL41 mutant ΔSma *(25)*.

The DHZ.1 double mutant and other double mutants deleted for different combinations of IE genes were examined for reduced toxicity in cell culture compared with both the first-generation single IE gene-deleted (ICP4⁻) paren-tal virus and the third-generation vectors deleted for three IE genes (THZ.1 and THZ.3). In this particular study, Vero cells were infected with the various viruses at a multiplicity of infection (MOI) of 1.0, and the number of viable cells was determined at 24 h post infection using trypan blue exclusion. In addition, we determined the number of apoptotic cells in culture using the FragEL (Roche, Indianapolis, IN) DNA fragmentation kit. This specific MOI was chosen since d120 uniformly kills cells at an MOI of 1 or greater. As shown in **Table 1**, the double DHZ.1 mutant is less cytotoxic than the SHZ.1 single IE gene mutant for Vero cells at an MOI of 1.0, although some toxicity still remains. Double IE gene deletion mutants lacking ICP22 and either ICP4 (DHZ.1) or ICP27 (DHZ.3) were less toxic than the ICP4/ICP27 (DHZ.4) double gene mutant. In addition, removal of three IE toxic genes (THZ.1) as well as other toxic gene targets such as vhs (THZ.3) further reduced HSV vec-tor toxicity to the point that it eventually approached that observed with mock-infected cells (**Table 1**). We have obtained similar results with these recombinant vectors both in various cell types in culture and in vivo *(103)*.

The cassette that was recombined into the ICP22 locus is unique in that it contains the *lacZ* gene construct flanked by two 8-bp recognition sites for restriction endonuclease *Pac*I that are not present at any other site in the viral genome. Thus, the expression cassette can be easily removed by digestion with the *Pac*I restriction enzyme followed by religation of the genome, yielding a recombinant with a single *Pac*I site that produces clear plaques on the 7b complementing cell line *(111)*. To introduce a new gene cassette efficiently into the ICP22 locus of this recombinant, the recombinant genome is cleaved with *Pac*I and used in marker transfer transfection assays with plasmid-con-taining sequences that span the *Pac*I site in ICP22 *(111)*. The desired recombi-nant will lack the HCMV-*lacZ* expression cassette and thus can be easily isolated as a clear plaque on a background of blue plaques produced by the parental virus. The recombination frequency obtained using this approach is 10-fold greater than that seen in standard marker rescue experiments *(111)* since the background generated by the parental viruses is greatly reduced owing to digestion of the parental viral DNA with *Pac*I. We have employed this method to generate a number of recombinants including a vector THD (**Fig. 4**) that expresses the full-length dystrophin gene product for gene therapy appli-cations to treat Duchenne muscular dystrophy *(1)*.

Table 1
Cytotoxicity of Recombinant HSV Vectors[a]

Vector	% Survival	% Apoptotic
Mock	100	0.3
SHZ.1(ICP4⁻)	52	15.6
DHZ.1(ICP4⁻/ICP22⁻)	80	3.6
DHZ.3(ICP22⁻/ICP27⁻)	85	2.5
DHZ.4(ICP4⁻/ICP27⁻)	75	6.2
THZ.1(ICP4⁻/ICP22⁻/ICP27⁻)	>100	0.6
THZ.3(ICP4⁻/ICP22⁻/ICP27⁻/UL41⁻)	>100	0.1

[a]Vero cells were either mock-infected or infected with the single IE gene mutant SHZ.1 (ICP4⁻), the double IE gene deletion mutants (DHZ.1, DHZ.3, and DHZ.4), or the triple IE gene (ICP4⁻/ICP22⁻/ICP27⁻) deletion mutants (THZ.1 and THZ.3) at an MOI = 1.0. The numbers of viable cells were counted at 1 day post infection by trypan blue exclusion. The number of cells in the culture undergoing apoptosis was determined using the FragEL detection kit (Roche). Subsequent deletion of the additional cytotoxic IE gene led to increased cell survival and a decrease in Vero cell apoptosis compared with the single IE gene deletion virus (SHZ.1). Deletion of ICP22 from the ICP4 backbone had the greatest effect on vector-induced toxicity. Deletion of UL41 from the triple IE gene-deleted virus backbone (THZ.3) only marginally affected toxicity.

2. Materials

1. Vero (African green monkey kidney, ATCC, cat. no. CCL81) cells. Other permissive cells such as HELs or BHKs, or complementing cell lines are required to propagate HSV-1 accessory or essential gene deletion viruses.
2. MEM/10% FCS: Eagle's modified essential medium (MEM) supplemented with nonessential amino acids, 100 U/mL penicillin G, 100 µg/mL streptomycin sulfate, 2 mM glutamine, and 10% fetal calf serum (FCS).
3. TBS, pH 7.5: 50 mM Tris-HCl, pH7.5, 150 mM NaCl, 1 mM EDTA.
4. Lysis buffer: 10 mM Tris-HCl, pH 8.0, 10 mM EDTA.
5. TE: 10 mM Tris-HCl, pH 8.0, 1 mM EDTA.
6. TE equilibrated phenol/chloroform/isoamyl alcohol (25:24:1 v/v).
7. 2× HBS: 20 mM HEPES, 135 mM NaCl, 5 mM KCl, 5.5 mM dextrose, 0.7 mM Na_2HPO_4, pH 7.05. Accurate pH of this solution is critical.
8. X-gal staining solution that contains the chromogenic substrate X-gal at a final concentration of 300 µg/mL in TBS containing 14 mM $K_4Fe(CN)_6$, 14 mM $K_3Fe(CN)_6$. X-gal is highly insoluble and must be dissolved in dimethyl formamide (DMF) prior to addition to the staining solution.
9. 1.0% methylcellulose overlay: Add 25 g methylcellulose to 100 mL phosphate-buffered saline (PBS), pH 7.5, in a 500-mL sterile bottle containing a stir bar. Autoclave the bottle on the liquids cycle for 45 min. After the solution cools, add 350 mL of MEM supplemented with nonessential amino acids, 100 U/mL penicillin G, 100 µg/mL streptomycin sulfate, and 2 mM glutamine. Mix well and

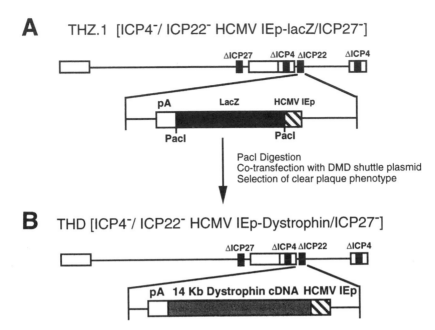

Fig. 4. Use of the *Pac*I method to generate a replication-defective HSV vector expressing the full-length dystrophin cDNA (DMD). The 14-kb dystrophin cDNA was cloned downstream from the HCMV IE promoter in the ICP22 shuttle plasmid that contains adequate HSV flanking DNA sequence, to allow efficient homologous recombination at the ICP22 gene locus of the vector. The expression cassette can then be introduced into the ICP22 locus of the THZ.1 vector by *Pac*I digestion of THZ.1 viral DNA and the subsequent cotransfection of the *Pac*I-digested THZ.1 DNA along with the shuttle plasmid containing the DMD expression cassette. **(A)** THZ.1 contains a *lacZ* expression cassette at the ICP22 locus of a vector deleted for ICP4, ICP27, and ICP22 and produces blue plaques following X-gal staining. **(B)** The new recombinant, THD, can be isolated following homologous recombination of the shuttle plasmid into the *Pac*I-digested THZ.1 genome by the identification of clear plaques. Following three rounds of limiting dilution analysis, the structure of THD can then be confirmed by Southern blot analysis *(1)*.

place the bottle on a stir plate at 4°C overnight. Once the methylcellulose has entered solution, add 50 mL of FCS.
10. 1× PBS, pH 7.5: 135 mM NaCl, 2.5 mM KCl, 1.5 mM KH$_2$PO$_4$, 8.0 mM Na$_2$HPO$_4$, pH 7.5.

3. Methods
3.1. Isolation of Viral DNA for Transfection

To engineer a new virus recombinant, plasmid is required containing the particular sequence of interest flanked by sufficient amounts of viral sequences homologous to the desired gene locus within the HSV-1 genome along with purified, infectious viral DNA. The quality and purity of these two reagents will determine the frequency of generating the desired recombinant virus. It is imperative that the sequence of interest in the transfer plasmid contain at least 500–1000 bp of flanking HSV-1 sequences to achieve a higher frequency of producing and isolating the recombinant. The quality of the viral DNA used in the transfections to synthesize new recombinants can be evaluated by two criteria: 1) whether the DNA is intact and at the proper concentration, as determined by Southern blot analysis *(112)*; or, more importantly, 2) whether the viral DNA is infectious and capable of producing an optimal number of plaques following transfection of 1 µg of viral DNA. We have optimized the protocol for the production of highly infectious viral DNA and routinely obtain preps in which 1 µg of purified viral DNA will yield 100–1000 plaques (*see* **Note 1**).

1. Infect a subconfluent to confluent monolayer of cells in a T150 tissue culture flask at an MOI of 3. The cell should have been split at both 1 and 3 days prior to infection.
2. Allow the infection to proceed for approximately 18–24 h depending on both the cell type and virus used. All cells should be rounded–up and still adherent to the flask, yet just about ready to detach.
3. Remove the cells by tapping the flask or use a cell scraper to dislodge the cells.
4. Pellet the cells for 5–10 m at 2060*g* at 4°C in a 15-mL conical polypropylene tube.
5. Wash the cells 1× with 10 mL of TBS (pH 7.5).
6. Lyse the cells in 5 mL of lysis buffer (10 m*M* Tris-HCl, pH 8.0, 10 m*M* EDTA) + 0.25 mg/mL Proteinase K + 0.6% sodium dodecyl sulfate (SDS).
7. Wrap the lid of the tube in parafilm and incubate the tube on a nutator platform at 37°C overnight.
8. Extract the suspension 2× with phenol/chloroform/isoamyl alcohol (25:24:1), being careful not to be too vigorous. However, it is important to invert the tube enough to achieve proper mixing of the phases. When removing the aqueous phase, remember to leave the interface behind, as the DNA stays at this point, thus increasing the overall yield.
9. Extract the aqueous phase 2× with chloroform, again being careful during mixing.
10. Remove the aqueous phase to a new tube, going as close to the interface as possible. The DNA present at the interface is extremely viscous and will enter the pipet as a visible slurry.
11. Add 2 vol of cold isopropanol to precipitate. Mix well.
12. The DNA can be spooled on a heat-sealed Pasteur pipet, or the mixture can be stored at –20°C overnight. If spooling the DNA, remove the spooled DNA and transfer the pipet to a new tube, breaking off the Pasteur pipet. Let dry overnight and then resuspend in 0.5–1.0 mL TE buffer or dH₂0 using wide-bore pipet tips (*see* **Note 2**).

At this point, it is important to determine the quality of the DNA prep (*see* **Note 3**). The DNA concentration was determined spectrophotometrically at 260 nm. Southern blot analysis of digests 1 μg quantities of the DNA prep with several diagnostic restriction enzymes should determine whether the DNA is intact. Transfection of 1 μg quantities of the DNA prep into permissive tissue culture cells will allow the determination of the number of infectious particles per 1 μg of DNA.

3.2. Construction of Recombinant Virus: Transfection of Plasmid/Viral DNA

Once the proper transfer plasmid has been constructed, by deleting the HSV-1 gene of interest yet maintaining sufficient HSV-1 flanking sequences to allow efficient recombination into the viral genome, it is then possible to marker-transfer these sequences into the virus, taking advantage of the highly active recombinational machinery of the virus (*see* **Note 4**). This is accomplished by transfecting both linearized plasmid and *Pac*I-digested purified viral DNA into permissive cells using the calcium phosphate method *(113,114)*. To readily detect the desired recombinants containing the desired therapeutic gene of interest, the target parental virus backbone should possess a reporter gene cassette (i.e., *lacZ* or green fluorescent protein [GFP]) at the desired site of recombination. Positive recombinants obtained from the transfection will produce clear plaques compared with the blue plaque phenotype of the parental virus. Alternatively, the introduction of a reporter gene cassette along with the therapeutic gene within the transfer plasmid will allow for more rapid identification and purification of the desired recombinant virus from the ensuing transfection reaction when the parental virus lacks a reporter gene.

1. Split Vero cells (for deletion of nonessential HSV-1 genes) or the specific complementing cell line (for deletion of essential HSV-1 gene[s]) 1:10 3 days prior to transfection.
2. At 1 day prior to transfecting, split the cells again and plate in 60-mm tissue culture dishes at a cell density of 1×10^6 cells/plate in MEM/10% FCS.
3. Linearize the plasmid construct at a restriction enzyme site between the flanking HSV-1 sequence and any *E. coli* DNA present within the bacterial vector (*see* **Note 3**).
4. Digest the purified viral DNA with *Pac*I to completion.
5. Make up the transfection mixture by adding 1–5 μg from the viral DNA prep (the amount of DNA that yields 200 or more plaques by transfection) to an amount of linearized plasmid DNA (or restriction fragment) equal to 10× and 50× genome equivalents of viral DNA. Viral DNA or plasmid DNA alone are used as controls for the transfection (*see* **Note 5**).

6. Add 600 μL of 2× HBS to each tube, mix, and place on ice for 20 min (*see* **Note 6**).
7. Add 41 μL of 2 *M* CaCl$_2$ dropwise, mixing gently. Set at room temperature for 20 min (see **Note 7**).
8. Aspirate the media from the plates and rinse 1x with 1 mL of 2× HBS.
9. Pipet transfection mixture up and down to break up large clumps of precipitate, add carefully to cell monolayers, and place plates at 37°C for 40 min in a CO$_2$ incubator.
10. Add 4 mL MEM/10% FCS per plate and place at 37°C for 4 h in a CO$_2$ incubator.
11. Aspirate media from plates and wash 1× with 1 mL of HBS, being careful not to cause the cell monolayer to lift off the plate.
12. Slowly and carefully add 2 mL of 20% glycerol (dilute 100% glycerol in 2× HBS) per plate and leave on cells at room temperature for exactly 4 min (*see* **Note 8**).
13. Carefully remove all of the glycerol shock solution by aspiration and wash the monolayer 3× with 2 mL of MEM/10% FCS. Be sure the monolayer remains intact.
14. Carefully add 4 mL of MEM/10% FCS and incubate the plates at 37°C in a CO$_2$ incubator.
15. Observe the plates twice daily under the microscope for the production of complete cytopathic effect (CPE), indicating the presence of infectious foci. This usually takes 3–5 days depending on the virus and cell type used to propagate the recombinant.
16. Once a majority of the cells have rounded up, remove the media and store it temporarily at 4°C. Isolate the virus from the cell pellet by three cycles of freeze-thawing and sonication followed by centrifugation at 913*g* for 5 min at 4°C. Then combine this supernate with the removed media and store at –80°C for use as a stock.

This stock of virus can now be analyzed for the presence of recombinants and purified as described in the next section.

3.3. Isolation and Purification of Recombinant Virus

The stock of virus obtained from the transfection can now be used to isolate the desired recombinant. It is necessary to isolate and purify the recombinant through three rounds of limiting dilution. This process can be greatly enhanced by the inclusion of a reporter gene cassette within the sequences to be recombined back into the viral genome or by the disruption of a reporter gene cassette following homologous recombination of the therapeutic gene cassette into the parental virus genome that already possesses the reporter cassette. We have optimized the limiting dilution procedure for the detection of recombinants expressing the β-galactosidase (*lacZ*) reporter gene.

3.3.1. Limiting Dilution Procedure for the Detection of Recombinants Expressing LacZ

The advantage of employing limiting dilution is that it does not require the standard plaque isolation technique in which single, well-isolated plaques are picked following agarose or methylcellulose overlay. Contamination of positive recombinants with parental virus represents a considerable problem using the standard plaque isolation procedure since it is difficult to find well-isolated plaques on a plate, or virus from an adjacent plaque gets sucked up along with that from the designated plaque in the act of picking a plaque through the agarose or methylcellulose overlay.

1. Titer the stock of recombinant virus from the transfection.
2. Add 30 plaque-forming units (pfu) of virus to 3 mL of 1×10^6 cells in suspension (MEM/10% FCS) within a 15-mL conical polypropylene tube.
3. Wrap the lid of the tube with parafilm, and place the tube on a nutator rocker platform at 37°C for 1 h.
4. Following the 1-h period for virus adsorption, add 7 mL of fresh media, and plate 100 µL of the mixture in each well of a 96-well flat bottomed plate.
5. Incubate the plates at 37°C in a CO_2 incubator for 2–5 days until the appearance of plaques.
6. The plates are scored for wells containing only single plaques. Theoretically, by adding 30 pfu, it should be possible to obtain 30 of 96 wells that contain single plaques.
7. Transfer the media from each well into a new 96-well plate. Store the plate at −80°C for future use as a virus stock.
8. Stain each well of the plate by adding 100 µL of the X-gal staining solution using a multichannel pipetor (*see* **Note 9**).
9. Incubate the plates at 37°C in a CO_2 incubator overnight or for several hours until the appearance of readily detectable blue plaques.

Wells containing single blue or clear plaques are scored and the overall frequency of original recombination event is determined. Media from the frozen stock 96-well plate stored at −80°C can now be used in the next round of the limiting dilution procedure. With each subsequent round of limiting dilution, the ratio of desired recombinants within the population should approach 100%. We will routinely proceed through one additional round of limiting dilution after the ratio reaches 100% to ensure purity of the stock. At this point, the virus stock can be used to produce a midi-stock for the eventual preparation of a high-titer stock for general experimental use. At the same time, this stock is used to produce viral DNA to confirm the presence of the insert as well as the absence of the deleted sequences by Southern blot analysis *(112)* (*see* **Note 10**).

3.4. Preparation of High-Titer Stocks of Recombinant Virus

Prepare a midi-stock of recombinant virus from a monolayer of cells in a T25 tissue culture flask and obtain the titer of the stock for preparation of the final stock.

3.4.1. Titration of Virus Stock

This procedure can be used to obtain the titer of any size virus stock.

1. Seed 6-well tissue culture plates with $0.5-1.0 \times 10^6$ cells per well at 1 day prior to titration of the stock.
2. Prepare a series of 10-fold dilutions (10^{-2} to 10^{-10}) of the virus stock in 1 mL of cold MEM without serum.
3. Add 100 μL of each dilution to a near confluent monolayer of cells in a single well of a 6-well tissue culture plate (in duplicate).
4. Allow the virus to adsorb for a period of 1 h at 37°C in a CO_2 incubator. Rock the plates every 15 min to distribute the inoculum.
5. Aspirate off the virus inoculum, add 3 mL of 1.0% methycellulose overlay, and reincubate the plates for 3–5 days, until well-defined plaques appear.
6. Aspirate off the methycellulose, and stain with 1 mL of 1% crystal violet solution (in 50:50 methanol/dH_2O, v/v) for 5 min. The stain fixes the cells and virus. Alternatively, the wells can be stained using the X-gal solution if the virus contains a *lacZ* reporter cassette.
7. Aspirate off the stain, rinse gently with tap water to remove excess stain, and air-dry.
8. Count the number of plaques per well, determine the average for each dilution, and multiply by a factor of 10 to get the number of pfu/mL for each dilution. Multiply this number by 10 to the power of the dilution to achieve the titer in pfu/mL.

3.4.2. Virus Stock Preparation

The following procedure calls for preparing a virus stock from two roller bottles worth of cells; however, it can be scaled up or down depending on specific needs (*see* **Note 11**).

1. We routinely infect 5–10 roller bottles of cells with virus to produce a high-titer stock. Each roller bottle is infected with virus at an MOI of 0.01–0.05, depending on the particular vector backbone, in a volume of 20 mL of serum-free media (DMEM; Gibco-BRL, Life Technologies) for 2 h.
2. After the 2-h adsorption period, add 80 mL of media (DMEM/10% FCS; Gibco-BRL) to each roller bottle. Then observe the bottles by light microscopy to determine the optimal time for harvesting virus. This time has been shown to vary dramatically depending on the particular vector backbone. Generally, viral CPE

should be obvious, with the majority of the cells rounded up yet not detached from the roller bottle surface.

3. Harvest the cells by scraping the cell monolayer into the media.
4. Collect medium and subject it to low-speed centrifugation at 2060g for 5 min at 4°C.
5. Store the supernatant on ice. Resuspend the cell pellet in a minimal volume (2–5 mL) of serum-free media and subject it to three rounds of sonication to disrupt it and release cell-associated virus.
6. Following low-speed centrifugation (2060g for 5 min at 4°C) to remove the cellular debris, combine this supernatant with the supernatant from the first low-speed spin.
7. Clarify this preparation further using an additional low-speed spin at 10–17,000g for 5 min.
8. Then pellet the virus by centrifugation at 48,400g in 50-mL Oak Ridge tubes for 30 min at 4°C.
9. Resuspend the virus pellet isolated following a high-speed spin in 500 μL of serum-free media per roller bottle of virus-infected cells and mix with OptiPrep solution (22% OptiPrep, 0.8% NaCl, 10 mM HEPES, 1 mM EDTA). Centrifuge the sealed tubes at 645,000g for 3.5 h in a Beckman Vti65.2 or Vti90 rotor at 4°C.
10. Following the centrifugation period, harvest the virus from the gradient. We routinely detect three bands of varying intensity depending on the particular vector backbone on the gradient. Harvest the lower two bands: approximately 60% of the virus is present in the second band from the bottom of the gradient. Alternatively, fractionate the tubes and determine the titer of each fraction. Then aliquote the vector stocks and store at −80°C until use.

The titer of the stock can be determined by the virus titration procedure detailed above. If concentration is not required, the OptiPrep solution acts as a cryopreservative. However, the virus can be furthered concentrated by centrifugation in a microfuge at 15,115g and 4°C. Following centrifugation, the viral pellet is resuspended in a minimal volume of PBS containing glycerol or dimethyl sulfoxide (DMSO). It is necessary to add glycerol to a final concentration of 10% to the virus stock to cryopreserve the virus properly.

4. Notes

1. The DNA that is produced using the viral DNA isolation technique is not composed exclusively of viral DNA but also contains cellular DNA. The cellular DNA in this mixture acts as carrier when precipitating the DNA during the isolation, thus increasing the yield of DNA. In addition, the cellular DNA acts as carrier DNA during transfections and increases the overall efficiency of forming a precipitate, thereby increasing the chance of obtaining the desired recombinant. If necessary, pure viral DNA can be prepared from virus harvested solely from the

media of infected cells or from virus particles that have been gradient purified. The yield of DNA obtained in this instance is significantly reduced.

2. The use of wide-bore pipetman tips (Bio-Rad, Hercules, CA) will help prevent shearing of the viral DNA, thereby increasing the infectivity of the viral DNA prep.

3. The quality of the viral DNA preparation is crucial to the recombination frequency. Quality can be evaluated by Southern blot or by determining the number of infectious centers following transfection. We have found that some preparations will appear to be intact by Southern blot analysis, yet still contain a significant number of viral genomes that are nicked and thus are not infectious. It is important that the preparation yield 100–1000 plaques/μg of viral DNA.

4. The quality of the plasmid DNA also plays a role in the recombination frequency. Two to five hundred base pairs of HSV-1 flanking sequence is sufficient; however, ≥1 kb will dramatically increase the chance of isolating the desired recombinant. The size of the sequence to be inserted into the HSV-1 vector genome can affect the generation of the desired recombinant. It is possible for the virus to package up to an additional 10% of the genome, 15 kb for wild-type virus and potentially more for single and multiple gene-deleted viruses. If the insert is too large or contains sequences that affect the stability of the viral genome, part or all of the insert will be lost over time and it will not be possible to obtain a purified isolate of the desired recombinant. The specific gene locus targeted for insertion/deletion can affect the recombination event. Recombination into the repeat sequences can yield a mixture of viruses containing insertion into one or both copies of the gene. The recombinational machinery of the virus can convert an isolate with a single copy into a recombinant with inserts in both copies. The same mechanism can also produce virus lacking the insert in both copies (i.e., wild-type virus). Southern blot analysis can confirm whether the insert is present in 0, 1, or 2 copies.

5. It is important to linearize the plasmid construct before transfection to increase the recombination frequency compared with that obtained with uncut supercoiled plasmid. Digestion of the plasmid to release the insert, followed by purification of the restriction fragment, does not increase the recombination frequency. Although the frequency is the same, use of purified fragment is superior since no chance exists for the insertion of plasmid vector sequences into the virus by semihomologous recombination.

6. The pH of the HBS transfection buffer (HEPES) is extremely crucial to the transfection efficiency. Depending on the cell type being transfected, other buffers such as BBS (BES) or PiBS (PIPES) may result in higher efficiencies.

7. Other transfection procedures can also be employed that will produce equivalent or greater recombination frequencies. Lipofectamine (Gibco-BRL) can be employed instead of the calcium phosphate method; however, other liposome preparations have not proved effective for the transduction of the large 152-kb HSV genome.

8. Glycerol or DMSO can be used to shock cells during transfection. The percentage of glycerol or DMSO used depends on the cell type being transfected. We have found that glycerol is less toxic than DMSO for Vero cells and that 20% glycerol produced the highest number of transformants with the lowest level of toxicity.

9. Bluo-gal can be substituted for X-gal in the X-gal staining solution. Although Bluo-gal is more costly than X-gal, it is superior since it produces a darker blue reaction product with reduced background staining of the cell monolayer and it has a greater solubility in DMF.

10. The viral DNA mini-prep procedure routinely yields enough DNA for three to five restriction enzyme digestions for Southern blot analysis.

11. Virus stocks should be maintained at a low passage. Use one vial of a newly prepared stock as a stock for preparing all future stocks. To reduce the chance of rescuing wild-type virus during the propagation of viruses carrying deletions of essential gene(s), stocks should be routinely prepared from single plaque isolates.

References

1. Akkaraju, G. R., Huard, J., Hoffman, E. P., et al. (1999). Herpes simplex virus vector-mediated dystrophin gene transfer and expression in MDX mouse skeletal muscle. *J. Gene Med.* **1,** 280–289.

2. Krisky, D. M., Marconi, P. C., Oligino, T. J., et al. (1998) Development of herpes simplex virus replication defective multigene vectors for combination gene therapy applications. *Gene Ther.* **5,** 1517–1530.

3. Dressler, G., Rock, D., and Fraser, N. (1987) Latent herpes simplex virus typy 1 DNA is not extensively methylated in vivo. *J. Gen. Virol.* **68,** 1761–1765.

4. Mellerick, D. M. and Fraser, N. (1987) Physical state of the latent herpes simplex virus genome in a mouse model system: evidence suggesting an episomal state. *Virology* **158,** 265–275.

5. Rock, D. and Fraser, N. (1985) Latent herpes simplex virus type 1 DNA contains two copies of the virion DNA joint region. *J. Virol.* **55,** 849–852.

6. Croen, K. D., Ostrove, J. M., Dragovic, L. J., Smialek, J. E., and Straus, S. E. (1987) Latent herpes simplex virus in human trigeminal ganlia. Detection of an immediate early gene "anti-sense" transcript by in situ hybridization. *N. Engl. J. Med.* **317,** 1427–1432.

7. Deatly, A. M., Spivack, J. G., Lavi, E., and Fraser, N. W. (1987) RNA from an immediate early region of the HSV-1 genome is present in the trigeminal ganglia of latently infected mice. *Proc. Natl. Acad. Sci. USA* **84,** 3204–3208.

8. Gordon, Y. J., Johnson, B., Romanonski, E., and Araullo-Cruz, T. (1988) RNA complementary to herpes simplex virus type 1 ICP0 gene demonstrated in neurons of human trigeminal ganglia. *J. Virol.* **62,** 1832–1835.

9. Rock, D. L., Nesburn, A. B., Ghiasi, H., et al. (1987) Detection of latency-related viral RNAs in trigeminal ganglia of rabbits latently infected with herpes simplex virus type 1. *J. Virol.* **61,** 3820-3826.

10. Spivack, J. G. and Fraser, N. W. (1987) Detection of herpes simplex virus type 1 transcripts during latent infection in mice. *J. Virol.* **61,** 3841–3847.

11. Stevens, J. G., Wagner, E. K., Devi-Rao, G. B., Cook, M. L., and Feldman, L. T. (1987) RNA complementary to a herpesviruses α gene mRNA is prominent in latently infected neurons. *Science* **255,** 1056–1059.

12. Fareed, M. and Spivack, J. (1994) Two open reading frames (ORF1 and ORF2) within the 2.0-kilobase latency-associated transcript of herpes simplex virus type 1 are not essential for reactivation from latency. *J. Virol.* **68,** 8071–8081.

13. Hill, J. M., Sedarati, F., Javier, R. T., Wagner, E. K., and Stevens, J. G. (1990) Herpes simplex virus latent phase transcription facilitates in vivo reactivation. *Virology* **174,** 117–125.

14. Ho, D. Y. and Mocarski, E. S. (1989) Herpes simplex virus latent RNA (LAT) is not required for latent infection in the mouse. *Proc. Natl. Acad. Sci. USA* **86,** 7596–7600.

15. Javier, R. T., Stevens J. G., Dissette V. B., and Wagner, E. K. (1988) A herpes simplex virus transcript abundant in latently infected neurons is dispensible for establishment of the latent state. *Virology* **166,** 254–257.

16. Leib, D. A., Bogard C.L., Kosz-Vnenchak M., et al. (1989b) A deletion mutant of the latency-associated transcript of herpes simplex virus type 1 reactivates from the latent infection. *J. Virol.* **63,** 2893–2900.

17. Sedarati, F., Izumi, K. M., Wagner, E. K., and Stevens, J. G. (1989) Herpes simplex virus type 1 latency-associated transcript plays no role in establishment or maintenance of a latent infection in murine sensory neurons. *J. Virol.* **63,** 4455–4458.

18. Steiner, I., Spivack, J. G., Lirette, R. P., et al. (1989) Herpes simplex virus type 1 latency-associated transcripts are evidently not essential for latent infection. *EMBO J.* **8,** 505–511.

19. Roizman, B. and Sears, A. (1996) Herpes simplex viruses and their replication, in *Fields Virology* (Fields, B. N., et al., eds.), Lippincott-Raven. Philadelphia, pp. 2231–2295.

20. Spear, P. (1993a) Membrane fusion induced by herpes simplex virus, in *Viral Fusion Mechanisms* (Bentz, J., ed.), CRC, Boca Raton, pp. 201–232.

21. Spear, P. G. (1993b) Entry of alphaherpesviruses into cells. *Semin. Virol.* **4,** 167–180.

22. Kwong, A. D. and Frenkel, N. (1987) Herpes simplex virus-infected cells contain a function(s) that destablizes both host and viral mRNAs. *Proc. Natl. Acad. Sci. USA* **84,** 1926–1930.

23. Kwong, A. D., Kruper, J. A., and Frenkel, N. (1988) Herpes simplex virus virion host shutoff function. *J. Virol.* **62,** 912–921.

24. Oroskar, A. and Read, G. (1989) Control of mRNA stability by the virion host shutoff function of herpes simplex virus. *J. Virol.* **63,** 1897–1906.

25. Read, G. S. and Frenkel, N. (1983) Herpes simplex virus mutants defective in the virion-associated shutoff of host polypeptide synthesis and exhibiting abnormal synthesis of α (immediate early) viral polypeptides. *J. Virol.* **46,** 498–512.

26. Ace, C. I., McKee, T. A., Ryan, J. M., Cameron, J. M., and Preston, C. M. (1989) Construction and characterization of a herpes simplex virus type 1 mutant unable to transinduce immediate-early gene expression. *J. Virol.* **63,** 2260–2269.

27. Batterson, W. and Roizman, B. (1983) Characterization of the herpes simplex virion-associated factor responsible for the induction of alpha-genes. *J. Virol.* **46,** 371–377.

28. Campbell, M. E. M., Palfeyman J. W., and Preston, C. M. (1984) Identification of herpes simplex virus DNA sequences which encode a trans-acting polypeptide responsible for stimulation of immediate early transcription. *J. Mol. Biol.* **180,** 1–19.

29. Kristie, J. and Roizman, B. (1987) Host cell proteins bind to the *cis*-acting site required for virion-mediated induction of herpes simplex virus 1 alpha genes. *Proc. Natl. Acad. Sci. USA* **84,** 71–75.

30. McKnight, J. L. C., Kristie, T. M., and Roizman, B. (1987) Binding of the virion protien mediating α gene induction in herpes simplex virus 1-infected cells to its *cis* site requires cellular proteins. *Proc. Natl. Acad. Sci. USA* **84,** 7061–7065.

31. Post, L., Mackem, S., and Roizman, B. (1981) Regulation of alpha genes of herpes simplex virus: expression of chimeric genes produced by fusion of thymidine kinase with alpha gene promoters. *Cell* **24,** 555–565.

32. Newcomb, W. and Brown, J. (1994) Induced extrusion of DNA from the capsid of herpes simplex virus type 1. *J. Virol.* **68,** 443–440.

33. McGeoch, D. J., Dolan, A., Donald, S., and Rixon, F.J. (1985) Sequence determination and genetic content of the short unique region in the genome of herpes simplex virus type 1. *J. Mol. Biol.* **181,** 1–13.

34. McGeoch, D. J., Dolan, A., Donald, S., and Brauer, D. H. K. (1986) Complete DNA sequence of short repeat region in the genome of herpes simplex virus type 1. *Nucleic Acids Res.* **14,** 1727–1744.

35. McGeoch, D. J., Dalrymple, M. A., Davison, A. J., et al. (1988) The complete DNA sequence of the long unique region in the genome of herpes simplex virus type 1. *J. Gen. Virol.* **69,** 1531–1574.

36. McGeoch, D. J., Cunningham, C., McIntyre, G., and Dolan, A. (1991) Comparative sequence analysis of the long repeat regions and adjoining parts of the long unique regions in the genomes of herpes simplex viruses types 1 and 2. *J. Gen. Virol.* **72,** 3057–3075.

37. Gruenheid, S., Gatzke, L., Meadows, H., and Tufaro, F. (1993) Herpes simplex virus infection and propagation in a mouse L cell mutant lacking heparan sulfate proteoglycans. *J. Virol.* **67,** 93–100.

38. Herold, B., Visalli, R., Susmarski, N., Brandt, C., and Spear, P. (1994) Glycoprotein C-independent binding of herpes simplex virus to cells requires cell surface heparan sulfate and glycoprotein B. *J. Gen. Virol.* **75,** 1211–1222.

39. Laquerre, S., Argnani, R., Anderson, D. B., et al. (1998) Heparan sulfate proteoglycan binding by herpes simplex virus type 1 glycoproteins B and C which differ in their contribution to virus attachment, penetration, and cell-to-cell spread. *J. Virol.* **72,** 6119–6130.

40. Krummenacher, C., Nicola, A. V., Whitbeck, J. C., et al. (1998) Herpes simplex virus glycoprotein D can bind to poliovirus receptor-related protein 1 or herpesvirus entry mediator, two structurally unrelated mediators of virus entry. *J. Virol.* **72,** 7064–7074.

41. Montgomery, R. I., Warner, M. S., Lum, B. J., and Spear, P. (1996) G. Herpes simplex virus 1 entry into cells mediated by a novel member of the TNF/NGF receptor family. *Cell* **87**, 427–436.
42. Rux, A., Willis, S., Nicola, A. V., et al. (1998) Functional region IV of glycoprotein D from herpes simplex virus modulates glycoprotein binding to the herpesvirus entry mediator. *J. Virol.* **72**, 7091–7098.
43. Sarrias, M. R., Whitbeck, J. C., Rooney, I., et al. (1999) Inhibition of herpes simpex virus gD and lymphotoxin-alpha binding to HveA by peptide antagonists. *J. Virol.* **73**, 5681–5687.
44. Terry-Allison, T., Montgomery, R., Whitbeck, J., et al. (1998) HveA (herpesvirus entry mediator A), a coreceptor for herpes simplex virus entry, also participates in virus-induced cell fusion. *J. Virol.* **72**, 5802–5810.
45. Whitbeck, J., Muggeridge, M., Rux, A., et al. (1999) The major neutralizing antigenic site on herpes simplex virus glycoprotein D overlaps a receptor-binding domain. *J. Virol.* **73**, 9879–9890.
46. Cocchi, F., Lopez, M., Menotti, L., et al. (1998) The V domain of herpesvirus Ig-like receptor (HIgR) contains a major functional region in herpes simplex virus-1 entry into cells and interacts physically with the viral glycoprotein D. *Proc. Natl. Acad. Sci. USA* **95**, 15700–15705.
47. Geraghty, R. J., Krummenacher, C., Cohen, G. H., Eisenberg, R. J., and Spear, P. G. (1998) Entry of alphaherpesviruses mediated by poliovirus receptor-related protein 1 and poliovirus receptor. *Science* **280**, 1618–1620.
48. Krummenacher, C., Baribaud, I., Ponce De Leon, M., et al. (2000) Localization of a binding site for herpes simplex virus glycoprotein D on herpesivrus entry mediator C by using antireceptor monoclonal antibodies. *J. Virol.* **74**, 10863–10872.
49. Shukla, D., Dal Canto, M. C., Rowe, C. L., and Spear, P. G. (2000) Striking similarity of murine nectin-1alpha to human nectin-1alpha (HveC) in sequence and activity as a glycoprotein D receptor for alphaherpesvirus entry. *J. Virol.* **74**, 11773–11781.
50. Fuller, A. O. and Spear, P. G. (1985) Specificities of monoclonal and polyclonal antibodies that inhibit adsorption of herpes simplex virus to cells and lack of inhibition by potent neutralizing antibodies. *J. Virol.* **55**, 475–482.
51. Highlander, S. L., Sutherland, S. L., Gage, P. J., et al. (1987) Neutralizing monoclonal antibodies specific for herpes simplex virus glycoprotein D inhibit virus penetration. *J. Virol.* **61**, 3356–3364.
52. Ligas, M. and Johnson, D. (1988) A herpes simplex virus mutant in which glycoprotein D sequences are replaced by β-galactosidase sequences binds to but is unable to penetrate into cells. *J. Virol.* **62**, 1486–1494.
53. Nicola, A. V., Ponce de Leon, M., Xu, R., et al. (1998) Monoclonal antibodies to distinct sites on herpes simplex virus (HSV) glycoprotein D block HSV binding to HVEM.J. *J. Virol.* **72**, 3595–3601.
54. Cai, W., Gu, B., and Person, S. (1988) Role of glycoprotein B of herpes simplex virus type 1 in viral entry and cell fusion. *J. Virol.* **62**, 2596–2604.
55. Desai, P., Schaffer, P., and Minson, A. (1988) Excretion of non-infectious virus

particles lacking glycoprotein H by a temperature-sensitive mutant of herpes-simplex virus type 1: evidence that gH is essential for virion infectivity. *J. Gen. Virol.* **69,** 1147–1156.

56. Hutchinson, L., Browne, H., Wargent, V., et al. (1992) A novel herpes simplex virus glycoprotein, gL, forms a complex with glycoprotein H (gH) and affects normal folding and surface expression of gH. *J. Virol.* **66,** 2240–2250.

57. Hutchinson, L., Goldsmith, K., Snoddy, D., et al. (1992) Identification and characterization of a novel herpes simplex virus glycoprotein, gK, involved in cell fusion. *J. Virol.* **66,** 5603–5609.

58. Roizman, B. and Sears, A. E. (1990) Herpes simplex viruses and their replication, in *Field's Virology* (Fields, B. N., et al., eds.), Raven, New York, pp. 1795–1841.

59. Roizman, B. and Sears, A. E. (1993) Herpes simplex viruses and their replication, in *The Human Herpesviruses* (Roizman, B., Whitley, R. J., and Lopez, C., eds.), Raven, New York, pp. 11–68.

60. Honess, R. and Roizman, B. (1974) Regulation of herpes simplex virus macromolecular synthesis. I. Cascade regulation of the synthesis of three groups of viral proteins. *J. Virol.* **14,** 8–19.

61. Honess, R. W. and Roizman, B. (1975) Regulation of herpes virus macromolecular synthesis: sequential transition of polypeptide synthesis requires functional viral polypeptides. *Proc. Natl. Acad. Sci. USA* **72,** 1276–1280.

62. DeLuca, N.A. and Schaffer, P. A. (1985) Activation of immediate-early, early, and late promoters by temperature-sensitive and wild-type forms of herpes simplex virus type 1 protein ICP4. *Mol. Cell. Biol.* **5,** 1997–2008.

63. Dixon, R. A. F. and Schaffer, P. A. (1980) Fine-structure mapping and functional analysis of temperature-sensitive mutants in the gene encoding the herpes simplex virus type 1 immediate early protein VP175. *J. Virol.* **36,** 189–203.

64. O'Hare, P. and Hayward, G. (1985) Three *trans*-acting regulatory proteins of herpes simplex virus modulate immediate-early gene expression in a pathway involving positive and negative feed regulation. *J. Virol.* **56,** 723–733.

65. O'Hare, P. and Hayward, G. S. (1985) Evidence for a direct role for both the 175,000 and 110,000-molecular-weight immediate-early protein of herpes simplex vius in transactivation of delayed-early promoters. *J. Virol.* **53,** 751–760.

66. Preston, C. (1979) Abnormal properties of an immediate early polypeptide in cells infected with the herpes simplex virus type 1 mutant tsK. *J. Virol.* **32,** 357–369.

67. Sacks, W. R. and Schaffer, P. A. (1987) Deletion mutants in the gene encoding the herpes simplex virus type 1 immediate-early protein ICP0 exhibit impaired growth in cell culture. *J. Virol.* **61,** 829–839.

68. Stow, N. and Stow, E. (1986) Isolation and characterization of a herpes simplex virus type 1 mutant containing a deletion within the gene encoding the immediate early polypeptide Vmw 110. *J. Gen. Virol.* **67,** 2571–2585.

69. Watson, R. and Clements, J. (1980) A herpes simplex virus type 1 function continuously required for early and late virus RNA synthesis. *Nature* **285,** 329–330.

70. Holland, L. E., Anderson, K. P., Shipman, C., and Wagner, E. K. (1980) Viral

DNA synthesis is required for efficient expression of specific herpes simplex virus type 1 mRNA. *Virology* **101,** 10–24.

71. Mavromara-Nazos, P. and Roizman, B. (1987) Activation of herpes simplex virus 1 γ2 genes by viral DNA replication. *Virology* **161,** 593–598.

72. Cook, M. L. and Stevens, J. G. (1973) Pathogenesis of herpetic neuritis and ganglionitis in mice: evidence of intra-axonal transport of infection. *Infect. Immun.* **7,** 272–288.

73. Stevens, J.G. (1989) Human herpesviruses: a consideration of the latent state. *Microbiol. Rev.* **53,** 318–332.

74. Deshmane, S. L. and Fraser, N. W. (1989) During latency, herpes simplex virus type 1 DNA is associated with nucleosomes in a chromatin structure. *J. Virol.* **63,** 943–947.

75. Johnson, P. A., Miyanohara, A., Levine, F., Cahill, T., and Friedmann, T. (1992) Cytotoxicity of a replication-defective mutant herpes simplex virus type 1. *J. Virol.* **66,** 2952–2965.

76. Leiden, J., Frenkel, N., and Rapp, F. (1980) Identification of the herpes simplex virus DNA sequences present in six herpes simplex virus thymidine kinase-transformed mouse cell lines. *J. Virol.* **33,** 272–285.

77. DeLuca, N. A., McCarthy, A .M., and Schaffer, P. A. (1985) Isolation and characterization of deletion mutants of herpes simplex virus type 1 in the gene encoding immediate-early regulatory protein ICP4. *J. Virol.* **56,** 558–570.

78. Sacks, W., Greene, C., Aschman, D., and Schaffer, P. (1985) Herpes simplex virus type 1 ICP27 is essential regulatory protein. *J. Virol.* **55,** 796–805.

79. Samaniego, L., Webb, A. and DeLuca, N. (1995) Functional interaction between herpes simplex virus immediate-early proteins during infection: gene expression as a consequence of ICP27 and different domains of ICP4. *J. Virol.* **69,** 5705–5715.

80. Katan, M., Haigh, A., Verrijzer, C., Vliet, P. v. d., and O'Hare, P. (1990) Characterization of a cellular factor which interacts functionally with Oct-1 in the assembly of a multicomponent transcription complex. *Nucleic Acids Res.* **18,** 6871–6880.

81. Kristie, T. and Sharp, P. (1993) Purification of the cellular C1 factor required for the stable recognition of the Oct-1 homeodomain by herpes simplex virus α-transinduction factor (VP16). *J. Biol. Chem.* **268,** 6525–6534.

82. Werstuck, G. and Capone, J. (1993) An unusual cellular factor potentiates protein-DNA complex assembly Oct-1 and Vmw65. *J. Biol. Chem.* **268,** 1272–1278.

83. Wilson, A., LaMarco, K., Peterson, M., and Herr, W. (1993) The VP16 accessory protein HCF is a family of polypeptides processed from a large precursor protein. *Cell* **74,** 115–125.

84. Xiao, P. and Capone, J. (1990) A cellular factor binds to the herpes simplex virus type 1 transactivator Vmw65 and is required for Vmw65-dependent protein-DNA complex assembly with Oct-1. *Mol. Cell. Biol.* **10,** 4974–4977.

85. Gerster, T. and Roeder, R. (1988) A herpesvirus trans-activating protein interacts

with transcription factor OTF-1 and other cellular proteins. *Proc. Natl. Acad. Sci. USA* **85**, 6347–6351.

86. O'Hare, P. and Goding, C. (1988a) Herpes simplex virus regulatory elements and the immunoglobulin octamer domain bind a common factor and are both targets for virion transactivation. *Cell* **52**, 435–445.

87. O'Hare, P., Goding, C. and Haigh, A. (1988b) Direct combinational interaction between a herpes simplex virus regulatory protein and a cellular octamer binding factor mediates specific induction of virus immediate-early gene expression. *EMBO J.* **7**, 4231–4238.

88. Preston, C., Frame, M., and Campbell, M. (1988) A complex formed between cell components and an HSV structural polypeptide binds to a viral immediate early gene regulatory DNA sequence. *Cell* **52**, 425–434.

89. Stern, S., Tanaka, M., and Herr, W. The Oct-1 homeodomain directs formation of a multiprotein-DNA complex with the HSV transactivator VP16. *Nature* (1989) **341**, 624–630.

90. Cai, W. and Schaffer, P. A. (1992) Herpes simplex virus type 1 ICP0 regulates expression of immediate-early, early, and late genes in productively infected cells. *J. Virol.* **66**, 2904–2915.

91. Everett, R. D. (1987) The regulation of transcription of viral and cellular genes by herpesvirus immediate-early gene products. *Anticancer Res.* **7**, 589–604.

92. Gelman, I. H. and Silverstein, S. (1985) Identification of immediate-early genes from herpes simplex virus that transactivate the virus thymidine kinase gene. *Proc. Natl. Acad. Sci. USA* **82**, 5265–5269.

93. Quinlan, M. P. and Knipe, D. M. (1985) Stimulation of expression of a herpes simplex virus DNA-binding protein by two viral factors. *Mol. Cell. Biol.* **5**, 957–963.

94. Zhu, Q. and Courtney, R. J. (1994) Chemical cross-linking of virion envelope and tegument proteins of herpes simplex virus type 1. *Virology* **204**, 590–599.

95. Maul, G. and Everett, R. (1994) The nuclear location of PML, a cellular member of the C3HC4 zinc-binding domain protein family, is rearranged during herpes simplex virus infection by the C3HC4 viral protein ICP0. *J. Gen. Virol.* **75**, 1223–1233.

96. Chen, J. and Silverstein, S. (1992) Herpes simplex viruses with mutations in the gene encoding ICP0 are defective in gene expression. *J. Virol.* **66**, 2916–2927.

97. Samaniego, L., Wu, N., and DeLuca, N. A. (1997) The herpes simplex virus immediate-early protein ICP0 affects transcription from the viral genome and infected-cell survival in the absence of ICP4 and ICP27. *J. Virol.* **71**, 4614–4625.

98. Rice, S., Long, M., Lam, V., and Spencer, C. (1994) RNA polymerase II is aberrantly phosphorylated and localized to viral replication compartments following herpes simplex virus infection. *J. Virol.* **68**, 988–1001.

99. Hill, A., P. Jugovic, I. York, et al. (1995) Herpes simplex virus turns off the TAP to evade host immunity. *Nature* **375**, 411–415.

100. Hill, A. and Ploegh, H. (1995) Getting the inside out: the transporter associated with antigen processing (TAP) and the presentation of viral antigen. *Proc. Natl. Acad. Sci. USA* **92**, 341–343.

101.York, I., Roop, C., Andrews, D., et al. (1994) A cytosolic herpes simplex virus protein inhibits antigen presentation to CD8+ T lymphocytes. *Cell* **77,** 525–535.
102.Johnson, P., Wang, M., and Friedmann, T. (1994) Improved cell survival by the reduction of immediate-early gene expression in replication-defective mutants of herpes simplex virus type 1 but not by mutation of the viron host shutoff function. *J. Virol.* **68,** 6347–6362.
103.Krisky, D. M., Wolfe, D., Goins, W. F., et al. (1998) Deletion of multiple immediate early genes from herpes simplex virus reduces cytotoxicity and permits long-term gene expression in neurons. *Gene Ther.* **5,** 1593–1603.
104.Marconi, P., Krisky, D., Oligino, T., et al. (1996) Replication-defective HSV vectors for gene transfer in vivo. *Proc. Natl. Acad. Sci. USA* **93,** 11319–11320.
105.Samaniego, L. A., Neiderhiser, L., and DeLuca, N. A. (1998) Persistence and expression of the herpes simplex virus genome in the absence of immediate-early proteins. *J. Virol.* **72,** 3307–3320.
106.Wu, N., Watkins, S. C., Schaffer, P. A., and DeLuca, N. A. (1996) Prolonged gene expression and cell survival after infection by a herpes simplex virus mutant defective in the immediate-early genes encoding ICP4, ICP27, and ICP22. *J. Virol.* **70,** 6358–6368.
107.McCarthy, A. M., McMahan, L., andSchaffer, P. A. (1989) Herpes simplex virus type 1 ICP27 deletion mutants exhibit altered patterns of transcription and are DNA deficient. *J. Virol.* **63,** 18–27.
108.Fink, D., DeLuca, N., Goins, W., and Glorioso, J. (1996) Gene transfer to neurons using herpes simplex virus-based vectors. *Annu. Rev. Neurosci.* **19,** 265–287.
109.Oligino, T., Poliani, P. L., Marconi, P., et al. (1996) In vivo transgene activation from an HSV-based gene vector by GAL4:VP16. *Gene Ther.* **3,** 892–899.
110.Frenkel, N., Locker, H., Batterson, W., Hayward, G., and Roizman, B. (1976) Anatomy of herpes simplex DNA. VI. Defective DNA originates from the S component. *J. Virol.* **20,** 527–531.
111.Krisky, D. M., Marconi, P. C., Oligino, T., et al. (1997) Rapid method for construction of recombinant HSV gene transfer vectors. *Gene Ther.* **4,** 1120–1125.
112.Southern, E. M. (1975) Detection of specific sequences among DNA fragments separated by gel electrophoresis. *J. Mol. Biol.* **98,** 503–517.
113.Graham, F. L. and Van der Eb, A. J. (1973) A new technique for the assay of infectivity of human adenovirus 5 DNA. *Virology* **52,** 456–467.
114.Shapira, M., Homa, F. L., Glorioso, J. C., and Levine, M. (1987) Regulation of the herpes simplex virus type 1 late (γ2) glycoprotein C gene: sequences between base pairs −34 to +29 control transient expression and responsiveness to transactivation by the products of immediate early (α) 4 and 0 genes. *Nucleic Acids Res.* **15,** 3097–3111.

Index

From: *Methods in Molecular Medicine, vol. 69, Gene Therapy Protocols, 2nd ed.*
Edited by: J. R. Morgan © Humana Press Inc., Totowa, NJ

SeV, *see* Sendai virus
SFV, *see* Semliki forest virus
SFV-1, *see* Simian foamy virus-1
Simian foamy virus-1 (SFV-1),
    advantages in gene transfer, 319, 320
    gene transfer vectors,
        cell culture, 325
        development, 322, 323
        infection of nonadherent cells, 326,
            328
        materials for production, 325, 326
        packaging cell lines, 324–326
        plasmids, 325
        titration using β–galactosidase, 326–
            328
        transfection using Lipofectamine,
            326–329
        variations, 323, 324
    genome, 321
    replication, 321, 322
    safety, 320
Simian immunodeficiency virus (SIV),
    gene transfer vectors,
        components, 353, 354
        concentrating, 356, 358, 359
        envelope protein selection, 354
        Gag-Pol expression plasmid
            optimization, 355, 356
        materials for production, 356
        minimization of construct, 354, 355
        plasmids, 356, 358
        titration, 357–359
        transfection, 356–359
    genome structure, 351, 353
    replication, 353
    tropism and safety, 351
SIV, *see* Simian immunodeficiency virus
Skin substitute,
    demand, 203
    keratinocyte ex vivo retroviral gene
        transfer,
        acellular dermis preparation, 205, 206,
            209, 210, 215
        advantages of vector, 204
        applications, 203, 204
        cell culture, 204, 205, 208, 215

Skin substitute *(cont.)*,
    keratinocyte ex vivo retroviral gene
        transfer *(cont.)*,
        genetically modified skin substitute
            preparation, 206, 207, 210–212,
            215
        genetic modification of cells, 208,
            209, 215
        materials, 204–208, 214, 215
        transplantation, 207, 212–214
Solvoplex,
    advantages of gene delivery, 83, 84
    di-*n*-propylsulfoxide synthesis, 89, 90
    formulations and efficiencies of lung
        gene transfer, 84, 85
    intratracheal injection,
        mouse, 90–93
        rat, 91
    mechanism of transfection, 87
    plasmid preparation, 88, 89
    preparation, 90, 92
    readministration, 87
    reporter gene studies,
        β–galactosidase, 91–93
        green fluorescent protein, 92, 93
        luciferase, 91, 93
        overview, 86
    rodents for study, 88, 92
    solvents, 88
    storage, 86
    toxicity, 87

T

Titration,
    adeno-associated virus,
        herpes simplex virus-1 hybrid vectors,
            468, 473, 474
        recombinant vectors, 430, 431, 433,
            438–442
    adenovirus,
        helper-dependent vector system, 397,
            406, 412
        ovine vectors, 424, 425
    feline immunodeficiency virus vectors,
        347, 348, 347, 348
    herpes simplex virus-1 vectors, 497